Lecture Notes in Control and Information Sciences

Volume 495

This series reports new developments in the fields of control and information sciences—quickly, informally and at a high level. The type of material considered for publication includes:

1. Preliminary drafts of monographs and advanced textbooks
2. Lectures on a new field, or presenting a new angle on a classical field
3. Research reports
4. Reports of meetings, provided they are

   (a) of exceptional interest and
   (b) devoted to a specific topic. The timeliness of subject material is very important.

Indexed by EI-Compendex, SCOPUS, Ulrich's, MathSciNet, Current Index to Statistics, Current Mathematical Publications, Mathematical Reviews, IngentaConnect, MetaPress and Springerlink.

Andrew J. Kurdila · Andrea L'Afflitto ·
John A. Burns

Data-Driven, Nonparametric, Adaptive Control Theory

Springer

Andrew J. Kurdila
Department of Mechanical Engineering
Virginia Tech
Blacksburg, VA, USA

Andrea L'Afflitto
Department of Industrial and Systems Engineering
Virginia Tech
Blacksburg, VA, USA

John A. Burns
Department of Mathematics
Virginia Tech
Blacksburg, VA, USA

ISSN 0170-8643 ISSN 1610-7411 (electronic)
Lecture Notes in Control and Information Sciences
ISBN 978-3-031-78002-8 ISBN 978-3-031-78003-5 (eBook)
https://doi.org/10.1007/978-3-031-78003-5

Mathematics Subject Classification: 93C40, 93D21, 46E22, 47B32

This Springer imprint is published by the registered company Springer Nature Switzerland AG
The registered company address is: Gewerbestrasse 11, 6330 Cham, Switzerland

To Ava and Patrick. My beloved grandkids, may you never be an engineer.
—Papa

To Paolo and Elena. May they remember how fun it was to reach for dad's keyboards.
—Papà

To my wife Gail.
—John

Preface

This book presents cohesively a new branch of control theory concerned with regulating nonlinear plants affected by uncertainties lying in a reproducing kernel Hilbert space (RKHS, also known as a native space). Classical control techniques, such as control Lyapunov functions, backstepping control, and model reference adaptive control, and their robust counterparts are generalized to allow for considerably broader classes of uncertainties than state-of-the-art and produce, for the first time, ultimate bounds on the trajectory tracking error that are explicit functions of how well uncertainties are captured by an RKHS. Several deterministic and stochastic data-driven strategies are investigated to characterize the space of functional uncertainties and, hence, reduce the reliance on oracles, that is, information available a priori without a proper justification.

The intended audience of this manuscript includes engineers, scientists, mathematicians, academics, and researchers from a diverse collection of fields. The proposed discussions include topics that are both theoretical and practical. The authors expect that the material will be appealing to those faced with implementing control systems in practical applications, as well as researchers specializing in statistical and machine learning theory, the theory of RKHSs, or approximation theory. Because of this breadth of backgrounds, some presentations may appear overly tutorial for specialists in a particular field. An engineer with expertise in classical adaptive control theory may not be very familiar with power functions and the many zeros theorem, but be all too familiar with Lyapunov stability analysis, projection methods, gradient learning laws, and their variants. On the other hand, experts in learning theory, approximation theory, or the theory of RKHSs may find some of the error bounds that arise from the study of interpolation in native spaces fairly basic. This text seeks to inform all these communities on how techniques from native spaces can be leveraged in applications to estimation and control theory for deterministic nonlinear ordinary differential equations.

Numerous numerical examples are presented in this book and the MATLAB® codes implementing these examples are made available online.[1,2,3] The scope of these resources is to showcase the applicability of the proposed problems and allow readers with diverse levels of expertise to have a head-start in this topic that, in the authors' opinion, seems more complex than it ultimately is.

This book may be consulted in multiple ways. A canonical back-to-back approach will lead the reader to explore in great detail elements of functional analysis; native spaces theory; stability theory; nonlinear, robust, and adaptive control over RKHSs; and deterministic and stochastic data-driven methods on native spaces. This approach, however, may be overly tedious to those eager to put the proposed theory to the test. In this case, the interested reader is recommended to start reading this book from Chap. 5, skipping proofs, refer to the earlier chapters as recommended by us throughout the text or as needed after having consulted the detailed index of terms, and test the computer codes provided by the authors online. This more direct approach may require re-reading the material of this book with increasing levels of detail to appreciate this work to the fullest extent. This book may also serve as a textbook in a graduate-level class, as it currently occurs at Virginia Tech, or for guided study courses. Although the ideal student for such a course has already passed courses in nonlinear systems theory and functional analysis, the tutorial style of this work allows students with only one of these two prerequisites to profit from this work.

We hope wholeheartedly that the readers will enjoy this work as much as we enjoyed writing it.

Blacksburg, VA, USA
November 2024

Andrew J. Kurdila
Andrea L'Afflitto
John A. Burns

---

[1] https://lafflitto.com/Codes/Springer_RKHS_Book_RCLF_Codes.zip.

[2] https://lafflitto.com/Codes/Springer_RKHS_Book_MRAC_Codes.zip.

[3] https://lafflitto.com/Codes/Springer_RKHS_Book_Newton_basis_MRAC.zip.

Acknowledgements

The authors would like to extend their deepest gratitude to the collaborators and students who, over the years, contributed to the development of this new theoretical framework invented by the unique combination of their interests and expertise. Special thanks to Dr. Daniel J. Stilwell in the Department of Electrical and Computer Engineering at Virginia Tech for having helped brainstorm the earliest results on adaptive control and native spaces, and Dr. Kyriakos G. Vamvoudakis in the School of Aerospace Engineering at Georgia Tech and his former student Dr. Filippos Fotiadis for their contributions to the early developments of a learning-based approach to adaptive control on native spaces. The authors are also indebted to their advisees Mr. Haoran Wang, Mr. Shengyuan Niu, and Mr. Giorgio A. Orlando, and their former advisee Dr. Nathan R. Powell, for having helped proofread this work and having produced several numerical simulations. Finally, the authors would like to thank their home institution, Virginia Tech, and in particular, the Department of Mechanical Engineering, the Grado Department of Industrial and Systems Engineering, and the Department of Mathematics, for their indirect support of this conspicuous effort.

Contents

Chapter 1
Introduction

1.1 Scope of This Book

This book presents in a tutorial, systematic manner how the theory of reproducing kernel Hilbert spaces (RKHSs) can yield novel solutions and insights for robust and adaptive control problems that arise in the study of some types of uncertain dynamical systems. In particular, we treat finite-dimensional deterministic systems in continuous time that are defined in terms of nonlinear ordinary differential equations (ODEs), and we explore how the properties of an RKHS can be exploited when designing control Lyapunov functions, backstepping control systems, model reference adaptive control (MRAC) systems, and their robust counterparts for these plants.

The role of uncertainty in modeling complex phenomena has been studied for several decades. The study of uncertainty modeling is well-documented across many disciplines including economics, finance, physics, medicine, biology, applied mathematics, and engineering. Uncertainty quantification for complex systems is now a field of study unto itself [1, 2], and the theoretical tools that make up this discipline are mature and diverse. Within the discipline of control theory, uncertainty quantification is studied systematically in the sub-discipline of robust control theory; see, for example, [3].

However, of all the techniques that constitute the field of uncertainty quantification, when a dynamical system is analyzed using the mathematical theory of an RKHS, the modeling approach is likely stochastic. This is particularly true in approaches that either address or have foundations in, statistical or machine learning theory. Bayesian or Gaussian process estimation are two specific kinds of such methods [4, 5], and they have been topics of widespread interest and development over the past decade. These methods have evolved concurrently to the development of associated algorithms of statistical or machine learning theory. The general theory of such stochastic, learning-theory-centric approaches is discussed in popular treatises such as [4, 6–8]. These basic principles for probabilistic methods in an RKHS

A. J. Kurdila et al., *Data-Driven, Nonparametric, Adaptive Control Theory*,
Lecture Notes in Control and Information Sciences 495,
https://doi.org/10.1007/978-3-031-78003-5_1

have been used in several recent efforts that have framed adaptive control problems in a *nonparametric* setting [9–12], and a similar solely deterministic nonparametric analysis is described in works like [13–17].

This book builds on these approaches and summarizes how a *deterministic* analysis in a native space setting can give rise to new and interesting results for some generally nonlinear systems governed by ODEs. The results derived for closed-loop performance bounds sometimes qualitatively resemble those in the more familiar stochastic cases, and we discuss when this is indeed the case. However, many methods and conclusions in this work are unique to the deterministic setting. This work summarizes new strong bounds on performance in the setting of a general, deterministic, nonparametric control theory in RKHSs. These bounds on the controllers' performances are explicit in the approximation dimension, which is a first time in the context of deterministic control of nonlinear dynamical systems affected by uncertainties. We argue that the derived bounds are often stronger and simpler to state than many of the bounds achieved using a stochastic analysis in an RKHS framework. Remarkably, this book is concluded with an analysis of the proposed control systems assessing the functional uncertainties using tools from stochastic machine learning and data-driven methods. These results from a stochastic setting enable a more careful comparison of the numerous, original deterministic results introduced in this book and some of the recent adaptive controllers based on features of Gaussian processes in an RKHS.

One of the unique features of the general approach discussed in this manuscript is that a limiting distributed parameter system (DPS) is introduced at the beginning of the control design process. Of the aforementioned references on adaptive control, the authors are only aware of a similar explicit study of an associated infinite-dimensional system in [12, 15]. This DPS includes a partial differential equation (PDE), which characterizes the desired ideal control behavior, and which can be viewed as an ordinary differential equation in a Banach space. The desired performance for the limiting DPS is described in terms of *nonparametric uncertainty classes* that reside in a generally infinite-dimensional native space.

As discussed at length in this text, it is impossible to implement the controllers that define the limiting DPS since they contain unknown estimates of the functional uncertainty in infinite-dimensional function spaces. To overcome this issue, the functional uncertainty estimates in these limiting controllers are projected onto certain finite-dimensional spaces, and the projection error produces an effect comparable to that of unmatched uncertainties in classical finite-dimensional approach systems. In this context, the proposed control systems assure robustness to the unavoidable approximation error occurring when transitioning controllers predicated on RKHS systems into finite-dimensional spaces. Thus, this volume can be understood as the first book to formulate a *general nonparametric theory* of deterministic robust and adaptive control theory in which the *functional uncertainty* is measured in terms of its membership in an RKHS.

The proposed implementable feedback control laws resemble those of conventional control systems, with an important modification. The methods in this text introduce suitable Grammian matrices that are required to obtain consistent approximations in the RKHS. The derived performance bounds rely on these Grammian matrices, and this is a key difference between the eventual practical controllers in this text and their conventional counterparts. The Grammians are required to obtain explicit bounds on the performance of the implementable controllers. The family of individual robust controllers derived in this text can therefore be understood as modifications of well-known algorithms from conventional control implementations to "respect the RKHS" projection onto finite-dimensional spaces. This book discusses how popular robustifications in classical control theory, such as the σ-modification and e-modification or MRAC, as well as many others, can be extended to plant models whose uncertainties are expressed through RKHSs. For details on classical robust and adaptive control techniques, see [18–26].

1.2 The Need for an Infinite-Dimensional Framework

The topics of robust control in general, and robust adaptive control in particular, have been addressed by a plethora of researchers. Although many relevant aspects of the state-of-the-art are discussed at length in relevant chapters, a comprehensive account of these techniques would be beyond the scope of this work. Overall, it can be safely claimed that the overwhelming majority of existing robust parametric adaptive control techniques share the following traits:

1. the uncertain system is governed by a system of nonlinear ODEs;
2. uncertainties in the system of ODEs are parameterized by a finite number of unknown real parameters;
3. adaptive estimation and control laws are typically described in terms of estimates of the plant states and the adaptive gains estimating or mirroring the behavior of the unknown real parameters.

In this book, we refer to the theory and algorithms that share these features as *finite dimensional, Euclidean, or real parametric* estimation and control theory. We choose this characterization because the state estimates and control laws evolve in finite-dimensional spaces, whose dimensions are fixed a priori, the associated analysis of stability and convergence in time is carried out within these Euclidean spaces, and uncertainties are provided by a regressor vector, or an equivalent form, given a priori and a finite set of unknown coefficients.

The framework considered in this text does not fall neatly into this category of approaches since the limiting system, which characterizes the closed-loop system's ideal behavior, is a type of DPS. Theories for the control of deterministic [27] or stochastic [28] infinite-dimensional systems in continuous time have been developed over the past few decades. The supporting theory uses many constructs that are distinct from those encountered in the study of real parametric adaptive estimation and control theory. These constructs include the theory of Banach space-valued functions, Bochner integrals, the theory of Gelfand triples, and distributional or weak derivatives, to name a few. Systems governed by PDEs are common examples that give rise to these DPSs [28–30]. Other examples include systems defined by delay differential equations, systems that depend on nonlinear memory effects like hysteresis, and more general types of functional differential equations [31–37]. In any case, the control theory for a DPS can be substantially more abstract and, hence, more difficult than the finite-dimensional theory for ODEs.

Some might argue that the added abstraction or complexity that is often necessary for the study of a DPS limits its widespread use in practical situations. The popularity of the ODE methods described in texts like [18–22, 24, 38] arises because of their mathematical precision and widespread applicability. The approaches in these texts have been implemented in a vast number of real-time feedback control applications. For many engineers or scientists who are faced with the implementation of adaptive controllers, the additional structures such as Gelfand triples, Bochner integrals, Sobolev spaces of Banach space-valued functions [28–30], which are required to frame some problem as a DPS, may not seem worth the extra effort. Furthermore, DPSs are not suitable for hardware implementation. A DPS associated with a PDE must be approximated to obtain an ODE, which, in turn, must be discretized in time for digital implementation. For all of these reasons, the choice to attack a problem in the full generality of the theory for DPS may be complex, and this aspect should not be underestimated.

In this book, we argue that by studying data-driven estimation and control problems for nonlinear ODE systems in terms of a DPS, important additional insights are often obtained in comparison to studying these problems in Euclidean spaces of fixed dimension. For instance, many of the advantages of this viewpoint are associated with developing a further, rigorous understanding of the role of approximation and learning theory in the performance of methods for adaptive estimation and control. We can be more specific about this claim by considering some iconic examples. In classical MRAC systems, matched functional uncertainties are captured by a regressor vector that, in many cases, is constructed offline, and whose dimension is constant. The proposed framework allows a paradigm shift. If the functional uncertainty lives in an infinite-dimensional space, then it is possible to use different numbers of parameters to approximate the uncertainty, and the number of these parameters, the selection of the basis used for building approximations, and the selection of the hypothesis space from which to approximate the functional uncertainty are now design choices. If we design a collection of adaptive controllers using approximating spaces of increasing dimensions, the proposed framework allows us to answer the following key questions:

> How do the choices of the dimension of the approximating space, of the basis used in approximations, and of the functional space that contains the uncertainty and the associated choice of state space explicitly affect the performance of a closed-loop control scheme as a function of? How can we design approaches, possibly automated, to guide these design choices?

The proposed framework allows us to characterize how the family of all the coordinate implementations indexed by the dimension of the approximating space, the basis used in approximations, and the functional space that contains the uncertainty bundled together performs. In this fashion, we are defining a general theory of nonparametric control that has the same goals as the corresponding theory of nonparametric function estimation of learning and approximation theories [39, 40]. Ultimately, our goal is to fully understand the implications of nonparametric function estimation theory as a tool to develop a corresponding nonparametric control theory that studies controller performance.

None of the standard textbooks on real parametric adaptive estimation and control in [18, 20–22, 24] directly address the question of obtaining such explicit relationships, certainly not in a systematic way, since the stability and convergence analysis in time is performed under the assumption that the space of functional uncertainties affecting the ODE model is fixed a priori. References [18, 19] come closest to this goal, but still do not connect the definition of the infinite-dimensional space of functional uncertainties and the dimension of the finite-dimensional approximating space explicitly to the controller's performance in the form of rigorous bounds on the ultimate trajectory tracking error. This book seeks to fill this gap by connecting controller performance to functional uncertainty in various native spaces.

1.3 A Strategy at the Intersection of Three Disciplines

The strategy taken in this book synthesizes results from different technical fields, as depicted symbolically in Fig. 1.1. The three primary disciplines that inform our approach are dynamical systems and control theory, data-driven methods, and approximation theory. In the following, we summarize how each of these fields contributes to our overall philosophy in this book.

1.3.1 Dynamical System Theory

The approach taken in this text views control problems on ODEs as types of DPSs, albeit in a relatively simple form. We show that many of the most complex concepts from the theory of DPS can be avoided to develop a theory of nonparametric control of ODEs. Our approach is to preserve and explain which technical features in the

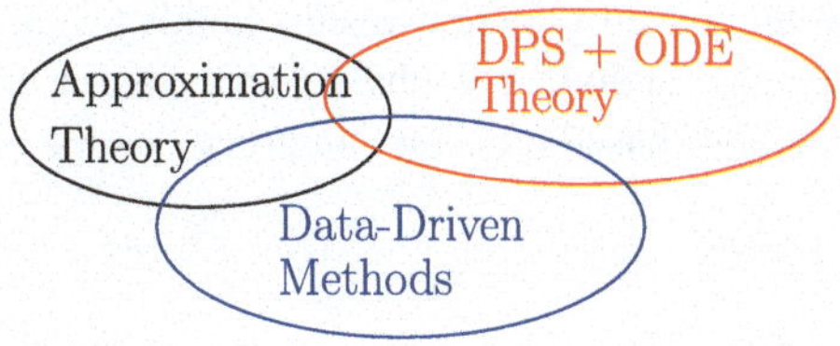

Fig. 1.1 The proposed robust and adaptive control framework lies at the intersection of three macro-areas, namely approximation theory, machine learning theory, and control theory

theory of infinite-dimensional DPS must be retained to have a useful theory to design control systems for ODEs.

At its foundation, the primary structural change between the control methods in Euclidean spaces and the methods presented in this text is that the uncertainty in the governing nonlinear ODEs is functional, not parametric. The adaptive gains, which in classical MRAC systems evolve in Euclidean spaces of finite fixed dimensions, are replaced by adaptive gains that evolve in infinite-dimensional hypothesis spaces chosen among RKHSs. The critical process is to extend the deterministic Lyapunov analyses in the finite-dimensional setting, which constitutes the backbone of the approaches in [18–24, 38], to the RKHS setting where we want to study performance over nonparametric functional uncertainty classes. However, despite this substantial paradigm shift, we have sought to conceptually maintain the same structures in both the control laws and the adaptive laws as in classical MRAC systems. The control systems presented in this book modify classical methods based on control Lyapunov functions, backstepping control, the dead-zone modification of MRAC and several of its smoothed variations, σ-modifications and e-modifications, the use of convex projection operators, error bounding methods, and adaptive error bounding methods. This way, those familiar with the deterministic Lyapunov analyses for these control systems originally derived for ODEs will recognize the general patterns by which they have been lifted easily to the setting of a DPS and functional uncertainty classes in an RKHS.

1.3.2 Data-Driven Methods

The explosion of interest and progress in machine and statistical learning theory in general and data-driven methods in particular, especially over the past decade, does not need any emphasis. The strength of these approaches in Bayesian estimation and estimation by Gaussian processes is attributable in no small part to the power of framing estimation in terms of RKHSs [4–8]. However, the overwhelming majority of such efforts in statistical learning theory involve samples generated by a discrete stochastic process or random walk. In this book, we achieve analogous general results for nonlinear, continuous time, deterministic dynamical systems. This result is obtained by incorporating popular tools in learning theory and data-driven methods over RKHSs into the deterministic framework of Lyapunov-based analysis and synthesis of control systems, with special emphasis on adaptive control systems. While most of the new results in this text are studied using a deterministic analysis, we also include a stochastic analysis for ease of comparison of the new results to methods based on Gaussian processes.

1.3.3 Approximation Theory

Approximation theory over RKHSs plays a critical role in the approach proposed in this book, which usually assumes that matched functional uncertainties are elements of an RKHS. Both the controlled plant and the adaptive laws define an infinite-dimensional system. Then, to enable a practical implementation, this system is projected onto finite-dimensional spaces. In this work, we study how the closed-loop control strategy performs for a wide class of systems based on the choice of hypothesis space in which the uncertainty lies; the definition of the admissible class of nonparametric functional uncertainty; the dimension of the finite-dimensional space employed for a finite-dimensional approximation; and the choice of basis that defines this space. These studies are at the core of modern approximation theory, machine learning theory, regression theory, Bayesian estimation theory, and Gaussian process estimation theory. However, they are substantially neglected in the literature on real parametric control in Euclidean spaces.

Many works on adaptive control systems in Euclidean spaces assume that some oracle provides a good choice of the approximating hypothesis space, that is, of the space in which the uncertainties affecting some dynamical models are assumed to lie. This oracle is embodied in the underlying uniform approximation assumption, whereby the matched uncertainties are well captured by a given regressor vector, irrespectively of the plant's initial condition and the closed-loop system's actual trajectory. By default, the approximation assumption is routinely stated with respect to the space of continuous functions of bounded support or the space of uniformly bounded functions over some subset of the state space that contains a trajectory of interest. This book introduces a general strategy that holds for a wide collection of hypothesis spaces that are RKHS. The topologies for these spaces are not equivalent, in contrast to the analysis of adaptive estimation and control in Euclidean spaces. The spaces in which the closed-loop system and adaptive laws evolve are therefore different and not generally equivalent.

The strategy taken in this book makes extensive use of analysis, estimation, and approximation in terms of scattered bases native spaces [41]. Instead of assuming that a rather weak form of the approximation assumption holds the space of continuous functions of bounded support or the space of uniformly bounded functions, we seek to enforce a stronger approximation assumption by the placement of centers of approximation of these scattered bases. In other words, the approximation requirements in the native space inform us on how to choose bases and where to locate them in the configuration space, that is, in the space where the plant dynamics evolve. Whenever possible, we tie the conclusions regarding the performance of methods for adaptive control to standard function spaces. For example, we relate the native spaces in which the functional uncertainty is contained to common spaces that characterize regularity like the Sobolev spaces over some subset of the state space.

1.4 Organization of This Book

The technical material in this text begins with a summary of the required mathematical background in Chap. 2. This introductory chapter covers some basic ingredients of real and functional analysis that are required in the proofs of stability and convergence of the controllers and learning schemes developed in the work.

Chapter 3 gives a self-contained description of native spaces for scalar-valued and vector-valued functions. Many dedicated texts, including [4–8, 41, 42], to name a few, present the general theory of native spaces. Chapter 3 extracts some key topics that appear in these references, or in the relatively recent body of research publications, and that are most relevant to the proposed problems of estimation and control for uncertain nonlinear systems. No new results are contained in Chap. 3. Chapter 3 begins with the study of real-valued native spaces and summarizes what are some of the more common choices for reproducing kernels in applications. A concise summary of standard approaches that characterize approximation errors is also included. The chapter concludes with a discussion of the theory of operator-valued kernels and vector-valued RKHS.

Chapter 4 presents an essential, self-contained account of the theory of dynamical systems and processes. This chapter summarizes some of the most popular definitions and tools from Lyapunov analysis for the study of the boundedness, stability, convergence, and ultimate boundedness of nonlinear dynamical systems in finite-dimensional spaces. This chapter also summarizes a few key definitions and theorems needed to analyze the infinite-dimensional systems that arise in limiting DPSs.

The core of the new material in this book begins with Chap. 5, which presents control systems for sets of deterministic, uncertain, nonlinear ordinary differential equations assuming that the uncertainties in the governing ODEs lie in a vector-valued RKHS induced by an operator-valued kernel. Initially, we extend the theory of robust control Lyapunov functions to affine-in-the-control dynamical systems affected by functionally matched uncertainties that reside in some vector-valued RKHS. Subsequently, we extend the classical backstepping control scheme to account for vector-valued uncertainties residing in some native space. Then, we present multiple solutions to the MRAC design problem for plants affected by parametric uncertainties and matched functional uncertainties that lie in some RKHS. The MRAC systems design problem is first solved in terms of a limiting DPS. Successively, to enable the implementation of this solution, finite-dimensional approximations of this limiting DPS are presented. These solutions include extensions of the classical dead-zone modification, e-modification of MRAC, and the use of convex projection operators. The treatise in Chap. 5 is concluded by two types of variable structure control systems, namely an error bounding method and and adaptive error bounding method. Numerical simulations show the applicability of the proposed results.

Chapter 6 continues the study of the methods introduced in Chap. 5, while focusing on how the dimensions of the approximating hypothesis space can be chosen, and, possibly, adapted to the controller's performance. Only deterministic approaches are studied in Chap. 6. The chapter begins by introducing three different architectures that merge elements of learning and control theory. These architectures include the learn-then-control (LTC), the switched learn-and-control (SLC), and the concurrent learn-and-control (CLC) approaches. This chapter also introduces the notion of almost approximation theory optimal (AAO) and ϵ-AAO adaptive controllers.

Chapter 7 further extends the deterministic analysis in Chap. 6 by introducing a stochastic structure. This chapter begins by reviewing the foundations of statistical and machine learning theory, with an emphasis on the empirical risk minimization problem and principles of Gaussian process estimation. Then, this chapter presents methods to characterize ultimate bounds on the trajectory tracking error of systems of ODEs controlled by adaptive control schemes, while the native space serving as the approximating hypothesis space is designed by choosing its centers randomly and accounting for the random nature of the uncertainties.

Chapter 8 reviews the key novelties introduced in the book. Emphasis is placed on identifying the limitations of the progress so far in the development of a general nonparametric adaptive control theory. Several open technical questions and potential venues for important future work are summarized.

References

1. Smith, R.C.: Uncertainty Quantification: Theory, Implementation, and Applications, vol. 12. SIAM, Philadelphia, PA (2013)
2. Sullivan, T.J.: Introduction to Uncertainty Quantification, vol. 63. Springer, New York, NY (2015)
3. Zhou, K., Doyle, J.C.: Essentials of Robust Control, vol. 104. Prentice Hall, Upper Saddle River, NJ (1998)
4. Berlinet, A., Thomas-Agnan, C.: Reproducing Kernel Hilbert Spaces in Probability and Statistics. Springer, New York, NY (2011)
5. Williams, C.K., Rasmussen, C.E.: Gaussian Processes for Machine Learning, vol. 2, no. 3. MIT Press, Cambridge, MA (2006)
6. Saitoh, S., Sawano, Y.: Theory of Reproducing Kernels and Applications. Springer, London, UK (2016)
7. Paulsen, V.I., Raghupathi, M.: An Introduction to the Theory of Reproducing Kernel Hilbert Spaces, vol. 152. Cambridge University Press, Cambridge, UK (2016)
8. Smola, B.S.A.J.: Learning with Kernels: Support Vector Machines, Regularization, Optimization, and Beyond. MIT Press, Cambridge, MA (2002)
9. Chowdhary, G., Kingravi, H.A., How, J.P., Vela, P.A.: Bayesian nonparametric adaptive control using Gaussian processes. IEEE Trans. Neural Netw. Learn. Syst. **26**(3), 537–550 (2015)
10. Rosenfeld, J.A., Kamalapurkar, R., Dixon, W.E.: State following (StaF) kernel functions for function approximation Part I: theory and motivation. In: American Control Conference, pp. 1217–1222. IEEE, Chicago, IL (2015)
11. Liu, M., Chowdhary, G., Da Silva, B.C., Liu, S.-Y., How, J.P.: Gaussian processes for learning and control: a tutorial with examples. IEEE Control Syst. Mag. **38**(5), 53–86 (2018)
12. Boffi, N.M., Tu, S., Slotine, J.-J.E.: Nonparametric adaptive control and prediction: Theory and randomized algorithms. J. Mach. Learn. Res. **23**(1), 12 841–12 886 (2022)
13. Farrell, J.A.: Stability and approximator convergence in nonparametric nonlinear adaptive control. IEEE Trans. Neural Netw. **9**(5), 1008–1020 (1998)

14. Chowdhary, G., How, J., Kingravi, H.: Model reference adaptive control using nonparametric adaptive elements. In: AIAA Guidance, Navigation, and Control Conference, p. 5038. Minneapolis, MN (2012)
15. Kurdila, A., Lei, Y.: Adaptive control via embedding in reproducing kernel Hilbert spaces. In: American Control Conference, pp. 3384–3389. IEEE, Florence, Italy (2013)
16. Kurdila, A.J., L'Afflitto, A., Burns, J.A., Wang, H.: Nonparametric adaptive control in native spaces: a DPS framework (Part I). Ann. Rev. Control **58**, 100969 (2024)
17. Kurdila, A.J., L'Afflitto, A., Burns, J.A., Wang, H.: Nonparametric adaptive control in native spaces: finite-dimensional implementations (Part II). Ann. Rev. Control **58**, 100969 (2024)
18. Farrell, J.A., Polycarpou, M.M.: Adaptive Approximation Based Control: Unifying Neural, Fuzzy and Traditional Adaptive Approximation Approaches, vol. 48. Wiley, Danvers, MA (2006)
19. Lavretsky, E., Wise, K.: Robust and Adaptive Control: With Aerospace Applications. Springer, London, UK (2013)
20. Ioannou, P.A., Sun, J.: Robust Adaptive Control. Courier Corporation, Mineola, NY (2012)
21. Krstic, M., Kanellakopoulos, I., Kokotovic, P.: Nonlinear and Adaptive Control Design. Wiley, New York, NY (1995)
22. Sastry, S., Bodson, M.: Adaptive Control: Stability, Convergence and Robustness. Courier Corporation, New York, NY (2011)
23. Tao, G.: Adaptive Control Design and Analysis, vol. 37. Wiley, Danvers, MA (2003)
24. Narendra, K.S., Annaswamy, A.M.: Stable Adaptive Systems. Dover, New York, NY (1989)
25. Slotine, J.-J.E., Li, W.: On the adaptive control of robot manipulators. Int. J. Robot. Res. **6**(3), 49–59 (1987)
26. Hovakimyan, N., Cao, C.: $\mathcal{L}_1$ Adaptive Control Theory: Guaranteed Robustness with Fast Adaptation. SIAM, Philadelphia, PA (2010)
27. Curtain, R., Zwart, H.: An Introduction to Infinite-Dimensional Linear Systems Theory. Springer (1995)
28. Da Prato, G., Zabczyk, J.: Stochastic Equations in Infinite Dimensions. Cambridge University Press, Cambridge, UK (2014)
29. Bensoussan, A., Da Prato, G., Delfour, M.C., Mitter, S.K.: Representation and Control of Infinite Dimensional Systems, vol. 2. Springer, New York, NY (2007)
30. Curtain, R.F., Zwart, H.: An introduction to infinite-dimensional linear systems theory. Springer, Berlin, Germany (2012)
31. Banks, H.T., Burns, J.A.: Hereditary control problems: numerical methods based on averaging approximations. SIAM J. Control Optim. **16**(2), 169–208 (1978)
32. Banks, H.T., Burns, J.A., Cliff, E.M.: Parameter estimation and identification for systems with delays. SIAM J. Control Optim. **19**, 791 (1981)
33. Burns, J.A., Herdman, T.L.: Adjoint semigroup theory for a class of functional differential equations. SIAM J. Math. Anal. **7**(5), 729–745 (1976)
34. Burns, J.A., Herdman, T.L., Stech, H.W.: Linear functional differential equations as semigroups on product spaces. SIAM J. Math. Anal. **14**, 98–116 (1983)
35. Delfour, M.C.: The largest class of hereditary systems defining a C_0 semigroup on the product space. Canadian J. Math. **32**(4), 969–978 (1980)
36. Hale, J.K., Lunel, S.M.V.: Introduction to Functional Differential Equations, vol. 99. Springer, New York, NY (2013)
37. Warga, J.: Optimal Control of Differential and Functional Equations. Academic Press, Cambridge, MA (2014)
38. Slotine, J.J.E., Li W et al.: Applied Nonlinear Control, vol. 199, no. 1. Prentice Hall (1991)
39. DeVore, R.A.: Nonlinear approximation. Acta Numerica **7**, 51–150 (1998)
40. Vapnik, V.N.: An overview of statistical learning theory. IEEE Trans. Neural Netw. **10**(5), 988–999 (1999)
41. Wendland, H.: Scattered Data Approximation, vol. 17. Cambridge University Press, Cambridge, UK (2004)
42. Aronszajn, N.: Theory of reproducing kernels. Trans. Am. Math. Soc. **68**, 337–404 (1950)

Chapter 2
Elements of Real and Functional Analysis

This chapter presents the reader with a compendium of key notions in real and functional analysis, which are needed throughout this book. Although it is expected that the reader has some prior knowledge of analysis, this chapter presents the material in a tutorial manner, starting with fundamental abstract spaces. Key function spaces, which are employed in multiple instances throughout this book, and their most relevant properties are discussed. An additional aim of this chapter is to present essential, perhaps known, tools consistently so that the reader need not refer to multiple sources, which may employ different notations and axiomatic structures. The interested reader is referred to specific points in classical references in multiple instances. The material presented here can be found in the references [1–6].

This chapter is organized as follows. Section 2.1 establishes fundamental notation, primarily in the context of set theory, commonly employed in this book. Section 2.2 provides essential elements of topological, metric, and inner product spaces. Duality pairings and Riesz representers, which are used abundantly in the following chapters, are introduced in Sect. 2.3. Finally, commonly used function spaces, with special emphasis on those employed in this book, are discussed in Sect. 2.4. Section 2.5 makes some closing remarks, and Sect. 2.6 proposes some homework problems.

2.1 Fundamental Notation

Mathematical notation is an essential tool for writing readable articles and books in engineering, mathematics, and the sciences. For this reason, we have attempted to provide consistent notation, use standard symbols, and give precise definitions of terms and concepts.

A. J. Kurdila et al., *Data-Driven, Nonparametric, Adaptive Control Theory*,
Lecture Notes in Control and Information Sciences 495,
https://doi.org/10.1007/978-3-031-78003-5_2

In this text, we use the symbol "$\triangleq$" to create definitions, so that $A \triangleq B$ is understood as "A is defined or given by B." Since many inequalities may contain constants whose values are not particularly noteworthy in subsequent arguments, we write $a \lesssim b$ if there is a constant $c > 0$, which does not depend on a or b, for which $a \leq c \cdot b$. It is important to note that c must be constant, that is, if a and b depend on some parameter, then c is independent of this parameter. The reverse inequality sign $\gtrsim$ is defined similarly. We use $a \approx b$ as shorthand for $a \lesssim b \lesssim a$.

The *empty set* is denoted by $\{\emptyset\}$. We denote the *set of real numbers* by $\mathbb{R}$, the *set of complex numbers* by $\mathbb{C}$, and the set of integers by $\mathbb{Z}$. The symbol $|\alpha|$ denotes the *absolute value* of the real number $\alpha \in \mathbb{R}$. For the complex number $\alpha = a + bi \in \mathbb{C}$, and $|\alpha| \triangleq \sqrt{a^2 + b^2}$ denotes its *modulus*. The *set of nonnegative real numbers* $[0, \infty)$ is denoted by $\mathbb{R}^+$. Likewise, we denote the *set of complex numbers with nonnegative real parts* by $\mathbb{C}^+ \triangleq \{\alpha = a + b\jmath : \text{Re}(\alpha) = a \geq 0\}$, where $\jmath \triangleq \sqrt{-1}$. Finally, the *set of nonnegative integers* is denoted by $\mathbb{Z}^+ \triangleq \{z \in \mathbb{Z} : z \geq 0\}$. The *natural numbers* $\mathbb{N}$ are the positive integers so that $\mathbb{N} \triangleq \mathbb{Z}^+ \backslash \{0\}$, or equivalently, $\mathbb{Z}^+ \triangleq \mathbb{N} \cup \{0\}$. A set $\mathbb{X}$ is *countable* if there is a one-to-one correspondence between all elements of $\mathbb{X}$ and the elements in a subset of $\mathbb{N}$.

The set of *positive real numbers* is denoted by $\mathbb{R}_+ \triangleq (0, \infty)$, and the *open right-half complex plane* is denoted by $\mathbb{C}_+ \triangleq \{\alpha = a + b\jmath : \text{Re}(\alpha) = a > 0\}$. The *imaginary axis* is the set $\mathbb{C}_0 \triangleq \{\alpha = a + b\jmath : \text{Re}(\alpha) = a = 0\}$.

Remark 2.1 It is important to note that the proposed notation has many variations and, depending on the particular scientific area, may or may not include the left boundary of these sets. We have tried to be consistent in that $\mathbb{R}^+ = [0, \infty)$ and $\mathbb{C}^+ = \{\alpha = a + b\jmath : \text{Re}(\alpha) = a \geq 0\}$ are closed, while $\mathbb{R}_+ = (0, \infty)$ and $\mathbb{C}_+ = \{\alpha = a + b\jmath : \text{Re}(\alpha) = a > 0\}$ are open sets.

The *real Cartesian space of dimension n* is defined by

$$\mathbb{R}^n \triangleq \left\{x = [x_1, x_2, \ldots, x_n]^{\mathrm{T}} : x_i \in \mathbb{R}, i = 1, 2, \ldots, n\right\}, \tag{2.1}$$

where ^{T} denotes the *transpose*. Therefore, $\mathbb{R}^n$ is the space of n-dimensional column vectors. Likewise, we set

$$\mathbb{C}^n \triangleq \left\{z = [z_1, z_2, \ldots, z_n]^{\mathrm{T}} : z_i \in \mathbb{C}, i = 1, 2, \ldots, n\right\}. \tag{2.2}$$

Let X and Y be two sets and $f(\cdot)$ be a function with *domain* $\mathscr{D}(f) \subseteq X$ and *range* $\mathscr{R}(f) \subseteq Y$. In this case, we write $f : \mathscr{D}(f) \subseteq X \to Y$. If $\mathscr{D}(f) = X$, then we write $f : X \to Y$. If $f : \mathscr{D}(f) \subseteq X \to Y$, then the *preimage of $A \subseteq Y$ under f* is defined as

$$f^{-1}(A) \triangleq \{x \in \mathscr{D}(f) : f(x) \in A\}. \tag{2.3}$$

2.2 Topological, Metric, Norm, and Inner Product Spaces

In this section, we present a very brief review of the basic abstract spaces needed in this text. The main definitions and results can be found in [5, Chap. 3], in [6, Chap. IV], and [7]. Generally speaking, an *abstract space* is a set having a certain amount of "algebraic and topological structure." For example, $\mathbb{R}^n$ is a linear vector space, which establishes its algebraic properties. The algebraic structure defines what operations are defined on the set. If a norm $\|\cdot\|$ is imposed on $\mathbb{R}^n$, then the pair $(\mathbb{R}^n, \|\cdot\|)$ becomes a normed linear space. The norm provides a topological structure to the set $\mathbb{R}^n$ so that concepts such as open sets, closed sets, continuity, and compactness can be defined. The following discussion is at the level of the well-known introductory books by Kreyszig in [3], or by Naylor and Sell in [7]. Therefore, the remainder of this section can be safely skipped by readers familiar with the basics of functional analysis and operator theory.

2.2.1 *Topological Spaces*

Many of the functional spaces considered in this book are topological spaces, which are the most basic abstract spaces we discuss. The definition of the topology enables concepts like continuity, compactness, and connectedness to be defined for the set X. Most of the proofs in the text do not make arguments that rely on such an abstract setting, although there are a few. For completeness, we just note the definition and intuition behind it, and the specific theorems used in Chaps. 5–7.

Definition 2.1 (*Topology on a Set*) Let $\mathbb{X}$ denote a set, and let τ denote a family of subsets of $\mathbb{X}$. Then, τ is a *topology on* $\mathbb{X}$ if

1. both the empty set $\{\emptyset\}$ and $\mathbb{X}$ are elements of τ,
2. any union of elements of τ is an element of τ, and
3. any intersection of finitely many elements of τ is an element of τ.

Given a topology τ on X, the members of τ are called *open sets* in $\mathbb{X}$. A subset of $\mathbb{X}$ is said *closed* if its complement is open. A subset of $\mathbb{X}$ may be open, closed, both open and closed, or neither. The empty set $\{\emptyset\}$ and $\mathbb{X}$ itself are both closed and open. A set that contains an open subset of $\mathbb{X}$ which contains a point x is called a *neighborhood of* x. A *connected set* cannot be represented as the union of two or more disjoint nonempty open subsets. A nonempty, open, and connected set is a *(topological) domain*.

Definition 2.2 (*Topological Spaces and Hausdorff Spaces*) If τ is a topology on $\mathbb{X}$, then the pair $(\mathbb{X}, \tau)$ is called a *topological space*. The space $(\mathbb{X}, \tau)$ is said to be a *Hausdorff space* (sometimes called a T_2 *space*) if, given any two distinct points $x \in \mathbb{X}$ and $y \in \mathbb{X}$, there exist neighborhoods U of x and V of y such that $U \cap V = \{\emptyset\}$.

In addition to open and closed sets, compact sets play a fundamental role in analysis. To define a compact set in a general topological space, the concepts of covers and open covers are needed.

Definition 2.3 (*Covers of Sets*) Let $(\mathbb{X}, \tau)$ be a topological space and $S \subseteq \mathbb{X}$. A *cover* of S is a collection of subsets $\mathscr{O} = \{U_\alpha \mid \alpha \in \mathscr{A}\}$ such that $S \subset \bigcup_\alpha U_\alpha$. The cover $\mathscr{O} = \{U_\alpha \mid \alpha \in \mathscr{A}\}$ is called a *finite cover* if it contains only a finite number of sets, i.e., $\mathscr{A} = \{1, 2, 3, \ldots, M\}$. A *subcover* of S is any subcollection $\mathscr{O}_{\text{sub}} = \{U_\beta \mid \beta \in \mathscr{B} \subseteq \mathscr{A}\} \subseteq \mathscr{O}$ that also covers S. A cover $\mathscr{O}$ is called *open* if each set $U_\alpha \in \mathscr{O}$ is open.

Definition 2.4 (*Compact Set*) Let $(\mathbb{X}, \tau)$ be a topological space and $S \subseteq \mathbb{X}$. The set $S \subseteq \mathbb{X}$ is *compact* if every open cover $\mathscr{O} = \{U_\alpha \mid \alpha \in \mathscr{A}\}$ of S has a finite subcover.

To ensure unique limits and other "regularity" properties, we shall require that all topological spaces $(\mathbb{X}, \tau)$ be Hausdorff spaces. Convergence in certain Hausdorff spaces can be described in terms of sequences and neighborhoods. A *sequential space* is a topological space whose topology can be completely characterized by its convergent/divergent sequences.

Definition 2.5 (*Convergent Sequence*) Let $\mathbb{X}$ be a Hausdorff space. A sequence $\{x_n\}$ converges to $x \in \mathbb{X}$ if for every open neighborhood U of x, there exists an integer N such that $\{x_n\} \in U$ for all $n \geq N$.

In a Hausdorff space, if a limit of the sequence exists, then it is unique.

Definition 2.6 (*Continuous Function*) Let $\mathbb{X}$, $\mathbb{Y}$ be Hausdorff spaces and $f : \mathbb{X} \to \mathbb{Y}$. Then, $f(\cdot)$ is *continuous* if any one of the following (equivalent) conditions is satisfied:

1. For every open set $O \subset \mathbb{Y}$, the preimage $f^{-1}(O) \subset \mathbb{X}$ is open in X.
2. For every closed set $C \subset \mathbb{Y}$, the preimage $f^{-1}(C) \subset \mathbb{X}$ is closed in X.

Remark 2.2 If $f : \mathbb{X} \to \mathbb{Y}$ is continuous and $\{x_n\} \to x \in \mathbb{X}$, then $\{f(x_n)\} \to f(x) \in \mathbb{Y}$. A function $f : \mathbb{X} \to \mathbb{Y}$ is *sequentially continuous* if the image of every convergent sequence is a sequence that converges to the image of the limit. Thus, every continuous function is sequentially continuous, but the converse is not always true. Any metric space is a sequential space (see [8]) and almost all the spaces we consider in this book are sequential.

2.2.2 Metric Spaces

The majority of function spaces in this text have more structure than topological spaces: they are metric spaces. Intuitively, the introduction of a metric endows the set $\mathbb{X}$ with a notion of distance between its elements, which are used to define the topology.

Definition 2.7 (*Metric Spaces*) A *metric space* is a pair $(\mathbb{X}, d)$, where $\mathbb{X}$ is a set and the *metric on* $\mathbb{X}$ is a real-valued function $d : \mathbb{X} \times \mathbb{X} \to \mathbb{R}$ such that

1. $d(x, y) = d(y, x) \geq 0$ for all $x, y \in \mathbb{X}$,
2. $d(x, y) = 0$ if and only if $x = y$, and
3. the *triangle inequality*

$$d(x, z) \leq d(x, y) + d(y, z), \quad \text{for all } \; x, y, z \in \mathbb{X} \tag{2.4}$$

is satisfied.

For simplicity, and as in the overwhelming majority of textbooks in functional analysis, nonlinear systems analysis, and control theory, we will often abuse notation and say that " $\mathbb{X}$ is a metric space" with the understanding that this refers to the pair $(\mathbb{X}, d)$. The metric on a set $\mathbb{X}$ is also referred to as the *distance* between any two points in $\mathbb{X}$. Having defined the notion of distance, we introduce the notion of open all.

Definition 2.8 (*Open Ball*) If $(\mathbb{X}, d)$ is a metric space and $r \in \mathbb{R}_+$, then the *open ball about* x_0 *of radius* r is defined

$$\mathcal{B}_r(x_0) \triangleq \{y \in \mathbb{X} : d(y, x_0) < r\}\,. \tag{2.5}$$

The notion of open ball then enables the following definitions.

Definition 2.9 (*Interior Point*) If $\mathscr{A} \subseteq \mathbb{X}$, then $x \in \mathscr{A}$ is an *interior point of* $\mathscr{A}$ if there exists a real number $r > 0$ such that $\mathcal{B}_r(x) \subset \mathscr{A}$. The set of all interior points of $\mathscr{A}$ is denoted by $\mathring{\mathscr{A}}$ and is called the *interior* of $\mathscr{A}$.

Definition 2.10 (*Open Set*) The subset $\mathscr{A}$ is *open* if every $x \in \mathscr{A}$ is an interior point of $\mathscr{A}$.

Definition 2.11 (*Bounded Set*) A set $\mathscr{A}$ is *bounded* if there is a positive number $m > 0$ such that $d(x, y) \leq m$ for all $x, y \in \mathscr{A}$. A set $\mathscr{A} \subseteq \mathbb{X}$ is *totally bounded* if for each $\varepsilon > 0$, there is a finite number of points $\{x_1, x_2, \ldots, x_m\} \subset \mathscr{A}$ such that $\mathscr{A} \subseteq \bigcup_{i=1}^{m} \mathcal{B}_\epsilon(x_i)$.

Definition 2.12 (*Neighborhood of a Point*) A set $\mathscr{N}$ is called a *neighborhood of* x if there exists a real number $r > 0$ such that $x \in \mathcal{B}_r(x) \subset \mathscr{N}$.

A large portion of the results presented in this book relies on complex structures most readers may be already familiar with in the context of the analysis on $\mathbb{R}^n$ and that in the following we present in a more general setting. To present such structures, firstly we introduce the notion of sequence and, hence, the notion of convergent sequence.

Definition 2.13 (*Sequence*) A *sequence in* $\mathbb{X}$ is an ordered set $\{x_n\} \subset \mathbb{X}$ that can be put in correspondence with the positive real integers, $n = 1, 2, 3, \ldots$.

Definition 2.14 (*Convergent Sequence*) The sequence $\{x_n\}$ converges to $x_0 \in \mathbb{X}$ if, for each $\epsilon > 0$, there exists an integer $N = N(\epsilon)$ such that, for all $n > N\epsilon)$, one has $d(x_n, x_0) < \epsilon$. In this case, we say that the sequence $\{x_n\}_{n\in\mathbb{N}}$ is *convergent* and write $x_n \to x_0$ as $n \to \infty$, or $\lim_{n\to\infty} x_n = x_0$.

This definition of convergence can be understood as a special case of convergence in Hausdorff topological spaces discussed in Sect. 2.2.1: a sequence $\{x_n\}_{n\in\mathbb{N}}$ is convergent if it is eventually contained in every neighborhood of x.

Definition 2.15 (*Cauchy Sequence*) A sequence in $\mathbb{X}$ is *Cauchy* if, for each $\epsilon > 0$, there exists an integer $N = N(\epsilon)$ such that for all $n > N(\epsilon)$ and $m > N(\epsilon)$, $d(x_n, x_m) < \epsilon$.

If we define the open sets of the topology to be the union of open balls $\mathscr{B}_r(x_0)$, then the metric space is a Hausdorff space. This means in particular that Theorem 2.6 holds in a metric space. In fact, we can replace all the appearances of "generalized sequence" in Theorem 2.6 by a "sequence," and the theorem holds in a form that is much easier to intuit. Every convergent sequence is Cauchy. However, the converse is not always true. Only special metric spaces $(\mathbb{X}, d)$ have this property which leads to the following definition.

Definition 2.16 (*Complete Metric Spaces*) A metric space $(\mathbb{X}, d)$ is *complete* if every Cauchy sequence $\{x_n\}$ is convergent. That is, there exists a $x_0 \in \mathbb{X}$ such that $\lim_{n\to\infty} x_n = x_0$.

The established notions of sequences and convergence of sequences lead to the following definition.

Definition 2.17 (*Limit Point*) Let $(\mathbb{X}, d)$ be a metric space and $\mathscr{A} \subset \mathbb{X}$. A point x_0 is called a *limit point of* $\mathscr{A}$ if there is a sequence $\{x_n\} \subset \mathscr{A}$ such that $\lim_{n\to\infty} x_n = x_0$.

In the following, we define several classes of sets, whose properties are directly linked to convergence properties of sequences.

Definition 2.18 (*Closed Set and Closure of a Set*) A set $\mathscr{A} \subset \mathbb{X}$ is *closed* if it contains all of its limit points. The *closure of* $\mathscr{A}$ is the set of all limit points of $\mathscr{A}$ and is denoted by $\bar{\mathscr{A}}$ so that $\mathscr{A} \subseteq \bar{\mathscr{A}} \subseteq \mathbb{X}$.

Definition 2.19 (*Boundary of a Set*) Given a set $\mathscr{A} \subset \mathbb{X}$, the *boundary* of $\mathscr{A}$ is defined as $\bar{\mathscr{A}} \setminus \mathscr{A}$.

It follows from Definition 2.18 that a set $\mathscr{A} \subseteq \mathbb{X}$ is closed if and only if $\mathscr{A} = \bar{\mathscr{A}}$.

Definition 2.20 (*Dense Set*) A subset $\mathscr{A} \subseteq \mathbb{X}$ is *dense in* $\mathbb{X}$ if $\bar{\mathscr{A}} = \mathbb{X}$.

Definition 2.21 (*Separable Space*) The metric space $(\mathbb{X}, d)$ is *separable* if there is a countable set $\mathscr{Q} \subseteq \mathbb{X}$ such that $\mathscr{Q}$ is dense in $\subseteq \mathbb{X}$.

Definition 2.22 (*Precompact and Compact Sets*) Let $\mathscr{A} \subseteq \mathbb{X}$, where $\mathbb{X}$ is a metric space. The set $\mathscr{A}$ is *precompact* if every sequence $\{x_n\} \subset \mathscr{A}$ contains a Cauchy subsequence $\{x_{n_k}\} \subset \mathscr{A}$.

It is easy to show that all precompact sets are bounded, but precompact sets do not have to be closed.

Definition 2.23 (*Locally Compact and Compact Metric Spaces*) A metric space $(\mathbb{X}, d)$ is *locally compact* if every closed and bounded subset is compact. A metric space $(\mathbb{X}, d)$ is *compact* if it is complete and totally bounded.

The next result recalls necessary and sufficient conditions for the compactness of sets. Usually, these conditions are more operative than Definition 2.4.

Theorem 2.1 *Let $\mathscr{A} \subseteq \mathbb{X}$, where $(\mathbb{X}, d)$ is a complete metric space. The set $\mathscr{A}$ is compact if and only if it is precompact and bounded. If, in addition, $(\mathbb{X}, d)$ is locally compact, then a set $\mathscr{A}$ is compact if and only if it is closed and bounded.*

Remark 2.3 There are some variations in the definition of what constitutes a precompact set in the literature. Some authors define a precompact (sometimes called *relatively compact*) set as one whose closure is compact. This can be shown to be equivalent to Definition 2.4 in terms of Cauchy sequences if $(\mathbb{X}, d)$ is a metric space.

Definition 2.24 (*Support of a Function*) Let $\mathbb{X}$ denote a metric space and $f : \mathscr{D}(f) \subseteq \mathbb{X} \to \mathbb{R}^m$ be a function. The *support* of f (also known as the *closed support of f*) is defined as

$$\operatorname{supp}(f) \triangleq \overline{\{x \in \mathscr{D}(f) \subseteq \mathbb{X} : f(x) \neq 0\}}.$$

Definition 2.25 (*Compact Support of a Function*) Let $\mathbb{X}$ denote a metric space. The function $f : \mathscr{D}(f) \subseteq \mathbb{X} \to \mathbb{R}^m$ is said to have *compact support* if $\operatorname{supp}(f)$ is compact.

2.2.3 Vector Spaces

In this book, vector spaces play a key role for their algebraic structure.

Definition 2.26 (*Real Vector Space*) A *real vector space over the field* $\mathbb{R}$ is a nonempty set $\mathbb{X}$ endowed with two operations, namely the *addition* of its elements and their *multiplication by scalars*. These two operations are such that

$$z \triangleq x + y \in \mathbb{X} \quad \text{for any} \quad x, y \in \mathbb{X}, \quad \text{and,}$$
$$z \triangleq \alpha \cdot x \in \mathbb{X} \quad \text{for any} \quad \alpha \in \mathbb{R}, x \in \mathbb{X}.$$

Furthermore, it holds that

$$(x + y) + z = x + (y + z), \quad \text{for any} \quad x, y, z \in \mathbb{X},$$

$$x + y = y + x,$$

there exists an element, denoted by $0 \in \mathbb{X}$, such that

$$x + 0 = x, \quad \text{for any} \quad x \in \mathbb{X},$$

there exists an element, denoted by $-x \in \mathbb{X}$ such that

$$x + (-x) = 0, \quad \text{for any} \quad x \in \mathbb{X},$$

and, finally,

$$\begin{aligned} \alpha(\beta x) &= (\alpha\beta)x, \quad \text{for any} \quad (\alpha, \beta, x) \in \mathbb{R} \times \mathbb{R} \times \mathbb{X}, \\ (\alpha + \beta)x &= \alpha x + \beta x. \end{aligned}$$

The elements of $\mathbb{X}$ are called *vectors*, and the elements of the field $\mathbb{R}$ are called *scalars*.

Thus, a vector space is a triplet $(\mathbb{X}, +, \cdot)$ given by a set, the operation "+" of addition of its elements, and the operation "$\cdot$" of multiplication of its elements by scalars. In this text, as in practically all texts, we just refer to the set $\mathbb{X}$ as a vector space instead of the triplet $(\mathbb{X}, +, \cdot)$. Furthermore, for brevity, and consistently with all relevant publications on this topic, we drop the multiplication operator "$\cdot$" in the product of scalars by vectors, except for very rare cases. Thus, we write αx instead of $\alpha \cdot x$ for $x \in \mathbb{X}$, $\alpha \in \mathbb{R}$. Finally, since all the vector spaces considered in this text are over the real field, we refer to the elements of $\mathbb{R}$ as scalars.

Definition 2.27 (*Linear Independence*) Let $\mathbb{X}$ denote a vector space over $\mathbb{R}$. The finite set $\{x_i\}_{i=1}^n \subset \mathbb{X}$ is *linearly independent* if

$$\sum_{i=1}^{n} \alpha_i x_i = 0 \tag{2.6}$$

if and only if $\alpha_1 = \ldots = \alpha_n = 0$. The finite collection $\{x_i\}_{i=1}^n$ is *linearly dependent* if it is not linearly independent. A general subset $S \subseteq \mathbb{X}$ is linearly independent if every finite set it contains is linearly independent. Otherwise, it is *linearly dependent*.

Definition 2.28 (*Finite-Dimensional Vector Spaces*) A vector space $\mathbb{X}$ is *finite-dimensional* if there is $n \in \mathbb{N}$ such that $\mathbb{X}$ contains a linearly independent set of n vectors and any set of $n + 1$ vectors in $\mathbb{X}$ is linearly dependent.

A vector space of dimension n may contain more than n elements. Thus, the following definition is useful.

Definition 2.29 (*Basis of a Vector Space*) A *(Hamel) basis* $\mathbb{B}$ for a vector space $\mathbb{X}$ is a linearly independent set $\mathbb{B} \subset \mathbb{X}$ such that any $x \in \mathbb{X}$ can be written as a finite linear combination of elements in $\mathbb{B}$. The collection of all the vectors in $\mathbb{X}$ that can be written as the sum of finite linear combinations of elements in $\mathbb{B}$ is the (finite linear) *span* of $\mathbb{B}$.

Example 2.1 A vector space extensively used in this book is $\mathbb{R}^n$. The most popular Hamel basis of $\mathbb{R}^n$ is the *canonical basis* is given by $\{e_1, \ldots, e_n\} \subset \mathbb{R}^n$, where

$$[e_1, \ldots, e_n] = \begin{bmatrix} 1 & \ldots & 0 \\ \vdots & \ddots & \vdots \\ 0 & \ldots & 1 \end{bmatrix} \in \mathbb{R}^{n \times n}, \tag{2.7}$$

that is, e_j comprises zeros in all entries, except for the jth entry that is equal to 1. △

Example 2.2 Here is another example that can help illustrate these definitions. Consider the sets

$$\mathbb{X} \triangleq \{v = \{v_i\}_{i=1}^{\infty} \mid v_i \in \mathbb{R}, v_i = 0 \text{ for all but a finite number of indices } i\},$$
$$\mathbb{B} \triangleq \{e_i = \{e_{i,k}\}_{k=1}^{\infty} \mid e_{i,i} = 1, e_{i,k} = 0 \text{ for } k \neq i, 1 \leq i < \infty, 1 \leq k < \infty\}.$$

The set $\mathbb{X}$ consists of real sequences that contain only a finite number of nonzero entries. This set, with the usual sum and product operations, is a vector space. If we add two elements of $\mathbb{X}$ or multiply an element of $\mathbb{X}$ by $\alpha \in \mathbb{R}$, then the resulting element is contained in $\mathbb{X}$ since it still has a finite number of nonzero terms. This means that the set $\mathbb{X}$ is closed under the addition of vectors and their multiplication by scalars, so $\mathbb{X}$ is a vector space. The subset $\mathbb{B} \subset \mathbb{X}$ contains what is known as the *canonical basis* for real-valued sequences. It is immediate that $\mathbb{B} \subset \mathbb{X}$, and that any finite collection of e_i in $\mathbb{B}$ is linearly independent. This means that $\mathbb{B} \subset \mathbb{X}$ is linearly independent, but it is not finite-dimensional. In fact, the span of all the elements in $\mathbb{B}$ is $\mathbb{X}$, and $\mathbb{B}$ is therefore an infinite-dimensional basis for $\mathbb{X}$. △

Definition 2.30 (*Vector Subspace*) A subset $V \subseteq \mathbb{X}$ of a vector space $\mathbb{X}$ is a *vector* (or *linear*) *subspace* of $\mathbb{X}$ if the operations of vector addition and multiplication by scalars defined on $\mathbb{X}$ are closed when restricted to V.

In our applications, vector spaces typically consist of well-known examples such as the Euclidean spaces $\mathbb{R}^n$, spaces of functions on a set, and bounded linear operators from one topological vector space to another. Section 2.4 describes several of the specific abstract vector spaces that consist of functions (or equivalence classes of functions) that are encountered in this text.

Definition 2.31 (*Linear Operators*) Given the vector spaces $\mathbb{X}$ and $\mathbb{Y}$, a *linear operator* $T : \mathscr{D}(T) \subseteq \mathbb{X} \to \mathbb{Y}$ is a function such that

1. $\mathscr{D}(T)$ is a linear subspace of $\mathbb{X}$, and
2. $T(\alpha x_1 + \beta x_2) = \alpha T(x_1) + \beta T(x_2)$ for all $x_1, x_2 \in \mathscr{D}(T)$ and $\alpha, \beta \in \mathbb{R}$.

The following notion extends the well-known definition of range and nullspace of matrices in $\mathbb{R}^{n\times m}$.

Definition 2.32 (*Range and Nullspace of a Linear Operator*) The *range* $\mathscr{R}(T)$ of a linear operator $T : \mathscr{D}(T) \subseteq \mathbb{X} \to \mathbb{Y}$ is the set $\mathscr{R}(T) \triangleq \{y = Tx \in \mathbb{Y} \mid x \in \mathscr{D}(T) \subseteq \mathbb{X}\}$. The *nullspace* $\mathscr{N}(T) \triangleq \{x \in \mathscr{D}(T) \mid Tx = 0\}$.

The nullspace and range of a linear operator $T : \mathscr{D}(T) \subseteq \mathbb{X} \to \mathbb{Y}$ are vector subspaces of $\mathbb{X}$ and $\mathbb{Y}$, respectively. If $\mathscr{D}(T) = \mathbb{X}$, then we write $T : \mathbb{X} \to \mathbb{Y}$. The *identity operator* over some vector space $\mathbb{X}$ is defined as $I_{\mathbb{X}} : \mathbb{X} \to \mathbb{X}$ such that

$$I_{\mathbb{X}}x = x, \quad \text{for all } x \in \mathbb{X}. \tag{2.8}$$

The right-hand side of (2.7) denotes the identity matrix in $\mathbb{R}^n$. If it is clear from the context, then we omit the subscript denoting the space on which the identity operator is defined.

In the following, we present the notions of composing operators, including the composition of operators with themselves, of eigenvalues and eigenvectors or linear operators, and of inverse operators.

Definition 2.33 (*Composition of Linear Operators*) Let $\mathbb{X}$, $\mathbb{Y}$, and $\mathbb{Z}$ be vector spaces. Given linear operators $T : \mathscr{D}(T) \subseteq \mathbb{X} \to \mathbb{Y}$ and $L : \mathscr{D}(L) \subseteq \mathbb{Y} \to \mathbb{Z}$, the composition operator $L \circ T$ is defined so that

1. $\mathscr{D}(L \circ T) = \{x \in \mathscr{D}(T) \mid Tx \in \mathscr{D}(L)\}$, and
2. for each $x \in \mathscr{D}(L \circ T)$, we define $(L \circ T)(x) = L(Tx)$.

Definition 2.34 (*Eigenvalues and Eigenvectors of Linear Operators*) Let $\mathbb{X}$ be a vector space over the of scalars $\mathbb{R}$, and let $T : \mathbb{X} \to \mathbb{X}$ be a linear operator. The vector $x \in \mathbb{X} \setminus \{0\}$ is an *eigenvector* of T if there exists $\lambda \in \mathbb{R}$ such that

$$Tx = \lambda x. \tag{2.9}$$

Definition 2.35 (*Inverse Operators*) Given the vector spaces $\mathbb{X}$ and $\mathbb{Y}$, the linear operator $T : \mathscr{D}(T) \subseteq \mathbb{X} \to \mathbb{Y}$ is *one-to-one* if, for each $y \in \mathscr{R}(T)$, there exist only one $x = x_y \in \mathscr{D}(T)$ such that $T(x) = T(x_y) = y$. If $T : \mathscr{D}(T) \subseteq \mathbb{X} \to \mathbb{Y}$ is one-to-one, then the *inverse operator* $T^{-1} : \mathscr{R}(T) \subseteq \mathbb{Y} \to \mathscr{D}(T) \subseteq \mathbb{X}$ is defined by $T^{-1}(y) = x_y$. The inverse operator T^{-1} of an operator T, when it exists, satisfies

$$\begin{aligned} T^{-1} \circ T &= I_{\mathscr{R}(T)}, \\ T \circ T^{-1} &= I_{\mathscr{D}(T)}, \end{aligned}$$

where $I_{\mathscr{D}(T)}$, $I_{\mathscr{R}(T)}$ are the identity operators on $\mathscr{D}(T)$, $\mathscr{R}(T)$, respectively.

Definition 2.36 (*Isomorphism*) A linear operator $T : \mathscr{D}(T) \subseteq \mathbb{X} \rightarrow \mathbb{Y}$ is an *isomorphism* if

1. $\mathscr{D}(T) = \mathbb{X}$,
2. $\mathscr{R}(T) = \mathbb{Y}$, and
3. $T^{-1} : \mathbb{Y} \rightarrow \mathbb{X}$ exists.

2.2.4 Normed Vector Spaces

In this section, we review normed vector spaces, which are obtained by endowing the algebraic structure associated with a vector space with a certain type of topological structure. This topology is created by introducing a norm, which can be interpreted as a type of "abstract length" of the elements in the vector space. Nearly all function spaces summarized later in Sect. 2.4, which are employed in this book, are types of normed vector spaces.

Because of the algebraic structure of vector spaces, it is possible to define a norm on it as follows.

Definition 2.37 (*Norms*) Let $\mathbb{X}$ be a real (complex) vector space. A *norm* on $\mathbb{X}$ is a function $\| \cdot \| : \mathbb{X} \rightarrow \mathbb{R}$ such that

(N1) $\|x\| \geq 0$,
(N2) $\|x\| = 0$ if and only if $x = 0$,
(N3) $\|\alpha x\| = |\alpha| \, \|x\|$ for all $\alpha \in \mathbb{R}$ ($\alpha \in \mathbb{C}$) and $x \in \mathbb{X}$, and
(N4) $\|x + y\| \leq \|x\| + \|y\|$ for all $x, y \in \mathbb{X}$.

Note that Definition 2.37 relies on the fact that it is possible to add elements of $\mathbb{X}$ together or to multiply them by a real scalar. Property (N2) is known as "definiteness," and stipulates that the norm of $x \in \mathbb{X}$ is zero only when $x = 0$. A *seminorm* satisfies (N1), (N3), and (N4) and dispenses with the requirement of definiteness.

Definition 2.38 (*Normed Vector Spaces*) A *normed linear space* is a pair $(\mathbb{X}, \|\cdot\|)$, where $\mathbb{X}$ is a vector space and $\|\cdot\|$ is a norm on $\mathbb{X}$.

Observe that if $(\mathbb{X}, \|\cdot\|)$ is a normed linear space, then its norm induces a "natural" metric on $\mathbb{X}$ by $d(x, y) = \|x - y\|$. Thus, all normed vector spaces are metric spaces for this natural choice. The *Euclidean norm* on $\mathbb{R}^n$ (or $\mathbb{C}^n$) is defined as

$$\|x\|_2 \triangleq \sqrt{\sum_{i=1}^{n} |x_i|^2}. \tag{2.10}$$

Thus, the pairs $(\mathbb{R}^n, \|\cdot\|_2)$ and $(\mathbb{C}^n, \|\cdot\|_2)$ are normed linear spaces, and $(\mathbb{R}^n, \|\cdot\|_2)$ is referred to as the *Euclidean space* of dimension $n > 0$.

Multiple norms can be imposed on the same space $\mathbb{X}$. For example, let $\mathbb{X} = \mathbb{R}^n$. The 1-*norm* on $\mathbb{R}^n$ is defined as

$$\|x\|_1 \triangleq \sum_{i=1}^{n} |x_i|, \quad \text{for all } x \in \mathbb{R}^n. \tag{2.11}$$

Alternatively, the *infinity-norm* on $\mathbb{R}^n$ is defined as

$$\|x\|_\infty \triangleq \max_{i \in \{1,\ldots,n\}} |x_i|, \quad \text{for all } x \in \mathbb{R}^n, \tag{2.12}$$

where x_i denotes the ith element of x. Thus, the pairs $(\mathbb{R}^n, \|\cdot\|_1)$ and $(\mathbb{R}^n, \|\cdot\|_\infty)$ are normed linear spaces, which are different from $(\mathbb{R}^n, \|\cdot\|_2)$. Indeed, the pairs $(\mathbb{R}^n, \|\cdot\|_2)$, $(\mathbb{R}^n, \|\cdot\|_1)$, and $(\mathbb{R}^n, \|\cdot\|_\infty)$ are different normed linear spaces since different norms are employed. In some cases, different norms can be equivalent as defined in the following.

Definition 2.39 (*Equivalent Norms*) Let $\mathbb{X}$ be a vector space and assume that $\|\cdot\|_\alpha$ and $\|\cdot\|_\beta$ are two norms on $\mathbb{X}$. These two norms are *equivalent* if there exist constants $c_1 > 0$ and $c_2 > 0$ such that

$$c_1 \|x\|_\alpha \le \|x\|_\beta \le c_2 \|x\|_\alpha, \quad \text{for all } x \in \mathbb{X}.$$

Example 2.3 The equivalence between the Euclidean norm and the infinity norm on $\mathbb{R}^n$ is proven by the fact that

$$\|x\|_\infty \le \|x\|_2 \le \sqrt{n}\|x\|_\infty, \quad \text{for all } x \in \mathbb{R}^n. \tag{2.13}$$

△

Normed vector spaces that have equivalent norms have the same topology. For example, when the norms are equivalent, the collection of all the convergent sequences contained in $(\mathbb{X}, \|\cdot\|_\alpha)$ is the same as that for $(\mathbb{X}, \|\cdot\|_\beta)$, and they have the same limiting values. In this book, we often just write

$$\|\cdot\|_\alpha \approx \|\cdot\|_\beta \quad \text{or} \quad (\mathbb{X}, \|\cdot\|_\alpha) \approx (\mathbb{X}, \|\cdot\|_\beta), \tag{2.14}$$

to designate such an equivalence. Note that the vector space $\mathbb{X}$ that goes into the definitions of normed vector spaces $(\mathbb{X}, \|\cdot\|_\alpha)$ and $(\mathbb{X}, \|\cdot\|_\beta)$ is the same, but these are not the same normed vector spaces since, in general, $\|x\|_\alpha \ne \|x\|_\beta$ for $x \in \mathbb{X}$. The norm equivalence, however, implies the topologies are the same, so a closed (or open) set in $(\mathbb{X}, \|\cdot\|_\alpha)$ is also a closed (or open) set in $(\mathbb{X}, \|\cdot\|_\beta)$. Connected sets in $(\mathbb{X}, \|\cdot\|_\alpha)$ are also connected in $(\mathbb{X}, \|\cdot\|_\beta)$. A sequence that converges to $x \in \mathbb{X}$ in the norm on $(\mathbb{X}, \|\cdot\|_\alpha)$ also converges to $x \in \mathbb{X}$ in the norm on $(\mathbb{X}, \|\cdot\|_\beta)$, and so forth.

While considerable effort may be required to establish the equivalence of generic normed spaces, it is well known that all norms on a finite-dimensional vector space

$\mathbb{X}$ are equivalent; see [3, 7]. In particular, since $\mathbb{R}^n$ and $\mathbb{R}^{n\times m}$ are finite-dimensional, all norms are equivalent on $\mathbb{R}^n$ and $\mathbb{R}^{n\times m}$. For these reasons, if clear from the context, we often express the Euclidean norm on $\mathbb{R}^n$ and $\mathbb{R}^{n\times m}$ as $\|\cdot\|$.

Remark 2.4 (*Comment on Notation*) Strictly speaking, a normed linear space is a pair $(\mathbb{X}, \|\cdot\|)$, where one specifies the (vector) space $\mathbb{X}$ and a norm $\|\cdot\|$ on $\mathbb{X}$. As in the almost totality of textbooks on linear algebra and control systems theory, if clear from the context, we abuse notation and state "$\mathbb{X}$ is a normed linear space" while meaning $(\mathbb{X}, \|\cdot\|)$.

Remark 2.5 (*Comment on Notation*) Unless a specific property is to be highlighted, we employ the generic symbol $\|\cdot\|$ to denote a norm. In this book, multiple normed vector spaces are often considered within the same reasoning. To disambiguate what space these norms refer to, we include the symbol of the corresponding space as a subscript of the symbol of the norm. For example, we denote a normed space as $(\mathbb{X}, \|\cdot\|_{\mathbb{X}})$ and, as already noted in Remark 2.4 we refer to $\mathbb{X}$ as "the normed space" and $\|\cdot\|_{\mathbb{X}}$ as its norm.

Definition 2.40 (*Bounded Linear Operators*) Let $\mathbb{X}$ and $\mathbb{Y}$ denote normed vector spaces. The linear operator $T : \mathbb{X} \to \mathbb{Y}$ is *bounded* if there exists $M > 0$ such that

$$\|Tx\|_{\mathbb{Y}} \leq M\|x\|_{\mathbb{X}}, \quad \text{for all } x \in \mathbb{X}. \tag{2.15}$$

Definition 2.41 (*Compact Linear Operators*) Let $\mathbb{X}$ and $\mathbb{Y}$ denote normed vector spaces. The linear operator $T : \mathbb{X} \to \mathbb{Y}$ is *compact* if it is bounded and the image of any bounded subset of $\mathbb{X}$ under T is relatively compact in $\mathbb{Y}$.

Definition 2.42 The *operator norm* $\|T\|$ of a bounded linear operator T is the smallest nonnegative constant M for which (2.15) holds.

It is straightforward to show that

$$\|T\| = \sup_{x\in\mathbb{X}, x\neq 0} \frac{\|Tx\|_{\mathbb{Y}}}{\|x\|_{\mathbb{X}}} = \sup_{x\in\mathbb{X}, \|x\|_{\mathbb{X}}=1} \|Tx\|_{\mathbb{Y}}, \tag{2.16}$$

and a linear operator between normed linear spaces is bounded if and only if it is continuous. Thus, in this setting, continuous linear operators and bounded linear operators are equivalent.

If the same norm is employed both at the numerator and at the denominator of the expression on the right-hand side of (2.16) (e.g., $\mathbb{X} = \mathbb{R}^n$, $\mathbb{Y} = \mathbb{R}^m$, and the Euclidean norm is employed both on $\mathbb{X}$ and on $\mathbb{Y}$), then the operator norm $\|\cdot\|$ is said to be *equi-induced*.

Example 2.4 Real-valued matrices, that is, elements of $\mathbb{R}^{n\times m}$ are among the most popular linear operators. For matrices, common operator norms are the (2, 2)-*equi-induced norm*

$$\|A\|_2 \triangleq \sup_{x\in\mathbb{R}^m, x\neq 0} \frac{\|Ax\|_2}{\|x\|_2}, \tag{2.17}$$

where $\|\cdot\|_2$ on the right-hand side of (2.17) is given by (2.10), and the (∞,∞)-*equi-induced norm*

$$\|A\|_\infty \triangleq \sup_{x\in\mathbb{R}^m, x\neq 0} \frac{\|Ax\|_\infty}{\|x\|_\infty}, \tag{2.18}$$

where $\|\cdot\|_\infty$ on the right-hand side of (2.18) is given by (2.12). △

Theorem 2.2 *Let* $\mathbb{X}$, $\mathbb{Y}$, *and* $\mathbb{Z}$ *denote normed vector spaces. If the operator norm* (2.16) *is equi-induced, then this norm is* submultiplicative, *that is,*

$$\|T_1 T_2\| \leq \|T_1\|\|T_2\| \tag{2.19}$$

for any $T_1 : \mathbb{X} \to \mathbb{Y}$ *and* $T_2 : \mathbb{Y} \to \mathbb{Z}$ *such that* $\mathscr{D}(T_1) \subseteq \mathscr{R}(T_2)$.

The operator norm can be used to define a normed linear space of bounded linear operators. Let $\mathbb{X}$ and $\mathbb{Y}$ be normed linear spaces. We denote by $\mathcal{L}(\mathbb{X},\mathbb{Y})$ the *normed linear space of bounded linear operators from* $\mathbb{X}$ *to* $\mathbb{Y}$ *equipped with the operator norm* given by (2.16). For brevity, we write $\mathcal{L}(\mathbb{X})$ instead of $\mathcal{L}(\mathbb{X},\mathbb{X})$.

Definition 2.43 (*Topological Dual Space*) The space $\mathcal{L}(\mathbb{X},\mathbb{R})$ of bounded (continuous) linear operators into the real numbers is called the *topological dual space* of $\mathbb{X}$ or, simply the *dual space* of $\mathbb{X}$.

For its relevance in this book, we write $\mathbb{X}' \triangleq \mathcal{L}(\mathbb{X},\mathbb{R})$ and elements of $\mathbb{X}'$ are called *bounded linear functionals on* $\mathbb{X}$.

Remark 2.6 (*Notational Warning*) The use of $\mathbb{X}'$ for the *topological* dual is not uniform in the literature. Some authors prefer to denote $\mathcal{L}(\mathbb{X},\mathbb{R})$ by $\mathbb{X}^*$. However, most modern references that deal with Sobolev spaces use the notation $\mathbb{X}'$ for the topological dual and $\mathbb{X}^*$ for the algebraic dual, which does not require linear functionals to be bounded.

Since normed vector spaces are metric spaces, we can apply Definitions 2.16 and 2.21 to normed linear spaces.

Definition 2.44 (*Banach Spaces*) A complete normed linear space is called a *Banach space*.

Example 2.5 One of the canonical examples of a Banach space is the sequence space ℓ^p for $1 \leq p \leq \infty$. Let $p \in \mathbb{R}_+$. The *sequence space* $\ell^p(\mathbb{R}^n)$ is the space of all sequences $\{x_n\}_{n\in\mathbb{N}} \subset \mathbb{R}^n$ such that $\sum_{i\in\mathbb{N}} \|x_i\|_{\mathbb{R}^n}^p$ exists and is finite. A norm on $\ell^p(\mathbb{R}^n)$ is given by

$$\|x\|_{\ell^p(\mathbb{R}^n)} \triangleq \begin{cases} \left(\sum_{i\in\mathbb{N}} \|x\|_{\mathbb{R}^n}^p\right)^{1/p} & \text{for } 1 \le p < \infty, \\ \sup_{i\in\mathbb{N}} \|x_i\|_{\mathbb{R}^n} & \text{for } p = \infty. \end{cases} \tag{2.20}$$

In the following, for brevity, we will write ℓ^p for $\ell^p(\mathbb{R}^n)$, unless the dimension of the vectors in the sequence needs to be specified. It can be proven that, endowed with the norm (2.20), $\ell^p(\mathbb{R}^n)$ is a Banach space [3, p. 61]. Lastly, we recall that if $p \in (0, 1)$, then ℓ^p is not a normed space, but it is a metric space with metric

$$d(x, y) = \sum_{n\in\mathbb{N}} |x_i - y_i| \quad \text{for all } x, y \in \ell^p.$$

△

As noted above, the space of bounded linear operators $\mathcal{L}(\mathbb{X}, \mathbb{Y})$ between two normed linear spaces $\mathbb{X}$ and $\mathbb{Y}$ is a normed linear space when equipped with the operator norm given by (2.16). The following result provides a condition that ensures that $\mathcal{L}(\mathbb{X}, \mathbb{Y})$ is also a Banach space. The proof of this theorem can be found in [6].

Theorem 2.3 (Completeness of $\mathcal{L}(\mathbb{X}, \mathbb{Y})$) *Let $\mathbb{X}$ and $\mathbb{Y}$ be normed linear space. If $\mathbb{Y}$ is a Banach space, then $\mathcal{L}(\mathbb{X}, \mathbb{Y})$ is a Banach space.*

Remarkably, in Theorem 2.3, the space $\mathbb{X}$ does not have to be a Banach space.

Definition 2.45 (*Compact Operator*) Let $\mathbb{X}$ and $\mathbb{Y}$ be normed vector spaces and let $T : \mathbb{X} \to \mathbb{Y}$ be a linear operator mapping bounded subset of $\mathbb{X}$ into relatively compact subsets of $\mathbb{Y}$. Then, the continuous bounded operator T is a *compact operator*.

2.2.5 Inner Product Spaces

The concept of inner products on linear spaces plays a fundamental role in this book. In particular, Hilbert spaces and their structure provide the framework that enables the methods and analysis presented in the following chapters. Intuitively, an inner product introduces a notion whereby if two elements of a vector space are perpendicular in some abstract sense, then their inner product is zero, and if two vectors are colinear in some abstract sense, then their inner product is maximized.

Definition 2.46 (*Inner Product on $\mathbb{R}^n$*) Let $\mathbb{X}$ be a real vector space. An *inner product* on $\mathbb{X}$ is a function $\langle\cdot,\cdot\rangle : \mathbb{X} \times \mathbb{X} \to \mathbb{R}$ satisfying:

- **(RI1)** $\langle x, x\rangle \ge 0$, for all $x \in \mathbb{X}$,
- **(RI2)** $\langle x, x\rangle = 0$ if and only if $x = 0$,
- **(RI3)** $\langle \alpha x, y\rangle = \alpha\langle x, y\rangle$ for all $\alpha \in \mathbb{R}$ and $x, y \in \mathbb{X}$,
- **(RI4)** $\langle x, y\rangle = \langle y, x\rangle$ for all $x, y \in \mathbb{X}$, and
- **(RI5)** $\langle x + z, y\rangle = \langle x, y\rangle + \langle z, y\rangle$ for all $x, y, z \in \mathbb{X}$.

If $\mathbb{X}$ is a complex linear space, the inner product is defined as follows.

Definition 2.47 (*Inner Product Spaces on* $\mathbb{C}^n$) Let $\mathbb{X}$ be a complex vector space. An *inner product* on $\mathbb{X}$ is a function $\langle \cdot, \cdot \rangle : \mathbb{X} \times \mathbb{X} \to \mathbb{C}$ satisfying:

(CI1) $\langle x, x \rangle$ is real-valued and $\langle x, x \rangle \geq 0$, for all $x \in \mathbb{X}$,
(CI2) $\langle x, x \rangle = 0$ if and only if $x = 0$,
(CI3) $\langle \alpha x, y \rangle = \alpha \langle x, y \rangle$ for all $\alpha \in \mathbb{C}$ and $x, y \in \mathbb{X}$,
(CI4) $\langle x, y \rangle = \overline{\langle y, x \rangle}$ for all $x, y \in \mathbb{X}$, and
(CI5) $\langle x + z, y \rangle = \langle x, y \rangle + \langle z, y \rangle$ for all $x, y, z \in \mathbb{X}$.

Definition 2.48 (*Inner Product Space*) An *inner product space* is a pair $(\mathbb{X}, \langle \cdot, \cdot \rangle)$ where $\mathbb{X}$ is a vector space and $\langle \cdot, \cdot \rangle$ is an inner product on $\mathbb{X}$.

An inner product on $\mathbb{X}$ induces a norm on $\mathbb{X}$ defined as

$$\|x\| \triangleq \sqrt{\langle x, x \rangle}, \quad \text{for all } x \in \mathbb{X}, \tag{2.21}$$

and we say that the norm is *compatible with the inner product.* We leave the reader with the task of proving that (2.21) is a norm. A solution to this relatively simple exercise can be found in classical books in functional analysis such as [3]. Thus, an inner product space is a normed linear space, and, hence, a metric space defined on $\mathbb{X}$. Moreover, inner product spaces have additional (geometric) structures and properties.

Theorem 2.4 *Let* $(\mathbb{X}, \langle \cdot, \cdot \rangle)$ *be an inner product space. Then, for all* $x, y \in \mathbb{X}$,

$$\langle x, y \rangle \leq \|x\| \|y\|, \qquad (\textit{Schwarz Inequality}) \tag{2.22a}$$

$$\|x + y\|^2 + \|x - y\|^2 = 2\left(\|x\|^2 + \|y\|^2\right), \qquad (\textit{Parallelogram Law}) \tag{2.22b}$$

$$\text{if } \langle x, y \rangle = 0, \text{ then } \|x + y\|^2 = \|x\|^2 + \|y\|^2. \qquad (\textit{Pythagorean Theorem}) \tag{2.22c}$$

Inner product spaces provide the framework needed to discuss the notion orthogonality. If $\langle x, y \rangle = 0$, then $x, y \in \mathbb{X}$ are *orthogonal.* This notion can be extended to arbitrary subsets of $(\mathbb{X}, \langle \cdot, \cdot \rangle)$.

Definition 2.49 (*Orthogonal Sets*) Let $(\mathbb{X}, \langle \cdot, \cdot \rangle)$ be an inner product space. Two sets $M \subseteq \mathbb{X}$ and $N \subseteq \mathbb{X}$ are *orthogonal* to one another if

$$\langle x, y \rangle = 0, \quad \text{for all } (x, y) \in M \times N. \tag{2.23}$$

Definition 2.49 can be applied to all partitions of a subset $M \subseteq \mathbb{X}$ over the inner product space $(\mathbb{X}, \langle \cdot, \cdot \rangle)$. In particular, if (2.23) is verified for all $(x, y) \in M \times M$ such that $x \neq y$, then the *vectors of* M *are orthogonal.* If $\|x\| = 1$ for all $M \subseteq \mathbb{X}$

and (2.23) is verified for all $(x, y) \in M \times M$ such that $x \neq y$, then the *vectors of M are orthonormal*. Since inner product spaces are metric spaces, we also apply the definition of completeness (see Definition 2.16) to inner product spaces.

Definition 2.50 (*Hilbert Spaces*) A complete inner product is called a *Hilbert space*.

Consistently with Remark 2.5, we make the following comment.

Remark 2.7 (*Comment on Notation*) Unless a specific property is to be highlighted, we employ the generic symbol $\langle \cdot, \cdot \rangle$ to denote an inner product. In this book, multiple inner product vector spaces are often considered within the same reasoning. To disambiguate what space these norms refer to, we include the symbol of the corresponding space as a subscript of the symbol of the norm. For example, we denote an inner product space as $(\mathbb{X}, \langle \cdot, \cdot \rangle_{\mathbb{X}})$ and we refer to $\mathbb{X}$ as "the inner product space" and $\langle \cdot, \cdot \rangle_{\mathbb{X}}$ as its inner product.

If one defines the *inner product* on $\mathbb{R}^n$ by

$$\langle x, y \rangle_{\mathbb{R}^n} = \sum_{i=1}^{n} x_i y_i, \quad \text{for all } x, y \in \mathbb{R}^n, \tag{2.24}$$

which is the most popular choice for an inner product on $\mathbb{R}^n$, then the pair $(\mathbb{R}^n, \langle \cdot, \cdot \rangle_{\mathbb{R}^n})$ is an inner product space. As already noted, the pair $(\mathbb{R}^n, \|\cdot\|_2)$, where $\|x\|_2 = \sqrt{\langle x, x \rangle_{\mathbb{R}^n}}$, is a normed linear space. Moreover, since $\mathbb{R}^n$ is finite-dimensional, these spaces are complete (see [9, p. 67]) so that $(\mathbb{R}^n, \|\cdot\|_2)$ is a Banach space and $(\mathbb{R}^n, \langle \cdot, \cdot \rangle_{\mathbb{R}^n})$ is a Hilbert space. Moreover, $(\mathbb{R}^n, \|\cdot\|_2)$ is locally compact, and, since all norms $\|\cdot\|$ on $\mathbb{R}^n$ are equivalent, $(\mathbb{R}^n, \|\cdot\|)$ is also locally compact.

One of the most popular choices for an inner product on $\mathbb{R}^{n \times m}$ is

$$\langle A, B \rangle_{\text{tr}} \triangleq \text{tr}\left(A^{\text{T}} B\right), \quad \text{for all } A, B \in \mathbb{R}^{n \times m}, \tag{2.25}$$

where $\text{tr} : \mathbb{R}^{a \times a} \to \mathbb{R}$ denotes the trace of its argument for any $a \in \mathbb{N}$. In this case, this inner product induces the *Frobenius norm*

$$\|A\|_{\text{Fr}} \triangleq \sqrt{\langle A, A \rangle_{\text{tr}}}, \quad \text{for all } A \in \mathbb{R}^{n \times m}. \tag{2.26}$$

The pair $\left(\mathbb{R}^{n \times m}, \|\cdot\|_{\text{Fr}}\right)$, is a normed linear space. Moreover, since $\mathbb{R}^{n \times m}$ is finite-dimensional, this space is complete. Moreover, $\left(\mathbb{R}^{n \times m}, \|\cdot\|_{\text{Fr}}\right)$ is locally compact.

Remark 2.8 (*Compactness and Finite-Dimensional Spaces*) The local compactness of $(\mathbb{R}^n, \|\cdot\|)$ implies that a set $\mathscr{A} \subset \mathbb{R}^n$ is compact if and only if it is closed and bounded. In particular, the closed unit ball $\bar{\mathcal{B}}_1(0) \triangleq \{x \in \mathbb{R}^n : \|x\| \leq 1\}$ is compact. This is true because the vector space $\mathbb{X}$ is finite-dimensional. If $\mathbb{X}$ is an infinite-dimensional vector space, then closed and bounded sets in the normed linear space $(\mathbb{X}, \|\cdot\|)$ may not be compact. More specifically, the closed unit ball in $(\mathbb{X}, \|\cdot\|)$ is compact if and only if $\mathbb{X}$ is a finite-dimensional vector space (e.g., see [6, Theorem 3.6]). This is an important technical point that, if ignored, can result in incorrect conclusions when applying Lyapunov and other direct methods; see Chap. 4. For instance a critical step in the classical proof of classical Lyapunov stability on $\mathbb{R}^n$ (see the proof of Theorem 4.1 in [10]) leverages the compactness (i.e., closedness and boundedness) of sub-level sets of Lyapunov functions.

Remark 2.9 It follows from Remark 2.8 and Definition 2.25 that the support of a function defined on $\mathbb{R}^n$ is compact if and only if it is closed and bounded.

To illustrate this point, consider the following example, inspired by [11, pp. 129–131], where the vector space $\mathbb{X}$ is a space of functions.

Example 2.6 Let $a < b$ be real numbers, let $[a, b] \subset \mathbb{R}$ be the closed interval $[a, b] \triangleq \{s : a \leq s \leq b\}$, and

$$C[a, b] \triangleq \{\varphi : [a, b] :\rightarrow \mathbb{R} : \varphi(\cdot) \text{ is a continuous function on } [a, b]\}. \tag{2.27}$$

Recall that $C[a, b]$ is a vector space with standard addition of functions defined by $[x(\cdot) + y(\cdot)](s) = x(s) + y(s)$ and scalar multiplication $[\alpha x(\cdot)](s) = \alpha x(s)$. Define the "sup norm"

$$\|x(\cdot)\|_\infty \triangleq \sup_{a \leq s \leq b} |x(s)| = \max_{a \leq s \leq b} |x(s)|. \tag{2.28}$$

The normed space $(C[a, b], \|\cdot\|_\infty)$ is a separable Banach space.

Observe that since $x(\cdot)$ is continuous on $[a, b]$, $f(\cdot) = |x(\cdot)|^2$ is also continuous and hence (Riemann) integrable. Moreover, $f(s) = |x(s)|^2 \geq 0$ for all $s \in [a, b]$. Thus, we define the 2-norm on $C[a, b]$ by

$$\|x(\cdot)\|_2 \triangleq \sqrt{\int_a^b |x(s)|^2 ds}. \tag{2.29}$$

To show that $\|x(\cdot)\|_2$ is a norm, one must verify conditions (N1)–(N4) are satisfied. Conditions (N1) and (N4) are obvious. The triangle inequality (N4) follows from Minkowski's inequality for integrals: if $1 \leq p < \infty$, then

$$\left(\int_a^b |f(s) + g(s)|^p ds\right)^{1/p} \leq \left(\int_a^b |f(s)|^p ds\right)^{1/p} + \left(\int_a^b |g(s)|^p ds\right)^{1/p}. \tag{2.30}$$

Establishing condition (N2) relies on the fact that if a continuous function $f(\cdot)$ satisfies $f(s) \geq 0$ for all $s \in [a, b]$ and $\int_a^b f(s)ds = 0$, then $f(s) = 0$ for all $s \in [a, b]$. Although $(C[a, b], \|\cdot\|_2)$ is a separable normed linear space, it is not complete and hence is not a Banach space. △

Example 2.7 Consider a special case of the sequence spaces ℓ^p discussed in Example 2.5. If $p = 2$, then ℓ^2 can be endowed with the inner product

$$\langle x, y\rangle_{\ell^2} \triangleq \sum_{n\in\mathbb{N}} x_n y_n, \quad \text{for all } x, y \in \ell^2, \tag{2.31}$$

and, with this inner product, ℓ^2 is a Hilbert space [3, p. 133]. △

Projection operators play a fundamental role in the theory of Hilbert spaces and approximations. Although projection operators can be defined in a very general setting, we focus on projection operators and orthogonal projections in inner product spaces. These projections play a key role in this book.

Definition 2.51 (*Projection and Orthogonal Projection*) Let $\mathbb{H}$ be an inner product space. A linear operator $\Pi : \mathbb{H} \to \mathbb{H}$ is a *projection* if $\Pi^2 = \Pi$. The projection Π is said to be an *orthogonal projection* if

$$\langle f, \Pi g\rangle_{\mathbb{H}} = \langle \Pi f, \Pi g\rangle_{\mathbb{H}} = \langle \Pi f, g\rangle_{\mathbb{H}}, \quad \text{for all } f, g \in \mathbb{H}. \tag{2.32}$$

Given the subset $\mathbb{H}_1$ of the Hilbert space $\mathbb{H}_2$, the *orthogonal complement to* $\mathbb{H}_1$ *in* $\mathbb{H}_2$ is defined as

$$\mathbb{H}_1^{\perp} \triangleq \{h_2 \in \mathbb{H}_2 : \langle h_2, h_1\rangle \text{ for some } h_1 \in \mathbb{H}_1\}. \tag{2.33}$$

Definition 2.51 can also be understood in terms of the adjoint operator $T^* : \mathbb{H}_2 \to \mathbb{H}_1$ of a bounded linear operator $T : \mathbb{H}_1 \to \mathbb{H}_2$. The adjoint operator T^* of the bounded linear operator T is the unique bounded linear operator that satisfies

$$\langle Tf, g\rangle_{\mathbb{H}_2} = \langle f, T^* g\rangle_{\mathbb{H}_1}, \quad \text{for all } (f, g) \in \mathbb{H}_1 \times \mathbb{H}_2. \tag{2.34}$$

The operator T is *self-adjoint* if $\mathbb{H}_1 = \mathbb{H}_2$ and $T = T^*$. It can be shown that Definition 2.51 is equivalent to stipulating that the conditions $\Pi^2 = \Pi = \Pi^*$ characterize orthogonal projections. In other words, a projection is an orthogonal projection if and only if it is self-adjoint. The *orthogonal projection error* $\epsilon_\Pi(f)$ is defined as

$$\epsilon_\Pi(f) \triangleq f - \Pi f = (I - \Pi)f, \quad \text{for all } f \in \mathbb{H}, \tag{2.35}$$

where $(I - \Pi)$ denotes the *complementary projection*. In several instances throughout this book, we write $\Pi_{\mathbb{H}}$ to remark on which Hilbert space underlies a projection operator and $\Pi_{\mathbb{H}^\perp}$ to remark on which Hilbert space underlies the corresponding complementary projection.

2.3 Duality Pairings and Riesz Representers

Let $(\mathbb{X}, \|\cdot\|_{\mathbb{X}})$ be a topological vector space, recall from Definition 2.43 that the topological dual space $\mathbb{X}' \triangleq \mathcal{L}(\mathbb{X}, \mathbb{R})$ is the space of bounded linear functionals on $\mathbb{X}$, and recall the notion of isomorphism introduced in Definition 2.36. Since $\mathbb{R}$ is a Banach space, it follows from Theorem 2.3 that $\mathbb{X}'$ is a Banach space. Although we know that $\mathbb{X}'$ is a set of continuous functionals, for applications, it is often very useful to have an equivalent representation of this space.

Definition 2.52 (*Congruent Spaces*) Let $(\mathbb{X}, \|\cdot\|_{\mathbb{X}})$ and $(\mathbb{Y}, \|\cdot\|_{\mathbb{Y}})$ be normed linear spaces, and let $T : \mathbb{X} \to \mathbb{Y}$ be an isomorphism. If T is bounded and $\|x\|_{\mathbb{X}} = \|Tx\|_{\mathbb{Y}}$ for all $x \in \mathbb{X}$, then T is called an *isometric isomorphism* and we say that $(\mathbb{X}, \|\cdot\|_{\mathbb{X}})$ is *congruent* to $(\mathbb{Y}, \|\cdot\|_{\mathbb{Y}})$.

Different descriptions and choices of notation are popular for describing dual spaces. To recall some of them, we introduce the notions of a *bilinear map* and the *duality pairing*.

Definition 2.53 (*Bilinear Maps*) Let $\mathbb{W}$, $\mathbb{V}$, and $\mathbb{X}$ be topological vector spaces over the field of scalars $\mathbb{R}$. A *bilinear map* is a function $B : \mathbb{W} \times \mathbb{V} \to \mathbb{X}$ such that, for all $w \in \mathbb{W}$, the map $B_w : v \mapsto B(w, v)$ is linear from $\mathbb{V}$ to $\mathbb{X}$, and, for all $v \in \mathbb{V}$, the map $B_v : w \mapsto B(w, v)$ is linear from $\mathbb{W}$ to $\mathbb{X}$.

If $\mathbb{W} = \mathbb{V}'$ and $\mathbb{X} = \mathbb{R}$, then we employ the duality pairing notation $\langle \cdot, \cdot \rangle_{V' \times V}$ defined by

$$\langle v^*, v \rangle_{\mathbb{V}' \times \mathbb{V}} \triangleq v^*(v) \in \mathbb{R} \tag{2.36}$$

for all $v^* \in \mathbb{V}'$ and $v \in \mathbb{V}$. The mapping $B : \mathbb{V}' \times \mathbb{V} \to \mathbb{R}$ defined as $B(\cdot, \cdot) \triangleq \langle \cdot, \cdot \rangle_{\mathbb{V}^* \times \mathbb{V}}$ is a bilinear mapping. If $\mathbb{V}$ is a Hilbert space, then the next result allows further exploring the relationship between the duality pairing notation and the inner product.

Theorem 2.5 (*Riesz Representation Theorem*, [12, Theorem 3.4]) *Let $\mathbb{H}$ be a Hilbert space and $h^* : \mathbb{H} \to \mathbb{R}$ be a bounded linear functional on $\mathbb{H}$ so that $h^* \in \mathbb{H}'$ belongs to the topological dual space of $\mathbb{H}$. Then, there is a unique $h \in \mathbb{H}$ such that $h^*(g) = \langle h, g \rangle_{\mathbb{H}}$ for every $g \in \mathbb{H}$. Furthermore, $\|h^*\|_{\mathbb{H}'} = \|g\|_{\mathbb{H}}$.*

Theorem 2.5 states that a Hilbert space $\mathbb{H}$ is isometrically isomorphic to its dual space $\mathbb{H}'$. The *Riesz map* $R : \mathbb{H}' \to \mathbb{H}$ is the linear operator that represents this relationship via

$$\langle h^*, g \rangle_{\mathbb{H}' \times \mathbb{H}} = \langle Rh^*, g \rangle_{\mathbb{H}} \triangleq \langle h, g \rangle_{\mathbb{H}} \quad \text{for all } h^* \in \mathbb{H}', g \in \mathbb{H}, \tag{2.37}$$

where $h \triangleq Rh^*$. In light of this observation, we recall the following definition.

Definition 2.54 (*Riesz Representers*) Let $\mathbb{H}$ be a Hilbert space and let $\mathbb{H}'$ denote the topological dual space. Given $h^* \in \mathbb{H}'$, the element $h \triangleq Rh^*$ that satisfies (2.37) is the *Riesz representer of* $h^* \in \mathbb{H}'$.

Since the norms on $\mathbb{H}'$ and $\mathbb{H}$ are related by

$$\|h^*\|_{\mathbb{H}'} = \|Rh^*\|_{\mathbb{H}} = \|h\|_{\mathbb{H}}, \quad \text{for all } h \in \mathbb{H}, \tag{2.38}$$

the Riesz operator R is an isometry and an isomorphism.

2.4 Commonly Used Function Spaces

Several well-known function spaces are needed in this book. These spaces primarily include certain spaces of continuous functions, measurable functions, and differentiable functions. We leave it as an exercise to prove that the space $\mathcal{F}(\mathbb{X}, \mathbb{U})$ of functions from $\mathbb{X} \triangleq \mathbb{R}^n$ into $\mathbb{U} \triangleq \mathbb{R}^m$ is indeed a vector space with the usual operations of pointwise addition of functions and multiplication of a function by a scalar.

2.4.1 *Continuous and Uniformly Continuous Functions*

The notion of continuous functions was introduced in Definition 2.6 in the context of topological spaces. As illustrated by Example 2.6, continuous functions provide a function space. Let $\Omega \subseteq \mathbb{X}$ and $f : \Omega \subseteq \mathbb{X} \to \mathbb{Y}$, where $\mathbb{X}$ and $\mathbb{Y}$ are topological spaces. We define

$$C(\Omega, \mathbb{Y}) \triangleq \{f : \Omega :\to \mathbb{Y} : f(\cdot) \text{ is a continuous function on } \Omega\}. \tag{2.39}$$

It is commonly assumed that Ω is a nonempty, open, and connected subset of $\mathbb{R}^n$, which is, by definition, a (topological) domain, and $\mathbb{Y} = \mathbb{R}^m$. Hence, $f : \Omega \subseteq \mathbb{X} \to \mathbb{Y}$ is a function between two metric spaces. In this case, we can provide another definition of continuous functions that is equivalent to Definition 2.6.

Definition 2.55 (*Continuous Functions on Metric Spaces*) Let $(\mathbb{X}, d_{\mathbb{X}})$ and $(\mathbb{Y}, d_{\mathbb{Y}})$ be Hausdorff metric spaces, let $f : \mathscr{D}(f) \subseteq \mathbb{X} \to \mathbb{Y}$ be a function, and let $\bar{x} \in \mathscr{D}(f)$. Then, $f(\cdot)$ is *continuous at* $\bar{x}$ if, for any $\varepsilon > 0$, there exists $\delta(\bar{x}, \varepsilon) > 0$ such that if $x \in \mathscr{D}(f)$ and $d_{\mathbb{X}}(x, \bar{x}) < \delta(\bar{x}, \varepsilon)$, then $d_{\mathbb{Y}}(f(x), f(\bar{x})) < \varepsilon$. A function that is *continuous on the set* $\Omega \subseteq \mathscr{D}(f)$ is continuous at all points in Ω.

If $\mathbb{Y}$ is a Banach space, then we define $C_b(\mathbb{X}, \mathbb{Y})$ to be the Banach space of *bounded continuous functions over* $\mathbb{X}$ endowed with the supremum norm

$$\|f\|_{C_b(\mathbb{X},\mathbb{Y})} \triangleq \sup_{x\in\mathbb{X}} \|f(x)\|_{\mathbb{Y}}. \tag{2.40}$$

If $\mathbb{Y} = \mathbb{R}^m$, then we write $(C_b(\mathbb{X}))^m$.

Definition 2.56 *(Uniformly Continuous Functions on Metric Spaces)* Let $(\mathbb{X}, d_{\mathbb{X}})$ and $(\mathbb{Y}, d_{\mathbb{Y}})$ be Hausdorff metric spaces and $f : \mathscr{D}(f) \subseteq \mathbb{X} \to \mathbb{Y}$ be a continuous function on Ω. If $\delta(\bar{x}, \varepsilon)$ in Definition 2.55 does not depend on $x \in \Omega$ (i.e., $\delta(\bar{x}, \varepsilon) = \delta(\varepsilon)$), then $f(\cdot)$ is *uniformly continuous on* Ω.

Next, we define the notion of restriction of a continuous function to a subset of its domain.

Definition 2.57 *(Restriction of a Continuous Function)* Given $\Omega \subseteq \mathbb{X}$ and $\Phi \subseteq \Omega$, the *classical restriction* of a continuous function $f(\cdot) \in C(\Omega, \mathbb{Y})$ to Φ is a well-defined function $f|_{\Phi}(\cdot) \in C(\Phi, \mathbb{Y})$ given by $f|_{\Phi}(x) = f(x)$ for $x \in \Phi \subset \Omega$. The *restriction* (or *classical trace*) *operator* is the linear operator $T_{\Phi} : C(\Omega, \mathbb{Y}) \to C(\Phi, \mathbb{Y})$ defined by $(T_{\Phi} f)(\cdot) = f|_{\Phi}(\cdot)$.

In the setting where $\Phi \subset \Omega$ and the function spaces consist of continuous functions, $(T_{\Phi} f)(\cdot) = f|_{\Phi}(\cdot)$ is the well-defined (classical) *trace operator*; see [13, 14]. However, one should also be aware that not all functions $f(\cdot) \in C(\Phi, \mathbb{Y})$ are restrictions of a function $g(\cdot) \in C(\Omega, \mathbb{Y})$.

Remark 2.10 The classical trace operator is well-defined only if $\Phi \subseteq \Omega \subseteq \mathbb{X}$. In particular, if Ω is an open set and $\Phi = \partial\Omega$ is the boundary of Ω, then $\Phi = \partial\Omega \not\subset \Omega$. In fact, $\partial\Omega \cap \Omega = \{\emptyset\}$ and the classical trace operator is not defined. However, if $\mathbb{X} = \mathbb{R}^n$, $\mathbb{Y} = \mathbb{R}^m$, and $\Omega \subseteq \mathbb{R}^n$ is a domain, then under certain conditions on Ω, the theory of Sobolev spaces can be used to define a *boundary trace operator* $T_{\partial\Omega} : C(\Omega, \mathbb{R}^m) \to C(\partial\Omega, \mathbb{R}^m)$. If Ω is closed (i.e., $\Omega = \bar{\Omega}$), then the boundary trace operator and the classical trace operator are the same.

If $\Omega \subseteq \mathbb{X}$, then the restriction operator $T_{\Omega} : C(\mathbb{X}, \mathbb{Y}) \to C(\Omega, \mathbb{Y})$ is well-defined. However, if $f(\cdot) \in C(\Omega, \mathbb{Y})$, then there may not be a function $g(\cdot) \in C(\mathbb{X}, \mathbb{Y})$ such that $g|_{\Omega}(\cdot) = f(\cdot)$. Let

$$V = \{f(\cdot) \in C(\Omega, \mathbb{Y}) : \exists\, g(\cdot) \in C(\mathbb{X}, \mathbb{Y}) \text{ such that } f(\cdot) = g|_{\Omega}(\cdot)\}. \tag{2.41}$$

We say that $\mathscr{E}_X : V \to C(\Omega, \mathbb{Y})$ is an *extension operator* to X if

$$T_{\Omega}\mathscr{E}_X f(\cdot) = f(\cdot) \quad \text{for all } f(\cdot) \in V. \tag{2.42}$$

Note that an extension operator $\mathscr{E}_X$ does not need to be unique, and, in general, $\mathscr{E}_X$ is not linear.

2.4.2 Continuous Embeddings

Continuous embeddings, which will be crucial to many of the proofs in this book, describe when one space is contained in another larger space, and when the norm in the smaller space dominates that of the larger space.

If $V \subseteq W$, then we use $\mathcal{I} : V \to W$ to represent the canonical identification or injection of V into W, where $\mathcal{I} : v \in V \mapsto i(v) = v \in W$. Thus, we introduce the notion of *continuous embedding*.

Definition 2.58 (*Continuous and Compact Embeddings*) A normed vector space $\mathbb{V}$ is *continuously embedded* in a normed vector space $\mathbb{W}$ if $\mathbb{V} \subseteq \mathbb{W}$ and there is a constant $C > 0$ such that $\|\mathcal{I}v\|_{\mathbb{W}} \triangleq \|v\|_{\mathbb{W}} \leq C\|v\|_{\mathbb{V}}$ for all $v \in \mathbb{V}$. The smallest such C for which this inequality holds is $\|\mathcal{I}\|$. The space $\mathbb{V}$ is *compactly embedded* in $\mathbb{W}$ if the continuous embedding $\mathcal{I}$ is a compact operator.

We write $V \hookrightarrow W$ whenever V is continuously embedded in W. In this book, we refer to the embedding function $\hookrightarrow$ as the *canonical injection*. If the normed vector space V is densely and continuously embedded in the normed vector space W, then we write $V \hookrightarrow\!\!\!\!\rightarrow W$. If we have multiple embeddings and must be explicit about to which embedding we refer, we write

$$V \overset{\mathcal{I}}{\hookrightarrow} W \quad \text{or} \quad V \overset{\mathcal{I}}{\hookrightarrow\!\!\!\!\rightarrow} W. \tag{2.43}$$

This notation is used to emphasize that the set V is contained in W, and the topology on V is stronger than that on W. In practical terms, for the analysis of our approximations and stability analysis in Chap. 4, the continuous embedding implies that $\mathbb{V}$-convergent sequences are $\mathbb{W}$-convergent too. This deduction is immediate since

$$\|v - v_\ell\|_W \leq \|\mathcal{I}\|\|v - v_\ell\|_V \tag{2.44}$$

for any sequence $\{v_\ell\}_{\ell\in\mathbb{Z}^+}$.

2.4.3 Absolutely Continuous and Lipschitz Continuous Functions

Other forms of continuity are given by absolutely continuous and Lipschitz continuous functions.

Definition 2.59 (*Absolutely Continuous Functions*) Let $I \subset \mathbb{R}$ be an interval. The function $f : I \to \mathbb{R}$ is *absolutely continuous on I* if for every $\varepsilon > 0$, there exists $\delta(\varepsilon) > 0$ such that a finite sequence of pairwise disjoint sub-intervals $(x_k, y_k) \subset I$ is such that $\sum_k |y_k - x_k| < \delta$, then $\sum_k |f(x_k) - f(y_k)| < \varepsilon$.

Definition 2.60 (*Lipschitz Continuous Functions*) Let $\mathbb{X}$ and $\mathbb{Y}$ be normed vector spaces. A function $f : \mathbb{X} \to \mathbb{Y}$ is *locally Lipschitz continuous at* $x_0 \in \mathbb{X}$ if there exists a neighborhood $\mathcal{N}$ of x_0 and $\gamma(x_0) > 0$ such that

$$\|f(x_1) - f(x_2)\|_{\mathbb{Y}} \leq \gamma(x_0)\|x_1 - x_2\|_{\mathbb{X}}, \quad \text{for all } x_1, x_2 \in \mathcal{N}. \tag{2.45}$$

The function $f(\cdot)$ is *Lipschitz continuous on* $\mathbb{X}$ if it is Lipschitz continuous at every point in $\mathbb{X}$. The function $f(\cdot)$ is *uniformly Lipschitz continuous on* $\mathbb{X}$ if it is Lipschitz continuous on $\mathbb{X}$ and there exists $\gamma > 0$ such that

$$\|f(x_1) - f(x_2)\|_{\mathbb{Y}} \leq \gamma\|x_1 - x_2\|_{\mathbb{X}}, \quad \text{for all } x_1, x_2 \in \mathbb{X}. \tag{2.46}$$

The subspace of *Lipschitz continuous functions from* $\mathbb{X}$ *into* $\mathbb{Y}$ is denoted by $C^{0,1}(\mathbb{X}, \mathbb{Y}) \subset C_b(\mathbb{X}, \mathbb{Y})$. We define a *seminorm* on $C^{0,1}(\mathbb{X}, \mathbb{Y})$ as

$$\|f\|_{C^{0,1}(\mathbb{X},\mathbb{Y})} \triangleq \inf \left\{ \gamma > 0 \;\middle|\; (2.45) \text{ holds} \right\},$$

and the *norm*

$$\|f\|_{C^{0,1}(\mathbb{X},\mathbb{Y})} \triangleq \|f\|_{C_b(\mathbb{X},\mathbb{Y})} + |f|_{C^{0,1}(\mathbb{X},\mathbb{Y})} \tag{2.47}$$

makes $C^{0,1}(\mathbb{X}, \mathbb{Y})$ a Banach space. If $\mathbb{Y} = \mathbb{R}$, then this space is just denoted by $C^{0,1}(\mathbb{X})$. If $f : I \to \mathbb{R}$ is Lipschitz continuous, where $I \subset \mathbb{R}$ is an interval, then f is absolutely continuous on I.

2.4.4 Classically Differentiable Functions

One way that additional regularity can be characterized is by membership of a function in spaces of classically differentiable functions. Let $f : \mathbb{R}^n \to \mathbb{R}$ be a function of n real variables. A concise description of various classical (partial) derivatives of $f(\cdot)$ over subsets $\Omega \subseteq \mathbb{R}^n$ can be given in terms of a multi-index given by the n-tuple $\alpha \triangleq (\alpha_1, \ldots, \alpha_n) \in (\mathbb{Z}^+)^n$. The *absolute value* of α is given by

$$|\alpha| \triangleq \alpha_1 + \alpha_2 + \cdots + \alpha_n. \tag{2.48}$$

We use the shorthand notation

$$D^\alpha f(x) \triangleq \frac{\partial^{|\alpha|} f(x)}{\partial x^\alpha} = \frac{\partial^{|\alpha|} f}{\partial x_1^{\alpha_1} \partial x_2^{\alpha_2} \cdots \partial x_n^{\alpha_n}} \tag{2.49}$$

for a *mixed derivative over* $\mathbb{R}^n$. If $\Omega \subseteq \mathbb{R}^n$ is a subset of $\mathbb{R}^n$, then we define $C^s(\Omega) \triangleq C^s(\Omega, \mathbb{R})$ for an integer smoothness index $s \in \mathbb{Z}^+$ as the set of functions that are

s-times continuously differentiable over Ω. The set $C_b^s(\Omega)$ is the Banach space that is obtained by endowing the set $C^s(\Omega)$ with the norm

$$\|f\|_{C_b^s(\Omega)} \triangleq \max_{0\leq|\alpha|\leq s} \|D^\alpha f\|_{C_b(\Omega)} \approx \sum_{|\alpha|=0}^{s} \|D^\alpha f\|_{C_b(\Omega)}. \tag{2.50}$$

The set of s-times classically differentiable and bounded $\mathbb{R}^m$-valued functions $f : \mathbb{R}^n \to Y = \mathbb{R}^m$ is denoted by $f \in (C_b^s(\Omega))^m$.

The space of all functions $f : \mathbb{R}^n \to \mathbb{R}$ such that $f \in C^s(\Omega)$ for all $s \in \mathbb{Z}^+$ is denoted by $C^\infty(\Omega)$. Sometimes, $C^\infty(\Omega)$ is called the *space of smooth functions*. The space $C_0^\infty(\Omega)$ is the set of all smooth functions that have compact support contained in Ω. Observe that if $f \in C_0^\infty(\Omega)$, then f is bounded on Ω.

Define $\mathcal{D}(\Omega) \triangleq C_0^\infty(\Omega)$ to be the *space of test functions* over the set Ω. The test functions $\mathcal{D}(\Omega)$ consist of smooth functions that have compact support contained in Ω. The topological dual $\mathcal{D}'(\Omega)$ of $\mathcal{D}(\Omega)$ is referred to as the *space of distributions*; see [14] for a discussion on the topology on $\mathcal{D}(\Omega)$ and on its dual space $\mathcal{D}'(\Omega)$. The test functions $\mathcal{D}(\Omega)$ and their dual spaces $\mathcal{D}'(\Omega)$ play a distinguished role in the study of Sobolev spaces and weak differentiability as discussed below.

2.4.5 Measurable and Integrable Functions

Commonly employed spaces, such as Bochner, Lebesgue, or Sobolev spaces, are not spaces of functions, but spaces of equivalence classes. To get a feel for the nature of equivalence classes of functions, we begin with a simple example provided by piecewise continuous functions.

Definition 2.61 ([15]) Let $I = [a, b]$ be a closed interval. A function $g : I \to \mathbb{R}$ is said to be *piecewise continuous* (PWC) on $[a, b]$ if

1. the function $g(\cdot)$ is bounded on I,
2. the right-hand limit $g\left(\hat{x}^+\right) = \lim_{x\to\hat{x}^+} g(x)$ exists (is finite) for all $\hat{x} \in [a, b)$,
3. the left-hand limit $g\left(\hat{x}^-\right) = \lim_{x\to\hat{x}^-} g(x)$ exists (is finite) for all $\hat{x} \in (a, b]$, and
4. there is a *finite* partition of $[a, b]$ given by $a = \hat{x}_0 < \hat{x}_1 < \hat{t}_2 < \ldots < \hat{x}_{p-1} < \hat{x}_p = b$, such that $g(\cdot)$ is continuous on each open subinterval $\left(\hat{x}_{i-1}, \hat{x}_i\right)$.

Let $\mathcal{PWC}(a, b)$ denote the vector space of PWC functions $g : [a, b] \to \mathbb{R}$. If $g \in \mathcal{PWC}(a, b)$, then $(\mathcal{R}) \int_a^b \|g(x)\|^2 \mathrm{d}x < \infty$, where we employ the notation $(\mathcal{R}) \int_a^b$ to emphasize this is the classical Riemann integral. It seems "natural" to try to define a norm on $\mathcal{PWC}(a, b)$ by

$$\|g\|_{\mathcal{PWC}(a,b)} \triangleq \left((\mathcal{R}) \int_a^b |g(x)|^2 \mathrm{d}x\right)^{1/2}. \tag{2.51}$$

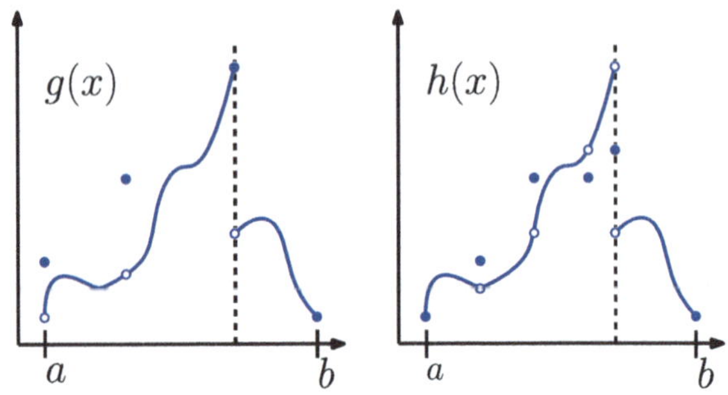

Fig. 2.1 The functions $g(\cdot)$ and $h(\cdot)$ are equal $e.f.$

However, the norm property (N2) in Definition 2.37 does not hold since any (nonzero) function $g(\cdot) \in \mathcal{PWC}(a, b)$ satisfying $g(x) = 0$ $e.f.$ will have zero Riemann integral, and (2.51) does not define a norm.

Definition 2.62 Given the vector space $\mathcal{PWC}(a, b)$, the functions $g \in \mathcal{PWC}(a, b)$ and $h \in \mathcal{PWC}(a, b)$ are equal except at a finite number of points (denoted by $g(x) = h(x)$ $e.f.$) if the set $\{x \in [a, b] : g(x) \neq h(x)\}$ consists of at most a finite number of points in $[a, b]$. If $g(x) = h(x)$ $e.f.$, then we say that g *is* $e.f.$-equivalent to h and write $g \sim h$. Furthermore, the *equivalence class* of g is the set

$$[g] \triangleq \{h \in \mathcal{PWC}(a, b) : \ h(x) = g(x) \ e.f. \} . \tag{2.52}$$

Figure 2.1 provides a graphical representation of functions equal except at a finite number of points. If follows from Definition 2.62 that $\sim$ is an equivalence relation on $\mathcal{PWC}(a, b)$. Furthermore, in light of (2.52), we define the space of equivalent classes

$$PWC(a, b) \triangleq \{[g] : g \in \mathcal{PWS}(a, b)\} . \tag{2.53}$$

With this notation, it is clear that the vector space of equivalent classes $PWC(a, b)$ is not the same as the space of functions $\mathcal{PWC}(a, b)$. However, now we can define a norm on $PWC(a, b)$ by

$$\|[g]\|_{PWS(a,b)} \triangleq \left((\mathcal{R}) \int_a^b |g(x)|^2 \mathrm{d}x \right)^{1/2} . \tag{2.54}$$

Observe that, if $h \in [g]$, then $\left((\mathcal{R}) \int_a^b |h(x)|^2 \mathrm{d}x \right)^{1/2} = \left((\mathcal{R}) \int_a^b |g(x)|^2 \mathrm{d}x \right)^{1/2}$ so the norm is well-defined and does not depend on which function $h \in [g]$ is used. Also, if h is any PWC function and $\left((\mathcal{R}) \int_a^b |h(x)|^2 \mathrm{d}x \right)^{1/2} = 0$, then $h(x) = 0$ $e.f.$. Thus, $[h] = [0]$ and condition (N2) in Definition 2.37 is satisfied.

While this example of $e.f.$-equivalence classes is intuitive, the standard setup for Lebesgue, Sobolev, or Bochner spaces uses a more general notion of equivalence classes. These equivalence classes are defined in terms of measures μ and μ-negligible sets. We first need to define a measure space, which begins with the notion of a σ-algebra.

Definition 2.63 (*σ-Algebra*) Let Ω be a nonempty set and let Σ be a family of subsets of Ω. The family Σ is called a *σ-algebra* on Ω, if

1. $\Omega \in \Sigma$,
2. Σ is closed under complementation, i.e., if $E \in \Sigma$, then so is its complement $\Sigma \backslash E \triangleq \{x \in \Sigma : x \notin E\}$, and
3. Σ is closed under countable unions, i.e., if $E_i \in \Sigma$, with $i \in \mathbb{N}$, then $\bigcup_{i=1}^{+\infty} E_i \in \Sigma$.

Definition 2.64 (*Measure and Measurable Sets*) A pair (Ω, Σ), where Ω is a set and Σ is a σ-algebra of subsets of Ω, is called a *measurable space*. Sets $E \in \Sigma$ are called *measurable sets*. A function $\mu : \Sigma \to [0, \infty]$ is a (countably additive) measure on (Ω, Σ) provided that

1. $\mu(\emptyset) = 0$, and
2. if $E_i \in \Sigma$ for each $i = 1, 2, \ldots$, are disjoint sets, then

$$\mu\left(\bigcup_{i=1}^{\infty} E_i\right) = \sum_{i=1}^{\infty} \mu(E_i). \tag{2.55}$$

If $\mu : \Sigma \to [0, \infty)$, then μ is called a *finite measure*. If Ω can be decomposed into a countable union of measurable sets of finite measure, then μ is called a *σ-finite measure*. The triplet $\Omega_\Sigma \triangleq (\Omega, \Sigma, \mu)$ is called a *measure space*. A set $E \in \Sigma$ is said to have *μ-measure zero* if $\mu(E) = 0$. Sets with μ-measure zero are called *null sets*. The measure μ is said to be a *complete measure* if $E \in \Sigma$ has μ-measure zero and $F \subseteq E$, then $F \in \Sigma$ and $\mu(F) = 0$.

If it is clear from the context, a μ-measure zero set is called a measure zero set. Measure spaces provide the framework to present elements of probability theory and stochastic processes.

Definition 2.65 (*Probability Space*) A *probability space* is a measure space (Ω, Σ, p), where the *probability measure* p is such that

1. $0 \leq p(E) \leq 1$, and
2. $p(\Omega) = 1$.

The set Ω is called the *sample space*. The σ-algebra Σ is called the *event space*. Any measurable subset of Ω is an *event*.

Definition 2.66 (*Measurable Functions and Random Variables*) Let $(\Omega_1, \Sigma_1, \mu_1)$ and $(\Omega_2, \Sigma_2, \mu_2)$ be measure spaces. A function $f : \Omega_1 \to \Omega_2$ is said to be a *measurable function* if, for every measurable set $S \in \Sigma_2$, the inverse image $f^{-1}(S) = \{x \in \Omega_1 : f(x) \in S\}$ is measurable, i.e., $f^{-1}(S) \in \Sigma_1$. If $(\Omega_1, \Sigma_1, \mu_1) = (\Omega, \Sigma, p)$ is a probability space, then a measurable function $w : \Omega \to \Omega_2$ is called a *random variable*.

2.4.6 Lebesgue Spaces

If $\Omega_\Sigma \triangleq (\Omega, \Sigma, \mu)$ is a measure space, then the measure $\mu(E)$ of a set $E \in \Omega$ can be thought of as a generalization of the notion of the size of the set E. For example, the Lebesgue measure generalizes the concepts of length, area, and volume of sets contained in $\mathbb{R}$, $\mathbb{R}^2$, and $\mathbb{R}^3$, respectively; see [2, Chap. I] and [5, Sect. 4.8]. The Lebesgue measure is a complete σ-finite measure.

Theorem 2.6 (Existence of the Lebesgue Measure on $\mathbb{R}^n$ [2, Theorem 1.35]) *There exist a σ -algebra Σ_L of subsets of $\mathbb{R}^n$ and a complete σ-finite measure λ such that the (Lebesgue) measure space $\mathbb{R}^n_L = (\mathbb{R}^n, \Sigma_L, \lambda)$ has the following properties:*

(a) every open set in $\mathbb{R}^n$ belongs to Σ_L,
(b) if $E = \{x \in \mathbb{R}^n : a_j \le x_j \le b_j, j = 1, 2, \dots, n\}$, then $E \in \Sigma_L$ and $\lambda(E) = \prod_{j=1}^n (b_j - a_j)$, and
(c) the measure is translation invariant: if $x \in \mathbb{R}^n$ and $E \in \Sigma_L$, then $x + E = \{x + y : y \in E\} \in \Sigma_L$ and $\lambda(x + E) = \lambda(E)$.

If $\Omega \subset \mathbb{R}^n$ is a nonempty open set, then the collection $\Sigma_{L\Omega} \triangleq \{F = \Omega \bigcap E : E \in \Sigma_L\}$ is a σ -algebra on Ω, and the Lebesgue measure λ restricted to $\Sigma_{L\Omega}$, defined by $\lambda(F) \triangleq \lambda(\Omega \bigcap E)$, is a complete measure. Thus, $\Omega_L \triangleq (\Omega, \Sigma_{L\Omega}, \lambda)$ is the natural measure space obtained by restricting the Lebesgue measure to Ω. The elements of Σ_L are called *Lebesgue measurable subsets* of $\mathbb{R}^n$ and λ is called the *Lebesgue measure* on $\mathbb{R}^n$.

Theorem 2.6 only proves the existence of the Lebesgue measure. A rigorous construction of this measure may be complex. Some classical textbooks devote considerable effort to the construction of this measure and the development of the corresponding Lebesgue integral. In this regard, we refer the reader to [2, 5].

Let $\Omega \subseteq \mathbb{R}^n$ be a measurable set. If $h : \Omega \to \mathbb{R}$ and $g : \Omega \to \mathbb{R}$ are measurable functions, then we say that g and h are equal almost everywhere (denoted by $g(x) = h(x)\, a.e.$) if the set $\{x \in \Omega : g(x) \ne h(x)\}$ has λ-measure zero. If $g(x) = h(x)\, a.e.$, then we write $g \sim h$. Since the Lebesgue measure is complete, it follows that $\sim$ is an equivalence relation on the set of all measurable functions $\mathcal{M}(\Omega) \triangleq \{f : \Omega \to \mathbb{R} : f \text{ is measurable}\}$. If $g \in \mathcal{M}(\Omega)$, then the *equivalence class* of g is

$$[g] \triangleq \{h \in \mathcal{M}(\Omega) : \; h(x) = g(x)\, a.e.\}\,. \tag{2.56}$$

Remark 2.11 (*Comment on Notation*) In this book, as in standard practice, we write f for the equivalence class $[f]$. In a few instances, for example, when we compare a special class of Hilbert spaces, namely native spaces, to Lebesgue or Sobolev spaces (see Sects. 3.5.2 and 3.5.3), we must be careful about whether the space is comprised of equivalence classes or not. In these cases, our notation is more explicit.

We assume that the reader has some familiarity with the notion of Lebesgue integral and its relationship to measure theory. The interested reader is referred to

[16–18] for a simple way to rigorously define the Lebesgue integral as a Riemann-type integral. The paper [17] defines the Lebesgue integral using only basic calculus. Moreover, Chap. 1 of [2] provides a basic introduction to the Lebesgue integral and its properties.

Definition 2.67 (*Lebesgue Integrable Functions*) Let $\Omega \subseteq \mathbb{R}^n$ be a measurable set and $h : \Omega \to \mathbb{R}$ be a measurable function. We say that $h(\cdot)$ is *Lebesgue integrable* on Ω if the Lebesgue integral of $h(\cdot)$ exists and is finite, i.e.,

$$\int_{\Omega} h(x)\mathrm{d}x = (\mathcal{L}) \int_{\Omega} h(x)\mathrm{d}x < \infty. \tag{2.57}$$

Remark 2.12 (*Comment on Notation*) In (2.51), we used $(\mathcal{R}) \int_a^b$ to emphasize that the integral is meant in the sense of Riemann. In (2.57), we used $(\mathcal{L}) \int_a^b$ to emphasize that the integral is meant in the sense of Lebesgue. Since in this book integrals are almost always meant in the sense of Lebesgue, we use the simpler notation $\int_a^b$ to denote Lebesgue integrals.

Given an integer p such that $1 \leq p < \infty$, the *space of measurable functions* is defined as

$$\mathcal{L}^p(\Omega, \mathbb{R}) \triangleq \left\{ f : \Omega \to \mathbb{R} : \int_{\Omega} |f(x)|^p \mathrm{d}x < \infty \right\}. \tag{2.58}$$

A measurable function $f : \Omega \to \mathbb{R}$ is *essentially bounded* if there exists a constant $b \geq 0$ such that $|f(x)| \leq b$ except on a set of measure zero. The essential supremum of $|f(x)|$ is defined as the smallest number $b_0 \geq 0$ such that the set $\{x \in \Omega : |f(x)| > b_0\}$ has measure zero and

$$\operatorname*{ess\,sup}_{x \in \Omega} |f(x)| = b_0. \tag{2.59}$$

Thus, for $p = \infty$, the space $\mathcal{L}^\infty(\Omega, \mathbb{R})$ is the space of essentially bounded functions

$$\mathcal{L}^\infty(\Omega, \mathbb{R}) \triangleq \left\{ f : \Omega \to \mathbb{R} : \operatorname*{ess\,sup}_{x \in \Omega} |f(x)| < \infty \right\}. \tag{2.60}$$

Definition 2.68 (*Lebesgue Spaces $L^p(\Omega, \mathbb{R})$*) Let $0 \leq p \leq \infty$. The Lebesgue space $L^p(\Omega, \mathbb{R})$ is the collection of all equivalence classes of functions in $\mathcal{L}^p(\Omega, \mathbb{R})$, given by

$$L^p(\Omega, \mathbb{R}) \triangleq \left\{ [f] : f \in \mathcal{L}^p(\Omega) \right\}, \tag{2.61}$$

where $f \sim g$ if and only if $f(x) = g(x)$ *a.e.* on Ω. We define the norm on $L^p(\Omega, \mathbb{R})$ by

$$\|f\|_{L^p(\Omega,\mathbb{R})} \triangleq \begin{cases} \left(\int_\Omega |f(x)|^p \mathrm{d}x\right)^{1/p} & \text{for } 1 \leq p < \infty, \\ \operatorname*{ess\,sup}_{x\in\Omega} |f(x)| & \text{for } p = \infty. \end{cases} \tag{2.62}$$

Remark 2.13 (*Extension to Vector-Valued Functions*) Definitions 2.67 and 2.68 can be extended to vector-valued functions $f : \Omega \to \mathbb{R}^m$, leading to the Lebesgue spaces $L^p(\Omega, \mathbb{R}^m)$ with $0 \leq p \leq \infty$. Moreover, these spaces can be extended to Banach space-valued functions by employing the *Bochner integral*, which is the Lebesgue integral for Banach space-valued functions. Specifically, if $f : \Omega \to \mathbb{Y}$, where $\mathbb{Y}$ is a Banach space with norm $\|\cdot\|_{\mathbb{Y}}$, then (2.62) becomes

$$\|f\|_{L^p(\Omega,\mathbb{Y})} \triangleq \begin{cases} \left(\int_\Omega \|f(x)\|_{\mathbb{Y}}^p \mathrm{d}x\right)^{1/p} & \text{for } 1 \leq p < \infty, \\ \operatorname*{ess\,sup}_{x\in\Omega} \|f(x)\|_{\mathbb{Y}} & \text{for } p = \infty. \end{cases} \tag{2.63}$$

The spaces $L^p(\Omega, \mathbb{Y})$ for $1 \leq p \leq \infty$ are Banach spaces. If $\mathbb{Y}$ is a Hilbert space with inner product $\langle\cdot,\cdot\rangle_{\mathbb{Y}}$, then $L^2(\Omega, \mathbb{Y})$ is a Hilbert space and

$$\langle f, g\rangle_{L^2(\Omega,\mathbb{Y})} \triangleq \int_\Omega \langle f(x), g(x)\rangle\rangle_{\mathbb{Y}} \mathrm{d}x. \tag{2.64}$$

Finally, the space $L^\infty(\Omega, \mathbb{Y})$ is not separable.

To emphasize on the measure μ employed to define the space $L^p(\Omega, \mathbb{Y})$, in some cases, we write $L^p_\mu(\Omega, \mathbb{Y})$. If $f : \Omega \to \mathbb{Y}$, then $\|f\|_{\mathbb{Y}} \leq \|f\|_{L^\infty(\Omega,\mathbb{Y})}$. If $f : \Omega \to \mathbb{Y}$ is such that $\|f\|_{L^p(\Omega,\mathbb{Y})} < \infty$, then it follows from (2.62) that f is bounded *a.e.* on Ω. For this reasons, if $\|f\|_{L^p(\Omega,\mathbb{Y})} < \infty$, then we say that f is *uniformly bounded* on Ω.

2.4.7 Weakly Differentiable Functions and Sobolev Spaces

Weak differentiability, which is instrumental in defining Sobolev space, provides another means of describing the regularity or smoothness of functions. Starting in Chap. 3, a special class of Hilbert spaces, known as RKHSs or native spaces, will play a preponderant role. We will frequently use native spaces that are equivalent to certain types of Sobolev spaces. For these reasons, we review the definition of Sobolev spaces $W^{s,p}(\Omega)$, where $\Omega \subseteq \mathbb{R}^n$. The interested reader is referred to [2, pp. 59–65] for additional details.

Definition 2.69 (*Weak Derivatives*) If $\alpha \in (\mathbb{Z}^+)^n$ is a multi-index, then $g : \mathbb{R}^n \to \mathbb{R}$ is the α-th weak derivative of $f : \mathbb{R}^n \to \mathbb{R}$ if

$$\int_{\mathbb{R}^n} g(x)\phi(x)\mathrm{d}x = (-1)^{|\alpha|} \int_{\mathbb{R}^n} f(x) D^{\alpha}\phi(x)\mathrm{d}x \tag{2.65}$$

for all test functions $\phi \in \mathcal{D}(\mathbb{R}^n)$, where $|\alpha|$ is given by (2.48).

Intuitively, a function is weakly differentiable if it satisfies an integration by parts formula when multiplied by smooth functions that have compact support in $\mathbb{R}^n$. Definition 2.69 implies that every classically differentiable function is weakly differentiable.

Having introduced the notion of weak derivatives, we now can define Sobolev spaces over subsets of $\mathbb{R}^n$. References [1] and [2, Chaps. 1–6] further discuss all the material presented here.

Definition 2.70 (*Real-Valued Sobolev Spaces, $s \in \mathbb{Z}^+$*) Let $\Omega \subseteq \mathbb{R}^n$, $s \in \mathbb{Z}^+$, and $1 \leq p \leq \infty$. The *real-valued Sobolev space* $W^{s,p}(\Omega) \triangleq W^{s,p}(\Omega, \mathbb{R})$ is the collection of real-valued functions in $L^p(\Omega)$ for which the αth weak derivative $D^{\alpha} f$ is an element of $L^p(\Omega)$ for each multi-index $\alpha \in (\mathbb{Z}^+)^n$ that satisfies $|\alpha| \leq s$. In particular,

$$W^{s,p}(\Omega) \triangleq \left\{ f(\cdot) \in L^p(\Omega) \mid D^{\alpha} f(\cdot) \in L^p(\Omega) \text{ for } 0 \leq |\alpha| \leq s \right\}. \tag{2.66}$$

Furthermore, we define $W_0^{s,p}(\Omega)$ as the closure of $C_0^{\infty}(\Omega)$ in the space $W^{s,p}(\Omega)$.

The integer s denotes the *smoothness index* of the Sobolev space since it measures the weak differentiability of function in $W^{s,p}(\Omega)$. Given $s \in \mathbb{Z}^+$, the Sobolev space $W^{s,p}(\Omega)$ is a Banach space with the norm

$$\|f\|_{W^{s,p}(\Omega)} \triangleq \begin{cases} \left(\sum_{|\alpha|=0}^{s} \|D^{\alpha} f\|_{L^p(\Omega)}^p\right)^{1/p} & \text{for } 1 \leq p < \infty, \\ \max_{0 \leq |\alpha| \leq k} \|D^{\alpha} f\|_{L^{\infty}(\Omega)} & \text{for } p = \infty. \end{cases} \tag{2.67}$$

For $s = 0$, Definition 2.70 reduces to the definition of Lebesgue space, and, hence,

$$W^{0,p}(\Omega) = L^p(\Omega). \tag{2.68}$$

We therefore interpret functions in the space $L^p(\Omega)$ as having no smoothness. For $s > 0$, we obtain different spaces that measure increasing levels of smoothness as $s \to \infty$.

The space $H^s(\Omega) \triangleq W^{s,2}(\Omega)$ is a Hilbert space with the inner product

$$\langle f, g \rangle_{H^s(\Omega)} \triangleq \langle f, g \rangle_{W^{s,2}(\Omega)} = \sum_{|\alpha|=0}^{s} \langle D^{\alpha} f, D^{\alpha} g \rangle_{L^p(\Omega)}. \tag{2.69}$$

A useful property of Sobolev spaces is that they define dense, nested, continuous embeddings,

$$W^{s+r,p}(\Omega) \hookrightarrow\!\!\!\rightarrow W^{s,p}(\Omega) \tag{2.70}$$

that hold for each pair of integer $(s, r) \in \mathbb{Z}^+ \times \mathbb{Z}^+$.

While the Definition 2.70 is for positive integer s, there are numerous ways to extend the notion of Sobolev spaces $W^{s,p}(\Omega)$ for all real $s \in \mathbb{R}^+$, including Fourier techniques, the theory of pseudodifferential operators, atomic decompositions, and several more approaches; [14, 19, Chap. 7]. Although the relationships among these equivalent definitions are not relevant to the scope of this book, the resulting spaces for real smoothness parameter s are needed. Various approximation theorems in this text make use of the Sobolev spaces $W^{s,p}(\Omega)$ for $s \in \mathbb{R}^+$. In the following, we present a definition for such spaces that has the advantage of being self-contained and concise. To this goal, let $s \geq 0$ be finite and not an integer. In this case, we define the norm over $W^{s,p}(\Omega)$as

$$\|f\|_{W^{s,p}(\Omega)} \triangleq \|f\|_{W^{\lfloor s \rfloor,p}(\Omega)} + \sum_{|\alpha|=\lfloor s \rfloor} \left(\int \int \frac{\|D^\alpha f(x) - D^\alpha f(y)\|^p}{\|x-y\|^{n+\{s\}p}} dx dy \right)^{1/p}, \tag{2.71}$$

where $s = \lfloor s \rfloor + \{s\}$, $\lfloor s \rfloor$ denotes the *integer part* and $0 < \{s\} < 1$ denotes the *remainder*. These considerations motivate the following definition.

Definition 2.71 (*Real-Valued Sobolev Spaces of Smoothness $s \in \mathbb{R}^+$*) Let $s \geq 0$ be a real number. The space $W^{s,p}(\Omega)$ is the collection of equivalence classes in $L^p(\Omega)$ for which the norm (2.71) is finite.

By definition, a domain $\Omega \subset \mathbb{R}^n$ is a nonempty open subset of $\mathbb{R}^n$. Under suitable (geometric) assumptions on Ω, the Sobolev space $W^{s,p}(\Omega)$ for $s \geq 0$ and $1 \leq p \leq \infty$ can be defined in terms of the restriction of functions in $W^{s,p}(\mathbb{R}^n)$. Furthermore, most of the embedding results employed in this book require that the domain Ω satisfy certain regularity properties. We provide a brief discussion of these properties. However, the interested reader may find a detailed presentation of these concepts and their role in the theory of Sobolev spaces in [2, pp. 81–87]. To begin this discussion, we refer to Definition 2.3 for the basic definitions of covers. If $\mathscr{O} = \{U_\alpha \mid \alpha \in \mathscr{A}\}$ is an open cover of a set $S \subset \mathbb{R}^n$, then $\mathscr{O}$ is *locally finite* if any compact set in $\mathbb{R}^n$ can intersect at most a finite number of elements of $\mathscr{O}$. As noted in [2], locally finite open covers must be countable so that $\mathscr{O} = \left\{U_j \mid j = 1, 2, 3, \ldots\right\}$. Although [2] presents six regularity properties that a domain $\Omega \subset \mathbb{R}^n$ may possess, we focus on conditions that are important for this book. Given a domain $\Omega \subset \mathbb{R}^n$ and a positive number $\delta > 0$, let

$$\Omega_\delta \triangleq \{x \in \Omega \mid \text{dist}(s, \partial\Omega) < \delta\}, \tag{2.72}$$

where

$$\mathrm{dist}(x, M) \triangleq \inf_{y \in \mathcal{M}} d(x, y). \tag{2.73}$$

defines the *distance between a point* $x \in \mathbb{R}^n$ *and a set* $\mathcal{M} \subset \mathbb{R}^n$, and $\partial\Omega$ denotes the boundary of Ω.

Definition 2.72 (*The Segment Condition*) A domain Ω satisfies the *segment condition* if every $x \in \partial\Omega$ has a neighborhood $\mathcal{N}_x$ and a nonzero vector y_x such that if $z \in \bar{\Omega} \bigcap \mathcal{N}_x$, then $z + ty_x \in 0 < t < 1$.

Let $v \in \mathbb{R}^n$ be a nonzero vector, let $\angle(x, v)$ be the angle between $x \in \mathbb{R}^n \setminus \{0\}$ and v, and let $\rho > 0$. The set

$$C \triangleq \left\{x \in \mathbb{R}^n \mid x = 0 \text{ or } \|x\| \leq \rho, \angle(x, v) \leq \kappa/2\right\} \tag{2.74}$$

is a *finite cone of height* ρ, axis direction v and aperture angle κ with vertex at the origin. Thus, $x + C = \{x + y \mid y \in C\}$ is a finite cone with vertex x.

Definition 2.73 (*Cone Condition*) A domain Ω satisfies the *cone condition* if there exists a finite cone C such that each $x \in \Omega$ is the vertex of a finite cone C_x contained in Ω and congruent to C.

Definition 2.74 (*Uniform Interior Cone Condition*) A domain Ω satisfies the *uniform interior cone condition* if there exists a finite cone C such that each $x \in \partial\Omega$ is the vertex of a finite cone C_x contained in Ω and congruent to C.

Finally, we present a regularity condition that is sufficient to prove the existence of the extension operators used in the native space-based control methods presented later in this book.

Definition 2.75 (*Uniform C^m-Regularity Condition*) A domain Ω satisfies the *uniform C^m-regularity condition* if there exists a locally finite open cover $\mathcal{O} = \left\{U_j \mid j = 1, 2, 3, \ldots\right\}$ of the boundary $\partial\Omega$, and a sequence Φ_j of m-smooth transformations mapping U_j onto the open unit ball $\mathcal{B}_1(0) = \{x \in \mathbb{R}^n \mid ||x|| < 1\}$ and having the inverse $\Psi_j = \Phi_j^{-1}$ such that

1. there is a finite number $k < \infty$ such that every subcollection $\mathcal{O}_k = \left\{U_j \in \mathcal{O}\right\}$ containing $p = k + 1$ members has empty intersection,
2. there is $\delta > 0$ such that $\Omega_\delta \subset \bigcup_{j=1}^{\infty} \Psi_j \left(\{y \in \mathbb{R}^n \,:\, ||y|| < 1/2\}\right)$,
3. $\Phi_j \left(U_j \bigcap \Omega\right) = \{y \in \mathcal{B}_1(0) \mid y_n > 0\}$ for each j, where y_n denotes the n-th component of the vector y, and
4. if $\left(\phi_{j,1}, \phi_{j,2}, \ldots, \phi_{j,n}\right)$ and $\left(\psi_{j,1}, \psi_{j,2}, \ldots, \psi_{j,n}\right)$ denote the components of Φ_j and Ψ_j, respectively, then there exists a finite constant M such that, for every multi-index α with $0 < |\alpha| \leq m$,

$$|D^\alpha \phi_{j,i}(x)| \leq M, \text{ for all } x \in U_j, \tag{2.75a}$$

$$|D^\alpha \psi_{j,i}(y)| \le M, \quad \text{for all } y \in \mathcal{B}_1(0). \tag{2.75b}$$

Remark 2.14 As noted in [2], the uniform C^m-regularity condition implies a strong local Lipschitz property and the cone conditions. Although the Sobolev embedding theorems only require the cone condition, establishing the existence of important extension operators requires stronger assumptions (at least a local Lipschitz condition). However, if Ω is a bounded domain, then the condition that Ω is a Lipschitz domain is sufficient for this book. The set $\Omega \subset \mathbb{R}^n$ is a *Lipschitz domain* if, for each point $x \in \partial\Omega$, there is a neighborhood $\mathcal{N}_x$ whose intersection with $\partial\Omega$ is the graph of a Lipschitz continuous function; see [2, p. 88]. Finally, we note that a bounded Lipschitz domain satisfies the interior cone condition; see [20, Lemma 1.3].

If the set Ω is sufficiently regular, then, for $s \in \mathbb{Z}^+$, it holds that

$$\|f\|_{W^{s,p}(\Omega)} = \begin{cases} \left(\sum_{|\alpha|=0}^{s} \|D^\alpha f\|_{L^p(\Omega)}^p\right)^{1/p} & \text{for } 1 \le p < \infty, \\ \max_{0 \le |\alpha| \le k} \|D^\alpha f\|_{L^\infty(\Omega)} & \text{for } p = \infty. \end{cases} \tag{2.76}$$

Corresponding changes can be made to (2.71) in the case $s \in \mathbb{R}^+$. The Sobolev spaces of negative order $-s < 0$ are the topological dual spaces $W^{-s,p}(\Omega) \triangleq (W^{s,p}(\Omega))^*$.

In this book, we use Sobolev spaces in specific instances in which they are examples of native spaces induced by a reproducing kernel. One complication arising in such an interpretation is that Sobolev spaces consist of equivalence classes, while, as discussed at length in Chap. 3, native spaces are collections of functions. The evaluation operator, which is formally defined in Chap. 3, is well-defined on functions in a native space, but it may not be well-defined on elements of a Sobolev space. The functions that reside in an equivalence class can have different values at points in the domain, and there is no way in general to define the evaluation function on such an equivalence class.

A critical step in concluding that certain Sobolev spaces are native spaces relies on the fact that some Sobolev spaces can be viewed as being continuously embedded in the bounded continuous functions, that is, $W^{s,p}(\Omega) \hookrightarrow C_b(\Omega)$. For $W^{s,p}(\Omega)$ to be continuously embedded in $C_b(\Omega)$, two conditions need to hold: (1) for each equivalence class $[f] \in W^{s,p}(\Omega)$, there is a representative $\hat{f} \in [f]$ that is a continuous function. This means that we can modify any function in the equivalence class $[f]$ on a set of measure zero to obtain the continuous function $\hat{f}$. In this sense, it is possible to view the equivalence classes as just a collection of (modifications on a set of measure zero of) continuous functions; (2) there is a constant $C > 0$ such that, for any such representative, $\hat{f}$, we have

$$\|\hat{f}\|_{C_b(\Omega)} \le C\|[f]\|_{W^{s,p}(\Omega)}. \tag{2.77}$$

The precise conditions that enable continuous embedding of $W^{r,p}(\Omega)$ in $C_b(\Omega)$ are contained in the famous Sobolev embedding theorem, a version of which is Theorem 2.12 in [2].

Theorem 2.7 (Sobolev Embedding Theorem) *Let Ω be a domain in $\mathbb{R}^n$. Let $1 \leq p < \infty$, $j \geq 0$, and $s \geq 1$ be integers.*

Part I. *Suppose that Ω satisfies the cone condition.*
Case A: *If either $s > n/p$ or $s = n$ and $p = 1$, then*

$$W^{j+s,p}(\Omega) \hookrightarrow C_b^j(\Omega) \tag{2.78}$$

and

$$W^{s,p}(\Omega) \hookrightarrow L^q(\Omega), \quad \text{for } p \leq q \leq \infty. \tag{2.79}$$

Case B: *If $s = n/p$, then*

$$W^{s,p}(\Omega) \hookrightarrow L^q(\Omega), \quad \text{for } p \leq q < \infty. \tag{2.80}$$

Case C: *If $s = n/p$, then*

$$W^{s,p}(\Omega) \hookrightarrow L^q(\Omega), \quad \text{for } p \leq q \leq p^* = \frac{np}{(n - sp)}. \tag{2.81}$$

Part II. *Suppose Ω satisfies the uniform C^s-regularity condition (2.75). Then, the target space $C_b^j(\Omega)$ in (2.78) can be replaced with the smaller space $C^j(\bar{\Omega})$. In particular, if $s > n/p > (s - 1)$, then*

$$W^{j+s,p}(\Omega) \hookrightarrow C^j(\bar{\Omega}). \tag{2.82}$$

Part III. *All the embeddings* (2.78)–(2.81) *hold for an arbitrary domain Ω when $W^{j+s,p}(\Omega)$ is replaced by $W_0^{j+s,p}(\Omega)$. In particular, if $s > n/p > (s - 1)$, then* **Part II** *implies the continuous embedding*

$$W_0^{j+s,p}(\Omega) \hookrightarrow C^j(\bar{\Omega}). \tag{2.83}$$

In summary, Theorem 2.7 guarantees that if the smoothness index s is large enough compared to the spatial dimension n, that is, if $s > n/p$, then the Sobolev space $W^{r,p}(\Omega)$ contains continuous, bounded functions. As noted in [14, p. 83], if Ω is a bounded subset of $\mathbb{R}^n$, then the cone condition is implied by the fact that the boundary of Ω is locally Lipschitz. There are many alternative versions of this theorem, which differ by regularity assumption on $\Omega \subset \mathbb{R}^n$.

Before closing this section that reviews scalar-valued spaces $W^{r,p}(\Omega)$ over subsets $\Omega \subset \mathbb{R}^n$, we state one version of the many-zeros theorem. This theorem plays a central role in the determination of rates of convergence of approximations in several estimation and control problems in Chaps. 3 and 5. A good overview and background for error analysis based on the many-zeros theorem can be found in [21, Sect. 11.6]. Alternative versions of this theorem are also described in [21–24], which shows that it holds over subsets of the Euclidean space or some compact Riemannian manifolds. The norms in the theorems in these references are bounded in terms of the fill distance $h_{\Xi_N,\Omega}$ of a set of N centers $\Xi_N \triangleq \{\xi_1, \ldots, \xi_N\} \subset \Omega \subseteq \mathbb{R}^n$ or of a compact Riemannian manifold. In its most general form, the *fill distance* is given by

$$h_{\Xi_N,\Omega} \triangleq \sup_{x\in\Omega} \min_{\xi\in\Xi_N} d_\Omega(x, \xi), \tag{2.84}$$

where d_Ω denotes a metric on Ω. For the case in this section, since $\Omega \subseteq \mathbb{R}^n$, the metric d_Ω is defined in terms of the Euclidean norm, that is, $d_\Omega(x, y) \triangleq \|x - y\|_2$.

Theorem 2.8 (Many-Zeros Theorem, [24]) *Let $\Omega \subseteq \mathbb{R}^n$ be a Lipschitz domain that satisfies the uniform interior cone condition (2.74), let $s \in \mathbb{N}$, let $k \in \mathbb{N}$ with $k < s$, and let $p \in \mathbb{N}$ with $1 \le p \le \infty$. If $p > 1$, then assume that $s > n/p$, and if $p = 1$, then assume that $s \ge n$. Let $\Xi_N \triangleq \{\xi_1, \ldots, \xi_N\} \subset \Omega$ be a collection of points for some $N \in \mathbb{N}$. Then, there exist constants $C_1, C_2 > 0$, which depend on k, m, p, Ω, and Ξ_N, but not N, such that, for every $f \in W^{s,p}(\Omega)$ such that $f|_{\Xi_N} = 0$, it holds that*

$$|f|_{W^{k,p}(\Omega)} \le C_1 h^{s-k}_{\Xi_N,\Omega} |f|_{W^{s,p}(\Omega)}, \tag{2.85}$$

$$\|f\|_{L^\infty(\Omega)} \le C_2 h^{s-n/p}_{\Xi_N,\Omega} |f|_{W^{s,p}(\Omega)}, \tag{2.86}$$

for all fill distances $h_{\Xi_N,\Omega}$ small enough.

This form of the many-zeros theorem describes the convergence in the "asymptotic regime" when $h_{\Xi_N,\Omega} \to 0$. That is, inequalities (2.85) and (2.86) hold for all $h_{\Xi_N,\Omega} < h_{cr}$ where h_{cr} is a fixed critical value of sample fill distance in Ω that depends on the problem data Ω and m, but not on N and Ξ_N. See [24] for the expressions of the constants C_1, C_2, h_{cr}, which we do not use in this book.

2.5 Concluding Remarks

This chapter provided a summary of elements of real and functional analysis pertinent to the scope of this book. Key topics covered in this chapter span a list of essential notation to a discussion on topological, metric, norm, and inner product spaces, the presentation of duality pairings and Riesz representers, an introduction to continuous embeddings, and, finally, a survey of classical function spaces.

The next chapter substantially leverages the content presented thus far, and will discuss in some depth a key topic for this book. In particular, Chap. 3 presents elements of native space theory. The presentation in Chap. 3 will refer to the material in this chapter from time to time.

2.6 Homework Problems

For an improved comprehension of the material exposed in this chapter, we recommend the following problems.

Exercise 2.1 Let $(\mathbb{X}, d)$ be a metric space. Show that any nonempty set $\mathcal{A} \subset \mathbb{X}$ is open if and only if it is a union of open balls. △

Exercise 2.2 Prove that the space of continuous functions $C[a, b]$ is complete. △

Exercise 2.3 Show that $\{x_1, \ldots, x_n\}$, where $x_j(t) = t^j$, is a linearly independent set in the space $C[a, b]$. △

Exercise 2.4 If Y and Z are subspaces of a vector space X, show that $Y \cap Z$ is a subspace of X, but $Y \cup Z$ need not be one. △

Exercise 2.5 Plot the unit spheres of radius 1 centered in the origin $\partial\mathcal{B}_1 \triangleq \{x \in \mathbb{R}^2 : \|x\|_p = 1\}$ for increasingly larger values of $p \in \mathbb{N}$, where

$$\|x\|_p \triangleq \left[|x_1|^p + |x_2|^p\right]^{\frac{1}{p}}, \quad \text{for any } x = [x_1, x_2]^\mathrm{T} \in \mathbb{R}^2. \tag{2.87}$$

Comment on your outcomes. △

Exercise 2.6 Show that equivalent norms on a vector space $\mathbb{X}$ induce the same topology for $\mathbb{X}$. △

Exercise 2.7 If $\|\cdot\|$ and $\|\cdot\|_0$ are equivalent norms on $\mathbb{X}$, show that the Cauchy sequences in $(\mathbb{X}, \|\cdot\|)$ and $(\mathbb{X}, \|\cdot\|_0)$ are the same. △

Exercise 2.8 Prove that $\mathbb{R}^n$ is not compact. △

Exercise 2.9 A metric space $\mathbb{X}$ is *locally compact* if every point of $\mathbb{X}$ has a compact neighborhood. Show that $\mathbb{R}^n$ is locally compact. △

Exercise 2.10 If $\mathbb{X}$ is a compact metric space and $M \subset \mathbb{X}$ is closed, show that M is compact. △

Exercise 2.11 Let X and Y be normed spaces. Show that a linear operator $T : X \to Y$ is bounded if and only if T maps bounded sets in X into bounded sets in Y. △

Exercise 2.12 Let $p : \mathbb{R}^n \to \mathbb{R}$ denote the *inner product* such that

$$p(x) = x^\mathrm{T} a, \qquad \text{for any } a \in \mathbb{R}^n. \tag{2.88}$$

Prove that p is linear and bounded. △

Exercise 2.13 Consider the space $C[a, b]$ and let

$$f_1(x) = \int_a^b x(t)y_0(t)\mathrm{d}t, \quad \text{for any } y_0 \in C[a, b] \tag{2.89}$$

$$f_2(x) = \alpha x(a) + \beta x(b), \quad \text{for any } \alpha, \beta \in \mathbb{R}. \tag{2.90}$$

Prove that both f_1 and f_2 are bounded linear operators. △

Exercise 2.14 Prove that the dual space of $\mathbb{R}^n$ is $\mathbb{R}^n$. △

Exercise 2.15 Prove that the dual space of ℓ^p is ℓ^q with $p \in (1, \infty)$ and $q \in \mathbb{R}$ such that $p^{-1} + q^{-1} = 1$. △

Exercise 2.16 If $\mathbb{X}$ is a normed space and $\dim(\mathbb{X}) = \infty$, show that the dual space $\mathbb{X}'$ is not identical with the algebraic dual space $\mathbb{X}*$. △

Exercise 2.17 Let $(\mathbb{X}, \langle \cdot, \cdot \rangle)$ is an inner product space. Assume that $\lim_{n\to\infty} x_n = x$, where $\{x_n\}_{n\in\mathbb{N}} \subset \mathbb{X}$ and $x \in \mathbb{X}$. Prove that if $\langle y, x_n \rangle = 0$ for all $n \in \mathbb{N}$, then $\langle y, x \rangle = 0$. △

Exercise 2.18 Let $f, g; [a, b] \to \mathbb{R}$ be absolutely continuous functions. Prove that

- $|f|$ is absolutely continuous.
- $\alpha f + \beta g$ is absolutely continuous for all $\alpha, \beta \in \mathbb{R}$.
- fg is absolutely continuous.
- If $|g(x)| \geq \delta > 0$ for all $x \in [a, b]$, then f/g is absolutely continuous. △

Exercise 2.19 Let $I \subset \mathbb{R}$ be an interval. Show that if $f : I \to \mathbb{R}$ is differentiable everywhere on I and $\dot{f}$ is bounded on I, then f is Lipschitz continuous on I. △

Exercise 2.20 Let $h : [-1, 1] \to \mathbb{R}$ be such that $h(x) = x^2 \sin(1/x)$ if $x \neq 0$, and $h(0) = 0$. Prove that h is Lipschitz on $[-1, 1]$. △

Exercise 2.21 Let A and B be measurable subsets of $\mathbb{R}^n$. Prove that

$$\mu(A \cup B) + \mu(A \cap B) = \mu(A) + \mu(B). \tag{2.91}$$

△

Exercise 2.22 Let $\{E_n\}_{n\in\mathbb{N}} \subset \mathbb{R}^n$ be a sequence of measurable sets such that $\mu(E_m \cap E_n) = 0$ for all $n \neq m$. Prove that

$$\mu\left(\bigcup_{n\in\mathbb{N}} E_n\right) = \sum_{n\in\mathbb{N}} \mu(E_n). \tag{2.92}$$

△

Exercise 2.23 Let A be any subset of $\mathbb{R}^n$, and let $\mathcal{L}(A) \triangleq \{E \cup A : E \text{ is measurable}\}$ be the restriction of all Lebesgue measurable sets to A. Show that $\mathcal{L}(A)$ is a σ-algebra on A. △

Exercise 2.24 Prove that the intersection of an arbitrary collection of σ-algebras on X is a σ-algebra on X. △

Exercise 2.25 Let $E \subseteq \mathbb{R}^n$ be a measurable set, and let $f, g, f_n : E \to \mathbb{R}$ be Lebesgue integrable functions. Prove that

- $f + g$ is Lebesgue integrable.
- fg is Lebesgue integrable.
- If $g(x) \neq 0$ a.e., then f/g is Lebesgue integrable.
- The function $|f|^p$ is Lebesgue integrable for each $p > 0$. △

Exercise 2.26 As a function of the parameters $\alpha, \beta \in \mathbb{R}$, determine for which $n \in \mathbb{N}$ and $p \in [1, \infty]$, the following functions belong to $W^{1,p}(\mathcal{B}_{1/2}(0))$:

- $|x|^\alpha$;
- $|\log |x||^\beta$;
- $|x|^\alpha \, |\log |x||^\beta$. △

References

1. Adams, R.A.: Sobolev Spaces. Academic Press (1975)
2. Adams, R.A., Fournier, J.J.F.: Sobolev Spaces, 2nd ed. Elsevier (2003)
3. Kreyszig, E.: Introductory Functional Analysis with Applications, vol. 17. Wiley (1991)
4. Naylor, A.W., Sell, G.R.: Linear Operator Theory in Engineering and Science. Springer, New York, NY (2000)
5. Taylor, A.E.: General theory of Functions and Integration. Dover Publications, New York, NY (1985)
6. Taylor, A.E., Lay, D.C.: Introduction to Functional Analysis. Krieger Publishing, Melbourne, FL (1986)
7. Naylor, A.W., Sell, G.R.: Linear Operator Theory in Engineering and Science. Springer Science and Business Media, New York, NY (1982)
8. Wilansky, A.: Topology for Analysis. Courier Corporation, Mineola, NY (2008)
9. Hewitt, E., Stromberg, K.: Real and Abstract Analysis: A Modern Treatment of the Theory of Functions of a Real Variable. Springer, Berlin, Germany (2013)
10. Haddad, W.M., Chellaboina, V.: Nonlinear Dynamical Systems and Control: A Lyapunov-Based Approach. Princeton University Press, Princeton, NJ (2008)
11. Kato, T.: Perturbation Theory for Linear Operators. Springer, Berlin, Germany (2013)
12. Conway, J.: A Course in Functional Analysis, series Graduate Texts in Mathematics. Springer, New York, NY (1994)
13. Brudnyi, A., Brudnyi, Y.: Methods of Geometric Analysis in Extension and Trace Problems, vols. 1, 102. Birkäuser, Baser, Switzerland (2011)
14. Adams, R.A., Fournier, J.J.: Sobolev Spaces. Elsevier, London, UK (2003)
15. Burns, J.A.: Introduction to the Calculus of Variations and Control with Modern Applications. CRC Press, Boca Raton, FL (2013)

16. McShane, E.J.: A Riemann-type Integral that Includes Lebesgue-Stieltjes, Bochner and Stochastic Integrals. American Mathematical Society, New York, NY (1969)
17. McShane, E.J.: A unified theory of integration. Am. Math. Mon. **80**, 349–359 (1973)
18. McShane, E.J.: Unified Integration. Academic Press, New York, NY (1983)
19. Triebel, H.: Theory of Function Spaces. Birkhäuser, Berlin, Germany (1983)
20. Fuselier, E., Wright, G.B.: Scattered data interpolation on embedded submanifolds with restricted positive definite kernels: Sobolev error estimates. SIAM J. Numer. Anal. **50**(3), 1753—1776 (2012)
21. Wendland, H.: Scattered Data Approximation, vol. 17. Cambridge University Press, Cambridge, UK (2004)
22. Fuselier, E., Wright, G.B.: Scattered data interpolation on embedded submanifolds with restricted positive definite kernels: Sobolev error estimates. SIAM J. Numer. Anal. **50**(3), 1753–1776 (2012)
23. Hangelbroek, T., Narcowich, F.J., Ward, J.D.: Kernel approximation on manifolds I: bounding the Lebesgue constant. SIAM J. Math. Anal. **42**(4), 1732–1760 (2010)
24. Hangelbroek, T., Narcowich, F., Ward, J.: Polyharmonic and related kernels on manifolds: interpolation and approximation. Found. Comput. Math. **12**, 625–670 (2012)

Chapter 3
Elements of Native Space Theory

While the review of the properties of general Hilbert spaces in Chap. 2 has been necessarily brief, this chapter studies more closely the subset of Hilbert spaces given by *reproducing kernel Hilbert spaces* (RKHS), also known as *native spaces*. The theory of RKHSs constitutes the theoretical foundation of many recent advances in statistical and machine learning theory, Bayesian estimation, and Gaussian process estimation, for example. Popular accounts of the general theory of RKHS can be found in [1–3]. References [4–6] provide a detailed treatment of learning theory in an RKHS, and [7] discusses how such approaches are pervasive in modern robotics.

This chapter is outlined as follows. Section 3.1 introduces scalar-valued native spaces, that is, native spaces that consist of scalar-valued functions defined over $\mathbb{X} \triangleq \mathbb{R}^n$. Section 3.2 discusses the problem of generating native spaces over subsets of $\mathbb{X} \triangleq \mathbb{R}^n$. In Sect. 3.3, we study real-valued kernels commonly employed to define scalar-valued RKHSs. Subsequently, in Sect. 3.4, we establish relationships between RKHSs and other spaces such as those of continuous functions and Sobolev spaces. Sections 3.6 and 3.7 leverage the discussion on scalar-valued RKHSs, introduce vector-valued RKHS, and present tools to compare these operators to one another. In Sect. 3.5, we establish relations between infinite-dimensional RKHS and finite-dimensional approximations. In particular, we present some key results for infinite-dimensional vector-valued RKHS and their finite-dimensional counterparts. Finally, Sect. 3.9 presents some concluding remarks and Sect. 3.10 recommends some homework problems.

The proposed discussion on vector-valued RKHSs is shorter than the discussion on scalar-valued RKHSs for multiple reasons. Firstly, it is possible to construct vector-valued RKHSs as Cartesian products of scalar-valued RKHSs. Thus, as shown in Sect. 3.8, several key properties of scalar-valued RKHSs can be ported more or less directly to vector-valued RKHS. Secondly, the theory for general vector-valued RKHSs is substantially more complex and less mature than that for scalar-valued RKHS. Finally, considering the complexity of this topic, we prefer a more tutorial,

A. J. Kurdila et al., *Data-Driven, Nonparametric, Adaptive Control Theory*,
Lecture Notes in Control and Information Sciences 495,
https://doi.org/10.1007/978-3-031-78003-5_3

hence, lengthier, approach to describe scalar-valued RKHSs. This presentation serves to build intuition for the more challenging case of vector-valued RKHSs.

3.1 Scalar-Valued Native Spaces

This section introduces scalar-valued RKHSs defined over an arbitrary set $\mathbb{X}$. For various applications discussed later in the text, several standard cases for the choice of $\mathbb{X}$ arise. Most frequently, we set $\mathbb{X} \triangleq \mathbb{R}^n$. Sometimes, we set $\mathbb{X} \triangleq \Omega \subset \mathbb{R}^n$. More rarely, $\mathbb{X}$ is a smooth Riemannian manifold $\mathcal{M} \subset \mathbb{R}^n$ of dimension n, or one of its k-dimensional, regularly embedded submanifolds with $n > k \geq 1$.

A technique for defining an RKHS starts with a Hilbert space $\mathbb{H}(\mathbb{X}, \mathbb{R})$ of $\mathbb{R}$-valued functions over a set $\mathbb{X}$ and shows that this space satisfies the *reproducing formula* or *reproducing property*.

Definition 3.1 (*Reproducing Property for Scalar-Valued Hilbert Spaces*) Let $\mathbb{H} \triangleq \mathbb{H}(\mathbb{X}, \mathbb{R})$ denote a real Hilbert space of scalar-valued functions defined over the set $\mathbb{X}$, with inner product $\langle \cdot, \cdot \rangle_{\mathbb{H}}$. If, for each $x \in \mathbb{X}$, there exists a function $\mathfrak{K}_x \in \mathbb{H}$ such that

$$f(x) = \langle f, \mathfrak{K}_x \rangle_{\mathbb{H}}, \quad \text{for all } f \in \mathbb{H}, \tag{3.1}$$

then $\mathbb{H}$ satisfies the *reproducing property*.

Definition 3.2 (*Scalar-Valued Reproducing Kernel Hilbert Space (RKHS)*) A real, scalar-valued Hilbert space $\mathcal{H}(\mathbb{X}, \mathbb{R})$ is a *scalar-valued Reproducing Kernel Hilbert Space* (RKHS), or, equivalently, a *scalar-valued native space*, if it satisfies the reproducing property (3.1).

Remark 3.1 (*Comment on Notation*) *Generic scalar-valued Hilbert spaces are denoted by* $\mathbb{H}(\mathbb{X}, \mathbb{R})$. *Given their key role, in this book, we denote scalar-valued RKHSs by* $\mathcal{H}(\mathbb{X}, \mathbb{R})$. *If clear from the context, we write* $\mathcal{H}$ *for* $\mathcal{H}(\mathbb{X}, \mathbb{R})$, *that is, we omit the domain and co-domain of the functions in the space* $\mathcal{H}$.

The next necessary and sufficient conditions provide an operative tool to verify whether some Hilbert space verifies the reproducing property. To state this result, note that it follows from the Definition 2.40 that a bounded linear functional $h^* \in \mathbb{H}'$ is a linear mapping $h : \mathbb{H} \to \mathbb{R}$ for which there is a constant $C > 0$ such that

$$|h^*(h)| \leq C\|h\|_{\mathbb{H}}, \quad \text{for all } h \in \mathbb{H}. \tag{3.2}$$

Furthermore, we introduce the following definition.

Definition 3.3 (*Evaluation Functional for Scalar-Valued Hilbert Spaces*) Given the Hilbert space $\mathbb{H}(\mathbb{X}, \mathbb{R})$, the *evaluation functional* $E_x : \mathbb{H} \to \mathbb{R}$ for a point $x \in \mathbb{X}$ is defined by the identity

$$E_x f \triangleq f(x) \in \mathbb{R}, \quad \text{for all } f \in \mathbb{H}. \tag{3.3}$$

The evaluation functional at x is always linear, but not necessarily bounded, in a generic Hilbert space of functions. However, as shown by the next result, in the case of RKHSs, evaluation functionals are bounded.

Theorem 3.1 ([8]) *The Hilbert space $\mathcal{H}$ is an RKHS if and only if all the linear evaluation functionals $E_x : \mathcal{H} \to \mathbb{R}$ are bounded for all $x \in \mathbb{X}$.*

Theorem 3.1 is equivalent to stating that, for each $x \in \mathbb{X}$, there exists a constant $C_x > 0$ such that

$$|E_x f| = |f(x)| \leq C_x \|f\|_{\mathcal{H}}, \quad \text{for all } f \in \mathcal{H}. \tag{3.4}$$

By the Riesz-representation theorem (Theorem 2.5), if the conditions of Theorem 3.1 are verified, then we are guaranteed the existence of an element $\mathfrak{K}_x \in \mathcal{H}$ such that

$$E_x f = f(x) = \langle \mathfrak{K}_x, f \rangle_{\mathcal{H}}, \quad \text{for all } x \in \mathbb{X}. \tag{3.5}$$

This characterization is usually one of the first theorems describing which particular Hilbert spaces are RKHSs [1–3]. However, in constructing useful scalar-valued RKHSs for applications, an alternative approach can often be more useful. The idea is to construct RKHSs by first defining an admissible *kernel function* $\mathfrak{K} : \mathbb{X} \times \mathbb{X} \to \mathbb{R}^+$ *over the set* $\mathbb{X}$ and subsequently defining an RKHS of functions generated by $\mathfrak{K}$. To illustrate this approach, we recall the following definitions of symmetry and positivity of kernels.

Definition 3.4 (*Scalar-Valued Symmetric Kernel*) A kernel function $\mathfrak{K} : \mathbb{X} \times \mathbb{X} \to \mathbb{R}^+$ is *symmetric* if $\mathfrak{K}(x, y) = \mathfrak{K}(y, x)$ for all $x, y \in \mathbb{X}$.

Definition 3.5 (*Scalar-Valued Kernel of Positive Type*) A kernel function $\mathfrak{K} : \mathbb{X} \times \mathbb{X} \to \mathbb{R}^+$ is *of positive type* if, for any integer $N \in \mathbb{N}$, any collection of N points

$$\Xi_N \triangleq \{\xi_i \in \Omega \mid 1 \leq i \leq N\} \subset \Omega \subset \mathbb{X}, \tag{3.6}$$

and any set of real coefficients $\{\alpha_1, \ldots, \alpha_N\} \subset \mathbb{R}$, it holds that

$$\sum_{i,j=1}^{N} \mathfrak{K}(\xi_i, \xi_j)\alpha_i\alpha_j \geq 0. \tag{3.7}$$

If the inequality in (3.7) is strict for any choice of N distinct centers in Ξ_N, then the kernel is of *strictly positive type*.

The points in Ξ_N are commonly called *centers*. It follows from Definition 3.5 that a kernel is of positive type if the matrix

$$\mathbb{K}_N \triangleq [\mathfrak{K}(\xi_i, \xi_j)] = [\langle \mathfrak{K}_{\xi_i}, \mathfrak{K}_{\xi_j} \rangle_{\mathcal{H}}] \in \mathbb{R}^{N \times N} \tag{3.8}$$

is positive semidefinite for any set of samples Ξ_N. By the reproducing property, the matrix defined in (3.8) is the *Grammian matrix* of the set of kernel basis functions $\{\mathfrak{K}_{\xi_i}\}_{1 \le i \le N}$ located at Ξ_N given by (3.6). Thus, a kernel is of *strictly* positive type if and only if the Grammian matrix for any collection of N distinct points is positive-definite.

Definition 3.6 (*Admissible Kernels and Mercer Kernels*) A real-valued function $\mathfrak{K} : \mathbb{X} \times \mathbb{X} \to \mathbb{R}^+$ is an *admissible kernel* if it is symmetric and of positive type. Additionally, if this kernel is continuous, then it is called a *Mercer kernel*.

In this book, we only consider continuous kernels, a few of which are summarized in Sect. 3.3 below. A theoretical discussion of real-valued Mercer kernels can be found in [6], which describes how they play a critical role in statistical and machine learning theory.

Once an admissible kernel $\mathfrak{K}$ is selected, we define the *kernel section located at* x as $\mathfrak{K}_x(\cdot) \triangleq \mathfrak{K}(x, \cdot)$. Often, a kernel section $\mathfrak{K}_x$ is used to define bases in building approximations, so we also refer to $\mathfrak{K}_x$ as the *kernel basis function centered at* x. We can think of an RKHS $\mathcal{H}$ as the space that is obtained from taking the limit of linear combinations of the "template" $\mathfrak{K}_x$ as the center x varies over $\mathbb{X}$. Thus, to make this intuition rigorous, an RKHS can be constructed as

$$\mathcal{H} \triangleq \overline{\text{span}\{\mathfrak{K}_x \mid x \in \mathbb{X}\}}, \tag{3.9}$$

where the closure is taken with respect to the candidate inner product defined as

$$\langle \mathfrak{K}_{x_1}, \mathfrak{K}_{x_2} \rangle_{\mathcal{H}} \triangleq \mathfrak{K}(x_1, x_2), \quad \text{for any } x_1, x_2 \in \mathbb{X}. \tag{3.10}$$

The space $\mathcal{H}$ defined according to (3.9) is the *native space induced by the kernel* $\mathfrak{K}$. Given a kernel $\mathfrak{K}$, (3.10) implies the norm bound

$$\|\mathfrak{K}_x\| = \sqrt{\mathfrak{K}(x, x)}, \quad \text{for all } x \in \mathbb{X}, \tag{3.11}$$

which is used frequently in our computations.

3.2 Subspaces Generated by Subsets and Restrictions to Subsets

In Sect. 3.1, we defined reproducing kernels over the entire set $\mathbb{X}$, and the corresponding RKHS $\mathcal{H}$ consists of real-valued functions defined over all of $\mathbb{X}$. RKHSs can be defined over a subset $\Omega \subset \mathbb{X}$ in multiple ways. If

$$\mathcal{H}_\Omega \triangleq \overline{\text{span}\{\mathfrak{K}_x \mid x \in \Omega\}} \subseteq \mathcal{H}, \tag{3.12}$$

then $\mathcal{H}_\Omega$ is the *closed subspace of* $\mathcal{H}$ *generated by the subset* $\Omega \subseteq X$. Since $\mathcal{H}_\Omega$ is contained in $\mathcal{H}$, all the functions in $\mathcal{H}_\Omega$ are defined over all of $\mathbb{X}$. The subspace $\mathcal{H}_\Omega \subset \mathcal{H}$ should not be confused with the restrictions of functions in $\mathcal{H}$ to the subset Ω. The space $\mathcal{H}_\Omega$ is equipped with the topology naturally inherited as a closed subspace of $\mathcal{H}$. We define $\Pi_\Omega : \mathcal{H} \to \mathcal{H}_\Omega$ as the $\mathcal{H}$-orthogonal projection onto $\mathcal{H}_\Omega$. The space $\mathcal{H}_\Omega$ is itself an RKHS, and the kernel that induces the native space $\mathcal{H}_\Omega$ is defined as

$$\mathfrak{K}_\Omega(x, y) \triangleq (\Pi_\Omega(\mathfrak{K}_x))(y) \triangleq \left\langle \Pi_\Omega \mathfrak{K}_x, \mathfrak{K}_y \right\rangle_{\mathcal{H}} = \left\langle \Pi_\Omega \mathfrak{K}_x, \Pi_\Omega \mathfrak{K}_y \right\rangle_{\mathcal{H}}. \tag{3.13}$$

A proof of the fact that (3.13) is a kernel is given by [1, Theorem 11]. A useful property of the subspace $\mathcal{H}_\Omega$ generated by the subset $\Omega \subseteq \mathbb{X}$ is that it defines a convenient orthogonal decomposition of $\mathcal{H}$, as described in the following result.

Theorem 3.2 *For any* $\Omega \subseteq \mathbb{X}$*, it holds that*

$$\mathcal{H} = \mathcal{H}_\Omega \oplus Z_\Omega, \tag{3.14}$$

where $\oplus$ *denotes the* orthogonal direct sum operator *and*

$$Z_\Omega \triangleq \{f \in \mathcal{H} : f(x) = 0 \textit{ for all } x \in \Omega \subset \mathbb{X}\}. \tag{3.15}$$

Per the definition of orthogonal direct sum operator, Theorem 3.2 implies that, for any $f \in \mathcal{H}$, there is a unique decomposition $f = h + z$ where $h \in \mathcal{H}_\Omega$ and $z \in Z_\Omega$, and moreover $h \perp z$.

It follows from Definition 2.57 that

$$\mathcal{R}_\Omega = T_\Omega(\mathcal{H}) = \{g : \Omega \to \mathbb{R} \ : \ g = T_\Omega f = f|_\Omega \text{ for some } f \in \mathcal{H}\}. \tag{3.16}$$

This space can be endowed with the norm

$$\|g\|_{\mathcal{R}_\Omega} \triangleq \inf \{\|f\|_{\mathcal{H}} : g = T_\Omega f = f|_\Omega \text{ for some } f \in \mathcal{H}\}. \tag{3.17}$$

The restriction operator T_Ω is a bounded linear operator from $\mathcal{H}$ to $\mathcal{R}_\Omega$, and in fact, $\|T_\Omega\| \le 1$. The norm in (3.16) on $\mathcal{R}_\Omega$ is referred to as the *trace norm* [9].

As one might expect, the RKHS $\mathcal{H}_\Omega$ generated by a subset $\Omega \subseteq \mathbb{X}$ that contains functions supported on $\mathbb{X}$ is closely related to the space of restrictions $\mathcal{R}_\Omega$ to the subset Ω. The next theorem explores the relationships between $\mathcal{H}$, $\mathcal{H}_\Omega$, and $\mathcal{R}_\Omega$.

Theorem 3.3 *Let the kernel function* $\mathfrak{K} : \mathbb{X} \times \mathbb{X} \to \mathbb{R}$ *induce the native space* $\mathcal{H}$*, and let* $\Omega \subseteq X$*.*

1. *The canonical extension operator* $\mathcal{E}_\Omega : \mathcal{R}_\Omega \to \mathcal{H}$

$$\mathcal{E}_\Omega \triangleq \left(T_\Omega|_{Z_\Omega^\perp}\right)^{-1} = \left(T_\Omega|_{\mathcal{H}_\Omega}\right)^{-1} \tag{3.18}$$

is well-defined. An inner product on the space of restrictions $\mathcal{R}_\Omega$ *is given by*

$$\langle f, g\rangle_{\mathcal{R}_\Omega} \triangleq \langle \mathcal{E}_S f, \mathcal{E}_S g\rangle_{\mathcal{H}}, \quad \textit{for all } f, g \in \mathcal{R}_\Omega, \tag{3.19}$$

which induces the norm in (3.17). Hence, $\mathcal{R}_\Omega$ *is an RKHS defined by the kernel* $\mathfrak{r} : \Omega \times \Omega \to \mathbb{R}$ *such that*

$$\mathfrak{r}(x, y) \triangleq \mathfrak{K}|_{\Omega\times\Omega}(x, y), \quad \textit{for all } x, y \in \Omega. \tag{3.20}$$

2. *For any* $g \in \mathcal{R}_\Omega$*, it holds that*

$$\mathcal{E}_\Omega g = \arg\min_{f\in\mathcal{H}} \{\|f\|_{\mathcal{H}} : g = T_\Omega f\}, \tag{3.21}$$

and, hence, the canonical extension operator $\mathcal{E}_\Omega$ *yields the minimum norm extension of* g.

3. *It holds that*

$$\Pi_\Omega = \mathcal{E}_\Omega T_\Omega \quad \textit{and} \quad \mathcal{E}_\Omega^* = T_\Omega. \tag{3.22}$$

Proof The result follows from [2, Sect. 2.2.1] or, equivalently, [3, Corollary 5.8] and is omitted for brevity. ∎

For additional discussions, the interested reader is also referred to [10].

3.3 Common Real-Valued Kernels on $\mathbb{R}^n$

The general construction outlined in Sect. 3.1 yields native spaces of real-valued functions over an arbitrary set $\mathbb{X}$. In many applications, the domain $\mathbb{X} \triangleq \mathbb{R}^n$ or some subset $\Omega \subset \mathbb{X} \triangleq \mathbb{R}^n$ are considered. In the following, we summarize a few of the common choices of the kernel $\mathfrak{K} : \mathbb{X} \times \mathbb{X} \to \mathbb{R}^+$ with $\mathbb{X} \triangleq \mathbb{R}^n$.

3.3.1 Globally Supported Kernels on $\mathbb{R}^n$

Many of the most common kernels $\mathfrak{K} : \mathbb{X} \times \mathbb{X} \to \mathbb{R}^+$ with $\mathbb{X} \triangleq \mathbb{R}^n$ are defined in terms of a *radial function* $\eta : [0, \infty) \to \mathbb{R}$. Then, the associated kernel is defined as

$$\mathfrak{K}(x_1, x_2) \triangleq \eta(\|x_1 - x_2\|_2), \quad \text{for all } (x_1, x_2) \in \mathbb{X} \times \mathbb{X}, \tag{3.23}$$

where $\|\cdot\|_2$ denotes the Euclidean norm on $\ell^2(\mathbb{R}^n)$. In this setting, the kernel section $\mathfrak{K}_x$ gives rise to approximations in terms of radial basis functions (RBFs).

Table 3.1 Radial functions η used to define kernels $\mathfrak{K}(x_1, x_2) \triangleq \eta(\|x_1 - x_2\|_2)$, for $x_1, x_2 \in \mathbb{X} = \mathbb{R}^n$. The symbol $K_{m-n/2}$ in the definition of Sobolev–Matérn kernels denotes the modified Bessel function of the second kind and $\lceil x \rceil$ denotes the ceiling function. Extracted from entries in Table 11.1 of [11] and Table 1 of [12]

Category	$\eta(r)$	Parameters
Gaussian	$e^{-\alpha r^2}$	$\alpha > 0$
Multiquadric	$(-1)^{\lceil \beta \rceil}(c^2 + r^2)^\beta$	$\beta > 0, \beta \notin \mathbb{N}$
Inverse multiquadric	$(r^2 + \beta^2)^\beta$	$\beta < 0$
Sobolev–Matérn	$r^{m-d/2} K_{m-n/2}(r)$	$2m > d$

Table 3.1 lists several popular choices of radial functions. In particular, Table 3.1 provides analytical expressions for the *Gaussian* (also known as *exponential*), *multiquadratic*, *inverse multiquadratic*, and *Sobolev–Matérn* kernels. Figures 3.1 and 3.2 provide graphical representations for the kernel functions deduced from the Gaussian and inverse multiquadratic radial basis functions.

The kernels defined using (3.23) and the radial functions in Table 3.1 share some important properties. First, these kernels are well-defined over $\mathbb{X} \triangleq \mathbb{R}^n$ for any dimension $n > 0$. Their simple closed-form expressions are readily implemented in applications. Moreover, for any dimension $n > 0$, these kernels are of strictly positive type on $\mathbb{R}^n$. This latter property leads to well-posed interpolation problems using any of these kernels for functions over any Euclidean space $\mathbb{R}^n$ for any dimension $n > 0$, a topic we discuss in Sect. 3.5.

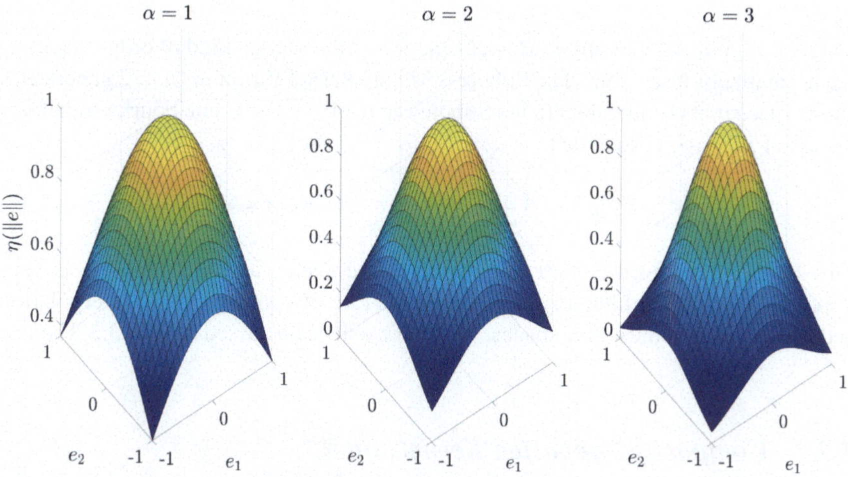

Fig. 3.1 Gaussian kernel function computed applying (3.23) with $e \triangleq x_1 - x_2$ and the Gaussian radial function shown in Table 3.1 for multiple values of α

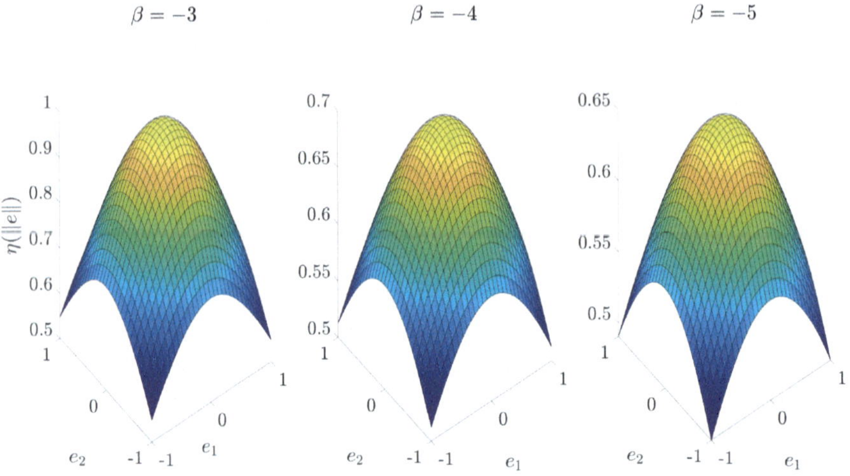

Fig. 3.2 Inverse multiquadratic kernel function computed applying (3.23) with $e \triangleq x_1 - x_2$ and the inverse multiquadratic radial function shown in Table 3.1 for multiple values of β

Of all the kernels in Table 3.1, the Sobolev–Matérn kernels are notable because the associated native spaces are Sobolev spaces. As explained in [11, Theorem 6.13] and [13], the Sobolev–Matérn kernel can also be written as

$$\begin{aligned} \mathfrak{K}_m(x_1, x_2) &\triangleq c_m \|x_1 - x_2\|_2^{m-n/2} K_{m-n/2}(\|x_1 - x_2\|_2) \\ &= c_m \|x_1 - x_2\|_2^{m-n/2} K_{n/2-m}(\|x_1 - x_2\|_2), \end{aligned} \tag{3.24}$$

where K_ν denotes the modified Bessel function of the second kind of order ν, and c_m is a normalization constant. The Sobolev–Matérn kernel $\mathfrak{K}_m$ for $m > n/2$ generates a native space that is equivalent to the Sobolev space $W^{m,2}(\mathbb{R}^n)$. The Fourier transform $\hat{\mathfrak{K}}_m(\omega)$ of $\mathfrak{K}_m(\cdot, 0)$ is such that

$$\hat{\mathfrak{K}}_m(\omega) = (1 + \|\omega\|_2^2)^{-m}, \quad \text{for any } \omega \in \mathbb{R}. \tag{3.25}$$

This Fourier transform is said to have algebraic decay, and this property is very important to some techniques for studying the rate of convergence of interpolation and projection in some native spaces. This topic will be discussed in Sect. 3.5.3 below.

3.3.2 Compactly Supported Kernels on $\mathbb{R}^n$

The radial functions in Table 3.1 yield kernels that have global support in $\mathbb{X} \times \mathbb{X}$ with $\mathbb{X} \triangleq \mathbb{R}^n$. In many applications, it is also advantageous to define kernels for $\mathbb{X} \triangleq \mathbb{R}^n$ that have compact support. The benefits that result from using compactly

Table 3.2 Compactly supported, minimal degree (Wendland) functions. In these expressions, $(x)_+ \triangleq x$ if $x \geq 0$, and $(x)_+ \triangleq 0$ if $x < 0$

Dimension	Function $\eta_{n,s}$	Smoothness
$n = 1$	$\eta_{1,0}(r) = (1-r)_+$	C^0
	$\eta_{1,1}(r) = (1-r)_+^3(3r+1)$	C^2
	$\eta_{1,2}(r) = (1-r)_+^5(8r^2+5r+1)$	C^4
$n \leq 3$	$\eta_{3,0}(r) = (1-r)_+^2$	C^0
	$\eta_{3,1}(r) = (1-r)_+^4(4r+1)$	C^2
	$\eta_{3,2}(r) = (1-r)_+^6(35r^2+18r+3)$	C^4
	$\eta_{3,3}(r) = (1-r)_+^8(32r^3+25r^2+8r+1)$	C^6
$n \leq 5$	$\eta_{5,0}(r) = (1-r)_+^3$	C^0
	$\eta_{5,1}(r) = (1-r)_+^5(5r+1)$	C^2
	$\eta_{5,2}(r) = (1-r)_+^7(16r^2+7r+1)$	C^4

supporting approximants are well known. These have been studied for decades in the development of finite element methods that approximate the solutions of PDEs, or of spline approximations of functions. For example, it follows from Definition 2.25 that functions having compact support in a domain Ω vanish on the boundary of the domain, and this property simplifies many arguments. Unfortunately, the situation in an RKHS is more involved if we want to define compactly supported kernels that are positive-definite on $\mathbb{R}^n$ for all choices of $n > 0$. The general problem of defining compactly supported kernels is discussed in [11, Chap. 9], and the interested reader is referred to this clear exposition for details. For the scope of this book, it suffices to note [11, Corollary 9.3], whereby a continuous, scalar-valued, compactly supported function cannot be positive-definite over the entire Euclidean space $\mathbb{R}^n$ for every choice of $n > 0$.

The author in [11] derives a family of compactly supported radial functions $\eta_{n,s}$ that are indexed by a dimension n and a smoothness index s; see Table 3.2 for some examples of such radial functions. These functions are sometimes referred to as *Wendland functions,* or *compact radial functions of minimal degree*. The Wendland function $\eta_{n,s}(\cdot)$ defines the kernel

$$\mathfrak{K}_{n,s}(x_1, x_2) \triangleq \eta_{n,s}(\|x_1 - x_2\|_2), \quad \text{for all } (x_1, x_2) \in \mathbb{R}^{\hat{n}} \times \mathbb{R}^{\hat{n}}, \tag{3.26}$$

that is positive-definite on $\mathbb{R}^{\hat{n}}$ for all $\hat{n} \leq n$. In (3.26), $\|\cdot\|_2$ denotes the ℓ^2-norm on $\mathbb{R}^{\hat{n}}$; see Example 2.5. This way, the radial function $\eta_{n,s}$ can be used to define a positive-definite kernel on any Euclidean space having dimension $\hat{n} \leq n$. The index s describes the global smoothness of the radial function $\eta_{n,s}$ in the sense that $\eta_{n,s} \in C^{2s}(\mathbb{R}^+)$. As discussed in [11, p. 201], the native space $\mathcal{H}_{n,s}$ that is defined in terms of $\eta_{n,s}$ is

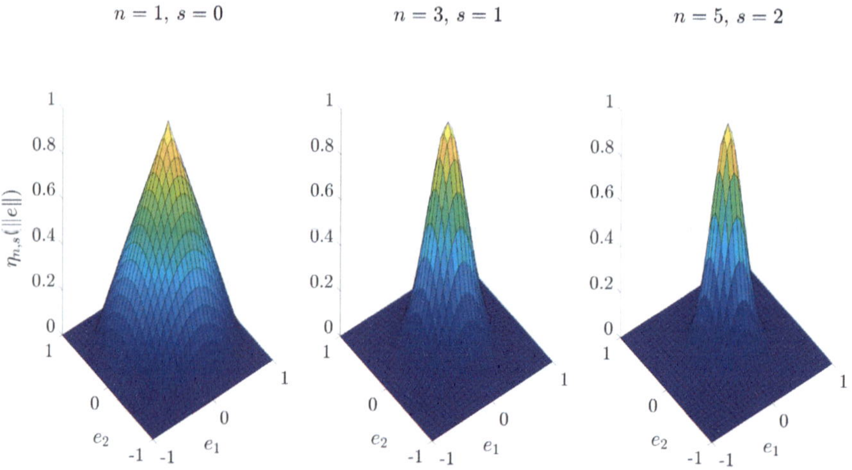

Fig. 3.3 Wendland kernel function computed applying (3.26) with $\eta_{n,s}$ as in Table 3.2 for multiple values of n and s and $e \triangleq x_1 - x_2$

norm-equivalent to the Sobolev space $W^{s+(n+1)/2,2}(\Omega)$ when $\Omega \subseteq \mathbb{R}^n$ has a Lipschitz boundary; see Definition 2.39 for the notion of norm equivalence. Figure 3.3 shows Wendland kernels.

3.4 Scalar-Valued Native Spaces and Other Function Spaces

The number of admissible kernels $\mathfrak{K} : \mathbb{X} \times \mathbb{X} \to \mathbb{R}$ that can be used to define a native space $\mathcal{H}$ of real-valued spaces of functions over an arbitrary set $\mathbb{X}$ is large. This diversity is one of the strengths that recommends using RKHSs in many applications. However, there are some drawbacks to having such a broad array of candidate kernels and native spaces. For any given application, different analysts may choose different kernels that induce different native spaces over $\mathbb{X}$. Exploring the relationship among different native spaces is far from a trivial exercise. The associated notions of convergence in one setting may not agree with our intuition regarding convergence in other settings. Looking ahead, when we study native space embedding for control theory in Chap. 5, the choice of the state space depends on the selection of the reproducing kernel. Since not all native spaces are equivalent to one another, arguments about stability and convergence depend on the choice of the kernel and its associated native space.

If we want to assess the performance of computational algorithms or control strategies formulated in a native space, then it is critical to calibrate how they compare to methods formulated in other spaces. In this section, we describe a few important examples of how native spaces can be related to other function spaces. We focus on comparing native spaces to one another. Then, we compare them to other common function spaces such as the continuous functions and Sobolev spaces of weakly differentiable functions.

3.4.1 Comparing Native Spaces to One Another

It follows from the definition of the reproducing kernel (Definition 3.2) or the constructive method captured by (3.9) that there are many ways to choose a kernel. If we are confronted with two different native spaces, it is often important in applications to know if one is contained in the other. Fortunately, there are rather simple means of establishing whether this is the case. The following result is a typical, well-known result of this type.

Theorem 3.4 ([3, Theorem 5.1]) *Suppose that $\mathfrak{K}_1$ and $\mathfrak{K}_2$ are the reproducing kernels that define the RKHSs $\mathcal{H}_1$ and $\mathcal{H}_2$, respectively, of real-valued functions over the same set $\mathbb{X}$. Then, $\mathcal{H}_1 \subseteq \mathcal{H}_2$ if and only if there is a constant $c > 0$ such that*

$$\mathfrak{K}_1(x, y) \leq c\mathfrak{K}_2(x, y), \quad \text{for all } x, y \in \mathbb{X}. \tag{3.27}$$

In this case, it holds that

$$\|f\|_{\mathcal{H}_2} \leq \sqrt{c}\|f\|_{\mathcal{H}_1}, \quad \text{for all } f \in \mathcal{H}_1. \tag{3.28}$$

Theorem 3.4 is equivalent to stating that the continuous embedding

$$\mathcal{H}_1 \overset{\mathcal{I}}{\hookrightarrow} \mathcal{H}_2 \tag{3.29}$$

is verified and the operator norm of the embedding is such that $\|\mathcal{I}\| \leq c$; see Definition 2.58.

3.4.2 Native Spaces and Continuous Functions

The following theorem gives a simple sufficient condition that guarantees that the real-valued native space $\mathcal{H}$ over a set $\mathbb{X}$ consists of continuous functions, that is, $\mathcal{H} \subseteq C(\mathbb{X})$. This is the only instance in the book where the native space $\mathcal{H}$ is generated by an admissible kernel that does not need to be continuous.

Theorem 3.5 ([2, 3, Theorems 2.3 and 2.17]) *Let $\mathfrak{K} : \mathbb{X} \times \mathbb{X} \to \mathbb{R}^+$ be the admissible kernel that generates the native space $\mathcal{H}$ of real-valued functions over the* *topological space $\mathbb{X}$. If $\mathfrak{K}$ is continuous, then all functions in $\mathcal{H}$ are continuous, that is, $\mathcal{H} \subseteq C(\mathbb{X})$.*

Proof Let $\mathcal{H}_0 \triangleq \text{span}\{\mathfrak{K}_x \ : \ x \in \mathbb{X}\}$. Since $\mathfrak{K}$ is continuous, each function in $\mathcal{H}_0$ is continuous. Furthermore, each function $f \in \mathcal{H}$ is the limit in the norm on $\mathcal{H}$ of a sequence of functions $\{f_k\}_{k\in\mathbb{N}} \subset \mathcal{H}_0$. However, norm convergence in a native space implies pointwise convergence since, for any $f \in \mathcal{H}$ and each $x \in \mathbb{X}$, if follows from (2.15), (2.22a), and (3.11) that

$$|f(x) - f_k(x)| = |\langle \mathfrak{K}_x, f - f_k\rangle_{\mathcal{H}}| \leq \|\mathfrak{K}_x\|\|f - f_k\| = \sqrt{\mathfrak{K}(x,x)}\|f - f_k\|. \tag{3.30}$$

Now, fix $\delta > 0$, and define the set

$$\mathcal{B}_{\delta^2}(x) \triangleq \left\{x \in \mathbb{X} \ : \ \mathfrak{K}(x,x) < \delta^2\right\}. \tag{3.31}$$

The fact that the mapping $\mathfrak{K}$ is continuous on $\mathbb{X} \times \mathbb{X}$ implies that the mapping $x \mapsto \mathfrak{K}(x,x)$ is continuous on $\mathbb{X}$. This means that $\mathcal{B}_{\delta^2}(x)$ is open in $\mathbb{X}$ since it is the inverse image of the open set $(-\infty, \delta^2) \subset \mathbb{R}$ under a continuous mapping. Thus, $\lim_\infty f_k = f$ uniformly on $\mathcal{B}_{\delta^2}(x)$. Since the uniform limit of continuous functions (taking values in a metric space) is continuous, the arbitrary function f is continuous in a neighborhood of x. ■

3.4.3 *Continuous Kernels Bounded on the Diagonal*

Theorem 3.5 only states that the native space $\mathcal{H}$ is a subset of the set of continuous functions $C(\mathbb{X})$ with $\mathbb{X}$ being a topological space. However, the functions in $\mathcal{H}$ are not required to be bounded on $\mathbb{X}$. If the diagonal of the kernel $\mathfrak{K}(x,x)$ is uniformly bounded, then we obtain much stronger results.

Definition 3.7 (*Kernels Uniformly Bounded on the Diagonal*) The kernel $\mathfrak{K} : \mathbb{X} \times \mathbb{X} \to \mathbb{R}^+$ is *uniformly bounded on the diagonal* if there is a constant $\bar{\mathfrak{K}} > 0$ such that

$$\|\mathfrak{K}_x\|^2_{\mathcal{H}} = \mathfrak{K}(x,x) \leq \bar{\mathfrak{K}}^2, \quad \text{for all } x \in \mathbb{X}. \tag{3.32}$$

The next result follows as a direct consequence of Definition 3.7.

Theorem 3.6 *If the kernel $\mathfrak{K}$ is uniformly bounded on the diagonal, then each function f in the induced RKHS $\mathcal{H}$ is bounded, and $\mathcal{H}$ is continuously embedded in the space of bounded continuous functions over $\mathbb{X}$, that is,*

$$\mathcal{H} \overset{I}{\hookrightarrow} C_b(\mathbb{X}), \tag{3.33}$$

with $I : \mathcal{H} \to C_b(\mathbb{X})$ being the canonical injection. In this case, $\|I\| \leq \bar{\mathfrak{K}}$.

Proof To prove the continuous embedding of $\mathcal{H}$ into $C_b(\mathbb{X})$, note that it follows from (2.40), (2.22a), (2.15), and (3.32) that

$$\|f\|_{C_b(\mathbb{X})} = \sup_{x\in\mathbb{X}} |\langle \mathfrak{K}_x, f\rangle_{\mathcal{H}}| \leq \sup_{x\in\mathbb{X}} \|\mathfrak{K}_x\|\|f\|_{\mathcal{H}} \leq \sup_{x\in\mathbb{X}} \sqrt{\mathfrak{K}(x,x)}\|f\|\mathcal{H} \leq \bar{\mathfrak{K}}\|f\|_{\mathcal{H}}. \tag{3.34}$$

This reasoning also implies that the family of evaluation functionals $\{E_x\}_{x\in\mathbb{X}}$ is uniformly bounded in operator norm, that is, $\|E_x\| \leq \bar{\mathfrak{K}}$. ■

The fact that (3.32) implies the uniform boundedness of the collection of all the evaluation functionals $\{E_x\}_{x\in\mathbb{X}}$ is an important property that is used in many of our applications. The previous argument for the continuous embedding of $\mathcal{H}$ into $C_b(\mathbb{X})$ can be further extended to infer that any generic Hilbert space $(\mathbb{H}, \langle\cdot,\cdot\rangle_{\mathbb{H}})$ that is continuously embedded in $C_b(\mathbb{X})$ is an RKHS. To show this result, let $\mathcal{I} : f \in \mathbb{H} \mapsto f \in C_b(\mathbb{X})$ be the canonical injection. In this case, the evaluation functional $E_x : \mathbb{H} \to \mathbb{R}$ is well-defined since $\mathbb{H} \subseteq C_b(\mathbb{X})$. We also have that

$$|E_x f| = |f(x)| \leq \|f\|_{C_b(\mathbb{X})} \leq \|\mathcal{I}\|\|f\|_{\mathbb{H}} \tag{3.35}$$

with $\|\mathcal{I}\|$ being the operator norm of the canonical injection $\mathcal{I} : \mathbb{H} \to C_b(\mathbb{X})$. Since (3.35) establishes that each evaluation functional E_x is a bounded operator, it follows from Theorem 3.1 that $\mathbb{H}$ is an RKHS.

3.4.4 Native Spaces and Sobolev Spaces

Although not all Sobolev spaces (see Definitions 2.70 and 2.71) with smoothness $s > 0$ are RKHSs, sufficiently smooth Sobolev spaces are native spaces. This fact is proven in the next theorem for Sobolev spaces over a subset $\Omega \subset \mathbb{X} \triangleq \mathbb{R}^n$, whose proof relies on the Sobolev embedding theorem (see Theorem 2.7). For the statement of this result, it is worthwhile recalling Definition 2.74.

Theorem 3.7 *Suppose that $s > n/2$, and let $\Omega \subset \mathbb{R}^n$ be a Lipschitz domain that satisfies the uniform interior cone condition. Then, $W^{s,2}(\Omega)$ is an RKHS.*

Proof The continuous embedding $W^{s,2}(\Omega) \hookrightarrow C_b(\Omega)$ for $s > n/2$ follows from the Sobolev embedding theorem (Theorem 2.7). The conclusion that $W^{s,2}(\Omega)$ is a native space now follows from the fact that $W^{s,2}(\Omega)$ is a Hilbert space continuously embedded in the space of bounded continuous functions, that is, $W^{s,2}(\Omega) \overset{\mathcal{I}}{\hookrightarrow} C_b(\Omega)$. Furthermore, it follows from (3.35) that $\|E_x\| \leq \|\mathcal{I}\|$, and every evaluation functional, is a bounded operator. We conclude that $W^{s,2}(\Omega)$ is an RKHS. ■

3.4.5 *Mercer Kernels and Smoothness Spaces*

As discussed in Chap. 2, one of the most important uses of Sobolev spaces $W^{s,p}(\Omega)$ is that they can be construed as a *smoothness scale* defined in terms of a real smoothness parameter $s \geq 0$. It holds that

$$W^{s_2,2}(\Omega) \subset W^{s_1,2}(\Omega) \subset L^2(\Omega) \tag{3.36}$$

whenever $s_1, s_2 \in [0, \infty)$ and $s_2 > s_1$. Since the smoothness index $s \in \mathbb{Z}^+$ can be interpreted as the number of weak derivatives that are guaranteed for an element of $W^{s,2}(\Omega)$, the functions in the space $W^{s,2}(\Omega)$ are increasingly smoother for larger values of s.

In this section, we review how to create a smoothness scale of RKHSs defined in terms of a smoothness parameter $s \geq 0$ using kernels $\mathfrak{K}$ of an RKHS $\mathcal{H}$. To this goal, we will employ the theory of integral operators associated with Mercer kernels. This section is important because one of the most powerful means of describing rates of convergence of approximations in a native space $\mathcal{H}$ exploits the theory of integral operators and Mercer kernels. One of the advantages of the theory of Mercer kernels is that it applies both to compact subsets Ω of $\mathbb{R}^n$ and, more generally, to closed subsets Ω of a smooth compact Riemannian manifold $\mathbf{M}$. A comprehensive discussion of the properties of Mercer kernels defined over a compact set is given in [6, 14]. An analysis of Mercer kernels over more general noncompact domains is presented in [15].

3.4.5.1 Spectral Decompositions for Mercer Kernels

In this section, we assume that (Ω, d_Ω) is a compact metric space (see Definition 2.23) and μ is a finite measure on Ω (see Definition 2.64). We also assume that the admissible kernel $\mathfrak{K} : \Omega \times \Omega \to \mathbb{R}$ is d_Ω-continuous and that the kernel $\mathfrak{K}$ is bounded on the diagonal. As proven in the previous section, it holds that $\mathcal{H} \hookrightarrow C_b(\Omega)$, $\|E_x\| \leq \bar{\mathfrak{K}}$ for all $x \in \Omega$, and the smallest constant $\bar{\mathfrak{K}}$ for which the norm bound inequality holds is the norm $\|\mathcal{I}\|$ of the canonical injection $\mathcal{I}$ of $\mathcal{H}$ into $C_b(\Omega)$. Thus, the embedding $\mathcal{I}$ is compact and $\mathcal{H}$ is separable [6, 14, 16]. The next result proves the compactness of the embedding (3.33).

Theorem 3.8 *Suppose that (Ω, d_Ω) is a compact metric space and the uniform bound in (3.32) holds for an admissible kernel that generates the native space $\mathcal{H}$. Then, the embedding $\mathcal{I} : \mathcal{H} \hookrightarrow C_b(\Omega)$ is compact.*

Proof Let $\{f_k\}_{k\in\mathbb{N}} \subset \mathcal{H}$ have the uniform bound $\|f_k\|_{\mathcal{H}} \le C$. Then, the family $\{f_k\}_{k\in\mathbb{N}}$ is equicontinuous since, for any f_k, it holds that

$$\begin{aligned} |f_k(x) - f_k(y)| &= |\langle \mathfrak{K}_x - \mathfrak{K}_y, f_k \rangle_{\mathcal{H}}| \\ &\le C\|\mathfrak{K}_x - \mathfrak{K}_y\|_{\mathcal{H}} \\ &= C\sqrt{\mathfrak{K}(x,x) - 2\mathfrak{K}(x,y) + \mathfrak{K}(y,y)}, \quad \text{for all } x, y \in \Omega. \quad (3.37) \end{aligned}$$

The right-hand side of (3.37) is a continuous function. Thus, for any $\epsilon > 0$, there is a $\delta = \delta(\epsilon)$ such that if $d_\Omega(x,y) \le \delta$, then $|f_k(x) - f_k(y)| \le \epsilon$ for all $k \in \mathbb{N}$. The collection $\{f_k\}_{k\in\mathbb{N}}$ is therefore equicontinuous, uniformly bounded, and defined on a compact metric space. By the Arzela–Ascoli theorem [17, Theorem 3.8], there is a subsequence of this family of functions that converges in the uniform norm on $C_b(\Omega)$. Hence, the embedding I is compact. ■

Since the measure μ is finite and $\|f\|^2_{L^2_\mu(\Omega)} \le \mu(\Omega)\|f\|^2_{C_b(\Omega)}$, the space of continuous bounded functions $C_b(\Omega)$ is continuously embedded in $L^2_\mu(\Omega)$. We denote this injection by $\mathcal{J} : C_b(\Omega) \to L^2_\mu(\Omega)$, and it follows from Theorem 3.6 that

$$\mathcal{H} \overset{I}{\hookrightarrow} C_b(\Omega) \overset{\mathcal{J}}{\hookrightarrow} L^2_\mu(\Omega). \quad (3.38)$$

Let $i_{\mathfrak{K}} \triangleq \mathcal{J} \circ I : \mathcal{H} \to L^2_\mu(\Omega)$ be the canonical injection of $\mathcal{H}$ into $L^2_\mu(\Omega)$. We can relate this injection to an integral operator from the identity

$$(i^*_{\mathfrak{K}} f)(x) = \langle \mathfrak{K}_x, i^*_{\mathfrak{K}} f \rangle_{\mathcal{H}} = \langle i_{\mathfrak{K}} \mathfrak{K}_x, f \rangle_{L^2_\mu(\Omega)} = \int_\Omega \mathfrak{K}(x,\xi) f(\xi) \mu(d\xi), \quad (3.39)$$

for all $f \in \mathcal{H}$. By definition, $i^*_{\mathfrak{K}} : L^2_\mu(\Omega) \to \mathcal{H}$. Moreover, since $i_{\mathfrak{K}} = \mathcal{J} \circ I$, and I is compact and $\mathcal{J}$ is bounded, the embedding $i_{\mathfrak{K}}$ is a compact operator (see Definition 2.58). We then define the integral operators

$$(T_{\mathfrak{K}} f)(x) \triangleq i^*_{\mathfrak{K}} i_{\mathfrak{K}} = \int_\Omega \mathfrak{K}(x,\xi) f(\xi) \mu(d\xi), \quad (3.40)$$

$$(L_{\mathfrak{K}} f)(x) \triangleq i_{\mathfrak{K}} i^*_{\mathfrak{K}} = \int_\Omega \mathfrak{K}(x,\xi) f(\xi) \mu(d\xi). \quad (3.41)$$

Remark 3.2 Although (3.40) and (3.41) have identical expressions, they capture different operators. Indeed, $T_{\mathfrak{K}} : \mathcal{H} \to \mathcal{H}$ and $L_{\mathfrak{K}} : L^2_\mu(\Omega) \to L^2_\mu(\Omega)$. Both the operators $T_{\mathfrak{K}}$ and $L_{\mathfrak{K}}$ are compact since they are the products of a compact and a bounded operator.

Theorem 3.9 *Suppose that the admissible kernel $\mathfrak{K}$ is continuous on the compact set Ω and satisfies the inequality in (3.32). The integral operators $i^*_{\mathfrak{K}}$, $L_{\mathfrak{K}}$, and $T_{\mathfrak{K}}$ are bounded, linear, compact operators from $L^2_\mu(\Omega) \to \mathcal{H}$, $L^2_\mu(\Omega) \to L^2_\mu(\Omega)$, and $\mathcal{H} \to \mathcal{H}$, respectively.*

Proof The proofs of these properties can be found in any of the references [6, 14, 16] and is omitted for brevity. ■

We next characterize the eigenvalues and eigenvectors of $L_{\mathfrak{K}} : L^2_\mu(\Omega) \to L^2_\mu(\Omega)$; see Definition 2.34.

Theorem 3.10 (Spectral Decomposition of $L_{\mathfrak{K}} : L^2_\mu(\Omega) \to L^2_\mu(\Omega)$, [6, 14, 16]) *If the Mercer kernel $\mathfrak{K} : \Omega \times \Omega \to \mathbb{R}$ satisfies the inequality (3.32), then the operator $L_{\mathfrak{K}} : L^2_\mu(\Omega) \to L^2_\mu(\Omega)$ is positive, self-adjoint, and compact. The eigenvalues $\{\lambda_k\}_{k\in\mathbb{N}}$ of $L_{\mathfrak{K}}$ are nonnegative $\lambda_1 \geq \lambda_2 \geq \cdots \geq 0$, and can only accumulate at zero. The collection of eigenvectors $\{[e_k]\}_{k\in\mathbb{N}}$ of $L_{\mathfrak{K}}$ are an L^2_μ-orthonormal basis for the closed subspace $(\mathcal{N}(L_{\mathfrak{K}}))^\perp \subseteq L^2_\mu(\Omega)$. Finally, $L_{\mathfrak{K}}$ can be represented as*

$$L_{\mathfrak{K}} f = \sum_{k=1}^{\infty} \lambda_k \left\langle [e_k], f \right\rangle_{L^2_\mu(\Omega)} [e_k] \quad \textit{strongly in } L^2_\mu(\Omega). \tag{3.42}$$

The notation $[e_k]$ in Theorem 3.10, which represents the equivalence class of eigenvectors, emphasizes the fact that the eigenvectors $[e_k]$ of $L_{\mathfrak{K}}$ are elements of $L^2_\mu(\Omega)$. This notation is commonly used in more general settings such as [15, 18–20], where it is critical to keep in mind that the eigenvectors in $L^2_\mu(\Omega)$ are equivalence classes. In fact, for Mercer kernels over a compact set Ω, much more can be said about the equivalence class of eigenvectors $[e_k] \in L^2_\mu(\Omega)$. This is accomplished using Nystrom's method for smoothing the equivalence class $[e_k]$; see [21]. Since $[e_k]$ is an eigenvector of $L_{\mathfrak{K}}$ in $L^2_\mu(\Omega)$, we can write

$$[e_k](x) = \frac{1}{\lambda_k}(L_{\mathfrak{K}}[e_k])(x) = \frac{1}{\lambda_k}\int_S \mathfrak{K}(x,\xi)[e_k](\xi)\mu(d\xi). \tag{3.43}$$

However, it holds that

$$|(L_{\mathfrak{K}}[e_k])(x) - (L_{\mathfrak{K}}[e_k])(y)| \leq \sup_{\eta\in\Omega} |\mathfrak{K}(x,\eta) - \mathfrak{K}(y,\eta)| \sqrt{\mu(S)} \|[e_k]\|_{L^2_\mu(\Omega)} \tag{3.44}$$

for any $x, y \in \Omega$. For this reason, it is always possible, with a Mercer kernel $\mathfrak{K}$ over a compact set Ω, to choose a continuous representative $e_k \in C_b(\Omega)$ of the equivalence class $[e_k] \in L^2_\mu(\Omega)$. We refer to (3.42) as a *spectral decomposition of* $L_{\mathfrak{K}}$. We conclude this section by summarizing some additional properties of Mercer kernels.

Theorem 3.11 (Mercer Kernel Expansions, [6, 16, 21]) *Suppose that Ω is compact, $\mathfrak{K}$ is an admissible Mercer kernel over $\Omega \times \Omega$ that generates the native space $\mathcal{H}$, μ is a finite measure on Ω, and $\{(\lambda_k, e_k)\}_{k\in\mathbb{N}}$ is the collection of eigenvalues and orthonormal eigenvectors of $L_{\mathfrak{K}} : L^2_\mu(\Omega) \to L^2_\mu(\Omega)$. Then, $\left\{(\lambda_k, \sqrt{\lambda_k} e_k)\right\}_{k\in\mathbb{N}}$ denotes the collection of eigenvalues and associated orthonormal eigenvectors of $T_{\mathfrak{K}} : \mathcal{H} \to \mathcal{H}$, and*

$$\mathfrak{K}(x, y) \triangleq \sum_{k \in \mathbb{N}} \lambda_k e_k(x) e_k(y), \tag{3.45}$$

and the convergence is absolute and uniform in $\Omega \times \Omega$.

3.4.5.2 Representations and Smoothness Spaces for Mercer Kernels

Given a spectral decomposition of a compact, self-adjoint, positive operator, such as $L_{\mathfrak{K}}$ in Theorem 3.10, we can also define spectral decompositions of all the operator powers $L^s_{\mathfrak{K}}$ for $s \in [0, \infty)$ via

$$L^s_{\mathfrak{K}} f = \sum_{k=1}^{\infty} \lambda^s_k \left([e_k], f\right)_{L^2_\mu(\Omega)} [e_k] \quad \text{strongly in } L^2_\mu(\Omega), \tag{3.46}$$

for all $f \in \mathscr{D}(L^s_{\mathfrak{K}}) \subseteq L^2_\mu(\Omega)$. We denote by $R^\infty_0 \subset \mathbb{R}^{\mathbb{N}}$ the collection of all sequences in $\mathbb{R}$ that are zero for all but a finite number of terms. We also define the finite linear span of the eigenfunctions $\{e_i\}_{i \in \mathbb{N}}$ as

$$E_0 \triangleq \left\{ f \in L^2_\mu(\Omega) : f = \sum_{i \in \mathbb{N}} c_i e_i, \{c_i\}_{i \in \mathbb{N}} \in R^\infty_0 \right\} \subset L^2_\mu(\Omega). \tag{3.47}$$

A candidate norm and an inner product over $\in E_0$ are given by

$$\|f\|_s \triangleq \left\| \sum_{i \in \mathbb{N}} c_i e_i \right\|_s = \sqrt{\sum_{i \in \mathbb{N}} \lambda_i^{-s} c_i^2} = \|\{\lambda^{-s/2} c_i\}_{i \in \mathbb{N}}\|_{\ell^2(\mathbb{N})}, \tag{3.48}$$

$$\begin{aligned} (f, g)_s &\triangleq \left(\sum_{i \in \mathbb{N}} c_i e_i, \sum_{i \in \mathbb{N}} d_i e_i \right)_s = \sum_{i \in \mathbb{N}} \lambda_i^{-s} c_i d_i \\ &= \left(\{\lambda_i^{-s/2} c_i\}_{i \in \mathbb{N}}, \{\lambda_i^{-s/2} d_i\}_{i \in \mathbb{N}} \right)_{\ell^2(\mathbb{N})}, \end{aligned} \tag{3.49}$$

respectively, for any $f, g \in E_0$.

We finally define the *power space* or *smoothness space* $\mathcal{H}^s(L_{\mathfrak{K}})$ *of smoothness index* $s > 0$ as the completion of set E_0 with the candidate norm (3.48).

Remark 3.3 (*Comment on Notation*) Later in this chapter, starting from Sect. 3.6, the expression $\mathcal{H}^m$ for some $m \geq 1$ will be employed to denote the vector-valued native spaces defined as the m-times Cartesian products of a scalar-valued RKHS $\mathcal{H}$, that is, $\mathcal{H}^m \triangleq \mathcal{H} \times \cdots \times \mathcal{H}$. This expression is used extensively in this book. This expression should not be confused with the expression for the power space $\mathcal{H}^s(L_{\mathfrak{K}})$ for some $s > 0$, which is primarily employed in Chap. 7. The meaning of these two similar expression will be apparent from the context.

The expression smoothness space is commonly used in approximation theory [22, 23]. The explanation of why this is an apt designation is as follows. Any $f \in L^2_\mu(\Omega)$ can be represented as $f = \sum_{i\in\mathbb{N}} c_i e_i$, with $\|f\|_{L^2_\mu(\Omega)} = \|\{c_i\}_{i\in\mathbb{N}}\|_{\ell^2(\mathbb{N})} \leq \infty$. If $f \in \mathcal{H}^s(L_{\mathfrak{K}})$, then the definition of the norm $\mathcal{H}^s(L_{\mathfrak{K}})$ requires that the weighted sequence $\{\lambda_i^{-s/2} c_i\}_{i\in\mathbb{N}}$ is square summable. The eigenvalues λ_i can be sorted to be decreasing for increasing values of the index i, which means that the weights $\lambda_i^{-s/2}$ are increasing as a function of the index i. It follows that for a function to be contained in $\mathcal{H}^s(L_{\mathfrak{K}})$, the sequence of coeffcients $\{c_i\}_{i\in\mathbb{N}}$ must converge faster to offset the growth of the weight $\lambda_i^{-s/2}$. Thus, as $s > 0$ increases, the sequence of the coefficients $\{c_i\}_{i\in\mathbb{N}}$ in the generalized Fourier series for $f = \sum_{i\in\mathbb{N}} c_i e_i$ must converge faster. The fact that such faster convergence of approximations corresponds to increased smoothness is a classical topic in approximation theory. In this sense, the description of the power space as an example of a smoothness space is consistent with intuition and with standard terminology in approximation theory.

The smoothness space $\mathcal{H}^s(L_{\mathfrak{K}})$ is a Hilbert space, although it may not necessarily be an RKHS [18, 19, 24]. This construction results in the scale of nested smoothness spaces

$$\mathcal{H}^{s_2}(L_{\mathfrak{K}}) \subset \mathcal{H}^{s_1}(L_{\mathfrak{K}}) \subset \ldots \mathcal{H}^1(L_{\mathfrak{K}}) \equiv \mathcal{H} \subset \ldots \subset \overline{\mathscr{R}(\mathcal{I})} \equiv \mathcal{H}^0(L_{\mathfrak{K}}) \subset L^2_\mu(\Omega), \tag{3.50}$$

where $s_2 > s_1 > 0$, $\mathcal{I} : \mathcal{H} \to L^2_\mu(\Omega)$ denotes the canonical injection of $\mathcal{H}$ into $L^2_\mu(\Omega)$, $\mathscr{R}(\mathcal{I})$ denotes the range space of $\mathcal{I}$, and the closure $\overline{(\cdot)}$ is taken with the norm of $L^2_\mu(\Omega)$. Sometimes, the space $\mathcal{H}^0$ is called the *sample space of the Gaussian process* $GP(\bar{m}, \mathfrak{K})$; see Chap. 7 for a more detailed discussion on this point.

Although the power space $\mathcal{H}^s(L_{\mathfrak{K}})$ may not be an RKHS for some $s \in (0, 1)$, the authors in [24] show that if the support of the measure μ is all of Ω and

$$\sum_{i\in\mathbb{N}} \lambda_i^s \phi_i(x) < \infty, \quad \text{for all } x \in \Omega, \tag{3.51}$$

then the power space $\mathcal{H}^s(L_{\mathfrak{K}})$ for $s \in (0, 1]$ is an RKHS for the kernel

$$\mathfrak{K}^s(x, y) \triangleq \sum_{i\in\mathbb{N}} \lambda_i^s e_i(x) e_i(y), \quad \text{for all } x, y \in \Omega. \tag{3.52}$$

The assumption whereby the support of μ is all of Ω ensures that $\mathscr{N}(L_{\mathfrak{K}}) = \{0\}$, which implies that

$$(\mathscr{N}(L_{\mathfrak{K}}))^\perp \equiv \overline{\mathscr{R}(\mathcal{I})} \equiv \mathcal{H}^0(L_{\mathfrak{K}}) \equiv L^2_\mu(\Omega). \tag{3.53}$$

Remark 3.4 The proposed construction of the power spaces $\mathcal{H}^s(L_{\mathfrak{K}})$ has been based on the operator $L_{\mathfrak{K}} : L^2_\mu(\Omega) \to L^2_\mu(\Omega)$, its powers $L^s_{\mathfrak{K}}$, and its $L^2_\mu(\Omega)$-orthonormal eigenvectors $\{e_i\}_{i\in\mathbb{N}}$. We can follow exactly the same process to define power spaces

$\mathcal{H}^s(T_{\mathfrak{K}})$ in terms of the operator $T_{\mathfrak{K}} : \mathcal{H} \to \mathcal{H}$ and its $\mathcal{H}$-orthonormal eigenvectors $\{\psi_i\}_{i\in\mathbb{N}}$. The definitions of the norms and inner products on the spaces $\mathcal{H}^s(T_{\mathfrak{K}})$ have the same form as in (3.48) and (3.49). However, we replace the eigenvectors e_i of $L_{\mathfrak{K}}$ with the eigenvectors ψ_i of $T_{\mathfrak{K}}$. It is important to realize that, in this case, the power spaces $\mathcal{H}^s(T_{\mathfrak{K}})$ for $s > 0$ are all contained in $\mathcal{H}$, that is,

$$\mathcal{H}^{s_2}(T_{\mathfrak{K}}) \subset \mathcal{H}^{s_1}(T_{\mathfrak{K}}) \subset \ldots \subset \mathcal{H}^0(T_{\mathfrak{K}}) \equiv \mathcal{H}, \tag{3.54}$$

for all $s_2 > s_1 > 0$. For additional details, see [24, Sect. 4.1] or [25].

When needed, we use the notation $\mathcal{H}^s(L_{\mathfrak{K}})$ or $\mathcal{H}^s(T_{\mathfrak{K}})$ to avoid ambiguities.

3.5 Approximation in Scalar-Valued Native Spaces

This section summarizes some characterizations of interpolation and projection approximations in scalar-valued native spaces. This material is crucial to the native space embedding methods of control introduced later in Chap. 5, and is necessarily brief. The interested reader is referred to [11] for a comprehensive account.

3.5.1 Representations via Interpolation and via Projection

In this section, we characterize the error induced when using a projection onto a finite-dimensional space to approximate an element of a generic Hilbert space and of an RKHS. This topic is of capital importance to appreciate the results in Chap. 5. To this goal, we discuss the differences between constructing approximations that interpolate samples and those that are derived in terms of orthogonal projection in a generic Hilbert space. Then, we prove that these approximations match in the case of RKHSs.

Throughout this section, the kernel $\mathfrak{K} : \mathbb{X} \times \mathbb{X} \to \mathbb{R}^+$ inducing the native space $\mathcal{H}$ is strictly positive on $\mathbb{X} \times \mathbb{X}$. Consider the finite collection of N distinct samples in the subset Ω given by (3.6), and define the associated finite-dimensional space

$$\mathcal{H}_N \triangleq \operatorname{span}\left\{\mathfrak{K}_{\xi_i} \ : \ \xi_i \in \Xi_N, \ 1 \le i \le N\right\} \subset \mathcal{H}. \tag{3.55}$$

The dimension of each of the spaces $\mathcal{H}_N$ is N, since strict positivity of the kernel implies that the Grammian matrix $\mathbb{K}_N$ defined in (3.8) is positive-definite for distinct centers in Ξ_N given by (3.6). Finally, Π_N denotes the $\mathcal{H}$-orthogonal projection onto $\mathcal{H}_N$, so that $\Pi_N : \mathcal{H} \to \mathcal{H}_N$.

3.5.1.1 Interpolation Approximations in a Generic Hilbert Space

Let $\mathbb{H}$ be a generic Hilbert space, which contains an unknown function $f \in \mathbb{H}$. We are tasked with building an approximation g_N of f that interpolates a given set of input/output measurements $\{(\xi_i, y_i)\}_{i=1}^N \triangleq \{(\xi_i, f(\xi_i))\}_{i=1}^N$; by stating that g_N interpolates the given data $\{(\xi_i, y_i)\}_{i=1}^N$, we mean that

$$g_N(\xi_i) = y_i = f(\xi_i), \quad \text{for all } 1 \le i \le N. \tag{3.56}$$

Next, we choose a linearly independent set of basis functions and associated finite-dimensional subspace $\mathbb{H}_N \subset \mathbb{H}$ as

$$\Phi_N \triangleq [\phi_1, \ldots, \phi_N]^{\mathrm{T}}, \tag{3.57}$$

$$\mathbb{H}_N \triangleq \text{span}\{\Phi_N\}, \tag{3.58}$$

and let $g_N = \sum_{j=1}^N \alpha_j \phi_j$. The real coefficients $\{\alpha_1, \ldots, \alpha_N\}$ must satisfy the linear system of equations

$$\sum_{j=1}^N \phi_j(\xi_i)\alpha_j = y_i, \quad \text{for all } i \in \{1 \ldots, N\}. \tag{3.59}$$

This process defines the *linear interpolation operator* $\mathcal{I}_N : \mathbb{H} \to \mathbb{H}_N$ by setting

$$\mathcal{I}_N f = g_N. \tag{3.60}$$

In a compact matrix form, (3.59) can be rewritten as

$$\mathbb{A}_N \boldsymbol{\alpha}_N = \boldsymbol{y}_N, \tag{3.61}$$

where $\mathbb{A}_N \triangleq [\phi_i(\xi_j)]$ denotes the *interpolation matrix*, $\boldsymbol{\alpha}_N \triangleq [\alpha_1, \ldots, \alpha_N]^{\mathrm{T}}$, and $\boldsymbol{y}_N \triangleq [y_1, \ldots, y_N]^{\mathrm{T}}$. This is a foundational problem in learning theory, which has been studied for decades. Conditions on this problem that ensure the well-posedness of this system of equations, and the fidelity of resulting interpolation approximations, are now classical topics of study [26].

3.5.1.2 Projection Approximations in a Generic Hilbert Space

It follows from Definition 2.51 and (2.35) that we can construct an approximation $f_N \triangleq \Pi_N f \in \mathbb{H}_N$ using the orthogonal projection $\Pi_N : \mathbb{H} \to \mathbb{H}_N$. The defining property of the orthogonal projection operator is that

$$\langle (I - \Pi_N) f, \phi_i \rangle_{\mathbb{H}} = 0, \quad \text{for all } i \in \{1, \ldots, N\}; \tag{3.62}$$

recall that orthogonal projection error $(I - \Pi_N)f$ is perpendicular to the subspace $\mathbb{H}_N$. If we set $f_N \triangleq \Pi_N f = \sum_{j=1}^{N} \beta_j \phi_j$, then the orthogonality conditions in (3.62) imply that the coefficients $\{\beta_1, \ldots, \beta_N\}$ that define the projection approximation f_N satisfy the matrix equations

$$\sum_{j=1}^{N} (\phi_i, \phi_j)_{\mathbb{H}} \beta_j = (\phi_i, f)_{\mathbb{H}}, \quad \text{for all } i \in \{1, \ldots, N\},$$

which can be rewritten as

$$\mathbb{G}_N \boldsymbol{\beta}_N = \bar{\boldsymbol{y}}_N, \tag{3.63}$$

where $\mathbb{G}_N \triangleq \left[(\phi_i, \phi_j)_{\mathcal{H}}\right]$ denotes the *Grammian matrix* of the basis $\Phi_N \subset \mathbb{H}$, $\boldsymbol{\beta}_N \triangleq [\beta_1, \ldots, \beta_N]^{\mathrm{T}}$, and $\bar{\boldsymbol{y}}_N \triangleq [(\phi_1, f)_{\mathbb{H}}, \ldots, (\phi_N, f)_{\mathbb{H}}]^{\mathrm{T}}$.

Remark 3.5 In a generic Hilbert space $\mathbb{H}$, the interpolation operator $\mathcal{I}_N$ and orthogonal projection operator Π_N for a subspace $\mathbb{H}_N \subseteq \mathbb{H}$ will generate different approximations, even for the same basis functions Φ_N and finite-dimensional subspace $\mathbb{H}_N$.

3.5.1.3 Projection and Interpolation Coincide in an RKHS

Remark 3.5 shows how, in general, an interpolation approximation is distinct from an orthogonal projection approximation. However, in an RKHS, these two approximations coincide for the finite-dimensional subspaces defined as in (3.55) in terms of the distinct set of centers and associated basis functions (3.57). This choice Φ_N is sometimes referred to as a *scattered basis* or a *basis of translates for the scattered centers in* Ξ_N.

To show that interpolation and orthogonal projection coincide for finite-dimensional subspaces defined in terms of a scattered basis, note that it follows from the definition of the projection error (2.35) that

$$\left((I - \Pi_N)f, \mathfrak{K}_{\xi_i}\right) = ((I - \Pi_N)f)(\xi_i) = 0, \quad \text{for all } \xi_i \in \Xi_N, \tag{3.64}$$

where we leveraged the reproducing formula (3.1) satisfied by all the bases $\{\mathfrak{K}_{\xi_i}\}_{i=1}^{N}$. Equation (3.64) implies that the orthogonal projection operator Π_N satisfies

$$(\Pi_N f)(\xi_i) = f(\xi_i), \quad \text{for all } i \in \{1, \ldots, N\}, \tag{3.65}$$

which is the defining property of the interpolation operator $\mathcal{I}_N$ in (3.60). Thus, for finite-dimensional subspaces defined in terms of scattered bases, it holds that $\mathcal{I}_N \equiv \Pi_N$. A similar reasoning allows us to prove that (3.63) and (3.61) are equivalent, that is, $\mathbb{G}_N = \mathbb{A}_N$, $\boldsymbol{y}_N = \bar{\boldsymbol{y}}_N$, and $\boldsymbol{\alpha}_N = \boldsymbol{\beta}_N$.

If $\mathcal{I}_N f = \Pi_N f = \sum_{j=1}^{N} \alpha_j \mathfrak{K}_{\xi_j}$, then it follows from the definition of kernel sections that the coefficients of the representation are such that

$$\sum_{i=1}^{N} \mathfrak{K}(\xi_i, \xi_j)\alpha_j = f(\xi_i), \quad \text{for all } j \in \{1, \ldots, N\}, \tag{3.66}$$

which can be rewritten in matrix form as

$$\mathbb{K}_N \boldsymbol{\alpha}_N = f_{\Xi_N}, \tag{3.67}$$

where $\mathbb{K}_N$ is given by (3.8) and $f_{\Xi_N} \triangleq [f(\xi_1), \ldots, f(\xi_N)]^{\mathrm{T}} \in \mathbb{R}^N$. In summary, for finite-dimensional subspaces defined in terms of scattered bases, we can write

$$\mathcal{I}_N f = \Pi_N f = \sum_{i=1}^{N} \sum_{k=1}^{N} \left[\mathbb{K}_N^{-1}\right]_{j,k} f(\xi_k)\, \mathfrak{K}_{\xi_j} = \left(\mathbb{K}_N^{-1} f_{\Xi_N}\right)^{\mathrm{T}} \mathfrak{K}_{\Xi_N}, \tag{3.68}$$

where $\mathfrak{K}_{\Xi_N} \triangleq \left[\mathfrak{K}_{\xi_1}, \cdots, \mathfrak{K}_{\xi_N}\right]^{\mathrm{T}}$.

The property whereby $\mathcal{I}_N \equiv \Pi_N$ is not shared by many of the other interpolation and projection operators defined using a host of other spaces of approximants, including those based on finite element, spline, or wavelet functions. This is an important theoretical and practical property of approximations built using scattered bases in kernel spaces.

The equivalence between projection and interpolation operators on RKHS has been proven assuming that the underlying reproducing kernel is of strictly positive type. The authors in [27] extend this result to problems in which the underlying Grammians are only positive semidefinite.

3.5.2 *Rates of Convergence in $L^\infty(\Omega)$ and the Power Function*

Section 3.5.1 discussed the mechanics of computing projections or interpolants in a scalar RKHS. For our discussions in the following chapters on the convergence of estimates of functions in ordinary differential equations, it is important to understand how rates of convergence can be studied for interpolation problems. This section presents the use of the power function for bounding the error in projections or interpolation problems; these results will be instrumental in deriving bounds on controller performance later in the text. Section 3.5.3 below presents alternatives to the results presented in this section that are based on versions of the many-zeros theorem (see Theorem 2.8). Throughout this section and the next one, we continue to assume that $\mathcal{H}$ is the native space induced by the real-valued kernel $\mathfrak{K} : \mathbb{X} \times \mathbb{X} \to \mathbb{R}$ with $\mathbb{X} \triangleq \mathbb{R}^n$, so that $\mathcal{H}$ consists of real-valued functions over $\mathbb{X} \triangleq \mathbb{R}^n$.

A few different versions of the definition of the power function are encountered in the literature. In this brief introduction, we follow the approach in [27], tailored to scalar-valued native spaces.

Definition 3.8 (*Power Function*) Let $U \subseteq \mathcal{H}$ be a closed subspace of $\mathcal{H}$ and denote by Π_U the $\mathcal{H}$-orthogonal projection onto U. The mapping $\mathcal{P} : \mathcal{H}' \to \mathbb{R}$ defined as

$$\mathcal{P}_U(h^*) \triangleq \sup_{f \neq 0} \frac{|h^*(f) - h^*(\Pi_U f)|}{\|f\|_{\mathcal{H}}}, \quad \text{for all } h^* \in \mathcal{H}', \tag{3.69}$$

denotes the *power function* $\mathcal{P}_U$ *associated with* U.

By definition, (3.69) is equivalent to

$$\mathcal{P}_U(h^*) = \|h^* \circ \Pi_{U^\perp}\|_{\mathcal{H}'} \quad \text{for all } h^* \in \mathcal{H}' \tag{3.70}$$

where $\Pi_{U^\perp} \triangleq I - \Pi_U$ denotes the complementary projection; see (2.35). The next result shows how the value $\mathcal{P}_U(h^*)$ of the power function can be concisely written in terms of the Riesz-representer h of h^*.

Theorem 3.12 *Let $h \in \mathcal{H}$ be the Riesz-representer of $h^* \in \mathcal{H}'$, so that $h^*(f) = \langle f, h\rangle_{\mathcal{H}}$ for all $f \in \mathcal{H}$. Then, it holds that*

$$\mathcal{P}_U(h^*) = \|(I - \Pi_U)h\|_{\mathcal{H}} = \|\Pi_{U^\perp}(h)\|_{\mathcal{H}}. \tag{3.71}$$

Proof It follows from (2.32) and (2.22a) that

$$\begin{aligned}
\sup_{f \neq 0} \frac{|h^*((I - \Pi_U)f)|}{\|f\|_{\mathcal{H}}} &= \sup_{f \neq 0} \frac{|\langle h, (I - \Pi_U)f\rangle_{\mathcal{H}}|}{\|f\|_{\mathcal{H}}} \\
&= \sup_{f \neq 0} \frac{|\langle (I - \Pi_U)h, f\rangle_{\mathcal{H}}|}{\|f\|_{\mathcal{H}}} \\
&\leq \|(I - \Pi_U)h\|.
\end{aligned}$$

The desired result now follows by noting that the maximum value of the Schwarz inequality occurs for $f = (I - \Pi_U)h$. ■

Definition 3.8 is given in terms of its action on an arbitrary element $h^* \in \mathcal{H}'$. In the applications that follow, when we derive rates of convergence of interpolation problems, it is most useful to choose $h^* \triangleq E_x$, the evaluation functional on $\mathcal{H}$ at $x \in \mathbb{X}$. In this case, the Riesz-representer h of $h^* \triangleq E_x$ is the kernel section $h \triangleq \mathfrak{K}_x$. We introduce the notation

$$\mathcal{P}_U(x) \triangleq \mathcal{P}_U(E_x) = \|(I - \Pi_U)\mathfrak{K}_x\| \tag{3.72}$$

in this special case. Since this choice is made extensively in this section, we note some of its properties.

Theorem 3.13 *Let $\mathfrak{K} : \mathbb{X} \times \mathbb{X} \to \mathbb{R}$ be the kernel that induces the scalar-valued native space $\mathcal{H}$ and let $U \subseteq \mathcal{H}$ be a closed subspace. Then,*

$$\mathcal{P}_U^2(x) = \mathfrak{K}(x, x) - \mathfrak{K}_U(x, x), \quad \textit{for all } x \in \mathbb{X}, \tag{3.73}$$

where $\mathfrak{K}_U(x, u) \triangleq (\Pi_U \mathfrak{K}_x)(u)$ is the kernel that induces the native space $U \subseteq \mathcal{H}$.

Proof First, recall that any closed subspace $U \subseteq \mathcal{H}$ of the native space $\mathcal{H}$ induced by the kernel $\mathfrak{K}$ is itself an RKHS with kernel $\mathfrak{K}_U$ given by $\mathfrak{K}_U(x, u) \triangleq (\Pi_U \mathfrak{K}_x)(u)$ for all $x, y \in \mathbb{X}$. [1]. Now, it follows from (3.5) that $\mathfrak{K}_x$ is the Riesz-representer of the evaluation functional E_x and it follows from Theorem 3.12 that

$$\begin{aligned}
\mathcal{P}_U^2(x) &= \|\mathfrak{K}_x - \Pi_U \mathfrak{K}_x\|_{\mathcal{H}}^2 \\
&= \langle \mathfrak{K}_x - \Pi_U \mathfrak{K}_x, \mathfrak{K}_x - \Pi_U \mathfrak{K}_x \rangle_{\mathcal{H}} \\
&= \mathfrak{K}(x, x) - 2(\Pi_U \mathfrak{K}_x)(x) + (\Pi_U \mathfrak{K}_x)(x) \\
&= \mathfrak{K}(x, x) - \mathfrak{K}_U(x, x), \quad \text{for all } x \in \mathbb{X}.
\end{aligned}$$

■

Theorem 3.13 justifies the definition of the power function of the finite-dimensional subspace $\mathcal{H}_N \triangleq \text{span}\{\mathfrak{K}_{\xi_i} : \xi_i \in \Xi_N\} \subset \mathcal{H}$ given by

$$\mathcal{P}_N(x) \triangleq \sqrt{\mathfrak{K}(x, x) - \mathfrak{K}_N(x, x)}, \tag{3.74}$$

where

$$\mathfrak{K}_N(x_1, x_2) \triangleq \mathfrak{K}_{\Xi_N}^{\mathrm{T}}(x_1)\mathbb{K}_N^{-1}\mathfrak{K}_{\Xi}(x_2), \quad \text{for all } x_1, x_2 \in \mathbb{X}, \tag{3.75}$$

denotes the reproducing kernel of $\mathcal{H}_N$, $\mathfrak{K}_{\Xi_N}(\cdot) \triangleq \left[\mathfrak{K}_{\xi_1}(\cdot), \cdots, \mathfrak{K}_{\xi_N}(\cdot)\right]^{\mathrm{T}}$, and $\mathbb{K}_N^{-1}$ denotes the inverse of the Grammian matrix (3.8).

Finally, we come to one of the critical results that gives a powerful characterization of the error in projection or interpolation in scalar-valued RKHSs.

Theorem 3.14 *Let $\mathfrak{K} : \mathbb{X} \times \mathbb{X} \to \mathbb{R}$ be the kernel that induces the scalar-valued RKHS $\mathcal{H}$ and let $U \subseteq \mathcal{H}$ be a closed subspace. A bound on the pointwise error is given by*

$$\begin{aligned}
|f(x) - (\Pi_U f)(x)| &\le \mathcal{P}_U(x)\|(I - \Pi_U)f\|_{\mathcal{H}} \\
&\le \mathcal{P}_U(x)\|f\|_{\mathcal{H}}, \quad \textit{for all } f \in \mathcal{H} \textit{and } x \in \mathbb{X}.
\end{aligned} \tag{3.76}$$

Proof It follows from (2.32) and (2.22a) that

$$\begin{aligned}|f(x) - (\Pi_U f)(x)| &= \left|\langle (I - \Pi_U)f, \mathfrak{K}_x\rangle_{\mathcal{H}}\right| \\ &= \left|\langle (I - \Pi_U)f, (I - \Pi_U)\mathfrak{K}_x\rangle_{\mathcal{H}}\right| \\ &\leq \|(I - \Pi_U)f\|_{\mathcal{H}}\|(I - \Pi_U)\mathfrak{K}_x\|_{\mathcal{H}} \\ &= \mathcal{P}_U(x)\|(I - \Pi_U)f\|_{\mathcal{H}} \\ &\leq \mathcal{P}_U(x)\|f\|_{\mathcal{H}}, \quad \text{for all } x \in \mathbb{X},\end{aligned}$$

for all $f \in \mathcal{H}$, and the result is proven. ■

This theorem is very general. It holds for any closed subspace U of the native space $\mathcal{H}$ induced by the kernel $\mathfrak{K}$. Furthermore, Theorem 3.14 gives a pointwise bound on the approximation error for the function $f \in \mathcal{H}$. Its most important application is whenever we set $U = \mathcal{H}_N$, the finite-dimensional subspace of $\mathcal{H}$. In this case, it follows from (3.74)–(3.76) that

$$|f(x) - (\Pi_N f)(x)| \leq \sqrt{\mathfrak{K}(x,x) - \mathfrak{K}_{\Xi_N}^{\mathrm{T}}(x)\mathbb{K}_N^{-1}\mathfrak{K}_{\Xi_N}(x)}\,\|(I - \mathbf{\Pi}_N)f\|_{\mathcal{H}}, \tag{3.77a}$$

$$\leq \sqrt{\mathfrak{K}(x,x) - \mathfrak{K}_{\Xi_N}^{\mathrm{T}}(x)\mathbb{K}_N^{-1}\mathfrak{K}_{\Xi_N}(x)}\,\|f\|_{\mathcal{H}}, \quad \text{for all } x \in \mathbb{X}, \tag{3.77b}$$

for all $f \in \mathcal{H}$.

Theorem 3.14 can be employed to derive explicit error bounds for interpolation problems over some bounded subset $\Omega \subset \mathbb{X} \triangleq \mathbb{R}^n$. Suppose that a set of centers Ξ_N given by (3.6) is used to define the space of approximants $\mathcal{H}_N$, and let Π_N be the $\mathcal{H}$-orthogonal projection onto $\mathcal{H}_N$. It follows from Theorem 3.14 that

$$|f(x) - (\Pi_N f)(x)| \leq \mathcal{P}_N(x)\|f\|_{\mathcal{H}}, \quad \text{for all } x \in \Omega \text{ and } f \in \mathcal{H}. \tag{3.78}$$

Pointwise rates of convergence over the bounded set Ω follow from this inequality, provided we can find upper bounds for the power function $\mathcal{P}_N(x)$ over $x \in \Omega$.

The bounds on the power function are often expressed in terms of the fill distance $h_{\Xi_N,\Omega}$ of the finite set of centers Ξ_N in the set of interest $\Omega \subseteq \mathbb{X} \triangleq \mathbb{R}^n$. Fill distances are defined as follows.

Definition 3.9 (*Fill Distance*) Let $\Xi_N \subset \Omega \subset \mathbb{X}$ given by (3.6) denote the set of centers defining $\mathcal{H}_N$. The *fill distance* is defined as

$$h_{\Xi_N,\Omega} \triangleq \sup_{y\in\Omega} \min_{\xi_i\in\Xi_N} \|y - \xi_i\|_2; \tag{3.79}$$

it is worthwhile comparing (3.79) with (2.84).

Several references, such as [11, 12], summarize upper bounds for the power function $\mathcal{P}_N(x)$ for a variety of common kernels. These bounds have the form

Table 3.3 Functions $\mathcal{N}(\cdot)$ that bound the power function as $\mathcal{P}_N(x) \lesssim \sqrt{\mathcal{N}(h_{\Xi_N,\Omega})}$ for $x \in \Omega$. In this table, $h \triangleq h_{\Xi_N,\Omega}$. The constant a is a generic constant that varies with the kernel choice. For additional details, see Table 3.1

Type	Kernel	$\mathcal{N}(\cdot)$
Gaussian	$e^{-\alpha r^2}$, $\alpha > 0$	$e^{-a\|\log h\|/h}$
Multiquadric	$(-1)^{\lceil\beta\rceil}(c^2+r^2)^\beta$, $\beta > 0, \beta \notin \mathbb{N}$	$e^{-a/h}$
Inverse multiquadric	$(c^2+r^2)^\beta$, $\beta < 0$	$e^{-a/h}$
Compactly supported functions	$\eta_{n,k}$	h^{2k+1}

$$\mathcal{P}_N(x) \lesssim \sqrt{\mathcal{N}(h_{\Xi_N,\Omega})} \tag{3.80}$$

for a function $\mathcal{N} : \mathbb{R}^+ \to \mathbb{R}^+$ that may depend on the dimension n, the set Ω, and the kernel $\mathfrak{K}$, but not on the samples Ξ_N.

Table 3.3 summarizes some expressions for the function $\mathcal{N}(\cdot)$; similar results can be found in Table 11.1 of [11] and Table 1 of [12]. Table 3.3 assumes that the set Ω exhibits some minimal notion of regularity. Specifically, it assumes that the set Ω satisfies the cone condition; see Definition 2.73 and [11] for the details. Combining (3.78) and (3.80), Table 3.3 provides an operative approach to bound the approximation error.

3.5.3 Rates of Convergence in $W^{s,2}(\Omega)$, $L^\infty(\Omega)$, and the Many-Zeros Theorem

In Sect. 3.5.2, rates of convergence of interpolants and projections in a native space were analyzed in terms of the power function. This is an effective method that has been frequently used for kernels $\mathfrak{K} : \mathbb{X} \times \mathbb{X} \to \mathbb{R}$ with $\mathbb{X} \triangleq \mathbb{R}^n$, or over regular subsets $\Omega \subset \mathbb{X} \triangleq \mathbb{R}^n$ [11, 12]. In this section, we summarize an alternate strategy for the estimation of errors that is based on the many-zeros theorem (Theorem 2.8) in native spaces that are Sobolev spaces over $\mathbb{R}^n$, over its subsets, or in native spaces that are continuously embedded in Sobolev spaces.

Let the domain $\Omega \subset \mathbb{X} \triangleq \mathbb{R}^n$ satisfy the regularity hypotheses of Theorem 2.8, let $\Xi_N \triangleq \{\xi_1, \ldots, \xi_N\} \subset \Omega$ denote a set of centers, let $\mathfrak{K} : \Omega \times \Omega \to \mathbb{R}$ denote the kernel that induces the native space $\mathcal{H}(\Omega)$, and let $\mathcal{H}_N \triangleq \text{span}\{\mathfrak{K}_\xi : \xi \in \Xi_N\}$ denote the space of approximants. If $\mathcal{H}(\Omega) \equiv W^{s,2}(\Omega)$ for a smoothness parameter $s > 0$, then the error in projection or interpolation in $W^{s,2}(\Omega)$ can be estimated as

$$\|(I - \Pi_N)f\|_{W^{k,2}(\Omega)} \lesssim h^{s-k}_{\Xi_N,\Omega}\|(I - \Pi_N)f\|_{W^{s,2}(\Omega)} \tag{3.81a}$$

$$\lesssim h^{s-k}_{\Xi_N,\Omega}\|f\|_{W^{s,2}(\Omega)}. \tag{3.81b}$$

The proof of a result stronger than (3.81) is provided in Sect. 3.8.2 below. The inequalities in (3.81) are the standard form of a *direct inequality*, or *Jackson inequality*, in approximation theory [26]; we make use of these inequalities in Chap. 5.

This strategy of bounding the error in projection using (3.81) relies on the ability to choose or identify in a closed form the kernel $\mathfrak{K} : \Omega \times \Omega \to \mathbb{R}$ that induces the native space $W^{s,2}(\Omega)$. In Sect. 3.4.4, we showed that every Sobolev space $W^{s,2}(\Omega)$ with $s > d/2$ is a native space by the Sobolev embedding theorem (see Theorems 2.7 and 3.7). However, determining a closed-form expression for the kernel over $\mathfrak{K} : \Omega \times \Omega \to \mathbb{R}$ that induces $\mathcal{H} \equiv W^{s,2}(\Omega)$ can be a challenge if $\Omega \subset \mathbb{R}^n$ is a generic set, or if Ω is unknown. In the following, we discuss an approach for practical implementations of this strategy that circumvents the task of finding a closed-form expression for the kernel over the subset Ω. This method is useful because it employs kernels and RKHSs over the larger ambient set $\mathbb{X}$ in which Ω is contained.

Assume that a radial function $\eta : \mathbb{R}^+ \to \mathbb{R}$ defines the kernel $\mathfrak{K} : \mathbb{X} \times \mathbb{X} \to \mathbb{R}$ as in (3.23); see Table 3.1 for a list of radial functions that generate kernels. At this point, it is important to emphasize that the kernel $\mathfrak{K}$ induces an RKHS $\mathcal{H}$ of functions over $\mathbb{X} \triangleq \mathbb{R}^n$, not just over Ω. Furthermore, recall that, by definition of the kernel section, $\mathfrak{K}_0(x) = \mathfrak{K}(x, 0)$. Finally, we denote the *Fourier transform* of the kernel section $\mathfrak{K}_0(x)$, $x \in \mathbb{X}$, by $\hat{\mathfrak{K}}_0(\xi)$, $\xi \in \mathbb{X}$. Thus, we introduce the following definition.

Definition 3.10 (*Algebraic Decay of Order s of the Fourier Transform of a Kernel*) The Fourier transform $\hat{\mathfrak{K}}_0(\xi)$ has *algebraic decay of order s* if there exist s, c_1, and $c_2 > 0$ such that

$$c_1 \left(1 + \|\xi\|_2^2\right)^{-s} \leq \hat{\mathfrak{K}}_0(\xi) \leq c_2 \left(1 + \|\xi\|_2^2\right)^{-s}, \quad \text{for all } \xi \in \mathbb{X}. \tag{3.82}$$

As shown in [10, 11], if the Fourier transform $\hat{\mathfrak{K}}(\xi)$, $\xi \in \mathbb{X}$, has algebraic decay with an exponent $s > n/2$, then

$$\mathcal{H}(\mathbb{R}^n) \approx W^{s,2}(\mathbb{R}^n), \tag{3.83}$$

where the constants of equivalency are the same c_1, c_2 as in (3.82), and

$$c_2 \|f\|^2_{W^{s,2}(\mathbb{R}^n)} \leq \|f\|^2_{\mathcal{H}(\mathbb{R}^n)} \leq c_1 \|f\|^2_{W^{s,2}(\mathbb{R}^n)}. \tag{3.84}$$

Again, let Ξ_N be the collection of N samples in $\Omega \subset \mathbb{X} \triangleq \mathbb{R}^n$, consider the space of approximants (3.55), and let $\Pi_N : \mathcal{H} \to \mathcal{H}_N$ be an orthogonal projection operator. In this case, since $\mathcal{H} \triangleq \mathcal{H}(\mathbb{R}^n)$ is a space of functions supported on $\mathbb{X} \triangleq \mathbb{R}^n$, the projection operator Π_N acts on functions f that are supported on *all of* $\mathbb{R}^n$. Since the coordinate realization of the operator Π_N only relies on samples $\Xi_N \subset \Omega$, the projection error $(I - \Pi_N)$ can also be interpreted as an operator on $W^{s,2}(\Omega)$ for which it remains true that $(I - \Pi_N) f|_{\Xi_N} = 0$. Applying the many-zeros theorem (Theorem 2.8), we obtain that

$$\|(I - \Pi_N) f\|_{W^{k,2}(\Omega)} \lesssim h^{s-k}_{\Xi_N,\Omega} \|(I - \Pi_N) f\|_{W^{s,2}(\Omega)}, \tag{3.85}$$

as in (3.81a). However, we cannot immediately obtain (3.81b) in the case at hand. While we know that $(I - \Pi_N)$ is a bounded linear operator from $\mathcal{H}(\mathbb{R}^n)$ to $\mathcal{H}(\mathbb{R}^n)$, since $\|(I - \Pi_N)\| \leq 1$, it is less obvious that $(I - \Pi_N)$ is a bounded operator from $W^{s,2}(\Omega)$ to $W^{s,2}(\Omega)$. This result is proven by the following lemma.

Lemma 3.1 ([28, Lemma 3.1]) *Consider the RKHS $\mathcal{H}$ given by (3.9), the finite-dimensional space (3.55), and the orthogonal projection operator $\Pi_N : \mathcal{H} \to \mathcal{H}_N$. If the set Ω is regular enough such that there exists a uniformly bounded extension operator $\bar{\mathcal{E}}_\Omega : W^{s,2}(\Omega) \to W^{s,2}(\mathbb{R}^n)$, then the projection error $(I - \Pi_N)$ can be interpreted as a bounded linear operator from $W^{s,2}(\Omega)$ to $W^{s,2}(\Omega)$.*

We have obtained the inequality in (3.81b). The relevance of this result lies in that, although we have not assumed that the kernel for $W^{s,2}(\Omega)$ is known in closed form, we obtain the same rate of convergence as if the kernel were known. Computations can therefore be carried out in terms of a kernel $\mathfrak{K}$ that is defined on all of $\mathbb{X} \triangleq \mathbb{R}^n$, and not necessarily on Ω.

The application of Lemma 3.1 is further facilitated whenever Ω has a sufficiently regular boundary. This result is captured by the following theorem.

Theorem 3.15 ([11, Corollary 10.48]) *Let $\mathfrak{K}(\cdot, \cdot)$ define an RKHS of functions defined over $\mathbb{R}^n$, assume that the kernel section $\mathfrak{K}_0(x) \triangleq \mathfrak{K}(x, 0)$ is contained in $L^1(\mathbb{R}^d)$, and assume that $\hat{\mathfrak{K}}_0(\xi)$, the Fourier transform of $\mathfrak{K}_0(x)$, has algebraic decay $(1 + \|\xi\|_2^2)^{-s}$ with an integer $s > n/2$. If the boundary of the domain $\Omega \subset \mathbb{R}^n$ is Lipschitz continuous, then*

$$\mathcal{H}(\Omega) \approx W^{s,2}(\Omega) \tag{3.86}$$

with equivalent norms, where $\mathcal{H}(\Omega)$ is the RKHS obtained by restriction of the functions in $\mathcal{H}(\mathbb{R}^n)$.

3.5.4 Numerical Examples

In this section, we describe some qualitative characteristics of approximations constructed in terms of scattered bases in a scalar-valued RKHS. In particular, we consider the problem of finding an approximation f_N of the unknown function f that is defined over a subset $\Omega \subset \mathbb{X}$. We assume that we are given M samples $\{(x_i, y_i)\}_{i=1}^M \subset \Omega \times \mathbb{R} \triangleq Z$, with the exact observations $y_i = f(x_i)$.

In the following examples, we let $f \equiv V^\star$ and construct approximations of

$$V^\star(x_1, x_2) \triangleq \frac{1}{2}x_1^2 + x_2^2; \tag{3.87}$$

this target function is the optimal value function for a feedback control problem studied in Sect. 6.2.2. We do not discuss the specifics of the control problem here but focus on how the definition of the subset of interest Ω and the samples affect the approximation process. We study four cases that illustrate how a priori knowledge of the set Ω, or knowing its regularity properties, can affect conclusions about the approximation of functions. We consider subsets $\Omega \subset \mathbb{X} \triangleq \mathbb{R}^2$ in the following typical scenarios:

Case 1: The subset is a regular domain, $\Omega \triangleq (-1, 1) \times (-1, 1) \subset \mathbb{X}$, and the samples are deterministically prescribed and quasi-uniform over Ω.

Case 2: The subset is a regular domain $\Omega \triangleq (-1, 1) \times (-1, 1) \subset \mathbb{X}$, and the samples $\{x_i\}_{i=1}^M$ are randomly generated by a probability measure μ on Ω.

Case 3: The subset $\Omega \triangleq \mathcal{M} \subset \mathbb{R}^2$ is a smooth compact regularly embedded Riemannian submanifold contained in $\mathbb{R}^2$, and the samples are deterministically prescribed and quasi-uniform over $\mathcal{M}$.

Case 4: The subset Ω is taken as the forward orbit $\Omega \triangleq \gamma^+(x_0)$, or the positive limit set $\Omega \triangleq \omega^+(x_0) \subset \mathbb{X}$, of the initial condition x_0 of a dynamical system on $\mathbb{X}$, as described in Chap. 4. The samples are deterministically prescribed.

In Cases 1–4, the approximations are constructed in terms of the Sobolev–Matérn kernel of order $\nu = 1.5$ defined over all of $\mathbb{X} \times \mathbb{X} \triangleq \mathbb{R}^2 \times \mathbb{R}^2$. This kernel is defined in closed form in Sect. 3.3.

3.5.4.1 Case 1: Deterministically Prescribed Samples on $\Omega \triangleq (-1, 1) \times (-1, 1)$

As depicted in Figs. 3.4 for $M = 9, 25, 49$, we pick the set of centers $\Xi_M \triangleq \{(x_i, y_i)\}_{i=1}^M$ uniformly over the set $\Omega \triangleq (-1, 1) \times (-1, 1) \subset \mathbb{X} \equiv \mathbb{R}^2$. The fill distance $h_{\Xi_M,\Omega} \approx 2/(\sqrt{M} - 1)$ decreases with M. We define the subspace

$$\mathcal{H}_M \triangleq \text{span}\left\{\mathfrak{K}_{x_i} \ : \ 1 \leq i \leq M\right\} \subset \mathcal{H}, \tag{3.88}$$

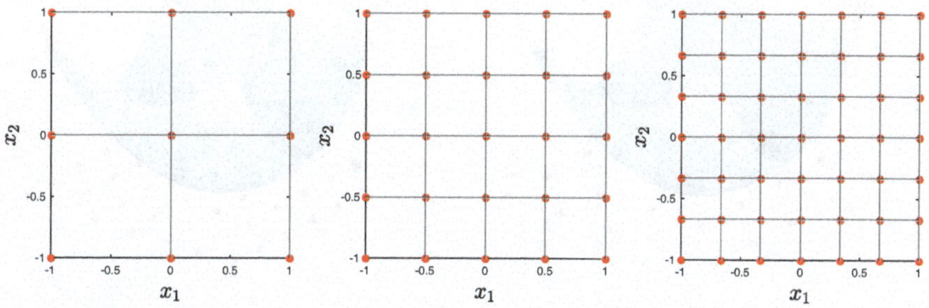

Fig. 3.4 Uniform grids for Case (1)

and the solution of (3.68) with $N = M$ generates coefficients $\Theta \triangleq [\theta_1, \ldots, \theta_M]^{\mathrm{T}}$ that define the approximation

$$V_M(x) \triangleq \Pi_M V^\star = \sum_{i=1}^{M} \theta_i \mathfrak{K}_{x_i}(x), \quad \text{for all } x \in \mathbb{X} \triangleq \mathbb{R}^2, \tag{3.89}$$

of $V^\star$. In (3.89), Π_M denotes the $\mathcal{H}$-orthogonal projection onto the subspace $\mathcal{H}_M \subset \mathcal{H}$ that is generated by $\Xi_M \subset \Omega$.

We emphasize that this estimate V_M of the target function $V^\star$ is *global*, that is, V_M is well-defined over the entire ambient space $\mathbb{X} \triangleq \mathbb{R}^2$. While Fig. 3.5 only plots the global function over Ω, since samples lie there, the approximation V_M is defined on all of $\mathbb{X} \triangleq \mathbb{R}^2$, and it satisfies the *global pointwise error bound*

$$|E_x(I - \Pi_M)V^\star| \leq \mathcal{P}_M(x)\|V^\star\|_{\mathcal{H}}, \qquad \text{for all } x \in \mathbb{X} \triangleq \mathbb{R}^2, \tag{3.90}$$

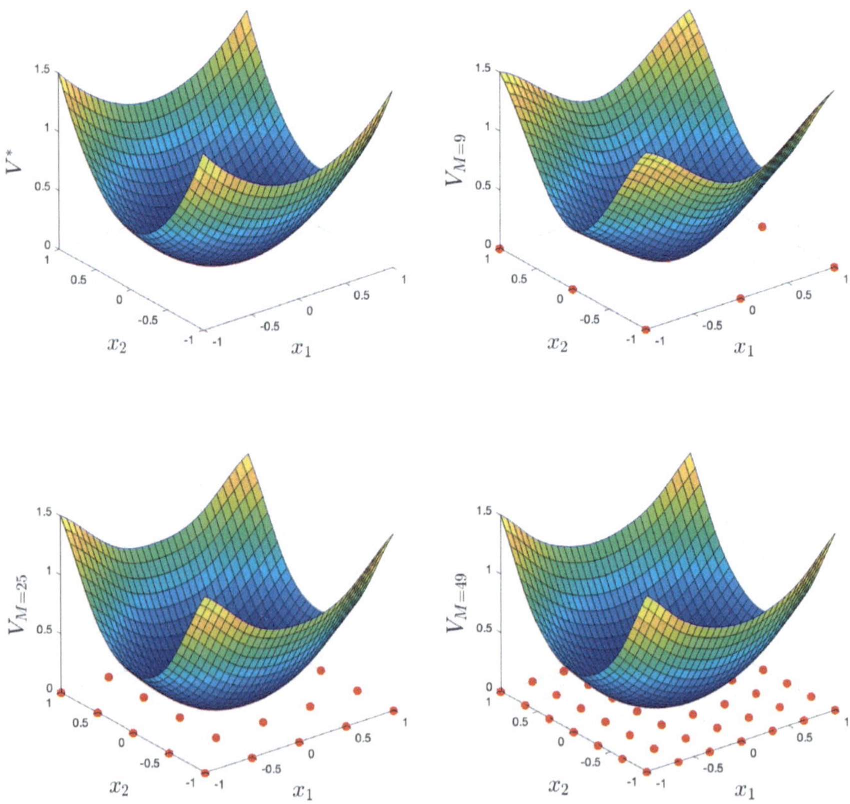

Fig. 3.5 Estimates V_M of $V^\star$, for $M = 9, 25, 49$

where $\mathcal{P}_M$ denotes the power function of the subspace $\mathcal{H}_M$ in $\mathcal{H}$, which is defined over all of $\mathbb{X}$; see (3.78). Since the samples are concentrated in $\Omega \subset \mathbb{X}$, the right-hand side of (3.90) does not need to be small for all $x \in \mathbb{X}$. Rather, as we choose a greater number M of samples in Ω, the right-hand side decreases for denser distributions of centers.

Figures 3.6–3.8 show various plots of the power function $\mathcal{P}_M$ of the subspace $\mathcal{H}_M \subset \mathcal{H}$. The power function $\mathcal{P}_M$ is defined over all of $\mathbb{X}$ since the kernel that defines $\mathcal{H}$ is defined over all of $\mathbb{X} \times \mathbb{X}$. Figures 3.6 and 3.7 show how, if we restrict ourselves to $\Omega \triangleq (-1, 1) \times (-1, 1)$, then the power function is progressively smaller by taking more samples that fill out the set Ω. Figure 3.8 expands the plotting region to $(-2, 2) \times (-2, 2)$ to show how the power function behaves outside of the domain Ω. This plot illustrates how choosing more samples in $\Omega = (-1, 1) \times (-1, 1)$ does not imply that the pointwise approximation error decreases to zero for points outside $\Omega \triangleq (-1, 1) \times (-1, 1)$. The global bound (3.90) still holds, but the power function does not decrease in regions far from the locations where samples are collected.

We can also study this behavior of the power function by considering restrictions of functions defined on all of the ambient space $\mathbb{X}$ to the subset $\Omega \subset \mathbb{X}$. As discussed in Sect. 3.2, if we are given a kernel $\mathfrak{K}$ defined on all of $\mathbb{X}$, then it is possible to define

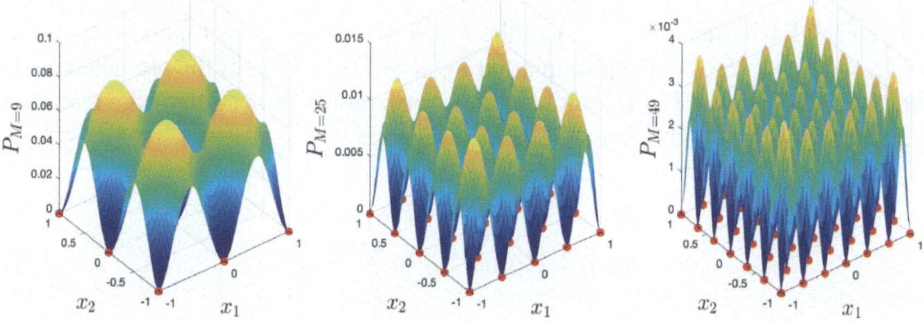

Fig. 3.6 Plot of the power function $\mathcal{P}_M(x)$ over $(-1, 1) \times (-1, 1)$

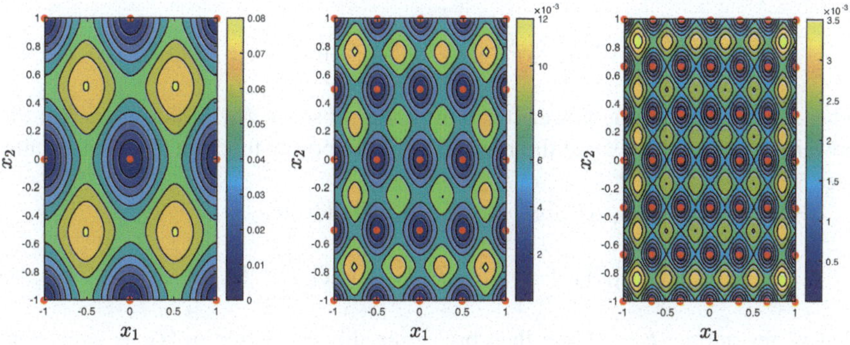

Fig. 3.7 Contour plot of the power function $\mathcal{P}_M(x)$ over $(-1, 1) \times (-1, 1)$

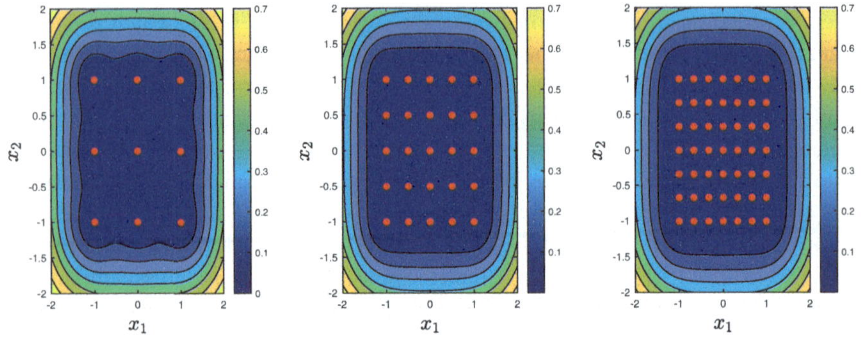

Fig. 3.8 Contour plot of the power function $\mathcal{P}_M(x)$ over $(-2, 2) \times (-2, 2)$

an associated RKHS $\mathcal{R}_\Omega$ that consists of the set of all restrictions of functions in $\mathcal{H}$ to the subset Ω. The reproducing kernel $\mathfrak{r} : \Omega \times \Omega \to \mathbb{R}$ of $\mathcal{R}_\Omega$ is just the restriction of the kernel $\mathfrak{K}$ originally defined on $\mathbb{X} \times \mathbb{X}$ to $\Omega \times \Omega$; see (3.20). Thus, we define the finite-dimensional space

$$\mathcal{R}_M \triangleq \text{span}\left\{\mathfrak{r}_{x_i} \; : \; 1 \leq i \leq M\right\} \subset \mathcal{R}_\Omega, \tag{3.91}$$

where each basis function $\mathfrak{r}_{x_i}$ is just the restriction of the basis function $\mathfrak{K}_{x_i}$ to the set Ω. We define the local approximation $V_{\Omega,M} \triangleq \Pi_{\Omega,M} V^\star|_\Omega$ of the restricted target function $V^*|_\Omega$, where $\Pi_{\Omega,M}$ denotes the $\mathcal{R}_\Omega$-orthogonal projection of the restricted target function $V^*|_\Omega$ onto $\mathcal{R}_M$. It is easy to show that

$$V_{\Omega,M} \triangleq \Pi_{\Omega,M} V^\star|_\Omega = \sum_{i=1}^{M} \theta_i r_{\xi_i}, \tag{3.92}$$

where the coefficients $\{\theta_1, \ldots, \theta_M\} \subset \mathbb{R}^M$ are the same as in (3.89). We also have the *local pointwise error estimate*

$$\left|E_x(I - \Pi_{\Omega,M})V^\star|_\Omega\right| \leq \mathcal{P}_{\Omega,M} \|V^\star|_\Omega\|_{\mathcal{R}_\Omega}, \quad \text{for all } x \in \Omega \subset \mathbb{X}, \tag{3.93}$$

where $\mathcal{P}_{\Omega,M}$ denotes the power function of the subspace $\mathcal{R}_M$ in the space $\mathcal{R}_\Omega$; the power function $\mathcal{P}_{\Omega,M}$ is just the restriction of the power function $\mathcal{P}_M$. For details, see (3.78).

Given the *trace* or *restriction operator* T_Ω, it holds that

$$T_\Omega \Pi_M f = \Pi_{\Omega,M} T_\Omega f, \quad \text{for all } f \in \mathcal{H}, \tag{3.94}$$

that is, restricting f to Ω and then projecting this restriction to $\mathcal{R}_M$ is equivalent to projecting f onto $\mathcal{H}_M$ and then restrict this projection ro $\mathcal{R}_M$. For these reasons, those familiar with the basics of the theory of RKHSs routinely use this simple relationship in practice, without pointing out all the details regarding the definitions of restriction operators, restricted spaces, and restricted power functions. We follow this convention in this text, and we are only explicit regarding restrictions when it is needed for a clear exposition.

The global and local pointwise error bounds in terms of the power function in (3.90) and (3.93), respectively, hold for any subset $\Omega \subset \mathbb{X}$, irrespectively of the regularity of the boundary of the set Ω. This is because Theorem 3.14 does not depend on the regularity of the subset $\Omega \subset \mathbb{X}$. However, in this example, the domain $\Omega \triangleq (-1, 1) \times (-1, 1)$ has a Lipschitz boundary, and it satisfies an interior cone condition. Thus, we can use the many-zeros theorem (Theorem 2.8) to derive upper bounds on the approximation error. If we know a closed-form expression of the reproducing kernel $\mathfrak{K}^s : \Omega \times \Omega \to \mathbb{R}$ for the Sobolev space $W^{s,2}(\Omega)$ with $s > n/2$, Theorem 2.8 guarantees that

$$\left|E_x(I - \Pi_M)V^\star|_\Omega\right| \le \|(I - \Pi_M)V^\star|_\Omega\|_{L^\infty(\Omega)} \lesssim h^{s-1/2}_{\Xi_M,\Omega}\|V^\star|_\Omega\|_{W^{s,2}(\Omega)}, \tag{3.95}$$

for all $x \in \Omega$; see (3.85). We emphasize that this bound requires that we know the closed-form expression for the kernel of $W^{s,2}(\Omega)$ over the set Ω. However, we override this requirement using Corollary 3.15. If we first choose a kernel $\mathfrak{K}$: $\mathbb{X} \times \mathbb{X} \to \mathbb{R}$ that generates $\mathfrak{K}_0(x) \triangleq \mathfrak{K}(x, 0)$ and satisfies the hypotheses of Corollary 3.15, then the restriction to Ω of this kernel still yields (3.85).

3.5.4.2 Case 2: Random Samples on $\Omega \triangleq (-1, 1) \times (-1, 1)$

In this example, we assume that the samples $\{x_i\}_{i=1}^M$ in the input/output measurements $\{(x_i, y_i)\}_{i=1}^M$ are generated by a random walk over the subset $\Omega \subset \mathbb{X}$. As already discussed, Ω is a domain (*i.e.*, it is an open and connected set), it has a Lipschitz boundary, and satisfies the interior cone condition. The random walk is determined by a probability measure μ on $\Omega \subset \mathbb{X}$. In this simple case, we still assume that the values $y_i = f(x_i)$ are measured without noise; this is a special case of the general

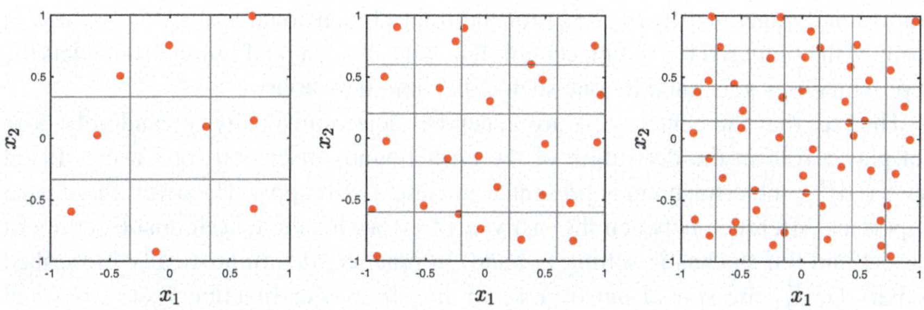

Fig. 3.9 Centers $\{x_i\}_{i=1}^M$ generated by a random walk, $M = 9, 25, 49$

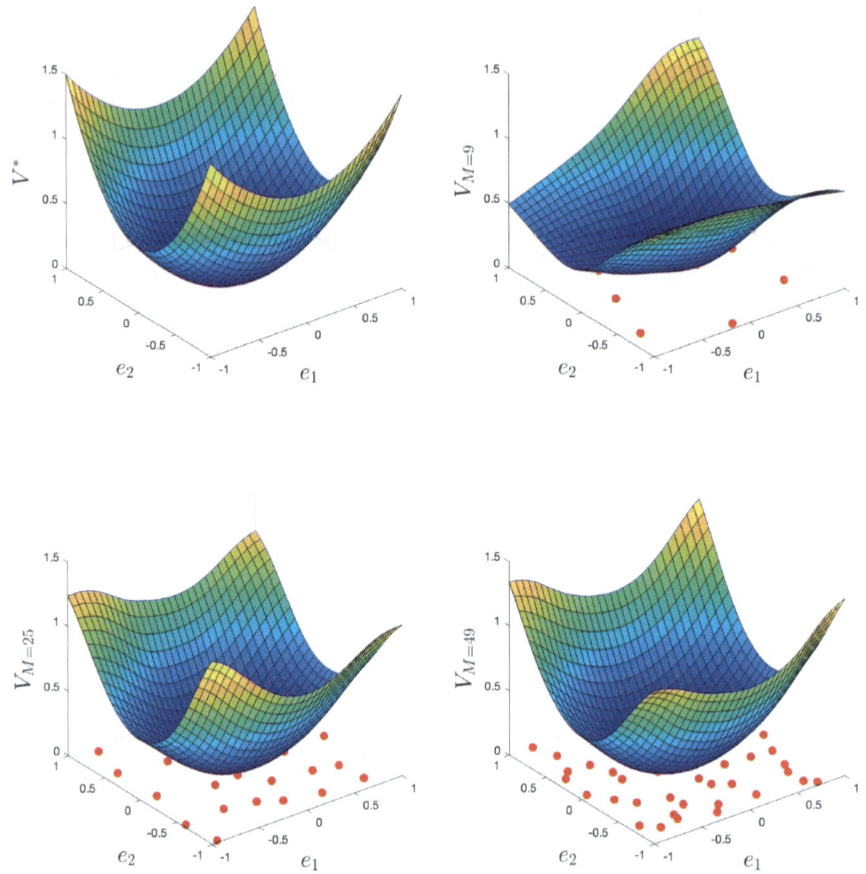

Fig. 3.10 Target function $V^{\star}$ and approximations determined by centers generated randomly with $M = 9, 25, 49$

problem described in Chap. 7. Examples of such centers are depicted in Fig. 3.9, which are generated using a uniform probability distribution over Ω. The approximations are generated from (2.32) for the finite-dimensional space $\mathcal{H}_M$ defined in terms of the scattered basis with centers $\Xi_M \triangleq \{x_1, \ldots, x_M\}$. Figure 3.10 depicts the approximations generated by one such collection of samples.

The fact that the centers Ξ_M are generated deterministically or randomly does not play a role in the derivation of the error bounds discussed for Case 1. Fixed $\{(x_i, y_i)\}_{i=1}^{M}$, the error bounds presented in Case 1 still apply. However, there is an important difference between the analysis of errors for the deterministic setting in Case 1 and the stochastic setting at hand. In practice, deterministically prescribed centers $\{x_i\}_{i=1}^{M}$ are spread out over Ω so that the power function $\mathcal{P}_N(x)$ is small over Ω. Typically, this is achieved by selecting centers for which the fill distance $h_{\Xi_M,\Omega}$ is small. Randomly prescribed centers $\{x_i\}_{i=1}^{M}$, however, do not guarantee that a small fill distance in Ω is attained or that the power function $\mathcal{P}_N(x)$ is small. In some cases, all centers may be concentrated near a single point in Ω, which is undesired. For this reason, in a probabilistic setting, indications of how likely it is that the approximation errors are small are provided by concentration inequalities. Let $Z \triangleq \Omega \times \mathbb{R}$ and $Z_M \triangleq \{(x_i, y_i)\}_{i=1}^M$. In the problem at hand, a concentration inequality is given by

$$Pr\left\{z_M \in Z^M : \|V^\star|_\Omega - V_{\Omega,M}\|_{L^\infty(\Omega)} \leq \epsilon_M \right\} \geq 1 - \delta_M \tag{3.96}$$

for two (small) constants $\epsilon_M, \delta_M > 0$. Intuitively, we think of ϵ_M as the desired error of approximation, and $1 - \delta_M$ as the probability of achieving the error ϵ_M, when using M samples. This inequality states how likely it is that the samples $z_M \in Z^M$ generate approximation errors less than ϵ_M. This kind of bounds describes the balance between achieving approximation errors bounded by ϵ_M and the size $1 - \delta_M$ of the set of samples that achieve such bounds. For additional details, see Sect. 7.6.

3.5.4.3 Case 3: Deterministically Prescribed Samples on a Manifold

In this case, samples of the unknown function f are concentrated on a subset $\mathcal{M} \subset \mathbb{X}$, which has zero measure in $\mathbb{R}^2$ and an empty interior as a subset of $\mathbb{R}^2$. This situation could arise, for example, in an adaptive tracking problem where the reference system generates a trajectory supported on a circle; see Chap. 5 for examples of these adaptive control problems.

For illustration, we selected the manifold $\mathcal{M}$ as the circle of radius one centered in the origin. Thus, the dimension of the state space $\mathbb{X} \triangleq \mathbb{R}^2$ is $n = \dim(\mathbb{X}) = 2$, and the dimension of the embedded manifold is $\dim(\mathcal{M}) = 1$. The manifold may be known or unknown. If the manifold $\mathcal{M}$ is known, then the same considerations as in Case 1 apply here. Examples of such a strategy can be found in [11, 29–32]. If the manifold $\mathcal{M}$ is unknown, then first we choose a kernel defined on $\mathbb{X} \times \mathbb{X}$. Then, we can define error bounds that are expressed in the native space defined by the restriction of the global kernel on $\mathbb{X} \times \mathbb{X}$ to $\mathcal{M} \times \mathcal{M}$, which is only used for theoretical purposes. This approach is used, for instance, in [10].

Figure 3.12 illustrates the global approximations V_M of $V^\star$ that are generated this way for $M = 9, 25, 49$, while the associated collections of centers are depicted in Fig. 3.11. From Fig. 3.12, we observe that the global estimates $V_M \not\to V^\star$ pointwise over Ω nor over $\mathbb{X}$ as $M \to \infty$. The global target function $V^\star$ is substantially different than the apparent global limit of V_M as $M \to \infty$. In fact, when we only collect samples over $\mathcal{M}$, the global approximations V_M obtained by orthogonal projection onto $\mathcal{H}_M$ converge to the minimal norm extension

$$V_M \to \mathcal{E}_\mathcal{M}(V^\star|_\mathcal{M}) \in \mathcal{H}_\mathcal{M} \subset \mathcal{H}, \quad \text{as } M \to \infty, \tag{3.97}$$

where $\mathcal{E}_\mathcal{M} : \mathcal{R}_\mathcal{M} \to \mathcal{H}_\mathcal{M}$ denotes the extension operator. This is because the minimal norm extension for these finite-dimensional approximations is the mapping

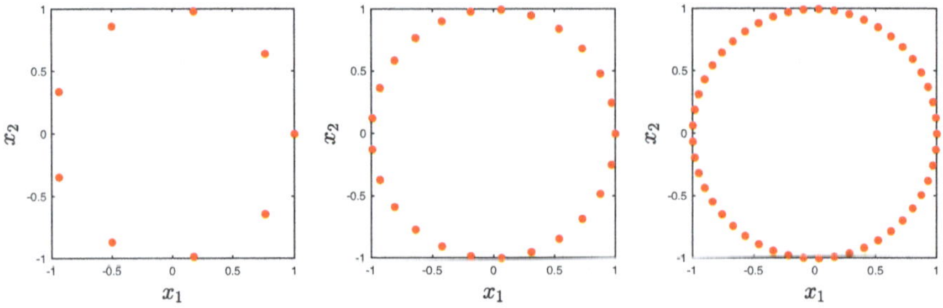

Fig. 3.11 Location of centers with $M = 9, 25, 49$

$$\mathcal{E}_{\mathcal{M}} : \sum_{i=1}^{M} \theta_i \mathfrak{r}_i \mapsto \sum_{i=1}^{M} \theta_i \mathfrak{K}_i, \tag{3.98}$$

where the coefficients are derived from the coordinate realizations of the orthogonal projections Π_N or $\Pi_{\mathcal{M},M}$; see [10] for a discussion on this fact, or consult Sect. 3.2. However, the error plots in Fig. 3.13 show that the restriction $V_{\mathcal{M},M}$ of V_M converges pointwise over $\mathcal{M}$ to the restriction $V^*|_{\mathcal{M}}$ as $M \to \infty$. The restricted approximation $V_{\mathcal{M},M}$ is a function just supported on $\mathcal{M}$, which is shown in Fig. 3.14. Finally, the globally defined value function V^*, the globally defined approximation V_M, and the restricted approximation $V_{\mathcal{M},M}$ are shown together in Fig. 3.15.

3.5.4.4 Case 4: Deterministically Prescribed Samples on a Forward Orbit or a Positive Limit Set

In this case, which arises naturally in control problems for which samples are collected along the trajectory of an ODE, samples reside in sets that are defined in terms of topological dynamics for ODEs. Suppose that $t \mapsto x(t) \in \mathbb{X} \triangleq \mathbb{R}^n$ is the trajectory of a time-invariant ODE (see Sect. 4.1 for a discussion on time-varying and time-invariant ODEs) starting at the initial condition $x_0 \in \mathbb{X} \triangleq \mathbb{R}^n$, and defined for all $t \in [0, \infty)$. The positive orbit $\gamma^+(x_0)$ and the positive limit set $\omega^+(x_0)$ are defined, respectively, as

$$\gamma^+(x_0) \triangleq \bigcup_{\tau \geq t_0} x(\tau), \tag{3.99}$$

$$\omega^+(x_0) \triangleq \left\{ y \ : \ \text{there exist } t_k \to \infty \text{ such that } y = \lim_{t_k \to \infty} x(t_k) \right\} \subset \mathbb{X} \triangleq \mathbb{R}^n. \tag{3.100}$$

To estimate the unknown function f over these sets, first we choose a reproducing kernel $\mathfrak{K}$ defined globally on $\mathbb{X} \times \mathbb{X}$. Since $\gamma^+(x_0) \subset \mathbb{X}$ and $\omega^+(x_0) \subset \mathbb{X}$, we can use the results of Sect. 3.2 to define the closed subspaces

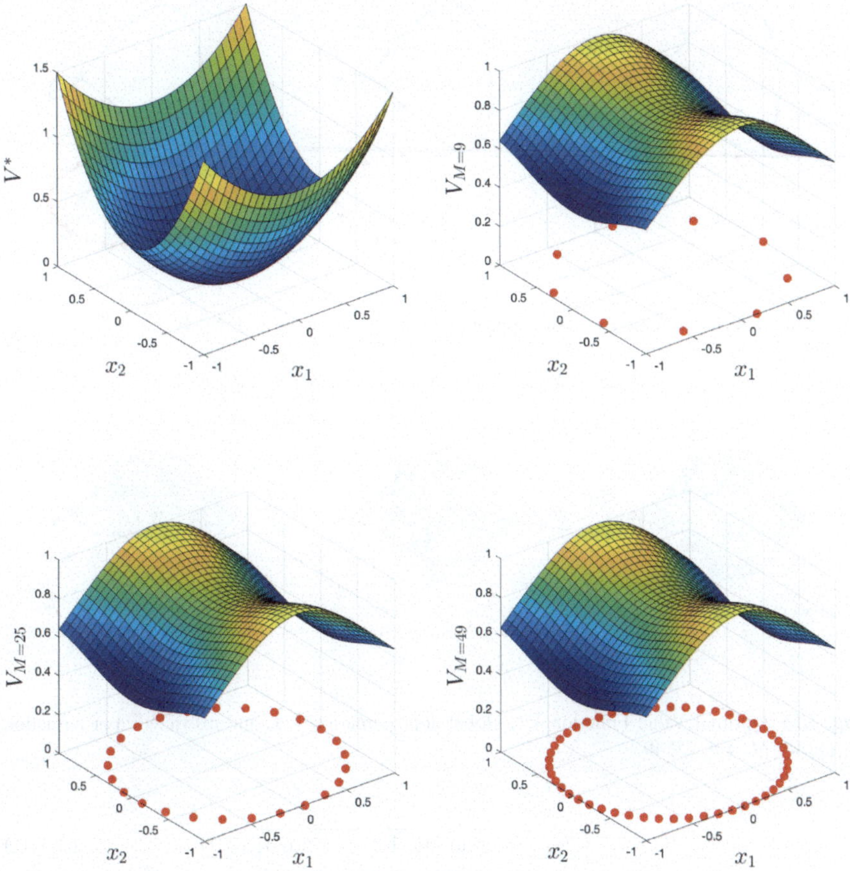

Fig. 3.12 Approximated Value Function and centers on the manifold $\mathcal{M}$

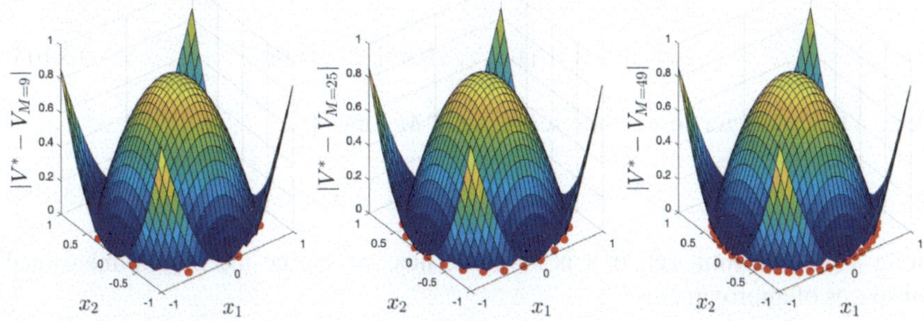

Fig. 3.13 Pointwise Magnitude of the Error $|V^\star(x) - V_M(x)|$, $x \in \Omega$, for $M = 9, 25, 49$

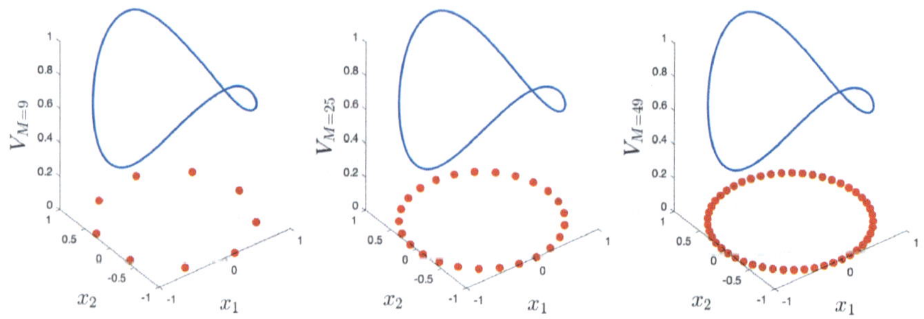

Fig. 3.14 Restricted Value Function Approximation $V_{\mathbf{M},M}$ for $M = 9, 25, 49$

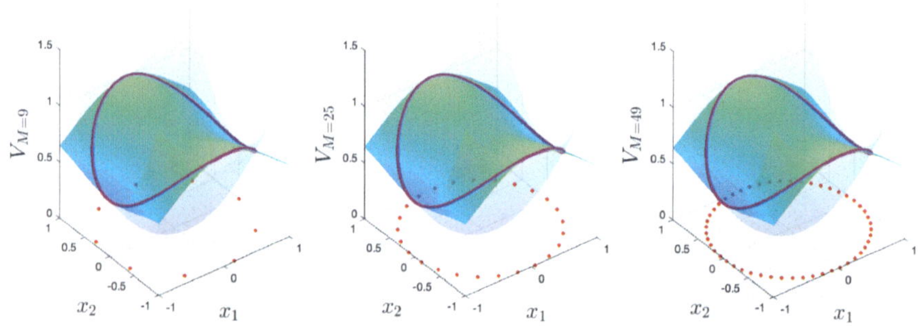

Fig. 3.15 Optimal Value Function $V^\star$, global approximation V_M, and restricted approximation $V_{\mathbf{M},M}$ for $M = 9, 25, 49$

$$\mathcal{H}_\gamma \triangleq \mathcal{H}_{\gamma^+(x_0)} \triangleq \overline{\text{span}\,\{\mathfrak{K}_x \mid x \in \gamma^+(x_0)\}} \subset \mathcal{H}, \tag{3.101}$$

$$\mathcal{H}_\omega \triangleq \mathcal{H}_{\omega^+(x_0)} \triangleq \overline{\text{span}\{\mathfrak{K}_x \mid x \in \omega^+(x_0)\}} \subset \mathcal{H}, \tag{3.102}$$

and then deterministically define samples as

$$\Xi_{\gamma,M} \triangleq \{x(t_1), \ldots, x(t_M)\} \subset \gamma^+(x_0) \tag{3.103}$$

along a trajectory for some finite sequence of M times $\{t_1, \ldots, t_M\} \subset [0, \infty)$, or

$$\Xi_{\omega,M} \triangleq \{\xi_1, \ldots, \xi_M\} \subset \omega^+(x_0), \tag{3.104}$$

along a positive limit set, if it is known. Thus, we define the finite-dimensional subspaces of approximants

$$\mathcal{H}_{\gamma,M} \triangleq \text{span}\{\mathfrak{K}_{x_i} \mid x_i \in \Xi_{\gamma,M}, 1 \le i \le M\} \subset \mathcal{H}_\gamma \subset \mathcal{H}, \tag{3.105}$$

$$\mathcal{H}_{\omega,M} \triangleq \text{span}\{\mathfrak{K}_{\xi_i} \mid \xi_i \in \Xi_{\omega,M}, 1 \le i \le M\} \subset \mathcal{H}_\omega \subset \mathcal{H}. \tag{3.106}$$

The construction of approximations using these spaces follows the same procedures as that in Cases 1–3. Figure 3.16 shows the data-driven centers $\Xi_{\gamma,M}$ for $M = 9, 25, 49$ that are selected along the trajectory $t \mapsto x(t)$, which is also depicted in the figure. Figure 3.17 shows the target function V^*, as well as the globally defined approximation V_M for $M = 9, 25, 49$ that are defined on $\mathbb{X} = \mathbb{R}^2$. As in Case 3, we do not expect that $V_M \to V^*$ in $L^\infty(\Omega)$ as $M \to \infty$. This fact is emphasized in Fig. 3.18, which shows the pointwise error $|V^*(x) - V_M(x)|$ for all $x \in \Omega$. However, we expect that this pointwise error $|V^*(x) - V_M(x)|$ converges pointwise to zero for all $x \in \gamma^+(x_0)$.

Figure 3.19 depicts three positive limit sets for autonomous systems; the same systems will be employed in Chap. 6. The positive limit set $\omega^+(x_0)$ is used to subselect the collection of centers $\Xi_{\gamma,M}$ shown in Fig. 3.20. These centers are used to define the subspaces $\mathcal{H}_{\omega,M}$, from which the approximations V_M in Fig. 3.21 are constructed for $M = 9, 25, 49$. The pointwise error plots in Fig. 3.22 demonstrates that, as $M \to \infty$, and the samples $\Xi_{\omega,M}$ become dense in $\omega^+(x_0)$, the pointwise error converges to zero in the norm $L^\infty(\omega^+(x_0))$.

3.6 Vector-Valued Native Spaces

3.6.1 Subspaces Generated by $\mathbb{X} \triangleq \mathbb{R}^n$

In this section, we present key elements of the theory of RKHSs that contain vector-valued functions. Let $\mathbb{X}$ denote the set over which functions are defined, and let these functions take values in a Hilbert space $\mathbb{U}$. By $\mathcal{F}(\mathbb{X}, \mathbb{U})$, we denote the vector space of all $\mathbb{U}$-valued functions over the set $\mathbb{X}$ with the usual vector space operations. In this section, we set $\mathbb{X} \triangleq \mathbb{R}^n$. Consistently with Remark 2.7, the inner product over $\mathbb{U}$ is denoted by $\langle \cdot, \cdot \rangle_{\mathbb{U}}$, and the inner product over $\mathbb{H}$ is denoted by $\langle \cdot, \cdot \rangle_{\mathbb{H}}$, and, consistently with Remark 2.5, the norms induced by $\langle \cdot, \cdot \rangle_{\mathbb{U}}$ and $\langle \cdot, \cdot \rangle_{\mathbb{H}}$ are denoted by $\|\cdot\|_{\mathbb{U}}$ and $\|\cdot\|_{\mathbb{H}}$, respectively.

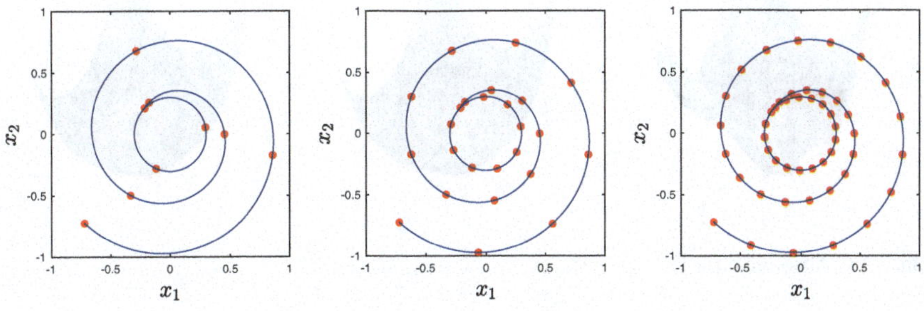

Fig. 3.16 Samples along a trajectory that define $\Xi_{\gamma,M}$ and $\mathcal{H}_{\gamma,M}$ for $M = 9, 25, 49$

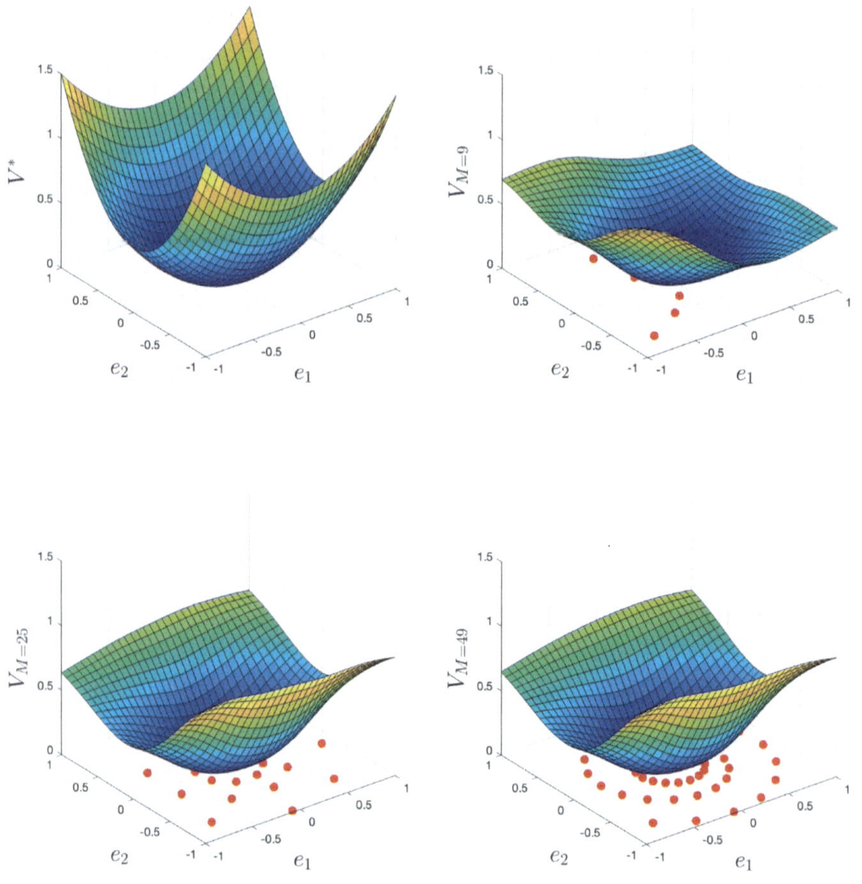

Fig. 3.17 Approximations V_M of V^* based on $\mathcal{H}_{\gamma,M}$, for $M = 9, 25, 49$

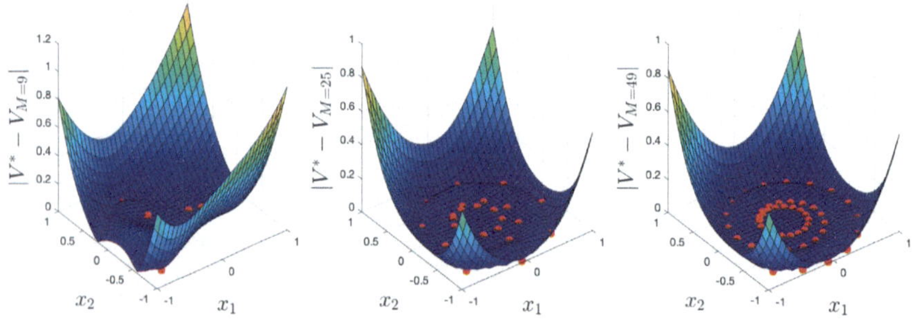

Fig. 3.18 Absolute error $|V^*(x) - V_{\gamma_M}(x)|$ for $x \in \Omega$, $M = 9, 25, 49$

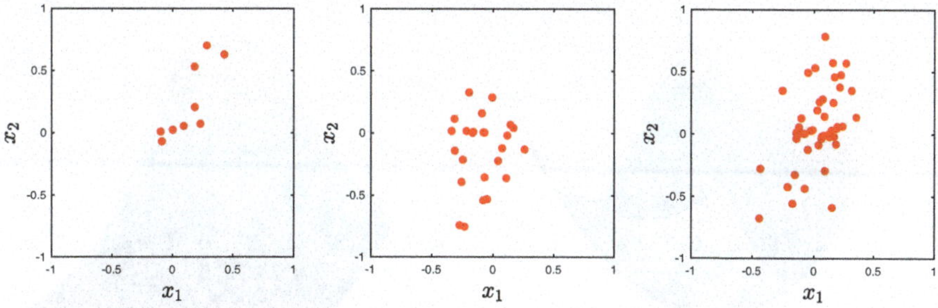

Fig. 3.19 Positive limit sets $\omega^+(x_0)$ from Chap. 6

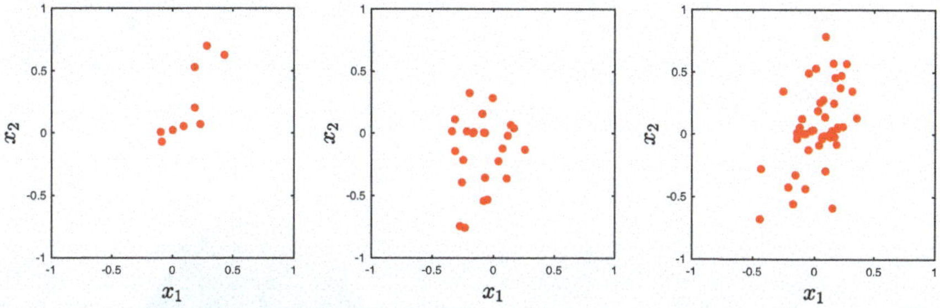

Fig. 3.20 Centers Ξ_{γ_M} extracted from the positive limit set $\omega^+(x_0)$

Before getting into the details regarding the definition and properties of vector-valued native spaces, we make one overall observation regarding terminology. One direct way to define reproducing kernel spaces $\mathcal{H} \subset \mathcal{F}(\mathbb{X}, \mathbb{U})$ of vector-valued functions is to let $\mathcal{H} \triangleq \mathcal{H}^m \triangleq \mathcal{H} \times \cdots \times \mathcal{H}$, with $\mathcal{H}$ a scalar-valued RKHS defined in terms of a scalar-valued reproducing kernel $\mathfrak{K} : \mathbb{X} \times \mathbb{X} \to \mathbb{R}$ and $\times$ denoting the *Cartesian product*. The theory in this section can be used to conclude that the vector-valued RKHSs $\mathcal{H} = \mathcal{H}^m$ is generated by the operator-valued kernel $\mathcal{K} : \mathbb{X} \times \mathbb{X} \to \mathcal{L}(\mathbb{U})$ that is diagonal, with $\mathcal{K}(x, y) \triangleq \mathfrak{K}(x, y) I_m$ for all $x, y \in \mathbb{X}$ and I_m the identity matrix on $\mathbb{U} \triangleq \mathbb{R}^m$. In this case, we say the vector-valued RKHS $\mathcal{H} \triangleq \mathcal{H}^m$ is *induced by a diagonal operator kernel*. In the general case, the vector-valued RKHS $\mathcal{H}$ is defined in terms of a non-diagonal operator kernel $\mathcal{K} : \mathbb{X} \times \mathbb{X} \to \mathcal{L}(\mathbb{U})$. In our study, it is often important to keep in mind whether a vector-valued RKHS is defined in terms of a diagonal or non-diagonal operator-valued kernel.

Scalar-valued RKHS were defined employing the reproducing formula (3.1). A similar condition can be used to characterize a $\mathbb{U}$-valued RKHS $\mathcal{H}$, albeit with a more general notion of a reproducing formula.

Definition 3.11 (*Reproducing Property for Vector-Valued Hilbert Spaces*) Let $\mathbb{H}(\mathbb{X}, \mathbb{U}) \subset \mathscr{F}(\mathbb{X}, \mathbb{U})$ denote a real, vector-valued Hilbert space of functions over the set $\mathbb{X}$. If, for each $x \in \mathbb{X}$, there is an operator $\mathcal{K}_x \in \mathcal{L}(\mathbb{U}, \mathbb{H})$ such that

$$\langle f(x), u \rangle_{\mathbb{U}} = \langle f, \mathcal{K}_x u \rangle_{\mathbb{H}}, \quad \text{for all } f \in \mathbb{H}(\mathbb{X}, \mathbb{U}), \tag{3.107}$$

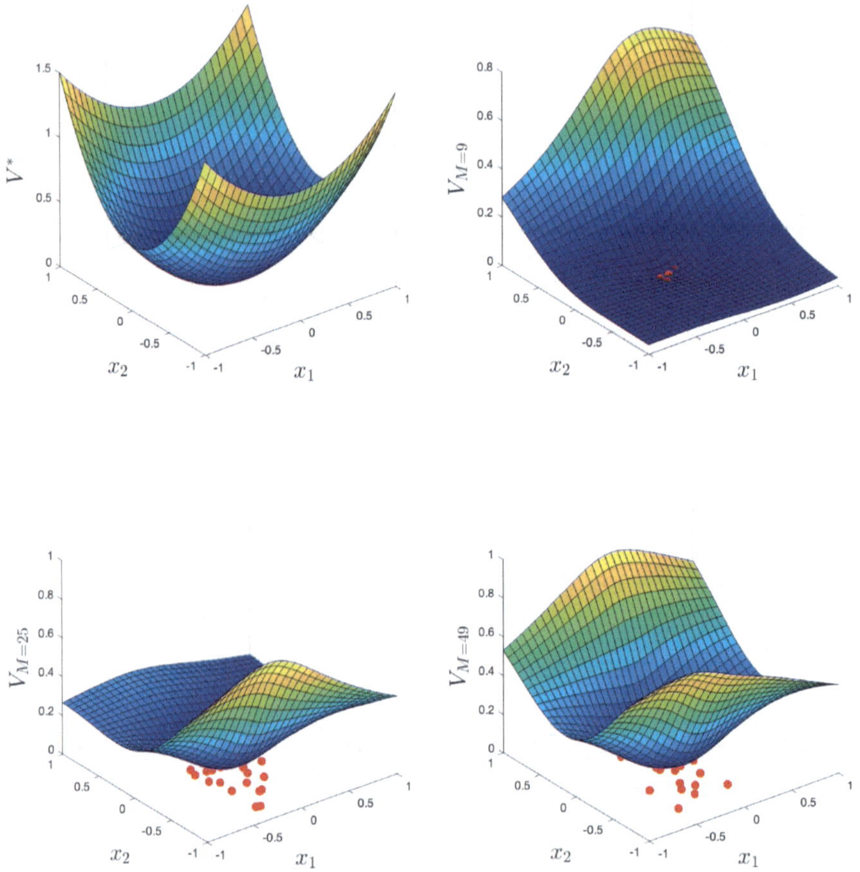

Fig. 3.21 Approximations V_M of $V^\star$ for $M = 9, 25, 49$

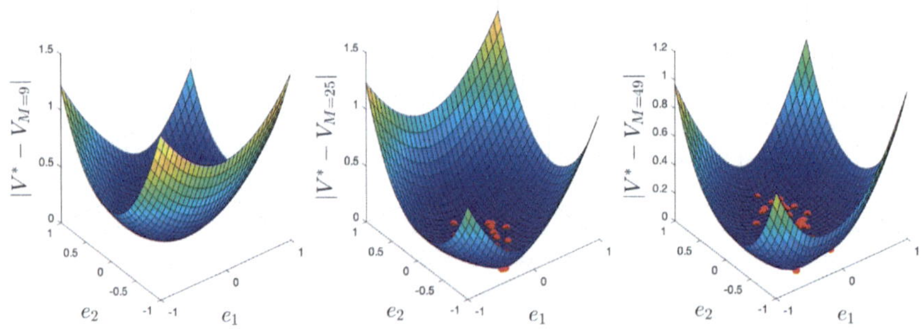

Fig. 3.22 Pointwise Approximation Error $|V^*(x) - V_M(x)|$ for $x \in \Omega$

then the *reproducing property for vector-valued Hilbert spaces* is satisfied.

Definition 3.12 (*Vector-Valued Reproducing Kernel Hilbert Space (RKHS)*) A real, vector-valued Hilbert space $\boldsymbol{\mathcal{H}}(\mathbb{X}, \mathbb{U})$ is a *vector-valued Reproducing Kernel Hilbert Space* (RKHS), or, equivalently, a *vector-valued native space*, if it satisfies the reproducing property (3.107).

Remark 3.6 (*Comment on Notation*) Generic vector-valued Hilbert spaces are denoted by $\mathbb{H}(\mathbb{X}, \mathbb{U})$. Given their key role, in this book, we denote vector-valued RKHSs by $\boldsymbol{\mathcal{H}}(\mathbb{X}, \mathbb{U})$. If clear from the context, we write $\boldsymbol{\mathcal{H}}$ for $\boldsymbol{\mathcal{H}}(\mathbb{X}, \mathbb{U})$, that is, we omit the domain and co-domain of the functions in the space $\boldsymbol{\mathcal{H}}$. It is also important to note the difference between the boldface symbol $\boldsymbol{\mathcal{H}}$ employed for vector-valued RKHSs and the symbol $\mathcal{H}$ employed for scalar-valued RKHSs in the first part of this chapter.

Consistently with Remark 2.5, the inner product over $\boldsymbol{\mathcal{H}}$ is denoted by $\langle \cdot, \cdot \rangle_{\boldsymbol{\mathcal{H}}}$ and the norm induced by $\| \cdot \|_{\boldsymbol{\mathcal{H}}}$. Not every $\mathbb{U}$-valued Hilbert space is a $\mathbb{U}$-valued RKHS. Furthermore, as in Sect. 3.1, we define the evaluation functional and provide a necessary and sufficient condition for a Hilbert space to be a vector-valued RKHs.

Definition 3.13 (*Evaluation Operator for a Vector-Valued Hilbert Spaces*) Given the Hilbert space $\mathbb{H}(\mathbb{X}, \mathbb{U})$ of vector-valued functions, the *evaluation operator* $E_x : \mathbb{H} \to \mathbb{U}$ for a point $x \in \mathbb{X}$ is defined so that

$$E_x f \triangleq f(x) \in \mathbb{U}, \quad \text{for all } f \in \mathbb{H}. \tag{3.108}$$

As for the scalar-valued case, the evaluation operator is linear, but not necessarily bounded in a generic Hilbert space of functions. The next result establishes a relationship between the boundedness of the evaluation functional and the reproducing property of a Hilbert space.

Theorem 3.16 *The Hilbert space $\boldsymbol{\mathcal{H}}$ is an RKHS if and only if all linear evaluation operators $E_x : \boldsymbol{\mathcal{H}} \to \mathbb{U}$ for $x \in \mathbb{X}$ are bounded.*

It follows from Theorem 3.16 that the Hilbert space $\boldsymbol{\mathcal{H}}$ is an $\mathbb{U}$-valued RKHS provided that, for each $x \in \mathbb{X}$, there exists a constant $C_x \geq 0$ such that

$$\|E_x f\|_{\mathbb{U}} \triangleq \|f(x)\|_{\mathbb{U}} \leq C_x \|f\|_{\boldsymbol{\mathcal{H}}}, \quad \text{for all } f \in \boldsymbol{\mathcal{H}}; \tag{3.109}$$

note the analogies between (3.109) and (3.4). As for the scalar-valued case, we refer to RKHSs that contain vector-valued functions simply as vector-valued RKHS.

The next examples provide important insights into the theory of vector-valued RKHS. The first of these examples correlates (3.107) with (3.109).

Example 3.1 If (3.109) is verified, then (3.107) is verified. For each fixed $x \in \mathbb{X}$ and $u \in \mathbb{U}$, define the linear functional $\lambda_{x,u} : f \mapsto (f(x), u)_{\mathbb{U}}$. If follows from (3.109) that, for each fixed $x \in \mathbb{X}$ and $u \in \mathbb{U}$, the linear functional $\lambda_{x,u}$ is bounded on $\mathcal{H}$ since

$$|\lambda_{x,u} f| = |\langle f(x), u \rangle_{\mathbb{U}}| = |\langle E_x f, u \rangle_{\mathbb{U}}| \leq C_x \|u\|_{\mathbb{U}} \|f\|_{\mathcal{H}}. \tag{3.110}$$

By the Riesz-representation theorem (Theorem 2.5), for the Hilbert space $\mathcal{H}$, fixed x, u, there is a representer $z = z(x, u) \in \mathcal{H}$ of the functional $\lambda_{x,u}$ such that $\lambda_{x,u} f \triangleq \langle E_x f, u \rangle_{\mathbb{U}} = \langle f, z(x, u) \rangle_{\mathcal{H}}$. However, by construction, $(x, u) \mapsto z(x, u)$ is a linear mapping in the variable u. Thus, there must exist some linear operator $\mathcal{K}_x$ such that $z(x, u) = \mathcal{K}_x u$. It only remains to show that $\mathcal{K}_x \in \mathcal{L}(\mathbb{U}, \mathcal{H})$. By the Riesz-representation theorem, we have

$$\begin{aligned}
\|\mathcal{K}_x u\|_{\mathcal{H}} &= \|z(x, u)\|_{\mathcal{H}} \\
&= \|\lambda_{x,u}\|_{\mathcal{H}'} \\
&= \sup_{f \neq 0} \frac{|\langle E_x f, u \rangle_{\mathbb{U}}|}{\|f\|_{\mathcal{H}}} \\
&\leq \|E_x\| \|u\|_{\mathbb{U}} \\
&\leq C_x \|u\|_{\mathbb{U}}, \quad \text{for all } (x, u) \in \mathbb{X} \times \mathbb{U}.
\end{aligned} \tag{3.111}$$

Thus, $\mathcal{K}_x$ is a bounded linear operator, that is, $\mathcal{K}_x \in \mathcal{L}(\mathbb{U}, \mathcal{H})$, and the reproducing formula in (3.107) holds for a bounded linear operator $\mathcal{K}_x \in \mathcal{L}(\mathbb{U}, \mathcal{H})$. △

The next example shows how, as expected, scalar-valued RKHSs are special cases of vector-valued RKHSs.

Example 3.2 Suppose that $\mathfrak{K} : \mathbb{X} \times \mathbb{X} \to \mathbb{R}$ is an admissible kernel that induces the native space $\mathcal{H}$ of real-valued functions. Choose the inner product space $\mathbb{U} \triangleq \mathbb{R}$ and define the operator $\mathcal{K}_x$ on $\mathbb{U}$ as the multiplication operator induced by $\mathfrak{K}_x$, that is,

$$\mathcal{K}_x u \triangleq \mathfrak{K}_x u, \quad \text{for all } x \in \mathbb{X} \text{ and } u \in \mathbb{U}.$$

In this case, it holds that

$$\|\mathcal{K}_x u\|_{\mathcal{H}} = \|\mathfrak{K}_x u\|_{\mathcal{H}} = \|\mathfrak{K}_x\|_{\mathcal{H}} |u| = \|\mathfrak{K}_x\|_{\mathcal{H}} \|u\|_{\mathbb{U}}, \quad \text{for all } u \in \mathbb{U} \text{ and } x \in \mathbb{X},$$

which implies that $\mathcal{K}_x \in \mathcal{L}(\mathbb{U}, \mathcal{H})$. Also, the $\mathbb{U}$-valued reproducing formula (3.108) is verified because, for each $x \in \mathbb{X}$, it holds that, $u \in \mathbb{U}$ and $f \in \mathcal{H}$

$$\langle f(x), u \rangle_{\mathbb{U}} = u \cdot f(x) = u \, \langle \mathfrak{K}_x, f \rangle_{\mathcal{H}} = \langle \mathfrak{K}_x u, f \rangle_{\mathcal{H}} = \langle \mathcal{K}_x u, f \rangle_{\mathcal{H}}. \tag{3.112}$$

Thus, any scalar-valued RKHS $\mathcal{H}$ can be viewed as a special case of the $\mathbb{U}$-valued RKHS with the choice $\mathbb{U} = \mathbb{R}$ and $\mathcal{K}_x u \triangleq \mathfrak{K}_x u$. △

The Cartesian product of scalar-valued RKHSs $\{\mathcal{H}_1, \ldots, \mathcal{H}_m\}$ produces a vector-valued RKHS $\mathcal{H}$ whose kernel $\mathcal{K}_x$ is diagonal, and each diagonal entry is given by the kernel underlying the corresponding scalar-valued RKHS. The next example shows the converse result, that is, that $\mathbb{U}$-valued RKHS with diagonal kernels are reducible to the Cartesian product of some scalar-valued RKHSs. Often, this example of a diagonal operator kernel can be a source of intuition for cases where the kernel $\mathcal{K}_x$ is not diagonal and $\mathcal{H}$ is not reducible to the Cartesian product of scalar-valued native spaces.

Example 3.3 Let $\mathbb{U} \triangleq \mathbb{R}^m$, and define the operator $\mathcal{K}_x : \mathbb{U} \to \mathcal{H}$ in terms of the component-wise multiplication operators

$$\mathcal{K}_x u \triangleq \begin{Bmatrix} \mathfrak{K}_{1,x} u_1 \\ \mathfrak{K}_{2,x} u_2 \\ \vdots \\ \mathfrak{K}_{m,x} u_m \end{Bmatrix} = \begin{bmatrix} \mathfrak{K}_{1,x} & 0 & \dots & & 0 \\ 0 & \mathfrak{K}_{2,x} & & & \vdots \\ \vdots & & & \ddots & 0 \\ 0 & 0 & \dots & 0 & \mathfrak{K}_{m,x} \end{bmatrix} \begin{bmatrix} u_1 \\ u_2 \\ \vdots \\ u_n \end{bmatrix}$$

for $x \in \mathbb{X}$ and $u = [u_1, \dots, u_m]^{\mathrm{T}} \in \mathbb{U} \triangleq \mathbb{R}^m$, where $\mathfrak{K}_{i,x} \triangleq \mathfrak{K}_i(x, \cdot)$ and $\mathfrak{K}_i$ denotes the kernel that generates the scalar-valued native space $\mathcal{H}_i$ for $i = 1, \dots m$. Thus, $\mathcal{K}_x \in \mathcal{L}(\mathbb{U}, \mathcal{H})$ and $\mathcal{K}_x$ satisfies the reproducing property in $\mathcal{H}$ for each $x \in \mathbb{X}$. $\triangle$

If $\mathcal{H}$ is a $\mathbb{U}$-valued RKHS, then the identity in (3.107) implies that

$$\mathcal{K}_x = (E_x)^* \triangleq E_x^*, \tag{3.113}$$

where $E_x^* : \mathbb{U} \to \mathcal{H}$ denotes the adjoint of the evaluation operator E_x at $x \in \mathbb{X}$. Indeed, by definition,

$$\langle E_x f, u \rangle_{\mathbb{U}} = \langle f, \mathcal{K}_x y \rangle_{\mathcal{H}} = \left\langle f, E_x^* u \right\rangle_{\mathcal{H}}, \quad \text{for all } x \in \mathbb{X} \text{ and } f \in \mathcal{H}. \tag{3.114}$$

Thus, the following definition is easily motivated.

Definition 3.14 (*Operator-Valued Canonical Kernel*) If $\mathcal{H}$ is a real, $\mathbb{U}$-valued RKHS, then the *canonical operator-valued kernel* $\mathcal{K} : \mathbb{X} \times \mathbb{X} \to \mathcal{L}(\mathbb{U})$ is defined as

$$\mathcal{K}(x, z) \triangleq E_x E_z^* = \mathcal{K}_x^* \mathcal{K}_z, \quad \text{for all } (x, z) \in \mathbb{X} \times \mathbb{X}. \tag{3.115}$$

Notably, operator-valued kernels are square matrices of dimensions $m \times m$. As for the scalar-valued RKHS, $\mathbb{U}$-valued RKHS can be conveniently constructed from admissible operator kernels. To show this result, first, we recall several definitions concerning operator-values kernels. Successively, we prove a result that allows us to construct $\mathbb{U}$-valued RKHSs and shows several of their properties.

Definition 3.15 (*Operator-Valued Symmetric Kernel*) The operator-valued kernel $\mathcal{K} : \mathbb{X} \times \mathbb{X} \to \mathcal{L}(\mathbb{U})$ is *symmetric* if

$$(\mathcal{K}(x_1, x_2))^* = \mathcal{K}(x_2, x_1) \quad \text{for all } x_1, x_2 \in \mathbb{X}. \tag{3.116}$$

Definition 3.16 (*Operator-Valued Kernel of Positive Type*) The operator-valued kernel $\mathcal{K} : \mathbb{X} \times \mathbb{X} \to \mathcal{L}(\mathbb{U})$ is of *positive type* if

$$\sum_{i,j=1}^{N} \left\langle \mathcal{K}(\xi_j, \xi_i) u_i, u_j \right\rangle_{\mathbb{U}} \geq 0 \tag{3.117}$$

If the inequality in (3.117) is strict for any choice of N distinct centers in Ξ_N, then the kernel is of *strictly positive type*. For any integer $N \in \mathbb{N}$, finite set $\{\xi_i\}_{i=1}^N \subset \mathbb{X}$ of points, and finite set $\{u_i\}_{i=1}^N \subset \mathbb{U}$ of elements.

Definition 3.17 (*Operator-Valued Admissible Kernel*) The operator-valued kernel $\mathcal{K} : \mathbb{X} \times \mathbb{X} \to \mathcal{L}(\mathbb{U})$ is *admissible* if it is symmetric and of positive type.

Thus, we can readily prove that canonical kernels are admissible.

Theorem 3.17 *The canonical kernel (3.115) is admissible.*

Proof The canonical kernel (3.115) is symmetric since $(\mathcal{K}(x, z))^* = (E_x E_z^*)^* = (E_z, E_x^*) = \mathcal{K}(z, x)$ for all $x, z \in \mathbb{X}$. This kernel is of positive type since, for any set of point $\{\xi_i\}_{i=1}^N \subset \mathbb{X}$ and any set of elements $\{u_i\}_{i=1}^N \subset \mathbb{U}$, it holds that

$$\sum_{i,j}^{N} \left\langle E_{\xi_j} E_{\xi_i}^* u_i, u_j \right\rangle_{\mathbb{U}} = \sum_{i,j}^{N} \left\langle E_{\xi_i}^* u_i, E_{\xi_j}^* u_j \right\rangle_{\mathbb{U}} = \left\| \sum_{i=1}^{N} \mathcal{K}_{\xi_i} u_i \right\|_{\mathbb{U}}^2 \geq 0.$$

■

The next example further elaborates on Example 3.2, and shows that, if $\mathbb{U} = \mathbb{R}$ and $\mathcal{K}_x y \triangleq \mathfrak{K}_x y$ for all $u \in \mathbb{U}$, then operator-valued symmetric kernels reduce to scalar-valued symmetric kernels.

Example 3.4 Consider the scalar kernel $\mathfrak{K} : \mathbb{X} \times \mathbb{X} \to \mathbb{R}$ that defines the scalar-valued native space $\mathcal{H}$, and, as in Example 3.2, let $\mathbb{U} = \mathbb{R}$ and $\mathcal{K}_x y \triangleq \mathfrak{K}_x y$ for all $u \in \mathbb{U}$. To verify the symmetry of the operator-valued kernel, note that

$$(\mathcal{K}(x, z))^* = \left(\mathcal{K}_z^* \mathcal{K}_x\right)^* = \mathcal{K}_z^* \mathcal{K}_x. \tag{3.118}$$

Thus, for any $u \in \mathbb{U}$, it holds that

$$(\mathcal{K}(x, z))^* u = \mathcal{K}_z^* \mathcal{K}_x u = E_z \mathfrak{K}_x u = \mathfrak{K}(z, x) u, \tag{3.119}$$

so that $(\mathcal{K}(x, z))^*$ is the multiplication by the kernel $\mathfrak{K}(z, x)$. The requirement whereby $(\mathcal{K}(x, z))^* = \mathcal{K}(z, x)$ is verified since

$$\mathfrak{K}(x,z)u_1u_2 = \langle \mathcal{K}(x,z)u_1, u_2\rangle_{\mathbb{U}} = \langle u_1, (\mathcal{K}(x,z))^* u_2\rangle_{\mathbb{U}} = \mathfrak{K}(z,x)u_1u_2$$

for all $x, z \in \mathbb{X}$ and $u_1, u_2 \in \mathbb{U} \triangleq \mathbb{R}$, and the result is proven. △

As in the study of scalar-valued native space $\mathcal{H}$, it is often more practical to start with an admissible kernel to define a $\mathbb{U}$-valued RKHS $\boldsymbol{\mathcal{H}}$. Once the mapping $\mathcal{K} : \mathbb{X} \times \mathbb{X} \to \mathcal{L}(\mathbb{U})$ is chosen, we define the operator $\mathcal{K}_x$ via the identity

$$(\mathcal{K}_x u)(z) = \mathcal{K}(z,x)u, \quad \text{for all } x, z \in \mathbb{X} \text{ and } u \in \mathbb{U}. \tag{3.120}$$

The construction of the operator $\mathcal{K}_x$ in this way from an admissible kernel $\mathcal{K}$ defines a unique $\mathbb{U}$-valued RKHS, as summarized in the next theorem.

Theorem 3.18 ([21, pp. 380–381, Proposition 2.3]) *Let $\mathcal{K} : \mathbb{X} \times \mathbb{X} \to \mathcal{L}(\mathbb{U})$ be an admissible operator-valued kernel over the set $\mathbb{X}$. Then, there is a unique vector-valued RKHS $\boldsymbol{\mathcal{H}}$ such that $\mathcal{K}_x \in \mathcal{L}(\mathbb{U}, \boldsymbol{\mathcal{H}})$ and the reproducing property (3.107) is verified. Furthermore, $\boldsymbol{\mathcal{H}}$ has the following properties:*

1. *The $\mathbb{U}$-valued RKHS $\boldsymbol{\mathcal{H}} \triangleq \boldsymbol{\mathcal{H}}(\mathbb{X}, \mathbb{U})$ is given by*

$$\boldsymbol{\mathcal{H}} = \overline{\text{span}\,\{\mathcal{K}_x u \mid x \in \mathbb{X}, u \in \mathbb{U}\}}, \tag{3.121}$$

where the closure is taken with respect to the inner product defined by

$$\langle f, g\rangle_{\boldsymbol{\mathcal{H}}} = \sum_{i,j=1}^{N} \alpha_i \beta_j \langle \mathcal{K}(a_i, b_j)v_j, u_i\rangle_{\mathbb{U}} \tag{3.122}$$

for any two functions f, g such that $f \triangleq \sum_{i=1}^N \alpha_i \mathcal{K}_{a_i} u_i$ and $g \triangleq \sum_{j=1}^N \beta_j \mathcal{K}_{b_j} v_j$, where $N \in \mathbb{N}$, $\{\alpha_i\}_{i=1}^N$ and $\{\beta_j\}_{j=1}^N \subset \mathbb{R}$ are real numbers, $\{a_i\}_{i=1}^N, \{b_j\}_{j=1}^N \subset \mathbb{X}$, and $\{u_i\}_{i=1}^N, \{v_j\}_{j=1}^N \subset \mathbb{U}$.

2. *For any $x \in \mathbb{X}$ and $f \in \boldsymbol{\mathcal{H}}$, it holds that*

$$\|f(x)\|_{\mathbb{U}} \le \sqrt{\|\mathcal{K}(x,x)\|_{\mathcal{L}(\mathbb{U})}}\,\|f\|_{\boldsymbol{\mathcal{H}}} \tag{3.123}$$

and

$$\|E_x\|_{\mathcal{L}(\boldsymbol{\mathcal{H}},\mathbb{U})} \le \sqrt{\|\mathcal{K}(x,x)\|_{\mathcal{L}(\mathbb{U})}}. \tag{3.124}$$

3. *Any $\|\cdot\|_{\boldsymbol{\mathcal{H}}}$-norm convergent sequence of functions $\{f_n\}_{n=1}^\infty$ is pointwise convergent in the norm $\|\cdot\|_{\mathbb{U}}$, that is, for each $x \in \mathbb{X}$, it holds that*

$$\|f(x) - f_n(x)\|_{\mathbb{U}} \to 0 \quad \text{as } n \to \infty. \tag{3.125}$$

This convergence is uniform on any subset $\Omega \subset \mathbb{X}$ for which there is a constant $C_\Omega > 0$ such that $\|\mathcal{K}(x,x)\|_{\mathcal{L}(\boldsymbol{\mathcal{H}})} \le C_\Omega$ for all $x \in \Omega$.

Proof The proof of this result is divided into three parts.

1. The proof of this part is outlined for a special case of higher interest to the scopes of this book. A complete proof can be found in [21, Proposition 2.3]. For the special case whereby $\mathbb{U} \triangleq \mathbb{R}^m$, the kernel $\mathcal{K}_x$ is diagonal as in Example 3.3, and $\mathcal{H} = \mathcal{H}_1 \times \cdots \times \mathcal{H}_m$ with each $\mathcal{H}_i$ a scalar-valued native space, (3.121) directly follows from (3.9) and (3.122) follows from (3.10).
2. It follows from the reproducing formula (3.107), (3.114), and (2.22a) that

$$|\langle f(x), u\rangle_{\mathbb{U}}| = |\langle \mathcal{K}_x u, f\rangle_{\mathcal{H}}| = |\langle u, E_x f\rangle_{\mathbb{U}}| \leq \|u\|_{\mathbb{U}}\|E_x f\|_{\mathbb{U}} \tag{3.126}$$

for all $x \in \mathbb{X}$, $u \in \mathbb{U}$, and $f \in \mathcal{H}$. By the Schwarz inequality (2.22a), the maximum value of the left-most term in (3.126) occurs when $f(x) = u$, which means that

$$\|f(x)\|_{\mathbb{U}}^2 = \|E_x f\|_{\mathbb{U}}^2 = |\left(E_x^* E_x f, f\right)_{\mathcal{H}}| \leq \|E_x^* E_x\|_{\mathcal{L}(\mathcal{H})}^2 \|f\|^2. \tag{3.127}$$

It follows from (3.115) that $\mathcal{K}(x, x) = E_x E_x^* = \mathcal{K}_x^* \mathcal{K}_x$, and the operator norms of $E_x E_x^* : \mathbb{U} \to \mathbb{U}$ and $E_x^* E_x : \mathcal{H} \to \mathcal{H}$ satisfy $\|E_x E_x^*\|_{\mathcal{L}(\mathbb{U})} = \|E_x^* E_x\|_{\mathcal{L}(\mathcal{H})}$. We conclude that (3.123) is verified. Finally, (3.124) follows from (3.123) and (2.15).
3. It follows from (2.22a) and (3.124) that, for any $x \in \mathbb{X}$ and $f \in \mathcal{H}$, it holds that

$$\begin{aligned}\|f(x) - f_n(x)\|_{\mathbb{U}} &= \|E_x(f - f_n)\|_{\mathbb{U}} \\ &\leq \|E_x\|\|f - f_n\|_{\mathcal{H}} \\ &\leq \sqrt{\|\mathcal{K}(x, x)\|}\|f - f_n\|_{\mathcal{H}}\end{aligned} \tag{3.128}$$

and $\|f - f_n\|_{\mathcal{H}} \to 0$ as $n \to \infty$. Thus, (3.125) is verified. ■

The statement of this theorem should be compared carefully to some of the key properties of scalar-valued RKHSs discussed in Sect. 3.1. For example, the sequence of inequalities in (3.128) should be compared with the scalar-valued case in (3.30).

3.6.2 Subspaces Generated by Restrictions to $\Omega \subset \mathbb{X}$

In Sect. 3.2, we summarized how, if we are given a scalar-valued kernel on the ambient space $\mathbb{X}$, we can use it to define various types of RKHS determined by a subset $\Omega \subset \mathbb{X}$. These considerations culminated in Theorem 3.3, which studied the relationships between such spaces of restrictions and spaces generated by a subset $\Omega \subset \mathbb{X}$. In analogy to the scalar-valued case, if an operator kernel $\mathcal{K}$ defines the RKHS $\mathcal{H} \triangleq \mathcal{H}(\mathbb{X}, \mathbb{U})$ of vector-valued functions, we define the closed subspace $\mathcal{H}_\Omega$ generated by $\Omega \subset \mathbb{X}$ to be

$$\mathcal{H}_\Omega \triangleq \overline{\text{span}\{\mathcal{K}_x u : u \in \mathbb{U}, x \in \Omega\}} \subset \mathcal{H}. \tag{3.129}$$

In this definition, the closure is taken with respect to the norm that $\mathcal{H}_\Omega$ inherits as a subspace of $\mathcal{H}$. As for scalar-valued kernels, we emphasize that the functions in $f \in \mathcal{H}_\Omega$ are mappings $f : \mathbb{X} \to \mathbb{U}$ defined on *all* of $\mathbb{X}$. Just as discussed earlier in Sect. 3.2, the closed subspace $\mathcal{H}_\Omega$ is an RKHS of vector-valued functions. It has the operator kernel

$$\mathcal{K}_{\Omega,x} \triangleq \Pi_\Omega \mathcal{K}_x \in \mathcal{H}_\Omega, \quad \text{for all } x \in \mathbb{X}, \tag{3.130}$$

where Π_Ω denotes the $\mathcal{H}$-orthogonal projection onto $\mathcal{H}_\Omega \subset \mathcal{H}$. This operator kernel satisfies the reproducing property since

$$\begin{aligned}\left\langle f, \mathcal{K}_{\Omega,x} u\right\rangle_{\mathcal{H}} &= \langle f, \Pi_\Omega \mathcal{K}_x u\rangle_{\mathcal{H}} = \left\langle \Pi_\Omega f, \mathcal{K}_{\Omega,x} u\right\rangle_{\mathcal{H}} \\ &= \langle f, \mathcal{K}_x u\rangle_{\mathcal{H}} = \langle E_x f, u\rangle_{\mathbb{U}} = \langle f(x), u\rangle_{\mathbb{U}}.\end{aligned}$$

By Proposition 7 and Corollary 2 of [33], we can define the subspace of functions in $\mathcal{H}$ that vanish on Ω as

$$\mathbf{Z}_\Omega \triangleq \{f \in \mathcal{H} : \ f(x) = 0 \text{ for all } x \in \Omega\} \subset \mathcal{H}, \tag{3.131}$$

and obtain the orthogonal decomposition

$$\mathcal{H} = \mathcal{H}_\Omega \oplus \mathbf{Z}_\Omega. \tag{3.132}$$

Alternatively, we can define the space of vector-valued restrictions $\mathcal{R}_\Omega$ to be

$$\mathcal{R}_\Omega \triangleq \{g : \Omega \to \mathbb{U} : g = T_\Omega f, \ f \in \mathcal{H}\}, \tag{3.133}$$

where T_Ω denotes the trace operator. The space of restrictions $\mathcal{R}_\Omega$ is an RKHS with the operator kernel $\mathcal{R} : \Omega \times \Omega \to \mathcal{L}(\mathbb{U})$ defined as

$$\mathcal{R}(x, y) \triangleq \mathcal{K}|_{\Omega\times\Omega}(x, y) \in \mathcal{L}(\mathbb{U}) \quad \text{for all } x, y \in \Omega. \tag{3.134}$$

The trace operator, when restricted to $\mathcal{H}_\Omega \subset \mathcal{H}$, is an isometry, that is,

$$\|f\|_{\mathcal{H}} = \|T_\Omega f\|_{\mathcal{R}}, \quad \text{for all } f \in \mathcal{H}_\Omega. \tag{3.135}$$

Thus, just as in the scalar-valued RKHS, we can identify $\mathcal{H}_\Omega$, which consists of functions defined on all of $\mathbb{X}$, with $\mathcal{R}_\Omega$, which contains functions defined only on $\Omega \subset \mathbb{X}$. The restricted trace operator $T_\Omega|_{\mathcal{H}_\Omega} : \mathcal{H}_\Omega \to \mathcal{R}_\Omega$, and its associated canonical extension operator $\mathcal{E}_\Omega \triangleq (T_\Omega|_{\mathcal{H}_\Omega})^{-1} : \mathcal{R}_\Omega \to \mathcal{H}_\Omega$, map these spaces isometrically onto one another.

3.7 Vector-Valued Native Spaces and Other Function Spaces

As discussed in Sect. 3.4 in the context of scalar-valued RKHSs, it is important to compare RKHSs induced by an operator kernel to other vector-valued function spaces. Since the theory of scalar-valued native spaces is more mature, there are currently more theorems that describe the situation for the scalar-valued case. However, if the vector-valued RKHS is the Cartesian product

$$\mathcal{H} = \mathcal{H}_1 \times \cdots \times \mathcal{H}_m, \tag{3.136}$$

where each $\mathcal{H}_i$ is a scalar-valued RKHS, then all the methods to relate the scalar-valued RKHS $\mathcal{H}_i$ to other common spaces of scalar-valued functions hold as discussed in Sect. 3.4.

In the following two sections, we describe two particular ways in which the strategies described in Sect. 3.4 can be extended or lifted to the case generated by operator-valued kernels.

3.7.1 *Comparing Vector-Valued Native Spaces to One Another*

Let $\mathbb{U}_1$ and $\mathbb{U}_2$ be Hilbert spaces, and let $\mathcal{K}_1 : \mathbb{X} \times \mathbb{X} \to \mathcal{L}(\mathbb{U}_1)$ and $\mathcal{K}_2 : \mathbb{X} \times \mathbb{X} \to \mathcal{L}(\mathbb{U}_2)$ be two admissible operator kernels that induce the native spaces $\mathcal{H}_1 \triangleq \mathcal{H}_1(\mathbb{X}, \mathbb{U}_1)$ and $\mathcal{H}_2 \triangleq \mathcal{H}_2(\mathbb{X}, \mathbb{U}_2)$, respectively. For any mapping $\phi : \mathbb{X} \to \mathcal{L}(\mathbb{U}_1, \mathbb{U}_2)$, we define an associated linear operator M_ϕ such that

$$(M_\phi f)(x) = \phi(x) f(x) \in \mathbb{U}_2, \quad \text{for all } f \in \mathcal{H}_1 \text{ and } x \in \mathbb{X}. \tag{3.137}$$

Note that, by construction, $M_\phi f \in \mathcal{F}(\mathbb{X}, \mathbb{U}_2)$, that is, $M_\phi f$, is an element of the vector space of functions over $\mathbb{X}$ that take values in $\mathbb{U}_2$; see Sect. 2.4. The operator M_ϕ is a multiplication operator since

$$M_\phi f \in \mathcal{H}_2 \subset \mathcal{F}(\mathbb{X}, \mathbb{U}_2). \tag{3.138}$$

As discussed in [34], $M_\phi \in \mathcal{L}(\mathcal{H}_1, \mathcal{H}_2)$. Of special interest is the case whereby $\mathbb{U} = \mathbb{U}_1 = \mathbb{U}_2$, that is, whenever functions in the vector-valued spaces $\mathcal{H}_1, \mathcal{H}_2$ take values in the same Hilbert space $\mathbb{U}$. In this case, the following result holds.

Theorem 3.19 ([3, Theorem 6.28]) *Let $\mathcal{K}_1 : \mathbb{X} \times \mathbb{X} \to \mathcal{L}(\mathbb{U})$ and $\mathcal{K}_2 : \mathbb{X} \times \mathbb{X} \to \mathcal{L}(\mathbb{U})$ be the operator kernels of the vector-valued RKHSs $\mathcal{H}_1$ and $\mathcal{H}_2$, respectively. Let $\phi : \mathbb{X} \times \mathbb{X} \to \mathcal{L}(\mathbb{U})$ be a linear mapping. Then, the following statements are equivalent:*

1. *The mapping ϕ induces a multiplication operator $M_\phi \in M(\mathcal{H}_1, \mathcal{H}_2)$.*
2. *The multiplication operator M_ϕ is a bounded linear operator, $M_\phi \in \mathcal{L}(\mathcal{H}_1, \mathcal{H}_2)$.*
3. *There exists a constant $c > 0$ such that*

$$\phi(x)\mathcal{K}_1(x, y)\phi(y)^* \leq c^2\mathcal{K}_2(x, y), \quad \textit{for all } x, y \in \mathbb{X}. \tag{3.139}$$

A consequence of Theorem 3.19 is the following result.

Theorem 3.20 ([3, Theorem 6.25]) *Let $\mathcal{K}_1 : \mathbb{X} \times \mathbb{X} \to \mathcal{L}(\mathbb{U})$ and $\mathcal{K}_2 : \mathbb{X} \times \mathbb{X} \to \mathcal{L}(\mathbb{U})$ be the operator kernels of the vector-valued RKHSs $\mathcal{H}_1$ and $\mathcal{H}_2$, respectively. If there exists a constant $c > 0$ such that*

$$\mathcal{K}_1(x, y) \leq c^2\mathcal{K}_2(x, y), \quad \textit{for all } x, y \in \mathbb{X}, \tag{3.140}$$

then, the continuous embedding

$$\mathcal{H}_1 \overset{i}{\hookrightarrow} \mathcal{H}_2 \tag{3.141}$$

is verified, and $\|i\| \leq c$, that is, the operator norm of the canonical embedding is bounded.

Theorems 3.19 and 3.20 provide relations between different vector-valued RKHSs and the underlying operator kernels. In particular, Theorem 3.20 provides sufficient conditions for some ordering of the RKHS induced by a continuous embedding.

3.7.2 Continuous Operator Kernels Bounded on the Diagonal

As discussed in Sect. 3.4.3, the analysis of native space embedding methods for scalar-valued RKHS $\mathcal{H}$ is simplified by assuming that the kernel $\mathfrak{K}(\cdot, \cdot)$ that defines $\mathcal{H}$ is bounded uniformly on the diagonal. As shown by Theorem 3.6, if the kernel function for scalar-valued RKHS is bounded on the diagonal, that is, if (3.32) is verified, then the continuous embedding (3.33) is verified. In the following, we generalize both Definition 3.7 and Theorem 3.6.

Definition 3.18 (*Uniformly Bounded on the Diagonal Kernel*) The operator-valued kernel $\mathcal{K} : \mathbb{X} \times \mathbb{X} \to \mathcal{L}(\mathbb{U})$ is *uniformly bounded on the diagonal* if there is a constant $\bar{\mathfrak{K}} > 0$ such that

$$\|\mathcal{K}(x, x)\|_{\mathcal{L}(\mathbb{U})} \leq \bar{\mathfrak{K}}^2, \quad \text{for all } x \in \mathbb{X}. \tag{3.142}$$

Theorem 3.21 *If the operator-valued kernel $\mathcal{K} : \mathbb{X} \times \mathbb{X} \to \mathcal{L}(\mathbb{U})$ is bounded uniformly on the diagonal and $\mathcal{H} \subset (C(\mathbb{X}))^m$, then $\mathcal{H}$ is continuously embedded in the space of bounded continuous functions, that is,*

$$\mathcal{H} \hookrightarrow (C_b(\mathbb{X}))^m. \tag{3.143}$$

Proof To prove the continuous embedding of $\mathcal{H}$ into $(C_b(\mathbb{X}))^m$, note that it follows from (2.22a), (3.115), and (3.142) that, for all $u \in \mathbb{U}$,

$$\begin{aligned} |\langle \mathcal{K}_x u, f\rangle_{\mathcal{H}}| &\leq \|\mathcal{K}_x u\|_{\mathcal{H}} \|f\|_{\mathcal{H}} \\ &< \sqrt{\langle \mathcal{K}_x u, \mathcal{K}_x u\rangle_{\mathcal{H}}} \|f\|_{\mathcal{H}} \\ &= \sqrt{\left\langle \mathcal{K}_x^* \mathcal{K}_x u, \alpha\right\rangle_{\mathcal{H}}} \|f\|_{\mathcal{H}}, \\ &= \sqrt{\langle \mathcal{K}(x,x)u, \alpha\rangle_{\mathcal{H}}} \|f\|_{\mathcal{H}} \\ &\leq \bar{\mathfrak{K}} \|u\|_{\mathbb{U}} \|f\|_{\mathcal{H}}, \end{aligned}$$

which implies that

$$\|\mathcal{K}_x\|_{\mathcal{L}(\mathbb{U},\mathcal{H})} \leq \bar{\mathfrak{K}}. \tag{3.144}$$

Furthermore, for any $f \in \mathcal{H}$, we have that

$$\begin{aligned} \|f(x)\|^2_{\mathbb{R}^m} &= \langle f(x), f(x)\rangle_{\mathbb{R}^m} \\ &= \langle E_x f, E_x f\rangle_{\mathbb{R}^m} \\ &= \left\langle \mathcal{K}_x^* f, \mathcal{K}_x^* f\right\rangle_{\mathcal{H}} \\ &\leq \|\mathcal{K}_x\|^2 \|f\|^2_{\mathcal{H}} \\ &\leq \bar{\mathfrak{K}}^2 \|f\|^2_{\mathcal{H}}, \end{aligned} \tag{3.145}$$

which completes the proof. ■

Note that, if the vector-valued native space $\mathcal{H}$ can be written as the Cartesian product (3.136) and the scalar-valued native space $\mathcal{H}$ is bounded on the diagonal by $\bar{\mathfrak{K}}$, then both (3.142) and (3.144) are readily verified. Indeed, for all $x \in \mathbb{X}$ and $u \in \mathbb{U}$, it holds that

$$\|\mathcal{K}_x u\|_{\mathcal{H}} = \|\mathfrak{K}_x I_m u\|_{\mathcal{H}} \leq \|\mathfrak{K}_x\|_{\mathcal{H}} \|u\|_{\mathbb{U}} \leq \bar{\mathfrak{K}} \|u\|_{\mathbb{U}} \tag{3.146}$$

so that $\|\mathcal{K}_x\|_{\mathcal{L}(\mathbb{U},\mathcal{H})} \leq \bar{\mathfrak{K}}$ in (3.144). Similar steps and analysis yield (3.142).

3.8 Approximations in Vector-Valued Native Spaces

Section 3.5 discussed approximation in scalar-valued native spaces $\mathcal{H}$. In this section, we address the same problem in the context of vector-valued spaces $\mathcal{H}$. Reference [27] discusses numerous properties of matrix kernels and the approximations that they generate. We primarily follow [27] in this section.

Similar to our approach to scalar-valued RKHS, we begin by building approximations of $f : \mathbb{X} \to \mathbb{U} \triangleq \mathbb{R}^m$ with a selection of distinct points or centers Ξ_N given by (3.6), where Ω is a compact subset of $\mathbb{X}$. It follows from (3.117) that the positivity of the matrix kernel $\mathcal{K}$ is equivalent to the positive semidefinitemess of the *generalized Grammian matrix*

$$\mathbb{K}_N \triangleq \begin{bmatrix} \mathcal{K}(\xi_1, \xi_1) & \mathcal{K}(\xi_1, \xi_2) & \cdots & \mathcal{K}(\xi_1, \xi_N) \\ \vdots & \vdots & \ddots & \vdots \\ \mathcal{K}(\xi_N, \xi_1) & \mathcal{K}(\xi_N, \xi_2) & \cdots & \mathcal{K}(\xi_N, \xi_N) \end{bmatrix} \in \mathbb{R}^{mN \times mN} \tag{3.147}$$

for any set of centers Ξ_N given by (3.6). If $m = 1$, then (3.147) reduces to the Grammian matrix (3.8). For this reason, we use the same notation for both matrices. The matrix operator kernel $\mathcal{K}$ is *positive-definite* if the generalized Grammian is a positive-definite matrix for any set of N distinct centers in $\Xi_N \subset \mathbb{X}$.

Given a collection of centers Ξ_N, we define the *space of approximants*

$$\mathcal{H}_{d(N)} \triangleq \text{span}\left\{\mathcal{K}_{\xi_i} e_j \mid \xi_i \in \Xi_N \text{ for } 1 \le i \le N, 1 \le j \le m\right\} \tag{3.148a}$$

$$= \text{span}\left\{\mathcal{K}_{\xi_i} u \mid \xi_i \in \Xi_N \text{ for } 1 \le i \le N, u \in \mathbb{U}\right\}, \tag{3.148b}$$

where $\{e_j\}_{j=1}^m$ denotes the *canonical basis* for $\mathbb{U} \triangleq \mathbb{R}^m$; see Example 2.1.

In this chapter, our approximation schemes in $\mathcal{H}$ always assume that the operator kernel $\mathcal{K}$ is positive-definite. This assumption does not prove to be overly restrictive in our applications, and it ensures that the family of functions $\{\mathcal{K}_{\xi_i} e_j : 1 \le i \le N, 1 \le j \le m\}$ is linearly independent and constitutes a basis for $\mathcal{H}_{d(N)}$. Note that when functions take values in $\mathbb{U} \triangleq \mathbb{R}^m$, the number of parameters $d(N) \triangleq \dim(\mathcal{H}_{d(N)}) = mN$ denotes the dimension of the space of approximants when using N centers for approximation. We suppose that the collection of all centers $\Xi \triangleq \bigcup_{N=1}^{\infty} \Xi_N$ is dense in the compact set $\Omega \subset \mathbb{X}$, that is, $\Omega \subseteq \overline{\bigcup_{N \in \mathbb{N}} \Xi_N} = \mathbb{X}$; see Definition 2.20.

In analogy to the situation for scalar-valued kernels, the subspace $\mathcal{H}_{d(N)}$ is a vector-valued native space that contains functions on $\mathbb{X}$ into $\mathbb{U}$. The kernel underlying $\mathcal{H}_{d(N)}$ is given by [27]

$$\mathcal{K}_N(x_1, x_2) \triangleq \mathcal{K}_{\Xi_N}(x_1)\mathbb{K}_N^{-1}\mathcal{K}_{\Xi_N}(x_2), \quad \text{for all } x_1, x_2 \in \mathbb{X}, \tag{3.149}$$

where $\mathbb{K}_N^{-1} \triangleq (\mathbb{K}_N)^{-1}$ denotes the inverse of the generalized Grammian (3.147), and

$$\mathcal{K}_{\Xi_N}(x) \triangleq \left[\mathcal{K}(x, \xi_1), \ldots, \mathcal{K}(x, \xi_N)\right] \in \mathbb{R}^{m \times mN}, \tag{3.150a}$$

$$= \left[\mathcal{K}_{\xi_1}(x), \ldots, \mathcal{K}_{\xi_N}(x)\right], \quad \text{for all } x \in \mathbb{X}. \tag{3.150b}$$

We define $\mathbf{\Pi}_N : \mathcal{H} \to \mathcal{H}_{d(N)}$ to be *$\mathcal{H}$-orthogonal projection onto $\mathcal{H}_{d(N)} \subseteq \mathcal{H}$*. As in the scalar case, the $\mathcal{H}$-orthogonal projection operator $\mathbf{\Pi}_N$ can be understood as an interpolation operator. Indeed, it follows from Definition 2.51 that

$$\left\langle (I - \mathbf{\Pi}_N)f, \mathcal{K}_{\xi_i}\alpha \right\rangle_{\mathcal{H}} = 0 \tag{3.151}$$

for each $\xi_i \in \Xi_N$ and $\alpha \in \mathbb{U}$. Furthermore, it follows from the reproducing formula (3.107) and (3.113) that

$$\left\langle (I - \mathbf{\Pi}_N)f, E^*_{\xi_i}\alpha \right\rangle_{\mathcal{H}} = 0, \tag{3.152a}$$

$$\left\langle E_{\xi_i}(I - \mathbf{\Pi}_N)f, \alpha \right\rangle_{\mathbb{U}} = 0, \quad \text{for all } \xi_i \in \Xi_N, \alpha \in \mathbb{U}, \tag{3.152b}$$

which imply that so the projection operator interpolates f with $f(\xi_i) = (\mathbf{\Pi}_N f)(\xi_i)$ at each center ξ_i.

The proposed analysis relies on the operator kernel being strictly positive-definite. In general, it is possible to perform a corresponding analysis when the generalized Grammian matrix is only positive semidefinite. In this case, the solution for projection operators, or interpolants, of the approximation problem can be expressed in terms of pseudoinverses of the generalized Grammian. A good discussion about this problem can be found in [27]. Chapter 6 of [3] provides an account of some of the subtleties that can arise in the study of interpolation problems in $\mathbb{U}$-valued native spaces for a general Hilbert space $\mathbb{U}$.

3.8.1 Rates of Convergence on $L^\infty(\Omega, \mathbb{U})$ and the Power Function

The fidelity of approximations obtained using the orthogonal projection $\mathbf{\Pi}_N$ are measured in a few different ways in this chapter. It is most common that the projection error is studied using the vector-valued power function.

Definition 3.19 (*Vector-Valued Power Function,* [27]) Consider the $\mathbb{U}$-valued RKHS given by (3.121) and the finite-dimensional $\mathbb{U}$-valued RKHS given by (3.148). The *vector-valued power function of the subspace $\mathcal{H}_{d(N)} \subseteq \mathcal{H}$ in the direction* $\alpha \in \mathbb{U} \triangleq \mathbb{R}^m$ is defined as

$$\mathcal{P}^{\alpha}_N(x) \triangleq \sqrt{\langle (\mathcal{K}(x,x) - \mathcal{K}_N(x,x))\alpha, \alpha \rangle_{\mathbb{U}}}, \tag{3.153}$$

where $\mathcal{K}(\cdot,\cdot)$ denotes the operator kernel that defines $\mathcal{H}$ and $\mathcal{K}_N(\cdot,\cdot)$ denotes the operator kernel that defines $\mathcal{H}_{d(N)}$.

Remarkably, if the operator-valued kernel $\mathcal{K}$ is bounded uniformly on the diagonal, then $\mathcal{P}^{\alpha}_N$ is bounded. This generalization of the scalar-valued power function (Definition 3.8) allows computing a bound on the approximation error $(I - \mathbf{\Pi}_N)f$ in a vector-valued native space that is analogous to (3.76).

Theorem 3.22 ([27, Corollary 2.11]) *Let $\mathcal{K} : \mathbb{X} \times \mathbb{X} \to \mathcal{L}(\mathbb{U})$ be the kernel that induces the vector-valued RKHS $\mathcal{H}$, where $\mathbb{U} \triangleq \mathbb{R}^m$. Then, a bound on the pointwise projection error is given by*

$$\left|\langle E_x(I-\boldsymbol{\Pi}_N)f,\alpha\rangle_{\mathbb{U}}\right| \le \mathcal{P}_N^{\alpha}(x)\|(I-\boldsymbol{\Pi}_N)f\|_{\mathcal{H}} \tag{3.154a}$$

$$\le \mathcal{P}_N^{\alpha}(x)\|f\|_{\mathcal{H}}, \quad \textit{for all } x\in\mathbb{X},\ f\in\boldsymbol{\mathcal{H}},\ \alpha\in\mathbb{U}. \tag{3.154b}$$

Rates of convergence for samples in the compact set $\Omega \subset \mathbb{X}$ can be obtained by taking the supremum of this pointwise error over Ω. In this case, Theorem 3.22 guarantees that

$$\|E_x(I-\boldsymbol{\Pi}_N)f\|_{\mathbb{U}} \le \sup_{\xi\in\Omega}\overline{\mathcal{P}}_N(\xi)\|(I-\boldsymbol{\Pi}_N)f\|_{\mathcal{H}}, \tag{3.155a}$$

$$\le \sup_{\xi\in\Omega}\overline{\mathcal{P}}_N(\xi)\|f\|_{\mathcal{H}}, \quad \text{for all } f\in\boldsymbol{\mathcal{H}}, \tag{3.155b}$$

with

$$\overline{\mathcal{P}}_N(x) \triangleq \sqrt{\|\mathcal{K}(x,x)-\mathcal{K}_N(x,x)\|}, \quad \text{for all } x\in\Omega. \tag{3.156}$$

If the vector-valued RKHS is defined using the Cartesian product of the same scalar-valued RKHS m-times, then the power function (3.74) can be leveraged. Indeed, if $\boldsymbol{\mathcal{H}} = \mathcal{H}^m$, then

$$\begin{aligned}\|E_x(I-\boldsymbol{\Pi}_N)f\|_{\mathbb{U}} &= \sqrt{\sum_{i=1}^{m}|E_x(I-\Pi_N)f_i|^2},\\ &= \mathcal{P}_N(x)\sqrt{\sum_{i=1}^{m}\|(I-\Pi_N)f_i\|_{\mathcal{H}}^2},\\ &= \mathcal{P}_N(x)\|(I-\boldsymbol{\Pi}_N)f\|_{\mathcal{H}},\\ &\le \mathcal{P}_N(x)\|f\|_{\mathcal{H}},\end{aligned} \tag{3.157}$$

where $\mathbb{U} \triangleq \mathbb{R}^m$ and $f \triangleq [f_1,\ldots,f_m]^{\mathrm{T}}$. In other words, if $\boldsymbol{\mathcal{H}} \triangleq \mathcal{H}^m$, then we can still use the power function $\mathcal{P}_N$ of the scalar-valued space to characterize errors in $\boldsymbol{\mathcal{H}}$.

Recall that the analysis of the rates of convergence of approximations in the scalar-valued space $\mathcal{H}$ using the power function $\mathcal{P}_N$ often employs upper bounds, as in Table 3.3, of the power function $\mathcal{P}_N$ in terms of the fill distance $h_{\Xi_N,\Omega}$ of the samples in Ω. In comparison to the scalar-valued case, to the authors' knowledge, the study of correspondingly general bounds on $\mathcal{P}_N^{\alpha}$ in the vector-valued case has not been yet performed in such a systematic fashion. In general, the study of vector-valued native spaces continues to be a topic much studied in machine learning theory, and the state-of-the-art is not as mature as in the study of the scalar-valued case.

3.8.2 Rates of Convergence in $\boldsymbol{W}^{s,2} \triangleq (W^{s,2}(\Omega))^m$ and the Many-Zeros Theorem

Convergence estimates for approximations based on the vector-valued power function $\mathcal{P}_N^\alpha$ defined in (3.153) are very general. Indeed, these estimates hold for $\mathbb{U}$-valued native spaces generated by any admissible operator-valued kernel $\mathcal{K}$. Similarly to Sect. 3.5.3, in the following, we use the many-zeros theorem (Theorem 2.8), adapted to the vector setting.

We begin by introducing the vector-valued Sobolev spaces $\boldsymbol{W}^{s,p}(\Omega)$ that consists of vector-valued functions $f \triangleq [f_1, \ldots, f_m]^{\mathrm{T}}$, where $f_i \in W^{s,p}(\Omega)$ for $1 \leq i \leq m$. Furthermore, we let the kernel function $\mathcal{K} : \Omega \times \Omega \to \mathcal{L}(\mathbb{U})$, where $\Omega \subset \mathbb{R}^n$, generate the RKHS $\mathcal{H}(\Omega, \mathbb{U})$ with $\mathbb{U} \triangleq \mathbb{R}^m$.

Theorem 3.23 *Suppose that the hypotheses of Theorem 2.8 are verified. If the $\mathbb{U}$-valued RKHS is continuously embedded in the vector-valued Sobolev space $\boldsymbol{W}^{k+s,p}(\Omega)$, that is,*

$$\mathcal{H} \triangleq \mathcal{H}(\Omega, \mathbb{R}^m) \hookrightarrow \boldsymbol{W}^{k+s,p}(\Omega) \triangleq (W^{k+s,p}(\Omega))^m, \tag{3.158}$$

then, there exist constants $C > 0$ and $h_0 > 0$ such that

$$\|f - \Pi_N f\|_{\boldsymbol{W}^{m,q}(\Omega)} \leq C h_{\Xi_N,\Omega}^{k+s-m-d(1/p-1/q)_+} \|f - \Pi_N f\|_{\boldsymbol{W}^{k+s,p}(\Omega)} \tag{3.159a}$$

$$\leq C h_{\Xi_N,\Omega}^{k+s-m-d(1/p-1/q)_+} \|f\|_{\mathcal{H}}, \tag{3.159b}$$

for all Ξ_N such that $h_{\Xi_N,\Omega} > h_0$.

Proof Let the entries of f and $\Pi_N f$ be given by $f \triangleq [f_1, \ldots, f_m]^{\mathrm{T}}$ and $\boldsymbol{\Pi}_N f \triangleq [\Pi_N f_1, \ldots \Pi_N f_m]^{\mathrm{T}}$, respectively. It follows from the scalar-valued many-zeros theorem (Theorem 2.8) that there are constants $C_i > 0$ and $h_{0,i} > 0, i = 1, \ldots, N$, such that

$$|f_i - \Pi_N f_i|_{W^{m,q}(\Omega)} \leq C_i h_{\Omega_N,\Omega}^{k+s-m-d(1/p-1/q)_+} |f_i - \Pi_N f_i|_{W^{k+s,p}(\Omega)}, \tag{3.160}$$

for all Ξ_N with $h_{\Xi_N,\Omega} \leq h_{0,i}$, where $(x)_+ \triangleq x$ if $x \geq 0$ and $(x)_+ \triangleq 0$ if $x < 0$. We add up all the seminorms in (3.160) to deduce (3.159), where the constant C arises from the constants C_i and h_i, the equivalence of norms on $\ell^2(\mathbb{R}^m)$, and the continuous embedding $\mathcal{H} \hookrightarrow \boldsymbol{W}^{k+s,p}(\Omega)$. ∎

Similarly to (3.81), (3.159) allows bounding the projection error in terms of the norm of f, or an upper bound thereof. This powerful result holds for any admissible kernel operator $\mathcal{K}$, even non-diagonal cases, as long as the embedding $\mathcal{H} \hookrightarrow W^{k+s,p}(\Omega)$ holds.

3.9 Concluding Remarks

This chapter presented systematically key topics in RKHS theory. The first part of this chapter has been dedicated to scalar-valued RKHSs, and the second part to vector-valued RKHSs. It is worthwhile to recall that the scope of this book is to present adaptive control systems for nonlinear plants, whose matched nonlinearities are captured by elements of some RKHSs. Thus, in this chapter, relevant properties of RKHSs have been presented and carefully proven.

It is apparent how the proposed discussion on scalar-valued RKHSs is considerably longer and more detailed than that on vector-valued RKHSs. This fact may appear dichotomic to the fact that Chap. 5 primarily introduces the control problem in the presence of vector-valued matched uncertainties. However, it is worthwhile to observe how the vector-valued RKHSs employed in this book are defined as Cartesian products of scalar-valued RKHSs, and, in this case, as discussed in Sects. 3.6 and 3.8, several key properties of scalar-valued RKHSs can be ported to the vector-valued case. Finally, some of the readers knowledgeable with classical adaptive control theory may not be familiar or as familiar with RKHS theory. Thus, the topic has been introduced in a lengthier, tutorial manner in the earlier parts of this chapter.

Although this chapter and Chap. 2 could be considered as a single unit, the next chapter partly disrupts this continuity to present key elements of the stability theory of nonlinear dynamical systems in both infinite and finite dimensions. Such a topic is necessary to address the core topic of this book, which is presented in Chap. 5.

3.10 Homework Problems

For an improved comprehension of the material exposed in this chapter, we recommend the following problems.

Exercise 3.1 Let $\mathfrak{K}_1, \mathfrak{K}_2 : \mathbb{X} \times \mathbb{X} \to \mathbb{R}^+$ denote admissible kernels. Prove that $\alpha\mathfrak{K}_1$ is an admissible kernel for any $\alpha > 0$. Prove that $(\mathfrak{K}_1 + \mathfrak{K}_2)$ is an admissible kernel. △

Exercise 3.2 Let $\mathbb{X}_1$ and $\mathbb{X}_2$ be subsets of the same set $\mathbb{X}$, let $A : \mathbb{X}_1 \to \mathbb{X}_2$ be continuous, and let $\mathfrak{K} : \mathbb{X}_2 \times \mathbb{X}_2 \to \mathbb{R}^+$ be a kernel function. Prove that $\mathfrak{K}(A(\cdot), A(\cdot))$ is a kernel on $\mathbb{X}_1$. △

Exercise 3.3 Let $\mathfrak{K}_1 : \mathbb{X}_1 \times \mathbb{X}_1 \to \mathbb{R}^+$ and $\mathfrak{K}_2 : \mathbb{X}_2 \times \mathbb{X}_2 \to \mathbb{R}^+$ be kernel functions, where $\mathbb{X}_1$ and $\mathbb{X}_2$ are subsets of the same set $\mathbb{X}$. Prove that $(\mathfrak{K}_1, \mathfrak{K}_2)$ is a kernel on $\mathbb{X}_1 \times \mathbb{X}_2$. △

Exercise 3.4 Prove that

$$\mathfrak{K}(x, y) \triangleq d\left(\langle x, y\rangle + c\right)^m \tag{3.161}$$

is a kernel on $\mathbb{R}^n$ for any $x, y \in \mathbb{R}^n$, $d \geq 1$, $m \geq 1$, and $c \geq 0$. △

Exercise 3.5 Prove that

$$\mathfrak{K}(x, y) \triangleq e^{\langle x, y \rangle}, \quad \text{for all } (x, y) \in \mathbb{R}^n \times \mathbb{R}^n, \tag{3.162}$$

is a kernel on $\mathbb{R}^n$. △

Exercise 3.6 Given the set of points $\{x_0, \ldots, x_k\} \subset \mathbb{R}$, known as *nodes*, the *Lagrange basis for polynomials of degree k* is given by the set $\{\ell_0(x), \ldots, \ell_k(x)\}$ such that

$$\ell_j(x) = \prod_{\substack{0 \le k \le m \\ m \ne j}} \frac{x - x_m}{x_j - x_m}, \quad \text{for all } x \in \mathbb{R}. \tag{3.163}$$

Show that the Lagrange basis for polynomials of degree k can be employed for approximation of functions via interpolation as discussed in Sect. 3.5.1. △

Exercise 3.7 The *Legendre polynomial* of degree n can be defined by means of *Rodrigues' formula*

$$P_n(t) \triangleq \frac{1}{2^n n!} \frac{\mathrm{d}^n}{\mathrm{d}t^n} \left(t^2 - 1\right)^n, \quad \text{for all } t \in [-1, 1]. \tag{3.164}$$

Prove that these polynomials are orthogonal to one another over the space $L^2[-1, 1]$ and can be employed for approximation of functions via projection as discussed in Sect. 3.5.1. △

Exercise 3.8 Let $\Omega = [-1, 1] \times [-1, 1] \subset \mathbb{R}^2$. Compute the fill distance for

$$\Xi_N = \left\{\xi_i = (-1)^i [i^{-1}, -i^{-1}]^\mathrm{T}, i = 1, \ldots, N\right\}, \tag{3.165}$$

$$\Xi_N = \left\{\xi_i = \left[\cos(i\alpha^{-1}\pi), \sin(i\alpha^{-1}\pi)\right]^\mathrm{T}, i = 1, \ldots, N\right\}, \tag{3.166}$$

$$\Xi_N = \left\{\xi_i = i^{-1}\left[\cos(i\alpha^{-1}\pi), \sin(i\alpha^{-1}\pi)\right], i = 1, \ldots, N\right\}, \tag{3.167}$$

where $\alpha \in [\pi/5, \pi/4]$. For each of these set of centers, evaluate the function $\mathcal{N}(\cdot)$ for all kernels in Table 3.3 and describe how such function varies. Comment on how the fill distance and $\mathcal{N}(\cdot)$ varies with N. △

Exercise 3.9 Let $\Omega = [-1, 1] \times [-1, 1] \subset \mathbb{R}^2$. Write a computer code that randomly distributed k points in Ω. Run this code 100 times for each k and characterize the fill distance as a function of $k = 2^p$ with $p \in \{0, 1, \ldots, 32\}$. Evaluate the function $\mathcal{N}(\cdot)$ for all kernels in Table 3.3 and describe how such function varies. △

Exercise 3.10 Reproduce the results in Sect. 3.5.4. △

References

1. Berlinet, A., Thomas-Agnan, C.: Reproducing Kernel Hilbert Spaces in Probability and Statistics. Springer, New York, NY (2011)
2. Saitoh, S., Sawano, Y.: Theory of Reproducing Kernels and Applications. Springer, London, UK (2016)
3. Paulsen, V.I., Raghupathi, M.: An Introduction to the Theory of Reproducing Kernel Hilbert Spaces, vol. 152. Cambridge University Press, Cambridge, UK (2016)
4. Williams, C.K., Rasmussen, C.E.: Gaussian Processes for Machine Learning, vol. 2, no. 3. MIT Press, Cambridge, MA (2006)
5. Smola, B.S.A.J.: Learning with Kernels: Support Vector Machines, Regularization, Optimization, and Beyond. MIT Press, Cambridge, MA (2002)
6. Cucker, F., Zhou, D.X.: Learning Theory: An Approximation Theory Viewpoint. Cambridge Press, Cambridge, UK (2007)
7. Thrun, S., Burgard, W., Fox, D.: Probabilistic Robotics. MIT Press, Cambridge, MA (2005)
8. Aronszajn, N.: Theory of reproducing kernels. Trans. Am. Math. Soc. **68**, 337–404 (1950)
9. Brudnyi, A., Brudnyi, Y.: Methods of Geometric Analysis in Extension and Trace Problems, vols. 1, 102. Birkäuser, Baser, Switzerland (2011)
10. Fuselier, E., Wright, G.B.: Scattered data interpolation on embedded submanifolds with restricted positive definite kernels: Sobolev error estimates. SIAM J. Numer. Anal. **50**(3), 1753–1776 (2012)
11. Wendland, H.: Scattered Data Approximation, vol. 17. Cambridge University Press, Cambridge, UK (2004)
12. Schaback, R.: Error estimates and condition numbers for radial basis function interpolation. Adv. Comput. Math. **3**, 251–264 (1994)
13. Narcowich, F.J., Ward, J.D., Wendland, H.: Sobolev error estimates and a Bernstein inequality for scattered data interpolation via radial basis functions. Construct. Approx. **24**(2), 175–186 (2006)
14. Baldassarre, L., Rosasco, L., Barla, A., Verri, A.: Vector field learning via spectral filtering. In: Joint European Conference on Machine Learning and Knowledge Discovery in Databases, pp. 56–71. Springer, Barcellona, Spain (2010)
15. Steinwart, I., Scovel, C.: Mercer's theorem on general domains: on the interaction between measures, kernels, and RKHSs. Construct. Approx. **35**, 363–417 (2012)
16. Smale, S., Zhou, D.-X.: Learning theory estimates via integral operators and their approximations. Construct. Approx. **26**(2), 153–172 (2007)
17. Conway, J.: A Course in Functional Analysis, Services. Graduate Texts in Mathematics. Springer, New York, NY (1994)
18. Steinwart, I.: Convergence types and rates in generic Karhunen-Loeve expansions with applications to sample path properties. Potential Anal. **51**(3), 361–395 (2019)
19. Fischer, S., Steinwart, I.: Sobolev norm learning rates for regularized least-squares algorithms. J. Mach. Learn. Res. **21**(205), 1–38 (2020)
20. Leiter, C., Lertvittayakumjorn, P., Fomicheva, M., Zhao, W., Gao, Y., Eger, S.: Towards explainable evaluation metrics for machine translation. J. Mach. Learn. Res. **25**(75), 1–49 (2024)
21. Carmeli, C., De Vito, E., Toigo, A.: Vector valued reproducing kernel Hilbert spaces of integrable functions and Mercer theorem. Anal. Appl. **4**(04), 377–408 (2006)
22. DeVore, R.A.: Nonlinear approximation. Acta Numerica **7**, 51–150 (1998)
23. Griebel, M., Oswald, P.: Convergence analysis of online algorithms for vector-valued kernel regression (2023). arXiv:2309.07779
24. Kanagawa, M., Hennig, P., Sejdinovic, D., Sriperumbudur, B.K.: Gaussian processes and kernel methods: a review on connections and equivalences (2018). arXiv:1807.02582
25. Rosasco, L., Belkin, M., De Vito, E.: On learning with integral operators. J. Mach. Learn. Res. **11**(2) (2010)

26. DeVore, R.A., Lorentz, G.G.: Constructive Approximation. Springer, Berlin, Germany (1993)
27. Wittwar, D., Santin, G., Haasdonk, B.: Interpolation with uncoupled separable matrix-valued kernels. Dolomites Res. Notes Approx. **11**, 23–39 (2018)
28. Narcowich, F., Ward, J., Wendland, H.: Sobolev bounds on functions with scattered zeros, with applications to radial basis function surface fitting. Math. Comput. **74**(250), 743–763 (2005)
29. Hangelbroek, T., Narcowich, F.J., Ward, J.D.: Kernel approximation on manifolds I: bounding the Lebesgue constant. SIAM J. Math. Anal. **42**(4), 1732–1760 (2010)
30. Hangelbroek, T., Narcowich, F., Ward, J.: Polyharmonic and related kernels on manifolds: interpolation and approximation. Found. Comput. Math. **12**, 625–670 (2012)
31. Maier, L.-B.: Ambient approximation on embedded submanifolds. Construct. Approx. **52**(2), 183–211 (2020)
32. Hangelbroek, T., Rieger, C.: Extending error bounds for radial basis function interpolation to measuring the error in higher order sobolev norms. Math. Comput. (2024)
33. Carmeli, C., De Vito, E., Toigo, A., Umanitá, V.: Vector valued reproducing kernel Hilbert spaces and universality. Anal. Appl. **8**(01), 19–61 (2010)
34. Barbian, C.: A characterization of multiplication operators on reproducing kernel Hilbert spaces. J. Oper. Theory 235–240 (2011)

Chapter 4
Elements of Dynamical Systems Theory

In this chapter, we present some fundamental results on dynamical systems and their relationships to well-posed ordinary differential equations (ODEs), and a few relevant notions concerning distributed parameter systems (DPS). As discussed in the introduction of this book, the general method of native space embedding for ordinary differential equations (ODEs) that is discussed in Chap. 5 proceeds in two parts. Initially, a "limiting system" is defined, which evolves in an infinite-dimensional state space. Subsequently, realizable control schemes are obtained by consistent approximation of the limiting system in finite dimensions. The theory of dynamical systems for ODEs presented in this chapter is used to study the systems that result from finite-dimensional approximations. A relatively smaller collection of results concerning the limiting DPS, which evolves in a generally infinite-dimensional configuration space, is presented.

This chapter is structured as follows. The notions of solution of ODEs and sufficient conditions for the existence of these solutions are presented in Sects. 4.1 and 4.2. The concepts of stability and attractivity, and, hence, asymptotic stability of equilibrium points of ODEs are presented in Sect. 4.3.1, and the concepts of boundedness and ultimate boundedness are discussed in Sect. 4.3.2. Lyapunov-like sufficient conditions for these desirable properties of equilibrium points and trajectories are outlined in Sects. 4.3.3–4.3.6. Section 4.4 presents control synthesis tools, namely control Lyapunov functions and robust control Lyapunov functions. The theory for infinite-dimensional state spaces in Sect. 4.5 builds upon concepts discussed in Chaps. 2 and 3, such as the C_0-semigroup, and are used to study the limiting DPS. This chapter is then concluded by Sect. 4.6, which collects some final remarks, and Sect. 4.7, which suggests some homework problems.

This chapter is not intended to comprehensively summarize theory and practice for finite and infinite-dimensional dynamical systems. The intent is to present coherently numerous concepts that are strategic to the scopes of this book in general and Chap. 5 in particular. Although throughout this chapter, we provide the reader with references

A. J. Kurdila et al., *Data-Driven, Nonparametric, Adaptive Control Theory*,
Lecture Notes in Control and Information Sciences 495,
https://doi.org/10.1007/978-3-031-78003-5_4

to specific discussions and examples in classical textbooks on nonlinear systems analysis, we encourage the interested readers to familiarize themselves with classical textbooks on ODEs, systems analysis, control synthesis in general, and adaptive control in particular such as such as [1–8]. The interested reader is also invited to consult additional, more advanced texts such as [9–22].

4.1 Well-Posedness of Ordinary Differential Equations

In this book, we primarily consider two classes of problems. The first class of problems is the *nonautonomous initial value problem (IVP)*

$$(IVP) \quad \begin{cases} (\Sigma) \;\; \dot{x}(t) = f(t, x(t)), \\ (IC) \; x(t_0) = x_0, \end{cases} \tag{4.1}$$

where $f : \mathbb{R} \times \Omega \to \mathbb{R}^n$, $\Omega \subseteq \mathbb{R}^n$ is open and connected, $x_0 \in \Omega$ denotes the *initial condition*, and $t_0 \geq 0$ denotes the *initial time*. In some cases, nonautonomous IVPs are also termed *time-varying IVPs* since time appears explicitly in the ODE governing the IVP and, in general, different initial times t_0 produce different solutions.

If $f(t, x) = f(x)$ is independent of t, then we consider the *autonomous IVP*

$$(IVP) \quad \begin{cases} (\Sigma) \;\; \dot{x}(t) = f(x(t)), \\ (IC) \; x(0) = x_0, \end{cases} \tag{4.2}$$

where $f : \Omega \to \mathbb{R}^n$ is the vector field on Ω, $\Omega \subseteq \mathbb{R}^n$ is open and connected, and $x_0 \in \Omega$. Note that, since f is time-independent, the initial time is set to 0. In some cases, autonomous IVPs are also termed *time-invariant IVPs*.

Usually, the solution of (4.1) (or (4.2)) is defined as a differentiable function $x(t)$ that satisfies (Σ) on an interval $\mathcal{I} \subseteq [t_0, \infty)$ containing t_0 and satisfies (IC). This is called a *classical* (or *strong*) *solution* to the IVP. If $f : \mathbb{R} \times \Omega \to \mathbb{R}^n$ is continuous in (t, x) and $x(\cdot)$ is a classical solution to the IVP (4.1), then $x(\cdot)$ satisfies the integral equation

$$x(t) = x_0 + \int_{t_0}^{t} f(s, x(s))\, ds, \quad \text{for all } t \in \mathcal{I}, \tag{4.3}$$

and the converse is true. An analogous integral equation can be defined for (4.2).

To deal with control systems, we need to allow for functions $f(t, x)$ that are not continuous in the pair (t, x). Thus, we need a *weaker* (*extended*) solution to the IVP. The following extension of the concept of a solution is due to Carathéodory and is discussed in [23–25]. For the statement of this definition, it is worthwhile recalling the notions of absolutely continuous functions and Lebesgue integrable functions presented in Sects. 2.4.3 and 2.4.5, respectively. The notions of Carathéodory conditions and Carathéodory solutions can be extended to the IVP (4.2); this exercise is left to the reader.

Definition 4.1 (*Carathéodory Conditions*) Let $D \subseteq \mathbb{R} \times \mathbb{R}^n$ be an open set. The function $f : D \to \mathbb{R}^n$ satisfies the *Carathéodory conditions on D* if

(C1) for $(t, \hat{x}) \in D$ and $\hat{x}$ fixed, the function $t \mapsto f(t, \hat{x})$ is measurable in t,
(C2) for $(\hat{t}, x) \in D$ and $\hat{t}$ fixed, the function $x \mapsto f(\hat{t}, x)$ is continuous in x, and
(C3) for each compact subset U of D, there exists a nonnegative real-valued function $m_U(\cdot)$, which is Lebesque integrable on U, and such that

$$\|f(t, x)\| \leq m_U(t), \quad \text{for all } (t, x) \in U, \tag{4.4}$$

almost everywhere in t.

Assuming the Carathéodory conditions hold on D, one can define what it means for $x(\cdot)$ to be a solution of the IVP given by (4.1).

Definition 4.2 (*Carathéodory Solutions*) Consider the IVP (4.1) and assume that f verifies the Carathéodory conditions. A *Carathéodory solution* to the IVP (4.1) is an absolutely continuous function $x : \mathcal{I} \to \mathbb{R}^n$ such that

(S1) $\mathcal{I} \subseteq [t_0, \infty)$ is a connected interval,
(S2) $t_0 \in \mathcal{I}$,
(S3) $x(t_0) = x_0$, and
(S4) Equation (4.3) is satisfied.

It is worthwhile recalling that there are other classes of solutions to IVPs, such as Filippov and Krasovskii solutions, or those generated by differential inclusions. These notions provide a rigorous account of solutions when the right-hand side $f(t, x)$ is even discontinuous in x or t. However, considering the scope of this book, we employ Carathéodory solutions in the study of all IVPs with some notable exception, which is duly noted. For this reason, in the remainder of this book, we refer to Carathéodory solution simply as solutions of IVPs.

Example 4.1 Consider the differential equation

$$\dot{x}(t) = [x(t)]^2, \quad x(0) = 1. \tag{4.5}$$

If $t \geq 0$, then any function of the form $\varphi(t) = \frac{1}{(1-t)}$ satisfies the differential equation. Thus, one might claim that $\varphi(t) = \frac{1}{(1-t)}$ with domain $D(\varphi) = \mathbb{R}^+ \setminus \{1\}$ is a solution to the IVP since $\varphi(\cdot)$ also satisfies $\varphi(0) = 1$. However, $D(\varphi)$ is not a connected interval so, by definition, $\varphi(\cdot)$ is not a solution to the IVP involving the ODE (4.5). Although the function $x(t) = \varphi(t), t > 1$, is a solution to the ODE, it does not satisfy the initial condition $x(0) = 1$ since $0 \notin (1, \infty)$. Finally, the function $x(t) = \varphi(t)$, $t < 1$, is the unique solution to the IVP with maximal interval of existence given by the connected interval $\mathcal{I}_{\max} = (-\infty, 1)$. △

Remark 4.1 Observe that a solution $x(t)$ is absolutely continuous (see Sect. 2.4.3) and, hence, is differentiable except on a set of measure zero. Thus, $\dot{x}(t)$ exists almost everywhere (see Sect. 2.4.6) and

$$\dot{x}(t) = f(t, x(t)), \quad t \in \mathcal{I} \text{ a.e.} \tag{4.6}$$

If $f(t, x)$ is continuous in t, then $\dot{x}(t)$ exists everywhere and $\dot{x}(t)$ is continuous. Therefore, in this case, $x(t)$ is a classical solution to the IVP. Also, it is important to note that, by definition, solutions are only defined on *connected* time intervals $\mathcal{I}$.

Local existence, uniqueness, and continuous dependence on initial data follow from Picard–Lindelöf type existence and uniqueness theorems for problems satisfying the Carathéodory conditions. In particular, if $f(t, x)$ satisfies a local Lipschitz condition, then a local solution exists, is unique, and can be extended to a maximal interval of existence; [24, Theorem 5.3] and [25, Theorem 3.9] are typical results. In the following, we recall one of the many results on the existence and uniqueness of solutions of IVPs. For the statement of this result, we write $x(t; t_0, x_0)$ instead of $x(t)$ to emphasize the dependence of solutions of the nonautonomous IVP (4.1) on both the initial time and initial conditions. In the case of the autonomous IPV (4.2), we write $x(t; x_0)$ instead of $x(t)$. Furthermore, we specialize the notion of the Lipschitz continuous function introduced in Definition 2.60.

Definition 4.3 (*Locally Lipschitz Continuous Functions*) Let $f : D \to \mathbb{R}^n$ and $D \subseteq \mathbb{R} \times \mathbb{R}^n$ be connected and such that $D = \mathcal{J} \times \Omega$, where $\mathcal{J} \subseteq \mathbb{R}$ is an interval and $\Omega \subseteq \mathbb{R}^n$. If, for each compact subset U of D, there exists a $c_U : \mathcal{J} \to \mathbb{R}^+$ that is Lebesque integrable on $\mathcal{J}$ and such that, for all $x, y \in \Omega$,

$$\|f(t, x) - f(t, y)\| \leq c_U(t)\|x - y\|, \quad t \in \mathcal{J} \text{ a.e.}, \tag{4.7}$$

then f is *locally Lipschitz continuous*.

Remark 4.2 (Comment on Terminology) In several publications, the notion of local Lipschitz continuity is referred to as Lipschitz continuity. In this book, to emphasize the fact that Definition 4.3 applies locally on compact sets, we explicitly distinguish between local and global Lipschitz continuity. This is the same convention as in popular treatises like [8].

Theorem 4.1 ([24, Theorem 5.3]) *Consider the IVP defined by* (4.1). *Suppose that* $f : D \to \mathbb{R}^n$ *satisfies the Carathéodory conditions on an open subset* $D \subseteq \mathbb{R} \times \mathbb{R}^n$ *and is locally Lipschitz continuous. Then, for any* $(t_0, x_0) \in D$, *there exists a unique solution* $x(t, t_0, x_0)$ *defined on an open connected set* $E \subseteq \mathbb{R} \times D$, *where* $\{t_0\} \times D \subset E$, *and* $x(t; t_0, x_0)$ *is continuous on* E.

Theorem 4.1 only guarantees the existence of solutions, locally in time. Thus, for each $(t_0, x_0) \in D$, solutions of the IVP may be extended to a maximum interval of existence $[t_0, t_{\max})$, where $t_{\max} \in (t_0, \infty)$. As shown by Example 4.1, some IVPs are defined on a proper subinterval of $[t_0, \infty)$ that contains t_0, and, hence, have a *forward finite escape time*. Existence of solutions on the semi-infinite time interval $[t_0, \infty)$ requires that additional properties be placed on $f(t, x)$. The following result, which is a direct consequence of [26, Theorem 3.1], provides sufficient conditions for the existence of solutions of an IVP on $[t_0, \infty)$.

Theorem 4.2 *Consider the IVP given by* (4.1)*. Suppose that* $f : D \to \mathbb{R}^n$ *satisfies the Carathéodory conditions on* $[t_0, \infty) \times \Omega$ *and is Lipschitz continuous. If there exist continuous functions* $\alpha, \beta : (t_0, \infty) \to \mathbb{R}_+$ *such that*

$$\langle f(t, x), x\rangle_{\mathbb{R}^n} \leq \alpha(t)\langle x, x\rangle_{\mathbb{R}^n} \text{ for all } t > t_0 \text{ and } x \in \Omega \text{ with } \|x - x_0\| > \beta(t), \tag{4.8}$$

then, for this value of t_0*, the solution* $x(t; t_0, x_0)$ *to* (4.1) *exists for all* $t \in [t_0, \infty)$.

Definition 4.4 (*Well-Posed IVPs*) The IVP problem (4.1) is *well-posed on* $[t_0, \infty)$ (or *forward complete on* $[t_0, \infty)$) if, for all $t_0 \in \mathbb{R}$ and $x_0 \in \Omega$, there is a unique solution $x(t; t_0, x_0)$ to the initial value problem (4.1) such that its maximal interval of existence contains $[t_0, \infty)$ and this solution depends continuously on the initial data (t_0, x_0).

In light of Definition 4.4, Theorem 4.2 provides sufficient conditions for the well-posedness of the IVP (4.1). As discussed in Sect. 2.2.4, all norms on $\mathbb{R}^n$ are equivalent. Therefore, the sufficient conditions provided by Theorem 4.2 are independent of the norm chosen to verify these sufficient conditions. The relevance of a result such as Theorem 4.2 is not to be underestimated. In the presence of a finite escape time, questions about the long-term behavior of solutions of IVPs, such as stability or ultimate boundedness, which are essential in this book as well as in substantially all the literature on control systems design, do not make sense. This is a reason why we are interested in IVPs that are well-posed on $[t_0, \infty)$ or forward complete.

Remark 4.3 (Choice of the State Space) So far, we have focused on ODEs whose dynamics evolve in the finite-dimensional space $\Omega \subseteq \mathbb{R}^n$. In this case, the *state space* is defined to be $\mathbb{X} \triangleq \mathbb{R}^n$. Although $\mathbb{X} \triangleq \mathbb{R}^n$ is a "natural" choice for the state space for ODE systems, there may be more than one suitable state space. To be useful as a model of a physical system, well-posedness requires that, for all physically relevant initial conditions x_0, the IVP has a unique solution and the solution depends continuously on (t_0, x_0). Consequently, the state space should contain all $x \in \mathbb{R}^n$. The issue of continuous dependence implies that one must also impose a metric (or norm) on the space $\mathbb{X}$. As noted, for ODE systems, well-posedness is independent of the norm on $\mathbb{R}^n$. For partial and delay differential equations and the limiting systems studied in Chap. 5, the choice of a suitable state space is not obvious. Moreover, there can be more than one practical choice of the state space which depends on the application to be addressed. In the following, we briefly touch on this issue when discussing infinite-dimensional processes and dynamical systems. A detailed discussion of this topic, with several practical examples, can be found in [26, Sect. 3.4].

4.2 Processes and Dynamical Systems on $\mathbb{R}^n$

In this section, we discuss how well-posed initial value problems define processes and dynamical systems on $\mathbb{R}^n$. In particular, a nonautonomous IVP gives rise to a process, while an autonomous IVP gives rise to a dynamical system.

Definition 4.5 (*Process over* $\mathbb{R}^n$) A *process* on $\mathbb{R}^n$ is a mapping $x : [t_0, \infty) \times \mathbb{R} \times \mathbb{R}^n \to \mathbb{R}^n$ such that

(P1) $x(t_0, t_0, x_0) = x_0$ for all $x_0 \in \mathbb{R}^n$,
(P2) $x(\tau + s, t_0, x_0) = x(\tau, s + t_0, x(s, t_0, x_0))$ for all $\tau \in [t_0, \infty)$, $s \in [0, \infty)$,
(P3) $x : (\cdot, t_0, x_0) : [t_0, \infty) \to \mathbb{R}^n$ is continuous (right continuous at $\tau = t_0$), and
(P4) $x(\tau, \cdot, \cdot) : \mathbb{R} \times \mathbb{R}^n \to \mathbb{R}^n$ is continuous for each $\tau \in [t_0, \infty)$.

Processes over $\Omega \subseteq \mathbb{R}^n$ generated by the IVP (4.1) can be computed by leveraging (4.3). Assume that the nonautonomous IVP problem defined by (4.1) is well-posed on $[t_0, \infty)$ with solution $x(t; t_0, x_0)$ for every initial condition $x_0 \in \Omega \subseteq \mathbb{R}^n$. The function $x : [0, \infty) \times \mathbb{R} \times \mathbb{R}^n \to \mathbb{R}^n$ defined for all $\tau \geq 0$ by

$$x(\tau, t_0, x_0) \triangleq x(\tau + t_0, t_0, x_0) = x_0 + \int_{t_0}^{t_0+\tau} f\left(\mu, x(\mu, t_0, x_0)\right) \mathrm{d}\mu \tag{4.9}$$

is the *process over* $\mathbb{R}^n$ *generated by the IVP* (4.1).

Remark 4.4 (Comment on Notation) Processes do not need to be defined by an IVP. However, in this book, we only focus on processes defined by an IVP. For this reason, consistently with the overwhelming majority of books in nonlinear systems analysis and control synthesis, we conflate the notion of Carathéodory solution of an IVP and processes generated by an IVP.

Definition 4.6 (*Dynamical System over* $\mathbb{R}^n$) A dynamical system on $\mathbb{R}^n$ is a mapping $x : [0, \infty) \times \mathbb{R}^n \to \mathbb{R}^n$ such that

(D1) $x(0, z) = z$ for all $z \in \mathbb{R}^n$,
(D2) $x(t + s, z) = x(t, x(s, z))$ for all $t, s \in [0, \infty)$,
(D3) $x : (\cdot, z) : [0, \infty) \to \mathbb{R}^n$ is continuous (right continuous at $t = 0$), and
(D4) $x(t, \cdot) : \mathbb{R}^n \to \mathbb{R}^n$ is continuous.

Assume that IVP problem defined by (4.2) is *well-posed on* $[0, \infty)$ with solution $x(t, x_0)$ for all initial conditions $x_0 \in \Omega \subseteq \mathbb{R}^n$. Since

$$x(t; x_0) = x_0 + \int_0^t f\left(x(\mu, x_0)\right) d\mu, \quad \text{for all } t \geq 0, \tag{4.10}$$

the function $x : [0, \infty) \times \mathbb{R}^n \to \mathbb{R}^n$ is the *dynamical system over* $\mathbb{R}^n$ *generated by the IVP* (4.2). In this sense, a dynamical system is a special process.

In some references, the study of dynamical systems is carried out using different, but ultimately equivalent, language. Some references use the language of semigroups to describe dynamical systems; we summarize this approach in Sect. 4.5.1 below. Such an approach is taken in [14, 18], for example, and a detailed account of the equivalence of these descriptions can be found in [26].

Remark 4.5 (Comment on Notation) In this book, to express that a Carathéodory solution of an IVP over the time interval $[t_0, \infty)$ is sought, we employ a short-hand notation so that problems in the same form as (4.1) are written as

$$\dot{x}(t) = f(t, x(t)), \quad x(t_0) = x_0, \quad t \geq t_0, \tag{4.11}$$

and problems in the same form as (4.2) are written as

$$\dot{x}(t) = f(x(t)), \quad x(0) = x_0, \quad t \geq 0. \tag{4.12}$$

Furthermore, consistently with substantially all books in nonlinear systems analysis and control synthesis, we refer to these IVPs as *dynamical systems*, hence conflating the notions of IVP, processes, and dynamical systems.

In the remainder of this book, the working hypotheses are that $t \mapsto f(\cdot, x)$ is continuous for all $x \in \mathbb{R}^n$ and both $x \mapsto f(t, \cdot)$ and $x \mapsto f(\cdot)$ are locally Lipschitz continuous over compact sets for all $t \in [t_0, \infty)$. There are only a few cases where this is well known not to be the case, such as dead-zone and variable structure methods. We comment on the additional technical and theoretical challenges that these methods can give rise to as they appear.

4.3 Lyapunov Stability, Boundedness, and Ultimate Boundedness

In this section, we present several classes of dynamical systems, namely Lyapunov stable, asymptotically stable, bounded, and uniformly ultimately dynamical systems. These types of systems are particularly popular because capture different desirable aspects of a "well-behaved" dynamical system.

4.3.1 Stability of Motion in $\mathbb{R}^n$

Engineering processes captured by means of ODEs often study motion near a fixed point, that is, a user-defined set condition. The idea is that if the process is moved away from this set point by a small perturbation, then it should not move too far from it. If possible, the process should also return to this set point. These fundamental, yet practical, ideas are captured by the following definitions.

Definition 4.7 (*Equilibrium Point*) Consider the nonlinear time-varying dynamical system (4.11). Any point $x_e \in \Omega \subseteq \mathbb{R}^n$ such that

$$0 = f(t, x_e), \qquad t \in [t_0, \infty) \tag{4.13}$$

is an *equilibrium point* of (4.11). Similarly, consider the nonlinear time-invariant dynamical system (4.12). Any point $x_e \in \Omega \subseteq \mathbb{R}^n$ such that

$$0 = f(x_e) \tag{4.14}$$

is an *equilibrium point* of (4.12).

Definition 4.8 (*Lyapunov Stability (Nonautonomous Systems)*) Consider the nonlinear time-varying dynamical system (4.11). The equilibrium point $x_e \in \Omega \subseteq \mathbb{R}^n$ is *Lyapunov stable* if for each $\varepsilon > 0$ such that $\mathcal{B}_\varepsilon(x_e) \subseteq \Omega$, there exists $\delta(\varepsilon, t_0) > 0$ such that if $\|x_0 - x_e\| < \delta(\varepsilon, t_0)$, then $\|x(t) - x_e\| < \varepsilon, t \geq t_0$.

It is important to notice how Lyapunov stability of an equilibrium point requires the existence of a solution of (4.11) for all $t \in [t_0, \infty)$. Indeed, a key step in the proofs of sufficient conditions to verify Lyapunov stability, such as those listed in Sect. 4.3.1 below, is the proof of the existence of solutions for all $t \geq t_0$.

Definition 4.9 (*Uniform Lyapunov Stability*) Consider the nonlinear time-varying dynamical system (4.11). The equilibrium point $x_e \in \mathbb{R}^n$ is *uniformly Lyapunov stable* if for each $\varepsilon > 0$ such that $\mathcal{B}_\varepsilon(x_e) \subseteq \Omega$, there exists $\delta(\varepsilon) > 0$ such that if $\|x_0 - x_e\| < \delta(\varepsilon)$, then $\|x(t) - x_e\| < \varepsilon, t \geq t_0$, for all $t_0 \geq 0$.

The classical ε-δ argument presented in Definitions 4.8 and 4.9 can be interpreted as continuity of solutions of (4.11) with respect to the initial condition x_0. In light of the semigroup property (i.e., (P4) in Definition 4.5), this notion also captures the idea whereby, to ensure a small displacement from this condition at times, the displacement from the equilibrium condition needs to be sufficiently small at t_0.

The only difference between Definition 4.8 and Definition 4.9 lies in that the former allows δ to be a function of the initial time $t_0 \geq 0$ and the latter does not. As discussed in [7, Chap. 4], non-uniform Lyapunov stability may be an overly weak condition for some problems of practical interest. As already noted, autonomous systems can be considered as a special case of nonautonomous systems. Thus, the following definition is easily justifiable in the light of Definition 4.9.

Definition 4.10 (*Lyapunov Stability (Autonomous Systems)*) Consider the nonlinear time-invariant dynamical system (4.12). The equilibrium point $x_e \in \mathbb{R}^n$ is *Lyapunov stable* if for each $\varepsilon > 0$, there exists $\delta(\varepsilon) > 0$ such that if $\|x_0 - x_e\| < \delta(\varepsilon)$, then $\|x(t) - x_e\| < \varepsilon, t \geq 0$. If all equilibrium points of (4.12) are Lyapunov stable, then (4.12) is Lyapunov stable.

Not all equilibrium points are Lyapunov stable. Indeed, if x_e is not a Lyapunov stable equilibrium point for (4.11) or (4.12), then x_e is an *unstable* equilibrium point. Next, we define the property whereby trajectories of a dynamical system converge to an equilibrium point.

Definition 4.11 (*Asymptotic Attractivity (Nonautonomous Systems)*) Consider the nonlinear time-invariant dynamical system (4.11). The equilibrium point $x_e \in \Omega \subseteq \mathbb{R}^n$ is *asymptotically attractive* if for every $t_0 \geq 0$, there exists $\delta(t_0) > 0$ such that $\|x_0 - x_e\| < \delta(t_0)$ implies that $\lim_{t\to\infty} \|x(t) - x_e\| = 0$. Furthermore, $x_e \in \Omega \subseteq \mathbb{R}^n$ is *uniformly asymptotically attractive* if there exists $\delta > 0$ such that $\|x_0 - x_e\| < \delta$ implies that $\lim_{t\to\infty} \|x(t) - x_e\| = 0$ uniformly in $t_0 \geq 0$. If $\Omega = \mathbb{R}^n$ and the equilibrium point $x_e \in \mathbb{R}^n$ is uniformly asymptotically attractive for all $x_0 \in \mathbb{R}^n$, then x_e is *globally asymptotically attractive*.

Definition 4.12 (*Asymptotic Attractivity (Autonomous Systems)*) Consider the nonlinear time-invariant dynamical system (4.12). The equilibrium point $x_e \in \Omega \subseteq \mathbb{R}^n$ is *asymptotically attractive* if there exists $\delta > 0$ such that $\|x_0 - x_e\| < \delta$ implies that $\lim_{t\to\infty} \|x(t) - x_e\| = 0$. If $\Omega = \mathbb{R}^n$ and the equilibrium point $x_e \in \mathbb{R}^n$ is asymptotically attractive for all $x_0 \in \mathbb{R}^n$, then x_e is *globally asymptotically attractive*.

Although Lyapunov stability and attractivity of some equilibrium points may seem equivalent concepts, the two notions are not correlated. The two-dimensional dynamical system

$$\dot{x}_1(t) = \frac{x_1^2(t)\,(x_2(t) - x_1(t)) + x_2^5(t)}{\left(x_1^2(t) + x_2^2(t)\right)\left[1 + \left(x_1^2(t) + x_2^2(t)\right)^2\right]}, \quad x_1(0) = x_{10}, \quad t \geq 0, \tag{4.15a}$$

$$\dot{x}_2(t) = \frac{x_2^2(t)\,(x_2(t) - 2x_1(t))}{\left(x_1^2(t) + x_2^2(t)\right)\left[1 + \left(x_1^2(t) + x_2^2(t)\right)^2\right]}, \quad x_2(0) = x_{20} \tag{4.15b}$$

provides a classical example of a system whose trajectories asymptotically converge to the equilibrium point $x_e = 0$ for all $x_0 \in \mathbb{R}^2$, but $x_e = 0$ is an unstable equilibrium point.

Definition 4.13 (*Asymptotic Stability (Nonautonomous Systems)*) Consider the nonlinear time-varying dynamical system (4.11). The equilibrium point $x_e \in \Omega \subseteq \mathbb{R}^n$ is *asymptotically stable* if it is both Lyapunov stable and asymptotically attractive. If all equilibrium points of (4.11) are asymptotically stable, then (4.11) is asymptotically stable. If $\Omega = \mathbb{R}^n$ and the equilibrium point $x_e \in \mathbb{R}^n$ is both Lyapunov stable and globally asymptotically attractive, then x_e is *globally asymptotically stable*.

Definition 4.14 (*Uniform Asymptotic Stability*) Consider the nonlinear time-varying dynamical system (4.11). The equilibrium point $x_e \in \mathbb{R}^n$ is *uniformly asymptotically stable* if it is both uniformly Lyapunov stable and uniformly asymptotically attractive. If all equilibrium points of (4.11) are uniformly asymptotically stable, then (4.11) is *uniformly asymptotically stable*. If $\Omega = \mathbb{R}^n$ and the equilibrium point $x_e \in \mathbb{R}^n$ is both Lyapunov stable and globally uniformly asymptotically attractive, then x_e is *globally uniformly asymptotically stable*.

Definition 4.15 (*Asymptotic Stability (Autonomous Systems)*) Consider the nonlinear time-invariant dynamical system (4.12). The equilibrium point $x_e \in \mathbb{R}^n$ is *asymptotically stable* if it is both Lyapunov stable and asymptotically attractive. If all equilibrium points of (4.12) are asymptotically stable, then (4.12) is *asymptotically stable*. If $\Omega = \mathbb{R}^n$ and the equilibrium point $x_e \in \mathbb{R}^n$ is both Lyapunov stable and globally asymptotically attractive, then x_e is *globally asymptotically stable*.

4.3.2 Boundedness and Ultimate Boundedness in $\mathbb{R}^n$

Aiming for Lyapunov or asymptotic stability of an equilibrium point or the entire system could be an overly ambitious goal, especially in the presence of external disturbances or other effects, which are not quantifiable at each time instant. For this reason, weaker conditions are commonly considered in the analysis of IVPs. In the following, we recall several variations of the notion of boundedness, that is, the property whereby small perturbations in the initial conditions induce small displacements of the system's trajectory from the equilibrium point at all times.

Definition 4.16 (*Uniform Boundedness*) Consider the nonlinear time-varying dynamical system given by (4.11) and let $x_e \in \Omega \subseteq \mathbb{R}^n$. The solutions of (4.11) are *uniformly bounded in a neighborhood of* x_e if there exists a constant $c > 0$, independent of $t_0 \geq 0$, and for every $a \in (0, c)$, there exists $\beta(a) > 0$, independent of t_0, such that if $\|x_0 - x_e\| < a$, then $\|x(t) - x_e\| < \beta(a)$ for all $t \geq t_0$.

Definition 4.17 (*Boundedness*) Consider the nonlinear time-invariant dynamical system given by (4.12) and let $x_e \in \Omega \subseteq \mathbb{R}^n$. The solutions of (4.12) are *bounded in a neighborhood of* x_e if there exists a constant $c > 0$ and for every $a \in (0, c)$, there exists $\beta(a) > 0$ such that if $\|x_0 - x_e\| < a$, then $\|x(t) - x_e\| < \beta(a)$ for all $t \geq 0$.

The notion of non-uniform boundedness can be extended to the time-varying dynamical system (4.11). However, this notion is not employed in this book and, hence, omitted. Definitions 4.16 and 4.17 can also be extended to any arbitrarily large neighborhood of x_e.

Definition 4.18 (*Global Uniform Boundedness*) Consider the nonlinear time-varying dynamical system given by (4.11) with $\Omega = \mathbb{R}^n$, and let $x_e \in \mathbb{R}^n$. The system (4.11) is *globally uniformly bounded* if for any $c > 0$, independent of $t_0 \geq 0$, there exists $\beta(c) > 0$, independent of t_0, such that if $\|x_0 - x_e\| < c$, then $\|x(t) - x_e\| < \beta(c)$ for all $t \geq t_0$.

Definition 4.19 (*Global Boundedness*) Consider the nonlinear autonomous dynamical system given by (4.12) with $\Omega = \mathbb{R}^n$, and let $x_e \in \mathbb{R}^n$. The system (4.12) is *globally bounded* if for any $c > 0$, there exists $\beta(c) > 0$ such that if $\|x_0 - x_e\| < c$, then $\|x(t) - x_e\| < \beta(c)$ for all $t \geq 0$.

Definitions 4.16–4.19 capture the notions of boundedness for dynamical systems. These notions are not to be confused with the notion of bounded functions presented in Sect. 2.4.6. In some applications, it may suffice to verify that if the system's initial condition is sufficiently close to some equilibrium point, then the system's trajectory is bounded after some finite time. Thus, Definition 4.16 can be relaxed as follows.

Definition 4.20 (*(Global) Uniform Ultimate Boundedness*) Consider the nonlinear time-varying dynamical system given by (4.11) and let $x_e \in \Omega$. The solutions of (4.11) are *uniformly ultimately bounded in a neighborhood of* x_e *with bound* b if there exist constants $b, c > 0$, independent of $t_0 \geq 0$, and for every $a \in (0, c)$, there exists $T(a, b) \geq t_0$, independent of t_0, such that if $\|x_0 - x_e\| < a$, then $\|x(t) - x_e\| < b$ for all $t \geq T(a, b)$. Finally, let $\Omega = \mathbb{R}^n$. If for any $c > 0$, independent of $t_0 \geq 0$, there exist both a constant $b > 0$ and $T(c, b) \geq t_0$, both independent of t_0, such that if $\|x_0\| < c$, then $\|x(t)\| < b$ for all $t \geq T(c, b)$, then the solutions of (4.11) are *globally uniformly ultimately bounded with bound* b.

Definition 4.21 (*(Global) Ultimate Boundedness*) Consider the nonlinear autonomous dynamical system given by (4.12) and let $x_e \in \mathbb{R}^n$. The solutions of (4.12) are *ultimately bounded in a neighborhood of* x_e if there exist constants $b, c > 0$ and for every $a \in (0, c)$, there exists $T(a, b) \geq 0$ such that if $\|x_0 - x_e\| < a$, then $\|x(t) - x_e\| < b$ for all $t \geq T(a, b)$. Finally, let $\mathbb{R}^n = \mathbb{R}^n$. If for any $c > 0$ there exist both a constant $b > 0$ and $T(c, b) \geq 0$ such that if $\|x_0\| < c$, then $\|x(t)\| < b$ for all $t \geq T(c, b)$, then the solutions of (4.12) are *globally ultimately bounded.*

Similar to the notion of Lyapunov stability of an equilibrium point and its more complex variations, the concepts of boundedness and ultimate boundedness presented herein require the existence of solutions for all $t \geq t_0 \geq 0$. A key step in the proofs of the sufficient conditions for boundedness and ultimate boundedness, such as those listed in Sect. 4.3.6 below, is the proof of the existence of solutions for all $t \geq t_0$.

Remark 4.6 (Convention on Equilibrium Points) It is common practice to assume, without loss of generality, that $x_e = 0$ is an equilibrium point of the dynamical systems (4.11) and (4.12), and we focus on the analysis of this equilibrium point only. This assumption can be easily verified by replacing x with $x - x_e$ and redefining $f(\cdot, \cdot)$ or $f(\cdot)$ accordingly. In light of this consideration, we always assume that $0 \in \Omega$.

4.3.3 Sufficient Conditions for Lyapunov Stability in $\mathbb{R}^n$

In this section, we recall popular sufficient conditions to verify the stability of equilibrium points. These conditions either stem from A. Lyapunov's original work or closely resemble those results. For the statement of these conditions, first, we recall the notions of positive-definite and nonnegative-definite functions.

Definition 4.22 The function $W : \mathbb{R}^n \to \mathbb{R}$ is *nonnegative-definite* if $W(0) = 0$ and $W(x) \geq 0$ for all $x \in \mathbb{R}^n$.

Definition 4.23 The function $W : \mathbb{R}^n \to \mathbb{R}$ is *positive-definite* if $W(0) = 0$ and $W(x) > 0$ for all $x \in \mathbb{R}^n \setminus \{0\}$.

Theorem 4.3 (Sufficient Condition for Lyapunov Stability [Nonautonomous Systems]) *Consider the nonlinear, time-varying dynamical system given by* (4.11), *and let* $x_e = 0$ *be an equilibrium point for* (4.11). *Let* $V : [t_0, \infty) \times \mathbb{R}^n \to \mathbb{R}$ *be continuously differentiable and such that*

$$V(t, 0) = 0, \quad \text{for all } t \in [t_0, \infty), \tag{4.16a}$$

$$W_1(x) \leq V(t, x), \quad \text{for all } (t, x) \in [t_0, \infty) \times \mathbb{R}^n, \tag{4.16b}$$

$$\frac{\partial V(t, x)}{\partial t} + \frac{\partial V(t, x)}{\partial x} f(t, x) \leq 0, \tag{4.16c}$$

where $W_1 : \mathbb{R}^n \to \mathbb{R}$ *is continuous and positive-definite. Then, the equilibrium point* $x_e = 0$ *of* (4.11) *is Lyapunov stable.*

Theorem 4.4 (Sufficient Condition for Uniform Lyapunov Stability) *Consider the nonlinear, time-varying dynamical system given by* (4.11), *and let* $x_e = 0$ *be an equilibrium point for* (4.11). *Let* $V : [t_0, \infty) \times \mathbb{R}^n \to \mathbb{R}$ *be continuously differentiable and such that*

$$W_1(x) \leq V(t, x) \leq W_2(x), \qquad (t, x) \in [t_0, \infty) \times \mathbb{R}^n, \tag{4.17}$$

where $W_1, W_2 : \mathbb{R}^n \to \mathbb{R}$ *are continuous and positive-definite, and* (4.16c) *is verified. Then, the equilibrium point* $x_e = 0$ *of* (4.11) *is uniformly Lyapunov stable.*

Theorem 4.5 (Sufficient Condition for Lyapunov Stability [Autonomous Systems]) *Consider the nonlinear, time-invariant dynamical system given by* (4.12), *and let* $x_e = 0$ *be an equilibrium point for* (4.12). *Let* $V : \mathbb{R}^n \to \mathbb{R}$ *be continuously differentiable, positive-definite, and such that*

$$\frac{\partial V(x)}{\partial x} f(x) \leq 0, \qquad x \in \mathbb{R}^n. \tag{4.18}$$

Then, the equilibrium point $x_e = 0$ *of* (4.12) *is Lyapunov stable.*

In the case of time-invariant systems, a positive-definite function $V : \mathbb{R}^n \to \mathbb{R}$ is known as a *Lyapunov function candidate*. In the case of time-varying systems, a function $V : [t_0, \infty) \times \mathbb{R}^n \to \mathbb{R}$ that verifies (4.16a) and (4.16b) is known as a *Lyapunov function candidate*. In the case of time-invariant systems, a Lyapunov function candidate that verifies (4.16c) is known as a *Lyapunov function*. In the case of time-varying systems, a Lyapunov function candidate that verifies (4.18) is known as a *Lyapunov function*. A Lyapunov function candidate $V : \mathbb{R}^n \to \mathbb{R}$ such that $\lim_{\|x\|\to\infty} V(x) = \infty$ is known as a *radially unbounded Lyapunov function candidate*. Theorem 4.5 is also known as the *Lyapunov direct method* for stability.

4.3.4 Sufficient Conditions for Attractivity in $\mathbb{R}^n$

In this section, we present sufficient conditions commonly used to predict asymptotic convergence of trajectories of dynamical systems. We begin with a classical tool from real analysis, Barbalat's lemma. For the statement of this result, we recall the notion of uniform continuity of a function provided in Definition 2.56.

Lemma 4.1 (Barbalat's lemma) *Let $g : [t_0, \infty) \to \mathbb{R}$ be uniformly continuous. If $\lim_{t\to\infty} \int_{t_0}^{t} g(\tau)\mathrm{d}\tau$ exists and is finite, then $\lim_{t\to\infty} g(t) = 0$.*

A direct consequence of Barbalat's lemma, which is of interest for adaptive control synthesis, is the following.

Lemma 4.2 *Let $g : [t_0, \infty) \to \mathbb{R}$ be continuously differentiable and assume that $\dot{g}(\cdot)$ is uniformly continuous. If $\lim_{t\to\infty} g(t)$ exists and is finite, then $\lim_{t\to\infty} \dot{g}(t) = 0$.*

Uniform continuity of functions is difficult to prove through a direct application of Definition 2.56. A sufficient condition to prove uniform continuity is the following.

Theorem 4.6 *Let $f : [t_0, \infty) \to \mathbb{R}$ be continuously differentiable. If $\dot{f} \in L^\infty([t_0, \infty), \mathbb{R})$, then f is uniformly continuous.*

As discussed at length in [8, Chap. 4], applied to autonomous ODEs, Barbalat's lemma provides weaker results than a more specialized result, namely LaSalle's invariance principle. To present this result, first, recall the notion of metric on a space $\mathbb{X}$ introduced in Definition 2.7 and the definition of distance from a set given by (2.73). Furthermore, we recall the notion of invariant and positively invariant sets.

Definition 4.24 (*Invariant Set*) Consider the nonlinear, time-invariant dynamical system given by (4.12), and the set $\mathcal{M} \subseteq \mathbb{R}^n$. If $x_0 \in \mathcal{M}$ implies that $x(t) \in \mathcal{M}$, $t \in \mathbb{R}$, then $\mathcal{M}$ is *invariant with respect to* (4.12).

Definition 4.25 (*Positively Invariant Set*) Consider the nonlinear, time-invariant dynamical system given by (4.12), and the set $\mathcal{M} \subseteq \mathbb{R}^n$. If $x_0 \in \mathcal{M}$ implies that $x(t) \in \mathcal{M}$, $t \geq 0$, then $\mathcal{M}$ is *positively invariant with respect to* (4.12).

Theorem 4.7 (LaSalle's Invariance Principle) *Consider the nonlinear, time-invariant dynamical system given by* (4.12)*, and let $\mathcal{M} \subset \mathbb{R}^n$ be a compact set that is positively invariant with respect to* (4.12)*. Let $V : \mathbb{R}^n \to \mathbb{R}$ be continuously differentiable and such that*

$$\frac{\partial V(x)}{\partial x} f(x) \leq 0, \qquad x \in \mathcal{M}. \tag{4.19}$$

Let $\mathcal{R} \triangleq \left\{x \in \mathcal{M} \mid \frac{\partial V(x)}{\partial x} f(x) = 0\right\}$, *and let* $\mathcal{N}$ *be the largest invariant set in* $\mathcal{R}$. *If* $x_0 \in \mathcal{M}$, *then* $\lim_{t\to\infty} \operatorname{dist}(x(t), \mathcal{N}) = 0$.

If the total derivative of a Lyapunov function along the trajectories of an autonomous dynamical system is nonpositive-definite, then LaSalle's invariance principle allows drawing conclusions on the asymptotic behavior of the system's trajectories; see [8, Chap. 4] for some examples. Replacing g with the total derivative of a Lyapunov function in Lemmas 4.1 and 4.2, Barbalat's lemma can be applied to analyze the asymptotic behavior of both time-varying and time-invariant dynamical systems. LaSalle's invariance principle, however, applies to time-invariant dynamical systems only. As discussed in [8, Chap. 4], in general, LaSalle's invariance principle provides stronger results than Barbalat's lemma.

4.3.5 Sufficient Conditions for Asymptotic Stability in $\mathbb{R}^n$

In this section, we present sufficient conditions for the asymptotic stability of nonlinear dynamical systems.

Theorem 4.8 *Consider the nonlinear, time-varying dynamical system given by* (4.11), *and let* $x_e = 0$ *be an equilibrium point for* (4.11). *Let* $V : [t_0, \infty) \times \Omega \to \mathbb{R}$ *be continuously differentiable and such that* (4.17) *is verified and*

$$\frac{\partial V(t,x)}{\partial t} + \frac{\partial V(t,x)}{\partial x} f(t,x) \leq -W(x), \qquad (t,x) \in [t_0, \infty) \times \Omega, \tag{4.20}$$

where $W : \Omega \to \mathbb{R}$ *is positive-definite. Then, the equilibrium point* $x_e = 0$ *of* (4.11) *is uniformly asymptotically stable. Furthermore, if* $\Omega = \mathbb{R}^n$, (4.17) *is verified by* $W_1(\cdot)$ *and* $W_2(\cdot)$ *radially unbounded, and* (4.20) *is verified by* $W(\cdot)$ *radially unbounded, then* $x_e = 0$ *is globally uniformly asymptotically stable.*

Theorem 4.9 *Consider the nonlinear, time-invariant dynamical system given by* (4.12), *and let* $x_e = 0$ *be an equilibrium point for* (4.12). *Let* $V : \Omega \to \mathbb{R}$ *be continuously differentiable, positive-definite, and such that*

$$\frac{\partial V(x)}{\partial x} f_a(x) < 0, \qquad x \in \Omega \setminus \{0\}. \tag{4.21}$$

Then, the equilibrium point $x_e = 0$ *of* (4.12) *is asymptotically stable. Furthermore, if* $\Omega = \mathbb{R}^n$ *and* $V(\cdot)$ *is radially unbounded, then* $x_e = 0$ *is globally asymptotically stable.*

Although straightforward to use, these conditions are too strict for problems of practical interest. Indeed, to prove the asymptotic stability of the equilibrium point of some dynamical system or, at least, the Lyapunov stability of this equilibrium point and the convergence of the controlled system's states, or part thereof, it is more common practice to resort to the sufficient conditions for Lyapunov stability discussed in Sect. 4.3.3 and the sufficient conditions for attractivity discussed in Sect. 4.3.4. For details, see [8, Chap. 4] to name a classical reference.

4.3.6 Sufficient Conditions for Ultimate Boundedness in $\mathbb{R}^n$

In the following, we present sufficient conditions for boundedness and ultimate boundedness of solutions of ODEs. For the statement of these conditions, we recall the notions of class $\mathcal{K}$ and $\mathcal{K}_\infty$ functions.

Definition 4.26 (*Class $\mathcal{K}$ and $\mathcal{K}_\infty$ Functions*) A continuous function $\alpha : [0, a) \to \mathbb{R}^+$, where $a \in (0, \infty]$, is a *class $\mathcal{K}$ function* if $\gamma(0) = 0$ and it is strictly increasing. A function $\alpha : \mathbb{R}^+ \to \mathbb{R}^+$ is a *class $\mathcal{K}_\infty$ function* if it is a class $\mathcal{K}$ function and $\lim_{x\to\infty} \alpha(x) = \infty$.

It is important to note how class $\mathcal{K}$ and class $\mathcal{K}_\infty$ functions are invertible. Furthermore, a radially unbounded Lyapunov function candidate $V : \mathbb{X} \to \mathbb{R}^+$, with $\mathbb{X} \triangleq \mathbb{R}^n$, can always be bounded by class $\mathcal{K}_\infty$ functions α_1 and α_2 such that

$$\alpha_1(\|x\|) \le V(x) \le \alpha_2(\|x\|), \quad \text{for all } x \in \mathbb{X}. \tag{4.22}$$

Theorem 4.10 (Sufficient Conditions for Boundedness) *Consider the nonlinear, time-varying dynamical system given by* (4.11), *and assume there exist a continuously differentiable function $V : [t_0, \infty) \times \Omega \to \mathbb{R}$ and class $\mathcal{K}$ functions $\alpha_1(\cdot)$ and $\alpha_2(\cdot)$ such that*

$$\alpha_1(\|x\|) \le V(t,x) \le \alpha_2(\|x\|), \qquad (t,x) \in [t_0, \infty) \times \Omega, \tag{4.23a}$$

$$\frac{\partial V(t,x)}{\partial t} + \frac{\partial V(t,x)}{\partial x} f(t,x) \le 0, \qquad (t,x) \in [t_0,\infty) \times \left(\Omega \setminus \mathcal{B}_\mu(0)\right), \tag{4.23b}$$

where $\mu > 0$ is such that $\mathcal{B}_{\alpha_1^{-1}(\eta)}(0) \subset \Omega$ for some $\eta \ge \alpha_2(\mu)$. Then, the solutions of (4.11) *are uniformly bounded in a neighborhood $\mathcal{N}$ of $x_e = 0$ such that $\mathcal{B}_\mu(0) \subseteq \mathcal{N} \subseteq \mathcal{B}_{\alpha_1^{-1}(\eta)}(0)$. Furthermore, for every $\delta \in (0, \gamma)$, if $\|x_0\| \le \delta$, then $\|x(t)\| \le \varepsilon$, where*

$$\varepsilon \triangleq \begin{cases} \alpha_1^{-1}(\alpha_2(\delta)), & \delta \in (\mu, \gamma), \\ \alpha_1^{-1}(\eta) & \delta \in (0, \mu], \end{cases} \tag{4.24}$$

where $\gamma \triangleq \sup\{r > 0 \mid \mathcal{B}_{\alpha_1^{-1}(\alpha_2(r))} \subset \Omega\}$. Finally, if $\Omega = \mathbb{R}^n$, both $\alpha_1(\cdot)$ and $\alpha_2(\cdot)$ are class $\mathcal{K}_\infty$ functions, and (4.23a) *and* (4.23b) *are verified, then the solutions of* (4.11) *are globally uniformly bounded, and for every* $x_0 \in \mathbb{R}^n$*, it holds that* $\|x(t)\| \leq \varepsilon$*, where* ε *is given by* (4.24) *and* $\delta = \|x_0\|$.

Theorem 4.11 (Sufficient Conditions for Uniform Ultimate Boundedness) *Consider the nonlinear, time-varying dynamical system given by* (4.11)*, and assume there exist a continuously differentiable function* $V : [t_0, \infty) \times \Omega \to \mathbb{R}$ *and class* $\mathcal{K}$ *functions* $\alpha_1(\cdot)$ *and* $\alpha_2(\cdot)$ *such that* (4.23a) *is verified and*

$$\frac{\partial V(t,x)}{\partial t} + \frac{\partial V(t,x)}{\partial x} f(t,x) \leq -W(x), \quad (t,x) \in [t_0, \infty) \times \left(\Omega \setminus \mathcal{B}_\mu(0)\right), \tag{4.25}$$

where $\mu > 0$ *is such that* $\mathcal{B}_{\alpha_1^{-1}(\eta)}(0) \subset \Omega$ *for some* $\eta > \alpha_2(\mu)$ *and* $W : \Omega \to \mathbb{R}$ *is continuous and such that* $W(x) > 0$ *for all* $x \in \Omega \setminus \mathcal{B}_\mu(0)$*. Then, the solutions of* (4.11) *are uniformly ultimately bounded with bound* $\alpha_1^{-1}(\eta)$*. Furthermore,* $\limsup_{t\to\infty} \|x(t)\| \leq \alpha_1^{-1}(\alpha_2(\mu))$*. Finally, if* $\Omega = \mathbb{R}^n$*, both* $\alpha_1(\cdot)$ *and* $\alpha_2(\cdot)$ *are class* $\mathcal{K}_\infty$ *functions, and* (4.23a) *and* (4.25) *are verified, then the solutions of* (4.11) *are globally uniformly ultimately bounded with bound* $\alpha_1^{-1}(\eta)$.

Although Theorems 4.10 and 4.11 apply to generic nonlinear, time-varying dynamical systems, their primary use consists in assessing the properties of systems affected by unknown external disturbances. These results are employed extensively in Chap. 5.

4.4 Control and Robust Control Lyapunov Functions in Euclidean Spaces

Sufficient conditions presented in Sects. 4.3.3–4.3.6 are instrumental in nonlinear systems analysis and can be employed to verify whether some control law meets user-defined performance in terms of boundedness of the systems' trajectories and asymptotic convergence of the controlled plant dynamics. Some experience in control systems design shows that, although there are several classes of solutions, there is no universal approach to designing feedback control systems for nonlinear plants, especially if affected by unknown external disturbances. Control Lyapunov functions (CLFs) and robust CLFs (RCLFs) provide a valuable tool that partly addresses the problem of designing a feedback-linearizing controller applicable to any nonlinear dynamical system.

To present CLFs, consider the nonlinear, controlled, time-invariant dynamical system given by

$$\dot{x}(t) = F(x(t), u(t)), \quad x(0) = x_0, \quad t \geq 0, \tag{4.26}$$

where $x : [0, \infty) \to \mathbb{R}^n$ denotes the *system's state*, $u : [0, \infty) \to \mathbb{R}^m$ denotes the *control input*, and $F : \mathbb{R}^n \times \mathbb{R}^m \to \mathbb{R}^n$ is Lipschitz continuous in the first argument and continuous in the second argument.

Definition 4.27 (*Control Lyapunov Function*) Consider the nonlinear, controlled, time-invariant system given by (4.26). A continuously differentiable, positive-definite, radially unbounded Lyapunov function candidate $V : \mathbb{R}^n \to \mathbb{R}$ such that

$$\inf_{u \in \mathbb{U}} \frac{\partial V(x)}{\partial x} F(x, u) < 0, \quad \text{for all } x \in \mathbb{R}^n \setminus \{0\}, \tag{4.27}$$

where $\mathbb{U} \subseteq \mathbb{R}^m$ is convex, is a *control Lyapunov function.*

The existence of CLF for a given dynamical system implies the existence of a feedback control law $\phi : \mathbb{R}^n \to \mathbb{R}^m$ such that

$$\frac{\partial V(x)}{\partial x} F(x, \phi(x)) < 0, \quad \text{for all } x \in \mathbb{R}^n \setminus \{0\}, \tag{4.28}$$

which, by Theorem 4.9, implies global asymptotic stability of the closed-loop system

$$\dot{x}(t) = F(x(t), \phi(x(t))), \quad x(0) = x_0, \quad t \geq 0. \tag{4.29}$$

The next result applies the notion of CLFs to a less general, and yet very powerful, dynamical model, namely

$$\dot{x}(t) = a(x(t)) + g(x(t))u(t), \quad x(0) = x_0, \quad t \geq 0, \tag{4.30}$$

where $x : [0, \infty) \to \mathbb{R}^n$ denotes the *system's state*, $u : [0, \infty) \to \mathbb{R}^m$ denotes the *control input*, and both $f : \mathbb{R}^n \to \mathbb{R}^n$ and $G : \mathbb{R}^n \to \mathbb{R}^{n \times m}$ are Lipschitz continuous. This affine-in-the-control system will be further analyzed in Chap. 5.

Theorem 4.12 ([7, Theorem 6.7]) *Consider the nonlinear, time-invariant, affine-in-the-control dynamical system* (4.30). *The continuously differentiable, radially unbonded function* $V : \mathbb{R}^n \to \mathbb{R}$ *is a control Lyapunov function for* (4.30) *if and only if*

$$\frac{\partial V(x)}{\partial x} f(x) < 0 \quad \textit{for all } x \in \mathcal{R}, \tag{4.31}$$

where $\mathcal{R} \triangleq \left\{ x \in \mathbb{R}^n : \frac{\partial V(x)}{\partial x} g(x) = 0 \right\}$.

In general, for systems in the same form as (4.26), which has a CLF, a feedback control law that guarantees global asymptotic stability of the closed-loop system by means of this CLF cannot be expressed explicitly. For systems in the same form as (4.30), which have a CLF, the associated control law can be written explicitly as

$$\alpha(x) = \begin{cases} -\left(c_0 + \frac{\alpha(x)+\left[\alpha^2(x)+\|\beta(x)\|^4\right]^{1/2}}{\|\beta(x)\|^2}\right)\beta(x) & \text{if } \beta(x) \geq 0, \\ 0 & \text{otherwise}, \end{cases} \tag{4.32}$$

where $c_0 \geq 0$ is user-defined, $\alpha(x) \triangleq \frac{\partial V(x)}{\partial x} f(x)$, and $\beta(x) \triangleq g^{\mathrm{T}}(x)\left[\frac{\partial V(x)}{\partial x}\right]^{\mathrm{T}}$.

RCLFs allow addressing a broader class of problems than CLFs. Indeed, consider the nonlinear, time-varying, perturbed dynamical system

$$\dot{x}(t) = F(x(t), u(t), w(t)), \quad x(0) = 0, \quad t \geq 0, \tag{4.33}$$

where $x : [0, \infty) \to \mathbb{R}^n$ denotes the *system's state*, $u : [0, \infty) \to \mathbb{R}^m$ denotes the *control input*, $w : [0, \infty) \to \mathbb{R}^p$ denotes a continuous-time *external disturbance*, and $F : \mathbb{R}^n \times \mathbb{R}^m \times \mathbb{R}^p$ is Lipschitz continuous in its first argument and continuous in the remaining ones. The external disturbance is assumed to be unknown and continuous in time.

Definition 4.28 (*Robust Control Lyapunov Function*) Consider the nonlinear, controlled, time-invariant system given by (4.33). A continuously differentiable, positive-definite, radially unbounded function $V : \mathbb{R}^n \to \mathbb{R}$ such that

$$\inf_{u\in\mathbb{U}} \sup_{w\in\mathbb{W}} \frac{\partial V(x)}{\partial x} F(x, u, w) < -W(x), \qquad \text{for all } x \in \mathcal{R}, \tag{4.34}$$

where $\mathbb{U} \subseteq \mathbb{R}^m$ is convex, $\mathbb{W} \subseteq \mathbb{R}^p$, $W : \mathbb{R}^n \to \mathbb{R}$ is positive-definite, $\mathcal{R} \triangleq \{x \in \mathbb{R}^n : V(x) > c\}$, and some $c > 0$, is a *robust control Lyapunov function*.

A more general definition of RCLFs can be found in [27, Chap. 3]. It follows from Theorem 4.11 that the existence of an RCLF implies the existence of a feedback control law $\alpha : \mathbb{R}^n \to \mathbb{R}^m$ such that the controlled dynamical system

$$\dot{x}(t) = F(x(t), \alpha(x(t)), w(t)), \quad x(0) = x_0, \quad t \geq 0 \tag{4.35}$$

is uniform ultimately bounded for all $w : [0, \infty) \to \mathbb{W}$.

4.5 Elements of Distributed Parameter Systems Theory

As discussed in the introduction to this chapter, the overall approach to nonparametric adaptive control described in Chap. 5 takes the form of a two-step process. Initially, an ideal nonparametric controller is defined. This ideal controller defines the nonparametric closed-loop system, which is called the limiting DPS and evolves in an infinite-dimensional state space. Afterward, realizable controllers are defined by constructing consistent approximations of the DPS. In this section, we present elements of dynamical systems theory for DPS, which are sufficient to comprehend the proposed analysis on the limiting DPS.

The idea of a process on $\mathbb{R}^n$ given in Definition 4.5 can be extended to a setting where the system evolves in a general metric space.

Definition 4.29 (*Process over a Metric Space* $\mathbb{Z}$) A *process* on the metric space $(\mathbb{Z}, d_{\mathbb{Z}})$ is a mapping $z : [t_0, \infty) \times \mathbb{R} \times \mathbb{Z} \to \mathbb{Z}$ such that

(P1) $z(t_0, t_0, z_0) = z_0$ for all $z_0 \in \mathbb{Z}$,
(P2) $z(\tau + s, t_0, z_0) = z(\tau, s + t_0, z(s, t_0, z_0))$ for all $\tau \in [t_0, \infty), s \in [0, \infty)$,
(P3) $z(\cdot, t_0, z_0) : [t_0, \infty) \to \mathbb{Z}$ is continuous on (t_0, ∞) and right continuous at $\tau = t_0$ on $[t_0, \infty)$, and
(P4) $z(\tau, \cdot, \cdot) : \mathbb{R} \times \mathbb{Z} \to \mathbb{Z}$ is continuous for each $\tau \in [t_0, \infty)$.

Remark 4.7 Continuity in Definition 4.29 is with respect to the metric $d_{\mathbb{Z}}$ on $\mathbb{Z}$. For instance, property **(P3)** requires that

$$t_k \to \tau \quad \text{implies} \quad d_{\mathbb{Z}}(z(t_k, t_0, z_0), z(\tau, t_0, z_0)) \to 0$$

as $k \to \infty$ for all $(t_0, z_0) \in [t_0, \infty) \times \mathbb{Z}$.

As for systems that evolve in Euclidean spaces $\mathbb{X} \triangleq \mathbb{R}^n$, processes over a metric space $\mathbb{Z}$ are sometimes referred to as a *nonautonomous dynamical systems* [28, 29].

Definition 4.6 can also be extended from the Euclidean space to metric spaces as follows.

Definition 4.30 (*Dynamical System over Metric Space* $\mathbb{Z}$) An *dynamical system* on the metric space $(\mathbb{Z}, d_{\mathbb{Z}})$ is a mapping $z : [0, \infty) \times \mathbb{Z} \to \mathbb{Z}$ such that

(D1) $z(0, z_0) = z_0$ for all $z_0 \in \mathbb{Z}$,
(D2) $z(t + s, z_0) = z(t, z(s, z_0))$ for all $t, s \in [0, \infty)$,
(D3) $z : (\cdot, z_0) : [0, \infty) \to \mathbb{Z}$ is continuous on (t_0, ∞) and right continuous at $\tau = t_0$ on $[t_0, \infty)$, and
(D4) $z(t, \cdot) : \mathbb{Z} \to \mathbb{Z}$ is continuous.

Again, it has become popular more recently to distinguish between processes and dynamical systems by referring to them as nonautonomous and autonomous dynamical systems, respectively [28, 29].

4.5.1 Strongly Continuous Linear Semigroups

In this section, we discuss some results in the theory of semigroups that are particularly useful to characterize dynamical systems over Banach spaces. The language of semigroups is used to express many important results in dynamical system theory in infinite dimensions, a few of which we need in Chaps. 5 and 6.

Definition 4.31 (*Strongly Continuous Linear Semigroups*) Let $\mathbb{Z}$ denote a Banach space, let $T : [0, \infty) \to \mathcal{L}(\mathbb{Z})$, and let I denote the identity operator on $\mathbb{Z}$. The family $\{T(t)\}_{t \geq 0}$ is a *strongly continuous semigroup* (also called a C_0*-semigroup*) on $\mathbb{Z}$ if

1. $T(0) = I$,
2. $T(t+s) = T(t)T(s)$ for all $t, s \in \mathbb{R}_+$, and
3. for every $z \in \mathbb{Z}$, $\lim_{t\to 0^-} \|T(t)z - z\| = 0$.

Remark 4.8 The first two properties of a C_0-semigroup are known as the *semigroup conditions*, while the third is the *continuity* or *C_0-condition*.

Definition 4.32 (*Infinitesimal Generators*) Let $\mathbb{Z}$ denote a Banach space and $T : [0, \infty) \to \mathcal{L}(\mathbb{Z})$ be such that $\{T(t)\}_{t\geq 0}$ is a C_0-semigroup on $\mathbb{Z}$. The *infinitesimal generator* of the semigroup $\{T(t)\}_{t\geq 0}$ is the linear operator $\mathcal{A} : \mathcal{D}(\mathcal{A}) \subset \mathbb{Z} \to \mathbb{Z}$, where

$$\mathcal{D}(\mathcal{A}) \triangleq \left\{ z \in \mathbb{Z} : \lim_{t\to 0^+} \frac{1}{t} \left(T(t) - I\right) z \text{ exists} \right\}, \tag{4.36}$$

and

$$\mathcal{A}z \triangleq \lim_{t\to 0^+} \frac{1}{t} \left(T(t) - I\right) z, \quad \text{for all } z \in \mathcal{D}(\mathcal{A}). \tag{4.37}$$

Infinitesimal generators are not necessarily defined over the entire space $\mathbb{Z}$ and may not be bounded. See the classical reference [30, 31] for good accounts of the nuances that arise in infinite dimensions. Next, we introduce the special class of C_0-semigroups. The next results provide a sufficient condition for the existence of an infinitesimal generator, which applies to uniformly continuous semigroups. This case is important to some results used in Chap. 5.

Definition 4.33 (*Uniformly Continuous Semigroups*) Let $\mathbb{Z}$ denote a Banach space and $T : [0, \infty) \to \mathcal{L}(\mathbb{Z})$ define a C_0-semigroup. The semigroup is *uniformly continuous* if

$$\lim_{t\to 0^+} \|T(t) - I\| = 0. \tag{4.38}$$

If $\{T(t)\}_{t\geq 0}$ is a uniformly continuous semigroup on $\mathbb{Z}$, then it is a C_0-semigroup on $\mathbb{Z}$, but not conversely.

Theorem 4.13 *Let $\mathbb{Z}$ denote a Banach space and $T : [0, \infty) \to \mathcal{L}(\mathbb{Z})$ be so that $\{T(t)\}_{t\geq 0}$ is a uniformly continuous semigroup on $\mathbb{Z}$. Then, there exists a bounded linear operator $\mathcal{A} : \mathbb{Z} \to \mathbb{Z}$ such that*

$$T(t) = e^{\mathcal{A}t} = \exp(\mathcal{A}t) \triangleq \sum_{k=0}^{\infty} \frac{\mathcal{A}^k t^k}{k!}, \quad \textit{for all } t \geq 0, \tag{4.39}$$

where the series converges in the uniform operator norm. Conversely, given bounded a bounded linear operator $\mathcal{A} : \mathbb{Z} \to \mathbb{Z}$, then (4.39) *defines a uniformly continuous semigroup on $\mathbb{Z}$.*

If $\mathcal{A} : \mathcal{D}(\mathcal{A}) \subset \mathbb{Z} \to \mathbb{Z}$ is the infinitesimal generator of the C_0-semigroup $\{T(t)\}_{t\geq 0}$, then the abstract IVP

$$\dot{z}(t) = \mathcal{A}z(t), \quad z(0) = z_0 \in \mathbb{Z}, \quad t \geq 0 \tag{4.40}$$

is well-posed and the (mild) solution is given by $z(t) = T(t)z_0$. Moreover, the non-homogeneous IVP initial value problem

$$\dot{z}(t) = \mathcal{A}z(t) + \mathcal{G}(t, z(t)), \quad z(t_0) = z_0 \in \mathbb{Z}, \quad t \geq t_0 \tag{4.41}$$

is well-posed, provided that $\mathcal{G} : [t_0, \infty) \times \mathbb{Z} \to \mathbb{Z}$, is continuous and satisfies a local Lipschitz condition in z. Solutions of (4.41) are characterized by the following result.

Theorem 4.14 (Nonlinear Variation of Parameters Formula, [18, Chap. 6, Theorem 1.4]) *Let $\mathcal{G} : [t_0, \infty) \times \mathbb{Z} \to \mathbb{Z}$ be continuous in t for $t \geq t_0 \geq 0$ and locally Lipschitz continuous in z, uniformly in t on bounded intervals. If $\mathcal{A} : \mathcal{D}(\mathcal{A}) \subset \mathbb{Z} \to \mathbb{Z}$ is the infinitesimal generator of a C_0-semigroup $\{T(t)\}_{t\geq t_0}$ on $\mathbb{Z}$, then, for every $z_0 \in \mathbb{Z}$, there exists a maximum interval of existence $[t_0, t_{\max})$ with $t_{\max} \leq \infty$ such that the initial value problem* (4.41) *has a unique mild solution $z(\cdot)$, and $z(\cdot)$ satisfies the* nonlinear variation of parameters formula

$$z(t) = T(t)z_0 + \int_{t_0}^{t} T(t-s)\mathcal{G}(s, z(s))\mathrm{d}s, \quad \textit{for all } t \in [t_0, t_{\max}). \tag{4.42}$$

Moreover, if $t_{\max} \in (t_0, \infty)$, then

$$\lim_{t \to t_{\max}^-} \|z(t)\|_{\mathbb{Z}} = \infty. \tag{4.43}$$

In the study of IVPs in Banach spaces, multiple notions of solution, such as strict, classical, weak, and mild, may be employed [18, 31, 32]. If $\mathbb{Z}$ is a finite-dimensional Banach space, these notions reduce to the classical solution of an IVP. Also, (4.42) agrees with the classical variation of parameters formula for $\mathbb{Z} = \mathbb{R}^n$. Finally, in this book in Chap. 5, the operator $\mathcal{A}$ is a bounded linear operator so that we have a uniformly continuous semigroup. In this case, the theory is considerably simpler than the general C_0 case. Chap. 2 in [33] discusses the processes generated by time-dependent bounded operators $\mathcal{A}(t)$.

4.5.2 *Topological Dynamics*

In this section, we present elements of topological dynamics needed for the scope of this book, specifically in our study of switched learn-and-control architectures in Chap. 6. For detailed discussions, the reader is referred to [26, 34].

Definition 4.34 (*Positive Orbit*) Let $z : [0, \infty) \times \mathbb{Z} \to \mathbb{Z}$ be a dynamical system on a metric space $\mathbb{Z}$. Define the operator $S(t) : \mathbb{Z} \to \mathbb{Z}$ so that $S(t)z_0 \triangleq z(t, z_0)$. The *positive orbit for the initial condition* $z_0 \in \mathbb{Z}$ is the set

$$\gamma^+(x_0) \triangleq \bigcup_{t \geq t_0} S(t)x_0. \tag{4.44}$$

Definition 4.35 (*Positive Limit Set*) Let $z : [0, \infty) \times \mathbb{Z} \to \mathbb{Z}$ be a dynamical system on a metric space $\mathbb{Z}$. Define the operator $S(t) : \mathbb{Z} \to \mathbb{Z}$ so that $S(t)z_0 \triangleq z(t, z_0)$. The *positive limit set* is defined as

$$\omega^+(x_0) \triangleq \Big\{ y \in \mathbb{Z} : y = \lim_{k \to \infty} S(t_k)x_0, \text{ for some monotonic sequence } \{t_k\}_{k \in \mathbb{N}} \text{ with } \lim_{k \to \infty} t_k = \infty \Big\}. \tag{4.45}$$

The following theorem applies to very general dynamical systems, under relatively weak hypotheses.

Theorem 4.15 ([26, Theorem 4.1]) *Let* $(\mathcal{M}, d_{\mathcal{M}})$ *be a complete metric space. Let* $z : [0, \infty) \times \mathbb{Z} \to \mathbb{Z}$ *be a dynamical system on a metric space* $\mathbb{Z}$*. Define the operator* $S(t) : \mathbb{Z} \to \mathbb{Z}$ *so that* $S(t)z_0 \triangleq z(t, z_0)$*. Suppose that the positive orbit* $\gamma^+(m_0)$ *is precompact, that is,* $\overline{\gamma^+(m_0)}$ *is compact. Then, the positive limit set* $\omega^+(m_0)$ *is nonempty, compact, invariant and connected. Moreover,*

$$\lim_{t \to \infty} \operatorname{dist}_{\mathcal{M}}(S(t)m_0, \omega^+(m_0)) = 0, \tag{4.46}$$

where $\operatorname{dist}(\cdot, \cdot)$ *is defined in (2.73).*

Definition 4.36 (*Limit Cycle*) Let $z : [0, \infty) \times \mathbb{Z} \to \mathbb{Z}$ be a dynamical system on a metric space $\mathbb{Z}$. A *limit cycle* is a closed curve $\Gamma \subset \mathbb{Z}$ such that Γ is the positive limit set of the positive orbit.

4.6 Concluding Remarks

This chapter presented key elements of nonlinear systems analysis needed for the discussions presented in Chap. 5. Special emphasis has been given to the notion of solution of IVPs, sufficient conditions for the existence and uniqueness of such solutions, various forms of stability and boundedness of solutions, and, finally, sufficient conditions to prove such forms of stability and boundedness. Essential elements of DPS theory have been reported to support key discussions later in this test.

The field of nonlinear dynamical systems analysis is so mature that it is impossible for a single book, let alone a single chapter, to be comprehensive. However, we have worked to be complete and summarize the analytical tools that are important to the rest of the book. Also, throughout this chapter, we referred the interested reader to specific sources for detailed discussions and examples.

4.7 Homework Problems

For an improved comprehension of the material exposed in this chapter, we recommend the following problems.

Exercise 4.1 The equations of motion of a planar pendulum of mass $m > 0$ and length $L > 0$ are given by

$$\begin{bmatrix}\dot\theta(t)\\ \ddot\theta(t)\end{bmatrix} = \begin{bmatrix}0 & 1\\ -c_1 & -c_2\end{bmatrix}\begin{bmatrix}\theta(t)\\ \dot\theta(t)\end{bmatrix} + \begin{bmatrix}0\\ 1\end{bmatrix}\left[-\frac{g}{L}\sin\theta(t) + f(\theta(t), \dot\theta(t))\right],$$
$$\begin{bmatrix}\theta(0)\\ \dot\theta(0)\end{bmatrix} = \begin{bmatrix}\theta_0\\ \theta_{\mathrm{d},0}\end{bmatrix}, \quad t \geq 0, \tag{4.47}$$

where $c_1, c_2 \in \mathbb{R}$ are unknown, $g > 0$ denotes the gravitational acceleration, and $f : \mathbb{R} \times \mathbb{R} \to \mathbb{R}$ is such that $f(0, 0) = 0$. Compute the equilibrium points of (4.47). △

Exercise 4.2 Prove that (4.47) is locally Lipschitz continuous in $\left[\theta, \dot\theta\right]^{\mathrm{T}}$. △

Exercise 4.3 Consider the nonlinear dynamical system

$$\dot e(t) = Ae(t) + B\left[u(t) + K^{\mathrm{T}}(t)\pi(t)\right], \quad e(t_0) = e_0, \quad t \geq t_0, \tag{4.48}$$

where $e : [t_0, \infty) \to \mathbb{R}^n$, $u : [t_0, \infty) \to \mathbb{R}^m$ is continuous and bounded, $\pi : [t_0, \infty) \to \mathbb{R}^N$ is continuous and bounded, $A \in \mathbb{R}^{n\times n}$, $B \in \mathbb{R}^{n\times m}$, and $K : [t_0, \infty) \to \mathbb{R}^{m\times N}$ is continuous and bounded. Prove that (4.48) is locally Lipschitz continuous in e. △

Exercise 4.4 Consider the nonlinear dynamical system given by (4.48) and

$$\dot{\hat K}(t) = -\Gamma\pi(t)e^{\mathrm{T}}(t)B, \quad \hat K(t_0) = K_0, \quad t \geq t_0, \tag{4.49}$$

where $\Gamma \in \mathbb{R}^{N\times N}$ is symmetric and positive-definite. Find the equilibrium points of these systems. △

Exercise 4.5 Consider the system given by (4.47) with $g = 0$ and $f(\theta, \dot\theta) = 0$ for all $(\theta, \dot\theta) \in \mathbb{R} \times \mathbb{R}$. Find its equilibrium points. Identify which of these points are Lyapunov stable, according to the values of c_1 and c_2. Can any of these points be globally asymptotically stable? △

Exercise 4.6 Consider the system given by (4.47) with $f(\theta, \dot{\theta}) = 0$ for all $(\theta, \dot{\theta}) \in \mathbb{R} \times \mathbb{R}$. Find its equilibrium points. Identify which of these points are asymptotically stable, according to the values of c_1 and c_2. Can any of these points be globally asymptotically stable? △

Exercise 4.7 Consider the system given by

$$\begin{bmatrix} \dot{\theta}(t) \\ \ddot{\theta}(t) \end{bmatrix} = \begin{bmatrix} 0 & 1 \\ -c_1 & -c_2 \end{bmatrix} \begin{bmatrix} \theta(t) \\ \dot{\theta}(t) \end{bmatrix} + \begin{bmatrix} 0 \\ 1 \end{bmatrix} \left[u(t) - \frac{g}{L} \sin \theta(t) \right],$$
$$\begin{bmatrix} \theta(0) \\ \dot{\theta}(0) \end{bmatrix} = \begin{bmatrix} \theta_0 \\ \theta_{\mathrm{d},0} \end{bmatrix}, \quad t \geq t_0, \tag{4.50}$$

where $u : [t_0, \infty) \to \mathbb{R}$ is continuous and bounded. Explain under which conditions (4.50) is bounded or ultimately bounded. △

Exercise 4.8 Consider the system given by (4.50) and design a control Lyapunov function and the corresponding feedback control law. △

Exercise 4.9 Consider the system given by

$$\begin{bmatrix} \dot{\theta}(t) \\ \ddot{\theta}(t) \end{bmatrix} = \begin{bmatrix} 0 & 1 \\ -c_1 & -c_2 \end{bmatrix} \begin{bmatrix} \theta(t) \\ \dot{\theta}(t) \end{bmatrix} + \begin{bmatrix} 0 \\ 1 \end{bmatrix} \left[u(t) - \frac{g}{L} \sin \theta(t) + w(t) \right],$$
$$\begin{bmatrix} \theta(0) \\ \dot{\theta}(0) \end{bmatrix} = \begin{bmatrix} \theta_0 \\ \theta_{\mathrm{d},0} \end{bmatrix}, \quad t \geq t_0, \tag{4.51}$$

where $w : [t_0, \infty) \to \mathbb{R}$ is continuous and bounded. Design a robust control Lyapunov function and the corresponding feedback control law. △

References

1. Farrell, J.A., Polycarpou, M.M.: Adaptive approximation based control: unifying neural, fuzzy and traditional adaptive approximation approaches, , vol. 48. Wiley, Danvers, MA (2006)
2. Lavretsky, E., Wise, K.: Robust and Adaptive Control: With Aerospace Applications. Springer, London, UK (2013)
3. Ioannou, P.A., Sun, J.: Robust Adaptive Control. Courier Corporation, Mineola, NY (2012)
4. Krstic, M., Kanellakopoulos, I., Kokotovic, P.: Nonlinear and Adaptive Control Design. Wiley, New York, NY (1995)
5. Sastry, S., Bodson, M.: Adaptive Control: Stability, Convergence and Robustness. Courier Corporation, New York, NY (2011)
6. Tao, G.: Adaptive Control Design and Analysis, vol. 37. Wiley, Danvers, MA (2003)
7. Haddad, W.M., Chellaboina, V.: Nonlinear Dynamical Systems and Control: A Lyapunov-Based Approach. Princeton University Press, Princeton, NJ (2008)
8. Khalil, H.K., Grizzle, J.W.: Nonlinear Systems, vol. 3. Prentice Hall, Upper Saddle River, NJ (2002)

9. Hovakimyan, N., Cao, C.: $\mathcal{L}_1$ Adaptive Control Theory: Guaranteed Robustness with Fast Adaptation. Society for Industrial and Applied Mathematics, Philadelphia, PA (2010)
10. Barreira, L., Valls, C.: Dynamical Systems: An Introduction. Springer Science and Business Media (2012)
11. Brunton, S.L., Kutz, J.N.: Data-driven Science and Engineering: Machine Learning, Dynamical Systems, and Control. Cambridge University Press (2022)
12. da Costa Campos, L.M.B.: Non-linear Differential Equations and Dynamical Systems. CRC Press, Boca Raton, FL (2019)
13. Dilão, R.: Dynamical System and Chaos: An Introduction with Applications. Springer, London, UK (2023)
14. Engel, K.-J., Nagel, R., Brendle, S.: One-parameter Semigroups for Linear Evolution Equations, vol. 194. Springer, London, UK (2000)
15. Hubbard, J.H., West, B.H.: Differential Equations: A Dynamical Systems Approach: Ordinary Differential Equations. Springer (2013)
16. Luo, A.: Bifurcation and Stability in Nonlinear Dynamical Systems. Springer, London, UK (2019)
17. Michel, A.N., Hou, L., Liu, D.: Stability of Dynamical Systems. Springer, London, UK (2008)
18. Pazy, A.: Semigroups of Linear Operators and Applications to Partial Differential Equations. Springer, New York, NY (2012)
19. Perko, L.: Differential Equations and Dynamical Systems, vol. 7. Springer Science and Business Media (2013)
20. Robinson, C.: Dynamical Systems: Stability, Symbolic Dynamics, and Chaos. CRC Press (1998)
21. Strogatz, S.H.: Nonlinear Dynamics and Chaos with Applications to Physics, Chemistry, and Engineering. CRC Press, Biology (2018)
22. Stuart, A., Humphries, A.R.: Dynamical Systems and Numerical Analysis, vol. 2. Cambridge University Press, Cambridge, UK (1998)
23. Coddington, E.A., Levinson, N.: Theory of Ordinary Differential Equations. McGraw-Hill, New York, NY (1955)
24. Hale, J.K.: Ordinary Differential Equations. Wiley, New York, NY (1969)
25. Reid, W.: Ordinary Differential Equations. Wiley, New York, NY (1971)
26. Walker, J.A.: Dynamical Systems and Evolution Equations: Theory and Applications. Plenum Press, New York (1980)
27. Freeman, R., Kokotovic, P.V.: Robust Nonlinear Control Design: State-Space and Lyapunov Techniques. Springer, New York, NY (2008)
28. Kloeden, P.E., Rasmussen, M.: Nonautonomous Dynamical Systems, no. 176. American Mathematical Society, Providence, RI (2011)
29. Bortolan, M.C., Carvalho, A.N., Langa, J.A.: Attractors Under Autonomous and Non-autonomous Perturbations, vol. 246. American Mathematical Society, Providence, RI (2020)
30. Curtain, R., Zwart, H.: An Introduction to Infinite-Dimensional Linear Systems Theory. Springer (1995)
31. Bensoussan, A., Da Prato, G., Delfour, M.C., Mitter, S.K.: Representation and Control of Infinite Dimensional Systems, vol. 2. Springer, New York, NY (2007)
32. Curtain, R.F., Zwart, H.: An Introduction to Infinite-Dimensional Linear Systems Theory, vol. 21. Springer, Berlin, Germany (2012)
33. Krein, S.G.: Linear Differential Equations in Banach Space. American Mathematical Society, Providence, RI (1972)
34. Saperstone, S.H.: Semidynamical Systems in Infinite Dimensional Spaces, vol. 37. Springer, New York, NY (2012)

Chapter 5
Native Space Embedding Control Methods

In this chapter, we synthesize elements of real and functional analysis, RKHS theory, and nonlinear systems theory presented in the previous chapters to design control systems for dynamical models affected by uncertainties characterized as elements of a vector-valued RKHS. In this approach, all the closed-loop systems determined by realizable feedback controllers evolve in finite-dimensional state spaces that can be interpreted as being contained in a single infinite-dimensional state space constructed using a vector-valued RKHS. Thus, we refer to these control techniques *native space embedding methods*. We begin in Sect. 5.1, where we discuss the uncertainty classes considered in this chapter. In Sect. 5.2, we consider affine-in-the-control dynamical systems and extend some oft-cited theory of robust control Lyapunov functions to account for operator-valued matched functional uncertainties in RKHSs. Subsequently in Sect. 5.3, we present model reference adaptive control (MRAC) systems subject to matched functional uncertainties. Since the adaptive gains corresponding to the functional uncertainties evolve over infinite-dimensional spaces, in Sect. 5.4, we discuss how such laws can be applied to problems of practical interest by projecting the adaptive gains into finite-dimensional RKHSs. To account for the resulting projection error, we present robustifications of the proposed MRAC laws based on evolutions of the dead-zone modification, σ-modification, and convex projection operator on RKHSs. Finally, in Sect. 5.5, we present adaptive variable structure methods to ensure uniform asymptotic trajectory tracking. Finally, in Sect. 5.6, we draw conclusions and highlight a few key points of this chapter, and in Sect. 5.7 we recommend several homework problems. Numerous numerical simulations prove the effectiveness and applicability of the proposed results.

A. J. Kurdila et al., *Data-Driven, Nonparametric, Adaptive Control Theory*,
Lecture Notes in Control and Information Sciences 495,
https://doi.org/10.1007/978-3-031-78003-5_5

5.1 Functional Uncertainty Classes in Vector-Valued RKHSs

Recently, in [1, 2], the authors detailed a general strategy for deterministic nonparametric control theory where the functional uncertainty in the governing nonlinear ordinary differential equations (ODEs) is contained in a native space $\mathcal{H}$ of scalar-valued functions. In this chapter, we describe in considerably greater detail how such a nonparametric control theory can be extended to the case when the uncertainty is contained in a native space $\mathcal{H}$ of $\mathbb{U}$-valued functions over $\mathbb{X} \triangleq \mathbb{R}^n$ that is induced by an operator kernel $\mathcal{K}$. At the end of the chapter, we summarize how the results of the case for vector-valued uncertainty yield the same qualitative conclusions as that for the scalar-valued case in [1, 2].

The analysis in this chapter of native space embedding for robust and adaptive control theory is formulated in certain types of nonparametric uncertainty classes, namely

$$C_R \triangleq \{f \in \mathcal{H} : \|f\|_{\mathcal{H}} \leq R\} \subset \mathcal{H}, \tag{5.1}$$

$$C_{N,\epsilon,R} \triangleq \{f \in \mathcal{H} : \|(I - \Pi_N)f\|_{\mathcal{H}} \leq \epsilon, \|f\|_{\mathcal{H}} \leq R\} \subset C_R \subset \mathcal{H}. \tag{5.2}$$

The orthogonal projection operator $\Pi_N : \mathcal{H} \to \mathcal{H}_{d(N)}$ in (5.2) verifies (2.32), and the space of approximants is given by (3.148), or, equivalently,

$$\mathcal{H}_{d(N)} \triangleq \text{span}\left\{\Phi_{d(N)}\right\}, \tag{5.3}$$

where $d(N) \triangleq mN$, N denotes the number of centers in $\Xi_N \subset \mathbb{X}$, m denotes the number of scalar-valued functions f_i in $f \triangleq [f_1, \ldots, f_m]^{\mathrm{T}} \in \mathcal{H}$,

$$\Phi_{d(N)} \triangleq \left\{\mathcal{K}_{\xi_i} e_j : 1 \leq i \leq N, 1 \leq j \leq m\right\}, \tag{5.4}$$

$\mathcal{K}_{\xi_i}$ verifies the reproducing property (3.107), ξ_i denotes the ith center in Ξ_N given by (3.6), and e_j denotes the jth canonical basis in $\mathbb{R}^m$ as defined in Example 2.1.

In the classical adaptive control theory, matched uncertainties are captured using a regressor vector and an unknown matrix, that is, $f(x) = \Theta^{\mathrm{T}}\Phi_N(x) \in \mathbb{R}^m$, where $\Theta \in \mathbb{R}^{p\times m}$ and $\Phi_N : \mathbb{X} \to \mathbb{R}^p$ denotes the regressor vector. In the following, we present an example concerning the rotational dynamics of a rigid body.

Example 5.1 The uncontrolled rotational dynamics of a rigid body are captured by

$$\dot{\omega}(t) = -I^{-1}\omega^{\times}(t)I\omega(t), \quad \omega(0) = \omega_0, \quad t \geq 0, \tag{5.5}$$

where $\omega : [0, \infty) \to \mathbb{R}^3$ denotes the *angular velocity* of the body, $(\cdot)^{\times}$ denotes the *cross product* operator, and $I \in \mathbb{R}^{3\times 3}$ is diagonal, positive-definite, and captures the *principal matrix of inertia*. Assuming that $I = \text{diag}(I_x, I_y, I_z)$ is unknown, (5.5) can be rewritten as

$$\dot{\omega}(t) = \Theta^{\mathrm{T}}\Phi(\omega(t)), \quad \omega(0) = \omega_0, \quad t \geq 0, \tag{5.6}$$

where $\Theta \triangleq -\mathrm{diag}\left((I_z - I_y), (I_x - I_z), (I_y - I_x)\right)$, $\Phi(\omega) \triangleq \left[\omega_z\omega_y, \omega_z\omega_x, \omega_y\omega_x\right]^{\mathrm{T}}$, and $\omega(t) = \left[\omega_x(t), \omega_y(t), \omega_z(t)\right]^{\mathrm{T}}$. $\triangle$

To encompass this classical approach to represent uncertainties, we define the real parametric uncertainty class for N parameters as

$$C_{\Phi_N,R'} \triangleq \left\{f = \Theta_N^{\mathrm{T}}\Phi_N \in \mathcal{H}_N : \Theta_N \in \mathbb{R}^p, \|\Theta_N\|_\infty \leq R'\right\} \subset \mathcal{H}, \tag{5.7}$$

where $\|\cdot\|_\infty$ is defined by (2.18), or, alternatively, we use the definition

$$C_{\Phi_N,R} \triangleq \left\{f = \Theta_N^{\mathrm{T}}\Phi_N \in \mathcal{H}_N : \|f\|_{\mathcal{H}} \leq R\right\} \subset C_{N,\epsilon,R} \subset \mathcal{H}. \tag{5.8}$$

It is easy to relate (5.7) and (5.8) owing to the equivalence of norms on a finite-dimensional space (for details, see Definition 2.39 and the subsequent discussion), and we use the definition of uncertainty class that is most convenient for a particular application at hand.

Note that parametric uncertainty classes, such as $C_{\Phi_N,R}$, have been used extensively in the basic theory for parametric adaptive control as described in the classical references [3–9]. For instance, methods based on the parameter projection operator in these texts ensure that parameter estimates are confined to a fixed convex subset of $\mathbb{R}^p$.

For the vector-valued RKHS $\mathcal{H}$ defined in terms of a general operator-valued kernel $\mathcal{K}$, it holds that

$$C_{\Phi_N,R} \subset C_{N,\epsilon,R} \subset C_R \subset \mathcal{H} \tag{5.9}$$

for any $R \in \mathbb{R}^+$, $\epsilon \in \mathbb{R}^+$, and $N \in \mathbb{N}$. Since the proposed control systems establish performance bounds over all systems affected by an uncertainty $f \in C_{N,\epsilon,R}$, the proposed bounds apply to the more common parametric uncertainty class $C_{\Phi_N,R}$. In general, $C_{N,\epsilon,R}$ is considerably larger than the parametric uncertainty class. This observation has important ramifications in many situations, such as in the study and further development of a *nonparametric theory of safety certificates* for adaptive control methods; see [8, 10] for a general discussion of safety in the adaptive control setting.

5.2 Nonlinear Robust Controllers for Uncertainty in an RKHS

In this section, we discuss how the native space embedding method in vector-valued native spaces induced by general operator-valued kernels can yield strong conclusions for some robust control approaches for classes of uncertain nonlinear ODEs, such as robust control Lyapunov functions and robust backstepping control. Although the arguments presented in this section do not use any of the principles of approximation summarized in Sect. 3.8, they demonstrate that there can be distinct advantages when casting a robust control problem in the framework of native space embedding. Before approaching this material, the reader is encouraged to review Sect. 4.4 for an introduction to CLFs and RCLFs, and the notions of uniform ultimate boundedness in Sect. 4.3.2.

5.2.1 Robust Control Lyapunov Functions Over Native Spaces

Consider the plant model

$$\dot{x}(t) = a(x(t)) + g(x(t))\left(u(t) + E_{x(t)}f\right), \quad x(0) = x_0, \quad t \geq 0, \tag{5.10}$$

where $x(t) \in \mathbb{X} \triangleq \mathbb{R}^n$ denotes the *state vector*, $u(t) \in \mathbb{U} \triangleq \mathbb{R}^m$ denotes the *control vector*, $a : \mathbb{X} \to \mathbb{X}$ denotes the *drift function*, $g : \mathbb{X} \to \mathbb{R}^{n\times m}$ denotes the *control influence operator*, and $f \in \mathcal{H} \triangleq \mathcal{H}(\mathbb{X}, \mathbb{U})$ denotes the functional uncertainty. The space $\mathcal{H} \triangleq \mathcal{H}(\mathbb{X}, \mathbb{U})$ is a native space that is induced by an operator-valued kernel and contains functions that map $\mathbb{U}$ into $\mathbb{X}$. We are interested in finding a feedback control law $\mu : \mathbb{X} \to \mathbb{U}$ such that the control input $u(t) \triangleq \mu(x(t))$, $t \geq 0$, drives the state $x(t) \to 0$ as $t \to \infty$, or, at least, assures uniform ultimate boundedness of the controlled trajectory.

It is not at all obvious how to construct an RCLF $V(\cdot)$ in closed form for a system as general as that in (5.10). However, there are many examples of Lyapunov functions, CLFs, or even RCLFs for low-dimensional ODEs in the classical treatises of real parametric adaptive control including [3, 5, 6, 8, 11]. In some cases, an RCLF is given for a model problem. In others, a Lyapunov function or a CLF is modified to yield an RCLF. In this brief section, we assume there exists a CLF $V(\cdot)$ and the associated feedback control law $\alpha(\cdot)$ for (5.10) with $f \equiv 0$; as discussed in [11], this is a standard assumption to deduce RCLFs.

The next theorem is a generalization to vector-valued uncertainties in a native space of Lemma 2.26 in [8]; the latter is a classical result that treats the case when $m = 1$, and the scalar functional uncertainty is defined in terms of a fixed finite number of regressors. For the statement of this result, let $V(\cdot)$ be a CLF of (5.10) with $f \equiv 0$. In this case, it follows from Definition 4.27 that there exist class $\mathcal{K}_\infty$ functions $\gamma_1(\cdot)$ and $\gamma_2(\cdot)$ such that

$$\gamma_1(\|x\|_{\mathbb{X}}) \leq V(x) \leq \gamma_2(\|x\|_{\mathbb{X}}), \quad \text{for all } x \in \mathbb{X}. \tag{5.11}$$

Furthermore, it follows from (4.27) that there exist a continuous, radially unbounded, positive-definite function $W : \mathbb{X} \to \mathbb{R}$ and a class $\mathcal{K}_\infty$ function $\gamma_3(\cdot)$ such that

$$\left\langle \left(\frac{\partial V(x)}{\partial x}\right)^{\mathrm{T}}, a(x) + g(x)\alpha(x) \right\rangle_{\mathbb{X}} \leq -W(x) \leq -\gamma_3(\|x\|_{\mathbb{X}}), \quad \text{for all } x \in \mathbb{X}, \tag{5.12}$$

where $\alpha : \mathbb{U} \to \mathbb{X}$ denotes a globally asymptotically stabilizing control law associated to $V(\cdot)$ such as, for instance, (4.32).

Theorem 5.1 *Consider the nonlinear dynamical model (5.10). Let $f \in \mathcal{H}(\mathbb{X}, \mathbb{U})$, where the RKHS $\mathcal{H}$ is induced by the matrix-valued kernel $\mathcal{K} : \mathbb{X} \times \mathbb{X} \to \mathcal{L}(\mathbb{U})$ and $\mathbb{U} = \mathbb{R}^m$ with $m \in \mathbb{N}$. Let $V(\cdot)$ denote a CLF for (5.10) with $f \equiv 0$ such that (5.11) and (5.12) are verified. Then, the feedback control*

$$\mu(x) \triangleq \alpha(x) - \beta\mathcal{K}(x, x)g^{\mathrm{T}}(x)\left[\frac{\partial V(x)}{\partial x}\right]^{\mathrm{T}}, \quad \textit{for all } x \in \mathbb{X}, \tag{5.13}$$

where $\beta > 0$, guarantees global uniform boundedness of (5.10) with $u(t) = \mu(x(t))$, $t \geq 0$, and

$$\limsup_{t\to\infty} \|x(t)\| \leq \gamma_1^{-1} \circ \gamma_2 \circ \gamma_3^{-1}\left(\frac{\|f\|_{\mathcal{H}}}{4\beta}\right), \tag{5.14}$$

where the class $\mathcal{K}_\infty$ functions $\gamma_1(\cdot)$, $\gamma_2(\cdot)$, and $\gamma_3(\cdot)$ verify (5.11) and (5.12).

Before proceeding with the proof of this result, we make a few observations to better understand the implications of casting this problem in terms of native space embedding.

Remark 5.1 (On the Uncertainty Parameterization) The approach in [8] assumes that the functional uncertainty is scalar with $m = 1$ and that the functional uncertainty is expressed in terms of a fixed, finite collection of basis functions, that is,

$$f(x) \triangleq \Theta_N^{\mathrm{T}}\Phi_N(x) = \sum_{k=1}^{N} \theta_k\phi_k(x), \quad \text{for all } x \in \mathbb{X}, \tag{5.15}$$

where $\Theta_N \triangleq [\theta_1, \ldots, \theta_N]^{\mathrm{T}}$ and $\Phi_N(x) \triangleq [\phi_1(x), \ldots, \phi_N(x)]^{\mathrm{T}}$. This means that the uncertainty is parameterized by a fixed, finite number $N > 0$ of real parameters $[\theta_1, \cdots, \theta_N]^{\mathrm{T}}$, and the uncertain function f is contained in $\Phi_N \triangleq \text{span}\,\{\phi_1, \ldots, \phi_N\}$. As discussed in Chap. 3, this assumption is pervasive in many studies of estimation and control problems for ODE systems.

Remark 5.2 (On the Feedback Function) With the parameterization of uncertainty in (5.15), the approach in [8] proposes a feedback controller of the form

$$\mu(x) \triangleq \alpha(x) - \beta\Phi_N^{\mathrm{T}}(x)\Phi_N(x)g^{\mathrm{T}}(x)\left[\frac{\partial V(x)}{\partial x}\right]^{\mathrm{T}}, \quad \text{for all } x \in \mathbb{X}, \tag{5.16}$$

which is qualitatively similar to (5.13). Note that the inner product $\Phi_N^{\mathrm{T}}(x)\Phi_N(x)$ in (5.16) is replaced with the operator kernel $\mathcal{K}(x, x) \in \mathcal{L}(\mathbb{U})$ in (5.13), *the latter of which depends neither on the basis nor the dimension* N.

Remark 5.3 (On the Controller's Performance and the Functional Uncertainty Class) Lemma 2.26 of [8] shows that, employing (5.16) with $m = 1$,

$$\limsup_{t\to\infty} \|x(t)\| \leq \gamma_1^{-1} \circ \gamma_2 \circ \gamma_3^{-1}\left(\frac{\|\Theta_N\|_\infty}{4\beta}\right). \tag{5.17}$$

If we assume that $\|\Theta_N\|_\infty \leq R$, then (5.17) reduces to

$$\limsup_{t\to\infty} \|x(t)\| \leq \gamma_1^{-1} \circ \gamma_2 \circ \gamma_3^{-1}\left(\frac{R}{4\beta}\right) \tag{5.18}$$

for all affine-in-the-control uncertain systems with $m = 1$ and functional uncertainty f residing in the parametric uncertainty class $C_{\Phi_N,R}$ defined in (5.7) that depends on the dimension N and on the choice of basis in Φ_N. Theorem 5.1, instead, allows for a broader uncertainty class. This result holds for all systems in which the uncertainty is contained in C_R defined in (5.1).

As shown in (5.9), the uncertainty class $C_{\Phi_N,R}$ is smaller than the uncertainty class C_R. A proof of this result that does not involve $C_{N,\epsilon,R}$ is the following. Suppose, without loss of generality, that the basis of N functions in $\Phi_N \triangleq [\phi_1, \ldots, \phi_N]^{\mathrm{T}}$ is contained in an infinite-dimensional scalar-valued native space $\mathcal{H}$ and satisfies $\|\phi_i\|_{\mathcal{H}} = 1$ for $1 \leq i \leq N$. Then, we have

$$C_{\Phi_N,R'} \subset C_R \tag{5.19}$$

for a constant $R' \triangleq R/\sqrt{\|\mathbb{K}_N\|N}$, where

$$\mathbb{K}_N \triangleq \left[\langle \phi_i, \phi_j\rangle_{\mathcal{H}}\right] \in \mathbb{R}^{N\times N} \tag{5.20}$$

denotes the Grammian matrix; the reader is encouraged to compare (5.20) and (3.8) and recognize their qualitative equivalence. The relationship (5.19) follows from the fact that, for any $f \in C_{\Phi_N,R'}$, it holds that $f = \Phi_N^{\mathrm{T}}\Theta_N$ for some $\|\Theta_N\|_\infty \leq R'$. Thus, it follows from (3.5), (2.22a), and (2.13) that

$$\begin{aligned}
\|f\|_{\mathcal{H}}^2 &= \Theta_N^{\mathrm{T}}\mathbb{K}_N\Theta_N \\
&\leq \|\Theta_N\|_2^2\|\mathbb{K}_N\|_2 \\
&\leq N\|\Theta_N\|_\infty^2\|\mathbb{K}_N\|_2 \\
&\leq N(R')^2\|\mathbb{K}_N\|_2 \\
&\leq R^2,
\end{aligned}$$

which implies that $f \in C_R$. However, $C_R \not\subset C_{\Phi_N, R'}$ for any N, R, $R' > 0$ when $\mathcal{H}$ is infinite dimensional. In light of (5.19), it is therefore important to appreciate that the ultimate bound captured by (5.14) is not tied to any basis to characterize the functional uncertainty, whereas the ultimate bounds captured by (5.17) or (5.18) do depend on the basis employed to capture the functional uncertainty. We now turn to the proof of Theorem 5.1.

Proof The derivative of the RCLF $V(\cdot)$ along the trajectories of (5.10) with $u(t) = \mu(x(t))$, $t \geq 0$, is given by

$$\dot{V}(x(t)) = \frac{\partial V(x(t))}{\partial x}(f(x(t)) + g(x(t))\alpha(x(t))) \tag{5.21}$$

$$- \beta \frac{\partial V(x(t))}{\partial x} g(x(t))\mathcal{K}(x(t), x(t))g^{\mathrm{T}}(x)\left[\frac{\partial V(x(t))}{\partial x}\right]^{\mathrm{T}}$$

$$+ \frac{\partial V(x(t))}{\partial x} g(x(t))E_{x(t)}f. \tag{5.22}$$

It follows from (4.27) that

$$\dot{V}(x(t)) = \frac{\partial V(x(t))}{\partial x}(f(x(t)) + g(x(t))\alpha(x(t))) \leq -W(x(t)), \quad t \geq 0, \tag{5.23}$$

where $W : \mathbb{X} \to \mathbb{R}$ is positive-definite and radially unbounded. Furthermore, it follows from (3.114) and (3.113) that

$$\frac{\partial V(x(t))}{\partial x} g(x(t))E_{x(t)}f = \left\langle g^{\mathrm{T}}(x)\left[\frac{\partial V(x(t))}{\partial x}\right]^{\mathrm{T}}, E_{x(t)}f\right\rangle_{\mathbb{R}^m}$$

$$= \left\langle E^*_{x(t)}g^{\mathrm{T}}(x)\left[\frac{\partial V(x(t))}{\partial x}\right]^{\mathrm{T}}, f\right\rangle_{\mathcal{H}}$$

$$= \left\langle \mathcal{K}_{x(t)}g^{\mathrm{T}}(x)\left[\frac{\partial V(x(t))}{\partial x}\right]^{\mathrm{T}}, f\right\rangle_{\mathcal{H}}, \quad t \geq 0, \tag{5.24}$$

and

$$\begin{aligned}
&\frac{\partial V(x(t))}{\partial x} g(x(t))\mathcal{K}(x(t), x(t))g^{\mathrm{T}}(x(t))\left[\frac{\partial V(x(t))}{\partial x}\right]^{\mathrm{T}} \\
&= \left\langle g^{\mathrm{T}}(x)\left[\frac{\partial V(x(t))}{\partial x}\right]^{\mathrm{T}}, \mathcal{K}(x(t), x(t))g^{\mathrm{T}}(x(t))\left[\frac{\partial V(x(t))}{\partial x}\right]^{\mathrm{T}}\right\rangle_{\mathbb{R}^m} \\
&= \left\langle g^{\mathrm{T}}(x)\left[\frac{\partial V(x(t))}{\partial x}\right]^{\mathrm{T}}, \mathcal{K}^*_{x(t)}\mathcal{K}_{x(t)}g^{\mathrm{T}}(x(t))\left[\frac{\partial V(x(t))}{\partial x}\right]^{\mathrm{T}}\right\rangle_{\mathbb{R}^m} \\
&= \left\langle \mathcal{K}_{x(t)}g^{\mathrm{T}}(x)\left[\frac{\partial V(x(t))}{\partial x}\right]^{\mathrm{T}}, \mathcal{K}_{x(t)}g^{\mathrm{T}}(x(t))\left[\frac{\partial V(x(t))}{\partial x}\right]^{\mathrm{T}}\right\rangle_{\mathcal{H}} \\
&= \left\| \mathcal{K}_{x(t)}g^{\mathrm{T}}(x)\left[\frac{\partial V(x(t))}{\partial x}\right]^{\mathrm{T}}\right\|^2_{\mathcal{H}}. \qquad (5.25)
\end{aligned}$$

Substituting (5.24) and (5.25) into (5.22), we obtain that

$$\begin{aligned}
\dot{V}(x(t)) \leq -W(x(t)) - \beta \left\| \mathcal{K}_{x(t)}g^{\mathrm{T}}(x(t))\left[\frac{\partial V(x(t))}{\partial x}\right]^{\mathrm{T}}\right\|^2_{\mathcal{H}} \\
+ \left\| \mathcal{K}_{x(t)}g^{\mathrm{T}}(x(t))\left[\frac{\partial V(x(t))}{\partial x}\right]^{\mathrm{T}}\right\|_{\mathcal{H}} \|f\|_{\mathcal{H}}, \quad t \geq 0,
\end{aligned}$$

and, completing the squares[1], we have that

$$\dot{V}(x(t)) \leq -W(x(t)) + \frac{\|f\|^2_{\mathcal{H}}}{4\beta}, \quad t \geq 0. \qquad (5.26)$$

Thus, it follows from (5.11) and (5.23) that the conditions of Theorem 4.11 are verified with $\mu = \gamma_3^{-1}\left(\frac{\|\Theta_N\|_\infty}{4\beta}\right)$, and the result is proven. ■

Example 5.2 To demonstrate the applicability of Theorem 5.1, consider the plant model

$$\dot{x}(t) = u(t) + E_{x(t)}f, \quad x(0) = x_0, \quad t \geq 0, \qquad (5.27)$$

which is in the same form as (5.10) with $n = 1$, $m = 1$, $a(x) = 0$ for all $x \in \mathbb{R}$, and $g(x) = 1$. If $f \equiv 0$, then (5.27) is stabilized, for example, by $\alpha(x) = -c_1 x$ for all $x \in \mathbb{R}$, and the associated CLF is

[1] For all $a \in \mathbb{R} \setminus \{0\}$ and $b, c, z \in \mathbb{R}$, it holds that $az^2 + bz + b = a(z + d)^2 + e$, where $d \triangleq \frac{b}{2a}$ and $e \triangleq c - d^2 a$. This identity is commonly known as *completion of squares*.

$$V(x) = \frac{1}{2}x^2, \tag{5.28}$$

which is a class $\mathcal{K}_\infty$ function. Thus, $\gamma_1(\|x\|_{\mathbb{X}}) = \gamma_2(\|x\|_{\mathbb{X}}) = V(x)$ for all $x \in \mathbb{R}$. Furthermore, if $f \equiv 0$, then (5.12) is verified with

$$W(x) = c_1 x^2, \quad \text{for all } x \in \mathbb{X}, \tag{5.29}$$

which is a class $\mathcal{K}_\infty$ function, and hence, $W(x) = \gamma_3(\|x\|_{\mathbb{X}})$. Thus, it follows from Theorem 5.1 that

$$\mu(x) = -c_1 x - \beta \mathfrak{K}(x, x)x, \quad \text{for all } x \in \mathbb{X} \tag{5.30}$$

guarantees uniform ultimate boundedness of the closed-loop plant trajectory, that is,

$$\limsup_{t \to \infty} \|x(t)\| \leq \sqrt{\frac{\|f\|_{\mathcal{H}}}{4c_1\beta}}. \tag{5.31}$$

To perform numerical simulations, we let $E_{x(t)} f = \left(\sum_{i=1}^{N} \theta_i \mathfrak{K}_{\xi_i}(x(t))\right) \cos t$, where $\Theta_N \triangleq [\theta_1, \ldots, \theta_N]^{\mathrm{T}}$. We also consider the Gaussian kernel

$$\mathfrak{K}(x_1, x_2) = \exp\left(-\frac{\|x_1 - x_2\|^2}{2l^2}\right), \quad (x_1, x_2) \in \mathbb{X} \times \mathbb{X}, \tag{5.32}$$

and the 3/2 Sobolev–Matérn kernel

$$\mathfrak{K}(x_1, x_2) = \left(1 + \frac{\sqrt{3}\|x_1 - x_2\|}{l}\right) \exp\left(-\frac{\sqrt{3}\|x_1 - x_2\|}{l}\right), \quad (x_1, x_2) \in \mathbb{X} \times \mathbb{X}, \tag{5.33}$$

with $l > 0$. Thus, we run 100 simulations choosing Θ_N randomly so that $\theta_i \in [-1, 1]$ and N randomly so that it is an integer between 5 and 15. The centers ξ_i are distributed on the interval $[-4, 4]$ at approximately equal distances with some random variability about the nominal position. Figures 5.1 and 5.2 show the performance of the control law (5.30). The MATLAB® codes generating these results can be found in [12]. △

Remark 5.4 It is interesting to observe how the Gaussian and the Sobolev–Matérn kernels employed in Example 5.2 are such that $\mathfrak{K}(x, x) = 1$ for all $x \in \mathbb{X}$. In fact, for any kernel defined in terms of a radial function, and in Table 3.1, it is possible to scale the kernel so that this is true. Thus, applied to Example 5.2, (5.13) reduces to a special case of the classical CLF law (5.16). An interesting fact is that this classical CLF law, by virtue of the nonparametric analysis above, guarantees stability over nonparametric uncertainty classes in any of the RKHS spaces $\mathcal{H}$ defined in terms

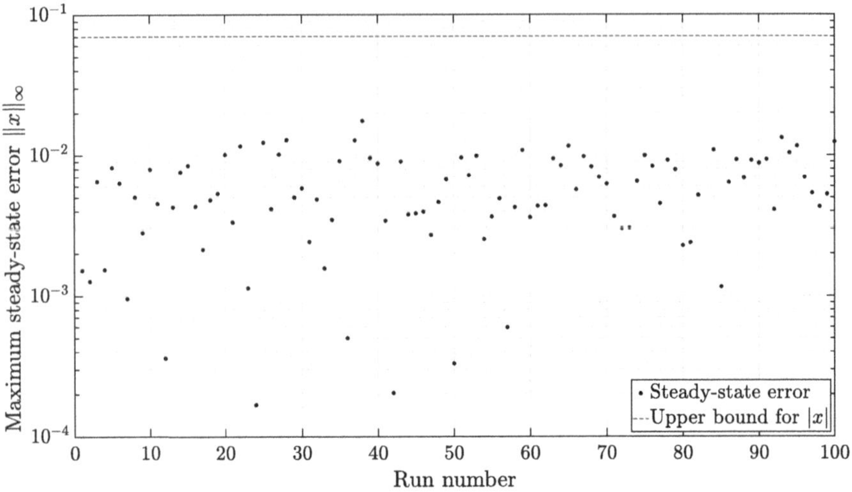

Fig. 5.1 Maximum steady state of 100 sample runs with the Gaussian kernel

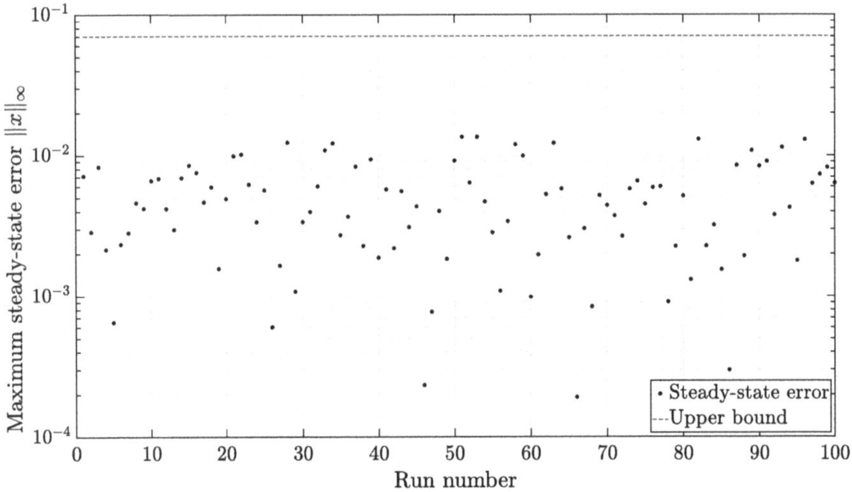

Fig. 5.2 Maximum steady state of 100 sample runs with the 3/2 Sobolev–Matérn kernel

of these radial kernels. This stability property applies to nonparametric uncertainties in spaces of analytic functions generated by the exponential kernel and an infinite family of Sobolev spaces of various smoothness s generated by the Wendland or Sobolev–Matérn kernels.

5.2.2 Robust Nonparametric Backstepping Over Native Spaces

In this section, we show how the preliminary results of Sect. 5.2.1 can leveraged to extend classical backstepping methods for robust nonlinear systems to the case when the uncertainty is contained in a native space. In particular, we consider a modification of the dynamical model (5.10) that is based on (2.292) of [8],

$$\dot{x}(t) = a(x(t)) + g(x(t))\xi(t) + E^1_{x(t)} f(\cdot, u(t)), \quad x(0) = x_0, \quad t \geq t_0, \tag{5.34a}$$

$$\dot{\xi}(t) = u(t) + E^2_{x(t)} h(\cdot, \xi(t), u(t)), \quad \xi(0) = \xi_0, \tag{5.34b}$$

where $x(t) \in \mathbb{X} \triangleq \mathbb{R}^n$, $u(t), \xi(t) \in \mathbb{U} \triangleq \mathbb{R}^m$, and $f : \mathbb{X} \times \mathbb{U} \to \mathbb{X}$ and $h : \mathbb{X} \times \mathbb{U} \times \mathbb{U} \to \mathbb{U}$ denote functional uncertainties. We assume that for each $\gamma \in \mathbb{U}$, $f(\cdot, \gamma) \in \mathcal{H}_1$, where $\mathcal{H}_1$ is an vector-valued RKHS induced by the operator kernel $\mathcal{K}^1 : \mathbb{X} \times \mathbb{X} \to \mathcal{L}(\mathbb{X})$. Similarly, we assume that for each $\gamma, \eta \in \mathbb{U}$, the function $h(\cdot, \gamma, \eta) \in \mathcal{H}_2$ is an vector-valued RKHS induced by the operator kernel $\mathcal{K}^2 : \mathbb{X} \times \mathbb{X} \to \mathcal{L}(\mathbb{U})$. The evaluation operators over $\mathcal{H}_1$ and $\mathcal{H}_2$ are denoted by E^1_x and E^2_x, respectively, for each fixed $x \in \mathbb{X}$. That is, $E^1_x f = f(x) \in \mathbb{X}$ for each $f \in \mathcal{H}_1$ and $x \in \mathbb{X}$, and $E^2_x h = h(x) \in \mathbb{U}$ for each $h \in \mathcal{H}_2$ and $x \in \mathbb{X}$.

The system given by (5.34) is in a cascaded form, which is suitable for backstepping control systems [8]. According to the backstepping design philosophy, $\xi(\cdot)$ in (5.34a) is interpreted as a pseudo-control input, and a control law for $\xi(\cdot)$ is designed so that $x(\cdot)$ attains the desired behavior, such as uniform ultimate boundedness despite uncertainties. Thus, $u(\cdot)$ is designed so that $\xi(\cdot)$ converges to its control law despite uncertainties.

As Sect. 5.2.1, a nonlinear controller for (5.34) begins by assuming that we have an RCLF for (5.34a). Thus, we assume that $V : \mathbb{X} \to \mathbb{R}$ is positive-definite and such that

$$\left\langle \left(\frac{\partial V(x)}{\partial x}\right)^{\mathrm{T}}, a(x) + g(x)\alpha(x) + E^1_x f(\cdot, \xi, u) \right\rangle_{\mathbb{X}} \leq -W(x) + b \tag{5.35}$$

for all $x \in \mathbb{X}$ and $u \in \mathbb{U}$, where $\alpha : \mathbb{X} \to \mathbb{U}$ denotes a continuously differentiable control law for $\xi(\cdot)$, $b > 0$ is constant, and $W : \mathbb{X} \to \mathbb{R}$ is continuous, positive-definite, and radially unbounded. Hence, the following result is applied to compute the control input $u(\cdot)$.

Theorem 5.2 (Robust Nonparametric Backstepping) *Consider the dynamical system given by (5.34), and suppose that (5.35) holds for the control Lyapunov function $V(\cdot)$. Then, the feedback control*

$$\mu(x,\xi) \triangleq -cz(x,\xi) + \frac{\partial \alpha(x)}{\partial x}(a(x)+g(x)\xi) - \frac{\partial V(x)}{\partial x} g(x) - \beta E_x^2 E_x^{2,*} z(x,\xi), \quad \textit{for all } (x,\xi) \in \mathbb{X}\times\mathbb{U}, \tag{5.36}$$

where $c > 0$ *and* $z(x,\xi) \triangleq \xi - \alpha(x)$*, guarantees global uniform boundedness of (5.34) with* $u(t) = \mu(x(t), \xi(t))$*,* $t \geq 0$*.*

Proof Consider the Lyapunov function candidate

$$\mathcal{V}(x,z) \triangleq V(x) + \frac{1}{2}\langle z, z\rangle_{\mathbb{U}}, \quad \text{for all } (x,z) \in \mathbb{X}\times\mathbb{U}. \tag{5.37}$$

The time derivative of $\mathcal{V}(\cdot,\cdot)$ along the trajectories of (5.34) with $u(t) = \mu(x(t), \xi(t))$, $t \geq t_0$, is given by

$$\begin{aligned}
\dot{\mathcal{V}}(x(t), z(t)) &= \left\langle \left(\frac{\partial V(x(t))}{\partial x}\right)^{\mathrm{T}}, a(x(t)) + g(x(t))(z(t)+\alpha(x(t))) + E_{x(t)}^1 f(\cdot, u(t))\right\rangle_{\mathbb{X}} \\
&\quad + \Big\langle z(t), u(t) + E_{x(t)}^2 h(\cdot, \xi(t), u(t)) \\
&\qquad - \frac{\partial \alpha(x(t))}{\partial x}\left(a(x(t)) + g(x(t))\xi(t) + E_{x(t)}^1 f(\cdot, u(t))\right)\Big\rangle_{\mathbb{U}} \\
&= \left\langle \left(\frac{\partial V(x(t))}{\partial x}\right)^{\mathrm{T}}, a(x(t)) + g(x(t))\alpha(x(t))) + E_{x(t)}^1 f(\cdot, u(t))\right\rangle_{\mathbb{X}} \\
&\quad - c\|z(t)\|_{\mathbb{U}}^2 - \beta\left(z(t), E_{x(t)}^2 E_{x(t)}^{2,*} z(t)\right)_{\mathbb{U}} \\
&\quad + \left\langle E_{x(t)}^{2,*} z(t), \frac{\partial \alpha(x(t))}{\partial x} h(\cdot, u(t)))\right\rangle_{\mathbb{U}} \\
&\quad + \left\langle z(t), E_{x(t)}^2 h(\cdot, \xi(t), u(t))\right\rangle_{\mathbb{U}} \\
&\leq -W(x(t)) - c\|z(t)\|_{\mathbb{U}}^2 - \beta\|E_{x(t)}^{2,*} z(t)\|_{\mathcal{H}_2}^2 \\
&\quad + \|E_{x(t)}^{2,*} z(t)\|_{\mathcal{H}_2}\left(\left\|\frac{\partial \alpha(x(t))}{\partial x} f\right\|_{\mathcal{H}_2} + \|h\|_{\mathcal{H}_2}\right) + b,
\end{aligned} \tag{5.38}$$

and the result now follows from Theorem 4.11. ■

The proposed robust backstepping approach can be then extended to dynamical systems in a cascaded form involving multiple pseudo-control inputs. This task, which can be performed by proceeding as in [8], is left as an exercise to the reader.

Example 5.3 To demonstrate the applicability of Theorem 5.2, consider the plant model

$$\dot{x}(t) = \xi(t) + E_{x(t)}^1 f(\cdot), \quad x(t_0) = x_0, \quad t \geq t_0, \tag{5.39a}$$

$$\dot{\xi}(t) = u(t), \quad \xi(t_0) = \xi_0, \tag{5.39b}$$

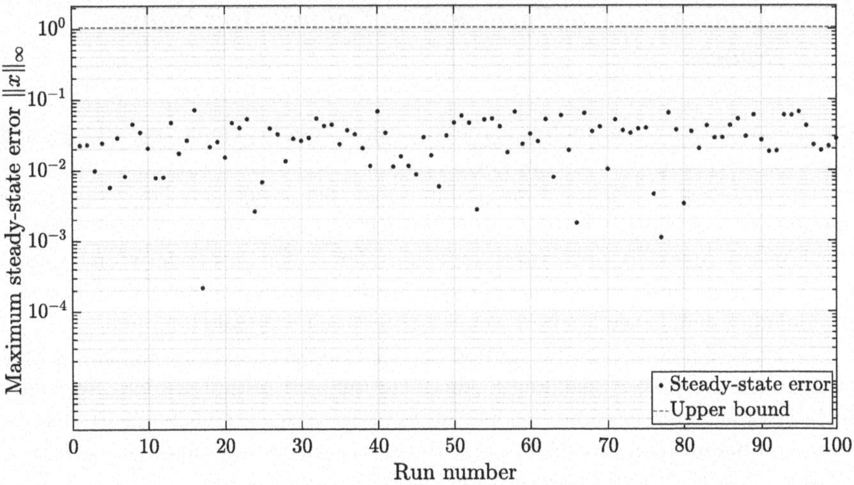

Fig. 5.3 Maximum steady state of 100 sample runs with the Gaussian kernel

which is in the same form as (5.34) with $n = 1$, $m = 1$, $a(x) = 0$ for all $x \in \mathbb{R}$, $g(x) = 1$, and $h = 0$. It is worthwhile noting how (5.39a) is in the same form as (5.27) and, hence, Example 5.2 can be leveraged to compute the control Lyapunov function for (5.39a) and the ideal feedback controller α in (5.35). Indeed, it follows from (5.26) that the feedback control law (5.13) guarantees that (5.35) is verified for this problem. Employing the same simulation parameters as in Example 5.2, we obtain the results shown in Figs. 5.3 and 5.4. △

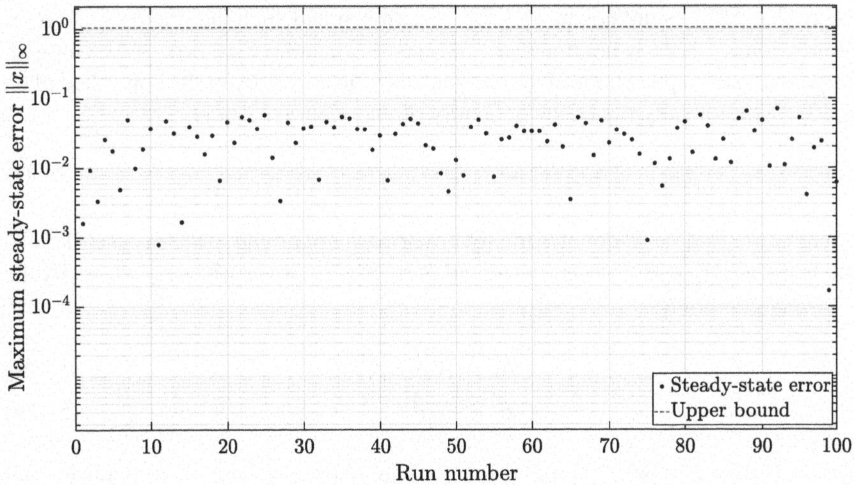

Fig. 5.4 Maximum steady state of 100 sample runs with the 3/2 Sobolev–Matérn kernel

5.2.3 *Remarks on Nonlinear Robust Control Over Native Spaces*

The proposed design of nonlinear robust control systems employing RCLFs and robust backstepping illustrates the powerful tools from RKHS theory that can be used to address classes of functional uncertainties that are considerably larger than those considered by existing nonlinear robust control systems. Indeed, in implementing the proposed approach, uncertainties are considered elements of an vector-valued RKHS, whereas classical approaches require some characterization of the functional uncertainties, and, hence, apply to smaller classes of uncertainties.

The analysis outlined in this section does not discuss how approximations play a role in native space embedding methods. In the next several sections, we study the role of the approximations in applications of native space embedding to adaptive control problems. In these problems, the functional uncertainty class $C_{N,\epsilon,R}$, defined in terms of errors of best approximation, plays a significant role.

5.3 MRAC Over Native Spaces of Vector-Valued Functions

In this section, we commence our study of MRAC systems design assuming that the matched functional uncertainty is captured by the vector-valued RKHS $\mathcal{H} \triangleq \mathcal{H}(\mathbb{X}, \mathbb{U})$ with $\mathbb{X} \triangleq \mathbb{R}^n$ and $\mathbb{U} \triangleq \mathbb{R}^m$. In particular, we consider nonlinear plants in the form

$$\dot{x}(t) = Ax(t) + B\left(u(t) + E_{x(t)}f\right), \quad x(t_0) = x_0, \quad t \geq t_0, \tag{5.40}$$

where $A \in \mathbb{R}^{n\times n}$ is unknown, $B \in \mathbb{R}^{n\times m}$ is known and such that the pair (A, B) is controllable, $f \in \mathcal{H}$ is unknown, $x : [t_0, \infty) \to \mathbb{X}$, and $u : [t_0, \infty) \to \mathbb{U}$. Our goal is to design an adaptive control system such that the trajectory of the controlled plant (5.40) asymptotically tracks the trajectory of the *reference model*

$$\dot{x}_r(t) = A_r x_r(t) + B_r r(t), \quad x_r(t_0) = x_{r,0}, \quad t \geq t_0, \tag{5.41}$$

where $r : [t_0, \infty) \to \mathbb{U}$ denotes the *reference command input*, which is user-defined and bounded, $A_r \in \mathbb{R}^{n\times n}$ is Hurwitz, and $B_r \in \mathbb{R}^{n\times m}$. The matrices A_r, B_r verify the *matching conditions*

$$A_r = A + B\alpha^{\mathrm{T}}, \tag{5.42a}$$

$$B_r = B\beta^{\mathrm{T}} \tag{5.42b}$$

for ideal gains $\alpha \in \mathbb{R}^{n\times m}$ and $\beta \in \mathbb{R}^{m\times m}$. The matrices α and β are defined ideal because α cannot be computed explicitly and there may be multiple β that verify (5.42b). If

$$u(t) = \alpha^{\mathrm{T}} x(t) + \beta^{\mathrm{T}} r(t) - E_{x(t)} f, \quad \text{for all } t \geq t_0,$$

then $\lim_{t\to\infty} \|x(t) - x_r(t)\| = 0$ uniformly in $t_0 \geq 0$. Unfortunately, since both α and f are unknown, this control input cannot be computed. Remarkably, the matching conditions (5.42) signify the assumption whereby there should be at least one ideal control input such that the trajectory controlled plant (5.40) can follow the trajectory of the reference model (5.41). If such conditions cannot be met, then the pursuit of an MRAC system would be fruitless.

The proposed RKHS embedding method for MRAC with vector-valued uncertainty is applied in two stages. In the first stage, which is addressed in this section, a *limiting distributed parameter system* (DPS), which evolves on an operator-valued space, is introduced. In the second stage, which is addressed in Sect. 5.4 below, consistent approximations of the limiting DPS generate finite-dimensional, realizable controllers.

Remark 5.5 (Comment on Notation) As noted in Remark 4.5, consistently with the literature in nonlinear systems analysis and control synthesis theory, in this book, we refer to initial value problems for which a Carathéodory solution is sought as dynamical systems. Starting this section, the design of control architectures involves multiple sets of differential equations, namely a dynamical system to control, e.g., (5.40), a reference model, e.g., (5.41), and IVPs that determine the dynamics of the gains needed to compute the control input, known as *adaptive laws*. Applying an adaptive control law, the dynamical system to control, the reference mode, and the adaptive laws become coupled, that is, are to be considered as a single dynamical model. Thus, to distinguish each set of IVPs, starting this chapter, we refer to the dynamical system to control as the *plant* or the *plant model*.

5.3.1 The Limiting Distributed Parameter System

Consider the plant model given by (5.40) and the reference model given by (5.41). The next theorem is the first key result of this section, and it provides an adaptive law so that the *limiting trajectory tracking error* $e(t) \triangleq x(t) - x_r(t), t \geq t_0$, asymptotically converges to zero despite the uncertainties in A, B, and $f \in \mathcal{H}$. For the statement of this result, let

$$\mathbb{Z} \triangleq \mathbb{X} \times \mathbb{R}^{n\times n} \times \mathbb{R}^{m\times m} \times \mathcal{H}, \tag{5.43}$$

and, for all $t \geq t_0$ consider the *adaptive laws*

$$\dot{\hat{\alpha}}(t) = -\Gamma_{\alpha} x(t) e^{\mathrm{T}}(t) P B, \quad \alpha(t_0) = \alpha_0, \tag{5.44a}$$

$$\dot{\hat{\beta}}(t) = -\Gamma_{\beta} r(t) e^{\mathrm{T}}(t) P B, \quad \beta(t_0) = \beta_0, \tag{5.44b}$$

$$\frac{\partial \hat{f}(t,\cdot)}{\partial t} = \Gamma_f \mathcal{K}_{x(t)} B^{\mathrm{T}} P e(t), \quad \hat{f}(t_0,\cdot) = f_0, \tag{5.44c}$$

where $\hat{\alpha} : [t_0, \infty) \to \mathbb{R}^{n \times m}$, $\hat{\beta} : [t_0, \infty) \to \mathbb{R}^{m \times m}$ denote *adaptive gain matrices*, $\hat{f} : [t_0, \infty) \times \mathbb{X} \to \mathcal{H}$ denotes the adaptive gain associated to the functional uncertainty, $\Gamma_{\alpha} \in \mathbb{R}^{n \times n}$, $\Gamma_{\beta} \in \mathbb{R}^{m \times m}$, and $\Gamma_f \in \mathbb{R}^{m \times m}$ denote the user-defined, symmetric, and positive-definite *adaptive rate matrices*, $P \in \mathbb{R}^{n \times n}$ denotes the symmetric, positive-definite solution of the *algebraic Lyapunov equation*

$$-Q = A_r^{\mathrm{T}} P + P A_{\mathrm{r}}, \tag{5.45}$$

and $Q \in \mathbb{R}^{n \times n}$ is symmetric, positive-definite, and user-defined. The adaptive laws for $\hat{\alpha}(t)$ and $\hat{\beta}(t)$ are the well-known real parametric adaptive learning laws for real matrices [4–6]. The adaptive law for $\hat{f}(t,\cdot)$ is a type of nonparametric adaptive learning law. At this point, we note that the adaptive law (5.44c) can be viewed as a partial differential equation (PDE), or as an ODE taking values in a Hilbert space, not as an ODE in some Euclidean space of fixed dimension.

Applying the *limiting control input*

$$u(t) = \hat{\alpha}^{\mathrm{T}}(t) x(t) + \hat{\beta}^{\mathrm{T}}(t) r(t) - E_{x(t)} \hat{f}(t,\cdot), \quad t \geq t_0, \tag{5.46}$$

the trajectory tracking error dynamics are given by

$$\dot{e}(t) = A_r e(t) - B\left(\tilde{\alpha}^{\mathrm{T}}(t) x(t) + \tilde{\beta}^{\mathrm{T}}(t) r(t) - E_{x(t)} \tilde{f}(t,\cdot)\right), \quad e(t_0) = x_0 - x_{r,0}, \tag{5.47}$$

where $\tilde{\alpha}(t) \triangleq \alpha - \hat{\alpha}(t)$, $\tilde{\beta}(t) \triangleq \beta - \hat{\beta}(t)$, and $\tilde{f}(t,\cdot) \triangleq f - \hat{f}(t,\cdot)$. Thus, the time-varying dynamical system given by the trajectory tracking error dynamics (5.47) and the adaptive laws (5.44) define a *limiting DPS* with a state $(e(t), \hat{\alpha}(t), \hat{\beta}(t), \hat{f}(t,\cdot))$ that evolves in $\mathbb{Z} \triangleq \mathbb{R}^n \times \mathbb{R}^{n \times m} \times \mathbb{R}^{m \times m} \times \mathcal{H}$. Note that $\left(e, \hat{\alpha}, \hat{\beta}, E_x \hat{f}\right) = (0, \alpha, \beta, E_x f)$ is the only equilibrium point of the nonlinear dynamical system given by (5.47) and (5.44). The next result proves the boundedness and asymptotic convergence of solutions of this system. For clarity of exposition, the proof of the existence of complete solutions of (5.47) and (5.44) is deferred to Sect. 5.3.2 below.

Theorem 5.3 *Consider the nonlinear plant dynamics (5.40), the reference model (5.41), the control input (5.46), and the adaptive laws (5.44). Assume that the matching conditions (5.42) are verified, and assume that the kernel $\mathcal{K}(\cdot,\cdot)$ that defines $\mathcal{H}$ is uniformly bounded on the diagonal. If the dynamical system given by (5.47) and*

(5.44) has a unique solution $(x, \tilde{f}) \in \mathbb{X} \times C^1([t_0, \infty), \mathcal{H})$ for every initial condition, then the trajectories of (5.47) and (5.44) are uniformly bounded and

$$\|x(t) - x_r(t)\|_{\mathbb{X}} \to 0 \quad as\ t \to \infty \tag{5.48}$$

uniformly in $t_0 \geq 0$.

Proof Consider the Lyapunov function candidate

$$V(e, \tilde{\alpha}, \tilde{\beta}, \tilde{f}) = \langle Pe, e\rangle_{\mathbb{R}^n} + \left\langle \Gamma_\alpha^{-1}\tilde{\alpha}, \tilde{\alpha}\right\rangle_{\mathrm{tr}} + \left\langle \Gamma_\beta^{-1}\tilde{\beta}, \tilde{\beta}\right\rangle_{\mathrm{tr}} + \left\langle \Gamma_f^{-1}\tilde{f}, \tilde{f}\right\rangle_{\mathcal{H}}, \quad (e, \tilde{\alpha}, \tilde{\beta}, \tilde{f}) \in \mathbb{Z}, \tag{5.49}$$

and note that, along the trajectories of (5.47) and (5.44), it holds that

$$\begin{aligned}
&\dot{V}\left(e(t), \tilde{\alpha}(t), \tilde{\beta}(t), \tilde{f}(t, \cdot)\right) \\
&= \langle P\dot{e}(t), e(t)\rangle_{\mathbb{R}^n} + \langle Pe(t), \dot{e}(t)\rangle_{\mathbb{R}^n} + 2\left(\left\langle \Gamma_\alpha^{-1}\tilde{\alpha}^{\mathrm{T}}(t), \dot{\tilde{\alpha}}(t)\right\rangle_{\mathrm{tr}} + \left\langle \Gamma_\beta^{-1}\tilde{\beta}(t), \dot{\tilde{\beta}}(t)\right\rangle_{\mathrm{tr}}\right. \\
&\quad \left. + \left\langle \Gamma_f^{-1}\tilde{f}(t, \cdot), \dot{\tilde{f}}(t, \cdot)\right\rangle_{\mathcal{H}}\right), \qquad (5.50) \\
&= -e^{\mathrm{T}}(t)Qe(t) \\
&\quad + 2\left[-\left\langle \tilde{\alpha}(t), x(t)e^{\mathrm{T}}(t)PB\right\rangle_{\mathrm{tr}} - \left\langle \tilde{\alpha}(t), \Gamma_\alpha^{-1}\dot{\hat{\alpha}}(t)\right\rangle_{\mathrm{tr}}\right] \qquad (5.51) \\
&\quad + 2\left[-\left\langle \tilde{\beta}(t), r(t)e^{\mathrm{T}}(t)PB\right\rangle_{\mathrm{tr}} - \left\langle \tilde{\beta}(t), \Gamma_\beta^{-1}\dot{\hat{\beta}}(t)\right\rangle_{\mathrm{tr}}\right] \qquad (5.52) \\
&\quad + 2\left[\left\langle \mathcal{K}_{x(t)}B^{\mathrm{T}}Pe(t), \tilde{f}(t, \cdot)\right\rangle_{\mathcal{H}} - \left\langle \tilde{f}(t, \cdot), \Gamma_f^{-1}\dot{\hat{f}}(t, \cdot)\right\rangle_{\mathcal{H}}\right], \qquad (5.53)
\end{aligned}$$

where we employed the property whereby $a^{\mathrm{T}}b = \mathrm{tr}\left(ba^{\mathrm{T}}\right)$ for any $a, b \in \mathbb{X}$. Thus, utilizing the adaptive laws (5.44), we deduce that

$$\dot{V}(e(t), \tilde{\alpha}(t), \tilde{\beta}(t), \tilde{f}(t, \cdot)) = -e^{\mathrm{T}}(t)Qe(t) \leq 0, \quad t \geq t_0 \tag{5.54}$$

along the trajectories of (5.47) and (5.44).

Since $\dot{V}(e(t), \tilde{\alpha}(t), \tilde{\beta}(t), \tilde{f}(t, \cdot)) \leq 0$ for all $t \geq t_0$, $V(e(t), \tilde{\alpha}(t), \tilde{\beta}(t), \tilde{f}(t, \cdot))$ is non-increasing, and since $V(\cdot, \cdot, \cdot, \cdot)$ is positive-definite, it follows from the monotone convergence theorem that $\lim_{t\to\infty} V\left(e(t), \tilde{\alpha}(t), \tilde{\beta}(t), \tilde{f}(t, \cdot)\right) = V_\infty$, where

$$V_\infty \leq V(e(t_0), \tilde{\alpha}(t_0), \tilde{\beta}(t_0), \tilde{f}(t_0, \cdot)). \tag{5.55}$$

The boundedness of $V(e(t), \tilde{\alpha}(t), \tilde{\beta}(t), \tilde{f}(t, \cdot))$ for all $t \geq t_0$ also implies that the states of (5.47) and (5.44) are bounded uniformly in $t_0 \geq 0$, that is, $e(\cdot) \in L^\infty\left(\mathbb{R}^+, \mathbb{R}^n\right)$, $\tilde{\alpha}(\cdot) \in L^\infty\left(\mathbb{R}^+, \mathbb{R}^{n\times n}\right)$, $\tilde{\beta}(\cdot) \in L^\infty\left(\mathbb{R}^+, \mathbb{R}^{m\times m}\right)$, and $\tilde{f}(\cdot) \in L^\infty\left(\mathbb{R}^+, \mathcal{H}\right)$. By integrating both sides of (5.54), we also conclude that $e(\cdot) \in L^2\left(\mathbb{R}^+, \mathbb{R}^n\right)$.

Next, to conclude the proof, we prove that $\dot{e}(\cdot) \in L^\infty\left(\mathbb{R}^+, \mathbb{X}\right)$. It follows from (5.47) that

$$\begin{aligned}\|\dot{e}(t)\|_{\mathbb{R}^n} \leq \|A_r\|\|e\|_{L^\infty(\mathbb{R}^+,\mathbb{X})} + \|\tilde{\beta}\|_{L^\infty(\mathbb{R}^+,\mathbb{R}^{m\times m})}\|r\|_{L^\infty(\mathbb{R}^+,\mathbb{U})} \\ + \|B\|\|e\|_{L^\infty(\mathbb{R}^+,\mathbb{X})}\|\tilde{\alpha}\|_{L^\infty(\mathbb{R}^+,\mathbb{R}^{n\times n})} \\ + \|B\|\|E_{x(t)}\|_{\mathcal{L}(\mathcal{H},\mathbb{U})}\|\tilde{f}(t,\cdot)\|_{\mathcal{H}}, \quad \text{for all } t \geq t_0.\end{aligned} \tag{5.56}$$

Now, $E_{x(t)} \in \mathcal{L}(\mathcal{H}, \mathbb{U})$ is a bounded operator per definition, and under the assumption that the kernel $\mathcal{K}$ is bounded on the diagonal, it follows from Example 3.1, (3.113), and the fact the adjoint of a bounded linear operator is bounded, that $\|E_{x(t)}\|_{\mathcal{L}(\mathcal{H},\mathbb{U})} \leq \bar{\mathfrak{K}}$. Furthermore, since $\|\tilde{f}(t,\cdot)\|_{\mathcal{H}} \leq \|\tilde{f}(t,\cdot)\|_{L^\infty(\mathbb{R}^+,\mathcal{H})}$ for all $t \geq t_0$ (see Sect. 2.4.6), we have proven that $\dot{e}(\cdot) \in L^\infty\left(\mathbb{R}^+, \mathbb{X}\right)$, and it follows from (4.6) that $e(\cdot)$ is uniformly continuous. Hence, $\dot{V}(e(t), \tilde{\alpha}(t), \tilde{\beta}(t), \tilde{f}(t,\cdot))$ is uniformly continuous for all $t \geq t_0$. Since $V(e(t), \tilde{\alpha}(t), \tilde{\beta}(t), \tilde{f}(t,\cdot))$ is bounded for all $t \in [t_0, \infty)$, it follows from Barbalat's lemma (Lemma 4.1 and [13]) that (5.48) is verified uniformly in $t_0 \geq 0$. ■

Remark 5.6 Before proceeding further, we note that in the proof of Theorem 5.3, we have applied the identity

$$\frac{d}{dt}\left\langle \tilde{f}(t,\cdot), \tilde{f}(t,\cdot)\right\rangle_{\mathcal{H}} = 2\left\langle \tilde{f}(t,\cdot), \dot{\tilde{f}}(t,\cdot)\right\rangle_{\mathcal{H}}. \tag{5.57}$$

If $\mathcal{H}$ is a finite-dimensional space, then (5.57) is always verified. More in general, if $\mathcal{H}$ is an operator-valued space, as assumed in this section, one needs to assume that $\tilde{f} \in C^1([0,\infty), \mathcal{H})$ for (5.57) to be verified. Indeed, under this assumption, we can interpret the inner product on the left-hand side of (5.57) as a composition of Fréchet differentiable operators. The term on the right-hand side of (5.57) then follows from the familiar composition law for taking the Fréchet derivative of a composition of Fréchet differentiable operators. Reference [14] discusses (5.57) for a more abstract DPS in the study of the so-called heat equation.

Theorem 5.3 proves that, applying the limiting control input (5.46) and the adaptive laws (5.44), the closed-loop plant trajectory is bounded at all times and asymptotically converges to the reference model's trajectory despite parametric uncertainties in α, β and *nonparametric uncertainty in* f, uniformly in the initial time. Although both (5.44a) and (5.44b) closely resemble those of classical MRAC [4, Chap. 9], (5.44c) is unique in comparison. Indeed, to the authors' knowledge, there does not exist an MRAC system that accounts for *nonparametric* matched uncertainties, such as f that appear in the limiting DPS, that are known to belong to some functional space, but whose representation is unknown. Thus, the framework introduced by Theorem 5.3 breaks from a conventional approach, whereby the matched uncertainty needs to be expressed in the form of the product of a regressor vector, which is either provided with a plant model or to be identified through neural networks or machine learning techniques [3, 4], and an unknown matrix.

Theorem 5.3 relies on the assumption whereby the nonlinear dynamical system is given by the limiting DPS in (5.47) and (5.44) is forward complete. In Sect. 5.3.2 below, we provide sufficient conditions for the existence of complete solutions of (5.47) and (5.44). The analysis is rather long, but it closely resembles the structure of arguments made in Chap. 4 for the forward completeness of ODEs.

Lastly, the limiting control input (5.46) is not implementable as-is. Indeed, $\hat{f}(t,\cdot)$, $t \geq t_0$, is the solution of a PDE and resides in an infinite-dimensional space. For this reason, we refer to (5.46) as the *limiting* feedback control input. Finite-dimensional approximations of $\hat{f}(t,\cdot)$, and, hence, of the limiting control input, which are implementable numerically, are discussed in Sect. 5.4 below.

5.3.2 Forward Completeness of the Limiting DPS

Theorem 5.3 assumes the existence of a complete solutions of (5.47) and (5.44), that is, of solutions over the interval $[t_0, \infty)$ for any initial condition. The next theorem provides sufficient conditions to guarantee the existence of solutions. For the statement of this result, note that the dynamical system given by (5.47) and (5.44) can be rewritten as

$$\dot{z}(t) = \mathcal{A}z(t) + \mathcal{G}(t, z(t)), \quad z(t_0) = z_0, \quad t \geq t_0, \tag{5.58}$$

where $z(t) \triangleq (e(t), \tilde{\alpha}(t), \tilde{\beta}(t), \tilde{f}(t)) \in \mathbb{Z}$, the operator matrix $\mathcal{A} \triangleq \operatorname{diag}\{A_r, 0, 0, 0\}$, and

$$\mathcal{G}(t, z) \triangleq \begin{bmatrix} -B\tilde{\alpha}^{\mathrm{T}}(x_r(t) + e) - B\tilde{\beta} + BE_{x_r(t)+e}\tilde{f} \\ \Gamma_\alpha(x_r(t) + e)e^{\mathrm{T}}PB \\ \Gamma_\beta r(t)e^{\mathrm{T}}PB \\ -\Gamma_f \mathcal{K}_{x_r(t)+e}B^{\mathrm{T}}Pe \end{bmatrix}$$

for each $z = (e, \tilde{\alpha}, \tilde{\beta}, \tilde{f}) \in \mathbb{Z}$.

Theorem 5.4 *Consider the closed-loop trajectory tracking error dynamics given by (5.58), and assume that the reference command input $t \mapsto r(t)$ is continuous and uniformly bounded. Furthermore, assume that the operator kernel $\mathcal{K}(x, y) \in \mathcal{L}(\mathbb{U})$ that defines the $\mathbb{U}$-valued native space $\mathcal{H}$ is such that*

1) The operator kernel $\mathcal{K}(\cdot,\cdot)$ is bounded on the diagonal in the sense that there is a constant $\bar{\mathfrak{K}} > 0$ such that

$$\|\mathcal{K}(x, x)\|_{\mathcal{L}(\mathbb{U})} \leq \bar{\mathfrak{K}}^2, \quad \textit{for all } x \in \mathbb{X}, \tag{5.59}$$

2) the mapping $(x, y) \mapsto \mathcal{K}(x, y)$ is Lipschitz continuous on $\mathbb{X}$, that is,

$$\|\mathcal{K}(x, y) - \mathcal{K}(p, q)\|_{\mathcal{L}(\mathbb{U})} \leq L(\|x - p\| + \|y - q\|) \tag{5.60}$$

for each $x, y \in \mathbb{X}$ and for all (p, q) in some neighborhood of (x, y), where $L(x, y) > 0$, and

3) the operator kernel $\mathcal{K}$ induces a native space $\mathcal{H}$ that is continuously embedded in the Lipschitz continuous functions

$$\mathcal{H} \hookrightarrow \left(C^{0,1}(\mathbb{X})\right)^m . \tag{5.61}$$

Then, for each initial condition $z_0 \in \mathbb{Z}$, there is a maximum interval of existence $[t_0, t_{\max})$ with $T_{\max} \in (t_0, \infty]$ such that (5.58) has a unique mild solution $z \in C([t_0, T_{\max}), \mathbb{Z})$. Furthermore, the map $z_0 \mapsto z$ is Lipschitz continuous from $\mathbb{Z}$ to $C([t_0, T_{\max}), \mathbb{Z})$. If $T_{\max} = \infty$ and $\mathcal{G} \in C^1([t_0, \infty) \times \mathbb{Z}, \mathbb{Z})$, then $z \in C^1([t_0, \infty), \mathbb{Z})$.

Before proceeding with the proof of Theorem 5.4, it is worthwhile making the following remark concerning assumptions (1)–(3) of Theorem 5.4.

Remark 5.7 (Comment on Assumption (1)) Let the operator-valued space $\mathcal{H}$ be the m-times Cartesian product of a scalar-valued space, that is, $\mathcal{H} \triangleq \mathcal{H}^m$, and let the scalar-valued kernel $\mathcal{H}$ be generated by an exponential, Sobolev–Matérn, inverse multiquadric, or Wendland kernel (see Sects. 3.3.1 and 3.3.2). Then Assumption (1) of Theorem 5.4 is verified.

Remark 5.8 (Comment on Assumption (2)) Equation (5.59) was already encountered as (3.142). If $\mathcal{H} \triangleq \mathcal{H}^m$, then Assumption 2) of Theorem 5.4 can be verified, for example, if the scalar-valued kernel $\mathfrak{K}(\cdot, \cdot)$ is contained in the space of Lipschitz continuous functions $C^{0,1}(\mathbb{X})$. The exponential, inverse multiquadric, sufficiently smooth Sobolev–Matérn, and sufficiently smooth Wendland kernels can be chosen to satisfy this property.

Remark 5.9 (Comment on Assumption (3)) If $\mathcal{H} = \mathcal{H}^m$, then Assumption (3) of Theorem 5.4 is verified for some common selections of the scalar-valued kernels. For example, the collection of Sobolev–Matérn kernels and Wendland kernels define native spaces $\mathcal{H}$ that are types of Sobolev spaces. These families can define Sobolev spaces of arbitrarily high order [15], and Sobolev spaces of sufficiently high order are continuously embedded in the spaces $C^{0,1}(\mathbb{X})$ of Lipschitz continuous functions [16] by a suitable Sobolev embedding theorem for Lipschitz spaces. See Theorem 2.7 for the more common embedding in $C_b(\mathbb{X})$.

We now turn to the proof of Theorem 5.4.

Proof The local existence and uniqueness result follows from Theorem 1.4 of Chap. 6 of [17]. To apply this result, we require that $e^{\mathcal{A}t} \triangleq \sum_{n=0}^{\infty} \frac{(\mathcal{A}t)^n}{n!}$ is a C_0-semigroup (see Definition 4.31) generated by $\mathcal{A}$; for each $z \in \mathbb{Z}$, the map $t \mapsto \mathcal{G}(t, z)$ is continuous over $[t_0, \infty)$; and, for each bounded subinterval of $[t_0, \infty)$, the map $z \mapsto \mathcal{G}(t, z)$ is Lipschitz continuous in bounded neighborhoods of $z = 0$, uniformly in time.

Since $\mathcal{A}$ is a bounded linear operator on $\mathbb{Z}$, it follows from Theorem 4.13 that the exponential $e^{\mathcal{A}t} \triangleq \sum_{n=0}^{\infty} \frac{(\mathcal{A}t)^n}{n!}$, $t \in [t_0, \infty)$, is a uniformly continuous semigroup (hence C_0) generated by $\mathcal{A}$. Thus, the first condition to apply Theorem 6.1.4 of [17] is satisfied.

Next, let $\mathcal{G} = [\mathcal{G}_1(t, z), \mathcal{G}_2(t, z), \mathcal{G}_3(t, z), \mathcal{G}_4(t, z)]^{\mathrm{T}}, (t, z) \in [t_0, \infty) \times \mathbb{Z}$, where $\mathcal{G}_i(t, z)$ denotes the ith component of $\mathcal{G}(t, z)$, $i = 1, \ldots, 4$. For each $z \in \mathbb{Z}$, the continuity in time of $\mathcal{G}_2(t, z)$ and $\mathcal{G}_3(t, z)$ is a direct consequence of the continuity of $x_r(\cdot)$ and $r(\cdot)$ on $[t_0, \infty)$. Furthermore, since $\tilde{f} \in \mathcal{H} \in (C^{1,0}(\mathbb{X}))^m \subseteq (C_b(\mathbb{X}))^m$, the composite term $E_{x_r(t)+e}$ is continuous for all $t \in [t_0, \infty)$, which, along with the continuity of $r(\cdot)$ and $x_r(\cdot)$, implies that $\mathcal{G}_1(t, z)$ is continuous in time. The bound

$$\|\mathcal{G}_4(t, z) - \mathcal{G}_4(\tau, z)\|_{\mathcal{H}} \quad \leq \|\mathcal{K}_{x_r(t)+e} - \mathcal{K}_{x_r(\tau)+e}\|_{\mathcal{L}(\mathbb{U},\mathcal{H})} \|B^{\mathrm{T}} P e\|_{\mathbb{U}} \tag{5.62}$$

holds for all $(t, \tau) \in [t_0, \infty) \times [t_0, \infty)$ and the continuity of the map $(x, z) \mapsto \mathcal{K}(x, z) \in \mathcal{L}(\mathbb{U})$ implies the continuity of the map $x \mapsto \mathcal{K}_x \in \mathcal{L}(\mathbb{U}, \mathcal{H})$. Since $x_r(\cdot)$ is continuous, it follows that $\mathcal{G}_4(t, z)$ is continuous in t for each fixed z. Thus, the second condition in Theorem 6.1.4 of [17] is satisfied.

Finally, we prove that, for all t in bounded subintervals of $[t_0, \infty)$, the map $z \mapsto \mathcal{G}(t, z)$ is Lipschitz continuous in bounded neighborhoods of $z = 0$, uniformly in time. For each bounded subinterval of $[t_0, \infty)$, the map $z \mapsto \mathcal{G}(t, z)$ is Lipschitz continuous in bounded neighborhoods of $z = 0$, if, for each $t > t_0$ and for each open ball $\mathcal{B}_R(0) \subset \mathbb{Z}$ of radius $R > 0$ centered at $z = 0$, there exists a constant $L_{t,R} > 0$ such that

$$\|\mathcal{G}(\tau, y) - \mathcal{G}(\tau, z)\|_{\mathbb{Z}} \leq L_{t,R} \|y - z\|_{\mathbb{Z}} \tag{5.63}$$

for all $\tau \in [t_0, t]$ and $y, z \in \mathcal{B}_R(0)$. To establish (5.63), let $y = \left(\tilde{y}, \tilde{c}, \tilde{d}, \tilde{g}\right) \in \mathcal{B}_R(0) \subset \mathbb{Z}$ and $z = \left(e, \tilde{\alpha}, \tilde{\beta}, \tilde{f}\right) \in B_R(0) \subset \mathbb{Z}$. Then, for each $t \in [t_0, \infty)$, $\mathcal{G}_3(t, \cdot)$ is Lipschitz continuous since

$$\begin{aligned} \|\mathcal{G}_3(\tau, y) - \mathcal{G}_3(\tau, z)\|_{\mathbb{R}^{m\times m}} &= \|\Gamma_\beta r(\tau)(e - \tilde{y})^{\mathrm{T}} P B\|_{\mathbb{R}^{m\times m}} \\ &\leq \Gamma_\beta r_{\max} \|PB\| \|(e - \tilde{y})\|_{\mathbb{X}} \end{aligned} \tag{5.64}$$

for all $\tau \in [t_0, t)$ and for all $(y, z) \in \mathbb{Z} \times \mathbb{Z}$, where $r_{\max} \geq 0$ denotes the upper bound on $\|r(\cdot)\|$. It follows from the triangle inequality (property (N4) of Definition 2.37) and the submultiplicative property of equi-induced norms (see Theorem 2.2) that, for each $t \geq t_0$,

$$\begin{aligned} &\|\mathcal{G}_2(\tau, y) - \mathcal{G}_2(\tau, z)\|_{\mathbb{R}^{n\times m}} \\ &\quad = \left\|\Gamma_\alpha \left((x_r(\tau) + \tilde{y})\tilde{y}^{\mathrm{T}} P B - (x_r(\tau) + e)e^{\mathrm{T}} P B\right)\right\|_{\mathbb{R}^{n\times m}} \\ &\quad \leq \|\Gamma_\alpha\| \|PB\| \left(x_{r,\max} \underbrace{\|e - \tilde{y}\|_{\mathbb{X}}}_{\text{Term 1}} + \underbrace{\left\|(\tilde{y}\tilde{y}^{\mathrm{T}} - ee^{\mathrm{T}})\right\|_{\mathbb{R}^{n\times n}}}_{\text{Term 2}} \right) \end{aligned} \tag{5.65}$$

for all $\tau \in [t_0, t)$ and for all $(y, z) \in \mathbb{Z} \times \mathbb{Z}$, where $\|x_r(t)\| \leq x_{r,\max}$, $t \geq t_0$, and the uniform boundedness of $x_r(\cdot)$ is guaranteed both by the Hurwitz property of A_r and the uniform boundedness of $r(\cdot)$. Term 1 in (5.65) is Lipschitz continuous. Term 2 in (5.65) is such that

$$\begin{aligned}\left\|(\tilde{y}\tilde{y}^{\mathrm{T}} - ee^{\mathrm{T}})\right\|_{\mathbb{R}^{n\times n}} &\leq \left\|\tilde{y}\tilde{y}^{\mathrm{T}} - e\tilde{y}^{\mathrm{T}} + e\tilde{y}^{\mathrm{T}} - ee^{\mathrm{T}}\right\|_{\mathbb{R}^{n\times n}}, \\ &\leq \left\|(\tilde{y} - e)\tilde{y}^{\mathrm{T}}\right\|_{\mathbb{R}^{n\times n}} + \left\|e(\tilde{y} - e)^{\mathrm{T}}\right\|_{\mathbb{R}^{n\times n}}, \\ &\leq 2R\|(e - \tilde{y})\|_{\mathbb{X}}, \end{aligned} \tag{5.66}$$

since $y, z \in B_R(0) \subseteq \mathbb{Z}$. Since both terms in (5.65) are Lipschitz continuous, for each bounded subinterval of $[t_0, \infty)$, $z \mapsto \mathcal{G}_2(\tau, z)$ is Lipschitz continuous in bounded neighborhoods of $z = 0$, uniformly in time. With regards to the term $\mathcal{G}_4(t, z)$, for each $t \in [t_0, \infty)$, we have that

$$\begin{aligned}&\|\mathcal{G}_4(\tau, y) - \mathcal{G}_4(\tau, z)\|_{\mathcal{H}} \\ &\quad\leq \|\Gamma_f\| \left\|\mathcal{K}_{x_r(\tau)+\tilde{y}} B^{\mathrm{T}} P\tilde{y} - \mathcal{K}_{x_r(\tau)+e} B^{\mathrm{T}} Pe\right\|_{\mathcal{H}} \\ &\quad\leq \|\Gamma_f\| \left(\bar{\mathfrak{K}}^2 \|B^{\mathrm{T}} P\| \|e - \tilde{y}\|_{\mathbb{X}}\right. \\ &\qquad \left. + \|B^{\mathrm{T}} P\|\ R\ \|\mathcal{K}_{x_r(\tau)+\tilde{y}} - \mathcal{K}_{x_r(\tau)+e}\|\right) \end{aligned} \tag{5.67}$$

for all $\tau \in [t_0, t)$ and for all for all $(y, z) \in B_R(0) \times B_R(0)$. Thus, we conclude that $z \mapsto \mathcal{G}_4(t, z)$ is Lipschitz continuous since the mapping $z \mapsto \mathcal{K}_z$ is assumed Lipschitz continuous in bounded neighborhoods of $z = 0$, uniformly in time. Finally, observe that, for each $t \in [t_0, \infty)$,

$$\begin{aligned}&\|\mathcal{G}_1(\tau, y) - \mathcal{G}(\tau, z)\|_{\mathbb{Z}} \\ &\quad\leq \|B\| \left(\|\tilde{c}^{\mathrm{T}}(x_r(\tau) + \tilde{y}) - \tilde{\alpha}^{\mathrm{T}}(x_r(\tau) + e)\|_{\mathbb{R}^n}\right. \\ &\qquad \left. + r_{\max}\|\tilde{\beta} - \tilde{d}\|_{\mathbb{R}^{m\times m}} + \|\tilde{f}(x_r(\tau) + e) - \tilde{g}(x_r(\tau) + \tilde{y})\|_{\mathbb{R}^m}\ \right) \\ &\quad\leq \|B\| \left(x_{r\max}\|\tilde{\alpha} - \tilde{c}\|_{\mathbb{R}^{n\times n}} + \|\tilde{c}^{\mathrm{T}}\tilde{y} - \tilde{\alpha}^{\mathrm{T}}e\|_{\mathbb{R}^n}\right. \\ &\qquad + \|\tilde{f}(x_r(\tau) + e) - \tilde{g}(x_r(\tau) + e)\|_{\mathbb{R}^m} \\ &\qquad \left. + \|\tilde{g}(x_r(\tau) + \tilde{y}) - \tilde{g}(x_r(\tau) - e)\|_{\mathbb{R}^m}\ \right) \end{aligned} \tag{5.68}$$

for all $\tau \in [t_0, t)$ and for all $(y, z) \in \mathcal{B}_R(0) \times \mathcal{B}_R(0)$. Now, similarly to (5.66), it follows that

$$\|\tilde{c}^{\mathrm{T}}\tilde{y} - \tilde{\alpha}^{\mathrm{T}}e\| \leq 2R\|e - \tilde{y}\|_{\mathbb{X}}, \tag{5.69}$$

since $y, z \in \mathcal{B}_R(0) \subseteq \mathbb{Z}$. Furthermore, for each $t \in [t_0, \infty)$, it holds that

$$\begin{aligned}\|\tilde{f}(x_r(\tau) + e) - \tilde{g}(x_r(\tau) + e)\| &\leq \|\tilde{f} - \tilde{g}\|_{C_b(\mathbb{X})} \\ &\leq \hat{C}\left\|\tilde{f} - \tilde{g}\right\|_{\mathcal{H}}, \end{aligned} \tag{5.70}$$

$$\begin{aligned}\|\tilde{g}(x_r(\tau)+\tilde{y})-\tilde{g}(x_r(\tau)-e)\| &\leq \|\tilde{g}\|_{(C^{0,1}(\mathbb{X}))^m}\|e-\tilde{y}\|_{\mathbb{X}} \\ &\leq \hat{D}\|\tilde{g}\|_{\mathcal{H}}\|e-\tilde{y}\|_{\mathbb{X}} \\ &\leq \hat{D}R\|e-\tilde{y}\|_{\mathbb{X}} \end{aligned} \tag{5.71}$$

for all $\tau \in [t_0, t)$ and for all $(y, z) \in \mathcal{B}_R(0) \times \mathcal{B}_R(0)$. In both (5.70) and (5.71), we have used the fact that $\mathcal{H}$ is continuously embedded in $(C_b(\mathbb{X}))^m$ and $(C^{0,1}(\mathbb{X}))^m$, respectively, that is, $\mathcal{H} \hookrightarrow (C_b(\mathbb{X}))^m$ and $\mathcal{H} \hookrightarrow (C^{0,1}(\mathbb{X}))^m$. The former embedding holds since the operator kernel is uniformly bounded on the diagonal, and the latter holds by assumption. These embeddings mean that there are positive constants $\hat{C}$ and $\hat{D}$ such that

$$\begin{aligned}\|q\|_{(C_b(\mathbb{X}))^m} &\leq \hat{C}\|q\|_{\mathcal{H}}, \\ \|q\|_{(C^{0,1}(\mathbb{X}))^m} &\leq \hat{D}\|q\|_{\mathcal{H}}\end{aligned}$$

for all $q \in \mathcal{H}$. Thus, (5.68)–(5.71) imply that the map $z \mapsto \mathcal{G}_1(t, z)$ is Lipschitz continuous in bounded neighborhoods of $z = 0$, uniformly in time, and the third condition to apply Theorem 6.1.4 of [17] is verified.

It follows from Theorem 1.4 in Chap. 6 of [17] that there exists a maximal time of existence $T_{\max} \in [t_0, \infty]$ such that a local mild solution $z \in C([t_0, T_{\max}), \mathbb{Z})$ exists and is unique. If $T_{\max} = \infty$, then there is a unique mild solution $z \in C([0, \infty), \mathbb{Z})$. We finally obtain that $z \in C^1([0, \infty), \mathbb{Z})$ from the assumption that $\mathcal{G}$ is C^1 and Theorem 1.5 of Chap. 6 in [17].

Using the same arguments as in [18, Theorem 3.4], the continuous dependence of solutions on initial conditions follows from Gronwall's inequality [18, Lemma A.1] and the Lipschitz conditions on $\mathcal{G}(\cdot, \cdot)$, and, hence, is omitted for brevity. ■

5.4 Finite-Dimensional Approximations of the Limiting DPS

As discussed in Sect. 5.3.1, the control input (5.46) is not implementable as-is because the term $\hat{f}(t, \cdot)$, $t \geq t_0$, resides in an infinite-dimensional space. In this section, we present finite-dimensional approximations of $\hat{f}(t, \cdot)$, $t \geq t_0$, and, hence, of the limiting control input. To this goal, we consider the finite-dimensional subspace (5.3), which can be equivalently defined as

$$\mathcal{H}_{d(N)} = \text{span}\,\{\mathcal{K}(\cdot, \xi_i)\,\alpha : \xi_i \in \Xi_N, \alpha \in \mathbb{U}\}, \tag{5.72}$$

and define the *approximation of* $\hat{f}(t, \cdot)$ as

$$\hat{f}_N(t, \cdot) \triangleq \sum_{i=1}^{N} \mathcal{K}(\cdot, \xi_i)\,\hat{\Theta}_i(t), \quad t \geq t_0, \tag{5.73}$$

where $\hat{\Theta}_i(t) \in \mathbb{U}$. In this case, the limiting control input (5.46) can be approximated by the finite-dimensional controller

$$u_N(t) = \hat{\alpha}_N^{\mathrm{T}}(t)x_N(t) + \hat{\beta}_N^{\mathrm{T}}(t)r(t) - E_{x_N(t)}\hat{f}_N(t,\cdot), \tag{5.74}$$

where $\hat{f}_N(t,\cdot) \in \mathcal{H}_{d(N)}$.

With the finite-dimensional subspace defined, we can use its corresponding projection operator $\mathbf{\Pi}_N$ to express the unknown vector nonlinearity as the sum of two orthogonal components, namely

$$E_{x(t)}f = E_{x(t)}\left(\mathbf{\Pi}_N f + (I - \mathbf{\Pi}_N) f\right), \qquad t \geq t_0, \tag{5.75}$$

where $\mathbf{\Pi}_N f$ denotes the *orthogonal projection* of an arbitrary function f residing in the infinite-dimensional space $\mathcal{H}$ onto the finite-dimensional space of approximants $\mathcal{H}_{d(N)}$, and $\mathbf{\Pi}_N f$ denotes the *projection error*. Furthermore, employing (5.74), the plant dynamics (5.40) reduce to

$$\dot{x}_N(t) = Ax_N(t) + B\left(u_N(t) + E_{x_N(t)}\mathbf{\Pi}_N f\right) + BE_{x_N(t)}(I - \mathbf{\Pi}_N) f, \\ x_N(t_0) = x_0, \quad t \geq t_0. \tag{5.76}$$

Thus, the trajectory tracking error dynamics are now defined as $e_N(t) \triangleq x_N(t) - x_r(t)$, $t \geq t_0$, and, letting $\tilde{f}_N(t,\cdot) \triangleq f - \hat{f}_N(t,\cdot)$, the trajectory tracking error dynamics are given by

$$\dot{e}_N(t) = A_r e_N(t) - B\left(\tilde{\alpha}_N^{\mathrm{T}}(t)x_N(t) + \tilde{\beta}_N^{\mathrm{T}}(t)r(t) - E_{x_N(t)}\mathbf{\Pi}_N\tilde{f}_N(t,\cdot)\right) \\ + BE_{x_N(t)}\underbrace{\left((I - \mathbf{\Pi}_N)\tilde{f}_N(t,\cdot)\right)}_{(I-\mathbf{\Pi}_N)f}, \quad e_N(t_0) = x_N(t_0) - x_r(t_0), \quad t \geq t_0. \tag{5.77}$$

Comparing (5.47) and (5.77), it appears that the trajectory tracking error dynamics obtained by employing a finite-dimensional approximation of the matched uncertainties result in an additional term, namely, $BE_{x_N(t)}((I - \mathbf{\Pi}_N)f)$, $t \geq t_0$, which can be interpreted as a nonparametric, external disturbance. This section presents several adaptive laws that account for this term.

5.4.1 The Dead-Zone Modification

To account for the projection error term in (5.77), in this section, we employ an approach inspired by the dead-zone modification of classical MRAC originally presented in [19]. We present two sets of robust MRAC systems. In Sect. 5.4.1.1, we develop the adaptive controller assuming that the native space of vector-valued functions is a Cartesian product $\mathcal{H} = \mathcal{H}^m$, where $\mathcal{H}$ denotes a scalar-valued RKHS. In this case, $\mathfrak{K}$ denotes the kernel generating $\mathcal{H}$ and the operator kernel for $\mathcal{H}$ is given by $\mathcal{K} = I_m \mathfrak{K}$. In Sect. 5.4.1.2, we address the case whereby $\mathcal{H}$ is induced by a generic operator kernel $\mathcal{K}$.

For the statement of the proposed results, let

$$\mathbb{Z}_N \triangleq \mathbb{X} \times \mathbb{R}^{n\times n} \times \mathbb{R}^{m\times m} \times \mathcal{H}_{d(N)}, \tag{5.78}$$

let $P \in \mathbb{R}^{n\times n}$ denote the symmetric, positive-definite solution of the algebraic Lyapunov equation (5.45) for some symmetric, positive-definite, user-defined matrix $Q \in \mathbb{R}^{n\times n}$, let $R > 0$ denote the radius of the uncertainty class C_R defined in (5.1), and let $\epsilon_N \triangleq \epsilon_N(f) \triangleq \|(I - \Pi_N)f\|_{\mathcal{H}}$ be the upper bound on the projection error in the uncertainty class $C_{N,\epsilon,R}$ in (5.2).

5.4.1.1 Operator-Value RKHS as the Cartesian Product of Scalar-Valued RKHSs

We define two different choices for minimum dead-zone width. If we assume the trajectory is contained in the user-defined set $\Omega_0 \supseteq \bigcup_{\tau\geq t_0>0} x_N(\tau)$, then let

$$d_{0,N} \triangleq \frac{2\|PB\|}{\lambda_{\min}(Q)} \sup_{\eta\in\Omega_0} \mathcal{P}_N(\eta)\epsilon_N \tag{5.79}$$

denote the *minimal dead-zone width*, where $\lambda_{\min}(\cdot)$ denotes the smallest eigenvalue of its argument; note that $d_{0,N}$ is finite since the power function $\mathcal{P}_N$ is uniformly bounded (see Sect. 3.8.1). In this case, let $d > d_{0,N}$ denote the *user-defined dead-zone width*. Although the closed-loop plant trajectory $x_N(\cdot)$ is only known a posteriori, that is, after the control input is applied, in most problems of practical interest, it is reasonable to assume that a subset of $\mathbb{X}$ where the closed-loop trajectory will lie can be estimated. If Ω_0 is considered unknown, but only assume that $f \in C_{R,\epsilon,N}$, then we define

$$d_\epsilon \triangleq \frac{2\|PB\|}{\lambda_{\min}(Q)} \bar{\mathfrak{K}} \epsilon \tag{5.80}$$

as the *minimal dead-zone width*, and $d > d_\epsilon$ as the *user-defined dead-zone width.*

We consider the adaptive laws

$$\dot{\hat{\alpha}}_N(t,\cdot) = \begin{cases} -\Gamma_\alpha x_N(t) e_N^{\mathrm{T}}(t) PB, & \|e_N(t)\| \geq d, \\ 0, & \|e_N(t)\| < d, \end{cases} \tag{5.81a}$$

$$\dot{\hat{\beta}}_N(t) = \begin{cases} -\Gamma_\beta r(t) e_N^{\mathrm{T}}(t) PB, & \|e_N(t)\| \geq d, \\ 0, & \|e_N(t)\| < d, \end{cases} \tag{5.81b}$$

$$\dot{\hat{f}}_N(t,\cdot) = \begin{cases} \Gamma_f \mathbf{\Pi}_N \mathcal{K}_{x_N(t)} B^{\mathrm{T}} P e_N(t), & \|e_N(t)\| \geq d, \\ 0, & \|e_N(t)\| < d, \end{cases} \tag{5.81c}$$

where $\hat{\alpha}_N(t_0) = \hat{\alpha}_{N,0}$, $\hat{\beta}_N(t_0) = \hat{\beta}_{N,0}$, $\hat{f}_N(t_0,\cdot) = \hat{f}_{N,0}$, and the adaptive rate matrices $\Gamma_\alpha \in \mathbb{R}^{n\times n}$, $\Gamma_\beta \in \mathbb{R}^{m\times m}$, and $\Gamma_f \triangleq \gamma_f I_m \in \mathbb{R}^{m\times m}$ are user-defined, symmetric, and positive-definite adaptive rate matrices.

Theorem 5.5 *Consider the closed-loop trajectory tracking error dynamics given by (5.58), its finite-dimensional counterpart given by (5.77), and the adaptive laws (5.81). Suppose that the assumptions of Theorems 5.3 and 5.4 are verified, assume that the closed-loop trajectory $t \mapsto x_N(t)$ is contained in Ω_0, assume that $f \in C_{N,\epsilon,R}$, and let $d \geq d_{0,N}$. Then, there exists $T \in (t_0,\infty)$ such that $\|e_N(t)\|_{\mathbb{X}} \triangleq \|x_N(t) - x_r(t)\|_{\mathbb{X}} \leq d$ for all $t \geq T$.*

Proof Consider the Lyapunov function candidate

$$\begin{aligned} &V(e_N, \tilde{\alpha}_N, \tilde{\beta}_N, \tilde{f}_N) \\ &= \langle P e_N, e_N\rangle_{\mathbb{R}^n} + \left\langle \Gamma_\alpha^{-1}\tilde{\alpha}_N, \tilde{\alpha}_N\right\rangle_{\mathrm{tr}} + \left\langle \Gamma_\beta^{-1}\tilde{\beta}_N, \tilde{\beta}_N\right\rangle_{\mathrm{tr}} \\ &\quad + \left\langle \Gamma_N^{-1}\tilde{f}_N(t,\cdot), \tilde{f}_N(t,\cdot)\right\rangle_{\mathcal{H}}, \quad (e_N, \tilde{\alpha}_N, \tilde{\beta}_N, \tilde{f}_N) \in \mathbb{Z}_N. \end{aligned} \tag{5.82}$$

If $\Omega_0 \supseteq \bigcup_{t\geq t_0} x_N(t)$, $d > d_{0,N}$, and $\|e_N(t)\|_{\mathbb{X}} \geq d \geq d_{0,N}$ for some $t \geq t_0$, then, it follows from (3.157) that along the trajectories of (5.77) and (5.81), it holds that

$$\begin{aligned} \dot{V}(t) &= -2\,\langle Q e_N(t), e_N(t)\rangle_{\mathbb{R}} \\ &\quad - 2\left\langle \tilde{\alpha}_N(t), x_N(t) e_N^{\mathrm{T}}(t) P B + \Gamma_\alpha^{-1}\dot{\hat{\alpha}}_N(t)\right\rangle_{\mathrm{tr}} \\ &\quad - 2\left\langle \tilde{\beta}_N(t), r(t) e_N^{\mathrm{T}}(t) P B + \Gamma_\beta^{-1}\dot{\hat{\beta}}_N(t)\right\rangle_{\mathrm{tr}} \\ &\quad + 2\left\langle \tilde{f}_N(t,\cdot), \mathbf{\Pi}_N \mathcal{K}_{x_N(t)} B^{\mathrm{T}} P e_N(t) - \Gamma_f^{-1}\dot{\hat{f}}_N(t,\cdot)\right\rangle_{\mathcal{H}} \\ &\quad + 2\left\langle \tilde{f}_N(t,\cdot), (I - \mathbf{\Pi}_N)\mathcal{K}_{x_N(t)} B^{\mathrm{T}} P e_N(t)\right\rangle_{\mathcal{H}} \\ &= -\langle Q e_N(t), e_N(t)\rangle_{\mathbb{X}} \\ &\quad + 2\left\langle B^{\mathrm{T}} P e_N(t), E_{x_N(t)}(I - \mathbf{\Pi}_N) f\right\rangle_{\mathbb{U}} \\ &\leq -\langle Q e_N(t), e_N(t)\rangle_{\mathbb{X}} \\ &\quad + 2\|PB\|\|e_N(t)\|_{\mathbb{X}}\|E_{x_N(t)}(I - \mathbf{\Pi}_N) f\|_{\mathbb{U}} \\ &\leq -\lambda_{\min}(Q)\|e_N(t)\|_{\mathbb{X}}\left(\|e_N(t)\|_{\mathbb{X}} - d_{0,N}\right) \\ &\leq -\lambda_{\min}(Q)\|e_N(t)\|_{\mathbb{X}}\left(\|e_N(t)\|_{\mathbb{X}} - d\right), \end{aligned} \tag{5.83}$$

where $\dot{V}(\cdot)$ represents $\dot{V}(e_N(\cdot), \tilde{\alpha}_N(\cdot), \tilde{\beta}_N(\cdot), \tilde{f}_N(t,\cdot))$ for notational brevity. Thus, the result follows from Theorem 4.11. ■

In the following, we present a result similar to Theorem 5.5 under less restrictive assumptions.

Theorem 5.6 *Consider the closed-loop trajectory tracking error dynamics given by (5.58), its finite-dimensional counterpart given by (5.77), and the adaptive laws (5.81). Suppose that the assumptions of Theorems 5.3 and 5.4 are verified, assume that* $f \in C_{N,\epsilon,R}$, *and let* $d \geq d_\epsilon$. *Then, there exists* $T \in (t_0, \infty)$ *such that* $\|e_N(t)\|_{\mathbb{X}} \triangleq \|x_N(t) - x_r(t)\|_{\mathbb{X}} \leq d$ *for all* $t \geq T$.

Proof The proof of this result is similar to that of Theorem 5.5 noting that, since $f \in C_{N,\epsilon,R}$ and $\mathcal{K}$ is uniformly bounded on the diagonal, $\|E_{x_N(t)}(I - \Pi_N) f\|_{\mathbb{U}} \leq \bar{\mathfrak{K}}^2 \epsilon$. ■

Similar to the classical dead-zone modification of MRAC, in order to apply Theorems 5.5 and 5.6, it is necessary to define a sufficiently large dead zone, that is, choose either $d \geq d_{0,N}$ or $d \geq d_\epsilon$. In practice, for fixed $\mathcal{H}_N$ and N, Theorems 5.5 and 5.6 should be applied iteratively, starting with a sufficiently large value of d, measuring the bound on the trajectory tracking error after a sufficiently long time, and then decreasing d until such a bound is no longer a decreasing function of d. If we were able to set $d = d_{0,N}$ or $d = d_\epsilon$, then we would obtain the ultimate bounds

$$\|x_N(t) - x_r(t)\|_{\mathbb{X}} \leq O\left(\epsilon_N \sup_{\eta \in \Omega} \mathcal{P}_N(\eta)\right), \tag{5.84a}$$

$$\|x_N(t) - x_r(t)\|_{\mathbb{X}} \leq O(\epsilon), \tag{5.84b}$$

respectively. Although (5.84a) yields a tighter performance bound, (5.84b) uses information that is more likely available.

Remark 5.10 The adaptive laws (5.81) are discontinuous, and, hence, may lead to chattering in the closed-loop plant dynamics. Since the discontinuities in (5.81) are of the second kind (i.e., jump discontinuities), then it is common practice to approximate these adaptive laws by means of continuous functions. The problem of approximating the step function has been studied extensively in the context of activation functions for neural networks; for details, see [20].

5.4.1.2 Generic Operator Kernels

The analysis in Theorems 5.5 and 5.6 assumed that $\mathcal{H} \triangleq \mathcal{H}^m$ and, hence, $\mathcal{K}(x, y) \triangleq \mathfrak{K}(x, y) I_m$ for all $x, y \in \Omega_0$, with I_m denoting the identity on $\mathbb{R}^m$ and $\mathfrak{K}$ the scalar-valued reproducing kernel of the RKHS $\mathcal{H}$ of real-valued functions over Ω_0. This choice is made only to guarantee the simple form of the error bound in (3.157), which is used to define the upper bounds on the time derivative of Lyapunov function in (5.83).

A modification of this argument for an RKHS $\mathcal{H}$ of $\mathbb{U}$-valued functions over Ω_0, which is induced by a generic admissible operator kernel $\mathcal{K} : \Omega_0 \times \Omega_0 \to \mathcal{L}(\mathbb{U})$, is presented here. For a generic operator kernel, (3.153) and (3.154) ensure that

$$\left|\langle E_x(I - \mathbf{\Pi}_N)f, \alpha\rangle_{\mathbb{U}}\right| \le \bar{\mathcal{P}}_N(x)\|\alpha\|_{\mathbb{U}}\|(I - \mathbf{\Pi}_N)f\|_{\mathcal{H}} \tag{5.85}$$

for all $x \in \Omega_0$, $\alpha \in \mathbb{U}$, and $f \in \mathcal{H}$, where

$$\bar{\mathcal{P}}_N(x) \triangleq \sqrt{\|\mathcal{K}(x, x) - \mathcal{K}_N(x, x)\|}; \tag{5.86}$$

see (3.156). If we use this more general bound, the conclusions of Theorems 5.5 and 5.6 hold replacing $\mathcal{P}_N$ with $\bar{\mathcal{P}}_N$.

5.4.1.3 Coordinate Implementation

In this section, we study the coordinate realizations of the adaptive law (5.81c) with $\mathcal{H}_{d(N)} \triangleq \mathcal{H}_N^m$, and a diagonal, positive-definite, adaptive rate matrix $\Gamma_f \triangleq \text{diag}\left(\gamma_f^1, \ldots, \gamma_f^m\right)$. Since the vector-valued RKHS $\mathcal{H}_{d(N)}$ is given by the Cartesian product of the scalar-valued RKHS $\mathcal{H}_N$ m times, it is possible to write $\hat{f}_N(t, \cdot)$ in the vector form

$$\hat{f}_N(t, \cdot) = \left[\sum_{j=1}^{N} \hat{\theta}_{N,j}^1(t)\mathfrak{K}_{\xi_{N,j}}, \ldots, \sum_{j=1}^{N} \hat{\theta}_{N,j}^m(t)\mathfrak{K}_{\xi_{N,j}}\right]^{\mathrm{T}} = \sum_{j=1}^{N} \mathcal{K}\left(\cdot, \xi_{N,j}\right) \hat{\Theta}_{N,j}(t), \tag{5.87}$$

where $\hat{\Theta}_{N,j}(\cdot) = \left[\hat{\theta}_{N,j}^1(\cdot), \ldots, \hat{\theta}_{N,j}^m(\cdot)\right]^{\mathrm{T}}$, $\mathfrak{K}_{\xi_{N,j}}(\cdot)$ denotes the kernel basis function centered at $\xi_{N,j}$, and $\hat{\theta}_{N,j}^l$ denotes the coefficient associated with the kernel basis function in the lth subspace of $\mathcal{H}$ with $l \in \{1, \ldots, m\}$. Taking only the lth component of $\hat{f}_N \triangleq \left[\hat{f}_N^1, \ldots, \hat{f}_N^m\right]^{\mathrm{T}}$, we have that

$$\frac{\partial \hat{f}_N^l}{\partial t}(t, \cdot) \triangleq \sum_{j=1}^{N} \dot{\hat{\theta}}_{N,j}^l(t)\mathfrak{K}_{\xi_{N,j}} = \gamma_f^l \Pi_N \mathfrak{K}_{x_N(t)}[B^{\mathrm{T}} P e_N(t)]_l$$
$$\text{for all } l \in \{1, \ldots, m\}, \text{ for all } t \ge t_0, \text{ and for each } i \in \{1, \ldots, N\}. \tag{5.88}$$

Here, $[B^{\mathrm{T}} P e_N(\cdot)]_l$ denotes the lth component of $B^{\mathrm{T}} P e_N(\cdot)$. Note that from the scalar reproducing property of the scalar RKHS, $E_{x_N(t)}^* \alpha = \mathfrak{K}_{x_N(t)}\alpha$ for any scalar $\alpha \in \mathbb{R}$. To calculate the projection Π_N of (5.88), we take the inner product of the above equation with an arbitrary $\mathfrak{K}_{\xi_{N,i}}$ for $1 \le i \le N$. This gives us

$$\sum_{j=1}^{N} \langle \mathfrak{K}_{\xi_{N,i}}, \mathfrak{K}_{\xi_{N,j}} \rangle_{\mathcal{H}} \dot{\hat{\theta}}^l_{N,j}(t) = \gamma^l_f \langle \mathfrak{K}_{\xi_{N,i}}, \Pi_N \mathfrak{K}_{x_N(t)} \rangle_{\mathcal{H}} [B^{\mathrm{T}} P e_N(t)]_l,$$
$$\text{for all } l \in \{1, \ldots, m\} \text{ for all } t \geq t_0, \text{ and for each } i \in \{1, \ldots, N\}. \quad (5.89)$$

Recall that we define the Grammian matrix of the scalar kernel $\mathfrak{K}$ to be $\mathbb{K}_N = [\langle \mathfrak{K}_{\xi_{N,i}}, \mathfrak{K}_{\xi_{N,j}} \rangle_{\mathcal{H}}]$. Thus, (5.89) simplifies to

$$\sum_{j=1}^{N} \mathbb{K}_{N,ij} \dot{\hat{\theta}}^l_{N,j}(t) = \gamma^l_f \mathfrak{K}\left(\xi_{N,i}, x_N(t)\right) [B^{\mathrm{T}} P e_N(t)]_l,$$
$$\text{for all } l \in \{1, \ldots, m\} \text{for all } t \geq t_0, \text{ and for each } i \in \{1, \ldots, N\}, \quad (5.90)$$

and, hence,

$$\dot{\hat{\theta}}^l_N(t) = \gamma^l_f \mathbb{K}_N^{-1} \begin{bmatrix} \mathfrak{K}\left(\xi_{N,1}, x_N(t)\right) \\ \vdots \\ \mathfrak{K}\left(\xi_{N,N}, x_N(t)\right) \end{bmatrix} [B^{\mathrm{T}} P e_N(t)]_l,$$
$$\text{for all } l \in \{1, \ldots, m\} \text{ and for all } t \geq t_0, \quad (5.91)$$

which is the adaptive law for lth entry of the vector of functions in $\hat{f}_N$.

Remark 5.11 (Comment on Computational Cost) It follows from (5.91) that, to implement the adaptive law for the adaptive gain corresponding to RKHS-embedded functional uncertainty, we must evaluate the effect of the kernel functions at each time step and at each center. While a large number of centers over a large subset of the state space would be ideal to increase the ability of the proposed method to capture matched uncertainties, this approach is likely to have a high computational cost. This problem, which applies to any of the control laws discussed in this chapter, is addressed in Chap. 6

5.4.1.4 Robustness to Exogenous Noise and Numerical Noise

Thus far, the model problem studied in this section did not consider any exogenous disturbances or numerical error introduced into the closed-loop control scheme due to the sensitivity of numerical calculations. In the context of classical MRAC theory, the problem of unmodeled external disturbances is well known. These effects may lead to the so-called *parameter drift*, that is, the effect whereby the trajectory tracking error asymptotically converges to zero while the adaptive gains, and, hence, the control input, diverge [4, Chap. 10]. Although the contribution of numerical errors is usually omitted in the study of real parametric adaptive control, it can be comparable to the effect of external disturbances, and it can be a significant factor in deriving performance bounds for nonparametric methods.

To account for these effects, consider the plant model

$$\dot{x}(t) = Ax(t) + B\left(u(t) + E_x(t)f\right) + \Gamma(t), \quad x(t_0) = x_0, \quad t \geq t_0, \tag{5.92}$$

where $\Gamma : [t_0, \infty) \to \mathbb{R}^n$ is an *unmatched uncertainty* that is uniformly bounded in time, that is,

$$\|\Gamma\|_{L^\infty([t_0,\infty),\mathbb{X})} \leq \gamma \tag{5.93}$$

for some $\gamma \geq 0$. We refer to $\Gamma(t)$, $t \geq t_0$, as an *unmatched uncertainty* because it does not act on the plant dynamics through the same channels as the control input $u(t)$. For this reason, the term $E_{x(t)}f$ in (5.40) and (5.92) is often termed as a matched uncertainty. Considering this plant model, proceeding as in the proof of Theorem 5.5 yields a bound on the derivative of the Lyapunov function along the trajectories of (5.77) and (5.81) given by

$$\dot{V}(t) \leq -\lambda_{\min}(Q)\|e_N(t)\| \left[\|e_N(t)\| - \left(d_{0,N} + \frac{2\gamma\|P\|}{\lambda_{\min}(Q)}\right)\right], \quad \text{for all } t \geq t_0, \tag{5.94}$$

whereas in Theorems 5.5 we choose $d \geq d_{0,N}$, to account for the unmatched disturbance $\Gamma(t)$, $t \geq t_0$, we define

$$\bar{d}_{0,N} \triangleq d_{0,N} + \frac{2\|P\|}{\lambda_{\min}(Q)}\gamma, \tag{5.95}$$

and let $d \geq \bar{d}_{0,N}$. Similarly, to apply Theorem 5.6, we choose $d \geq \bar{d}_\epsilon$, where

$$\bar{d}_\epsilon \triangleq d_\epsilon + \gamma. \tag{5.96}$$

Similar considerations on how to account for unmatched disturbances apply to the adaptive control techniques presented in the remainder of this chapter, and, hence, are omitted for brevity.

5.4.1.5 Numerical Example

This section summarizes some numerical results of the dead-zone algorithm with an example. Specifically, we consider the unstable plant model

$$\dot{x}(t) = \begin{bmatrix} 1 & -1 \\ 1 & 1 \end{bmatrix} x(t) + \left(u(t) + \frac{x_1(t)}{2\|x(t)\|} x(t) + \begin{bmatrix} -x_1\|x(t)\|^2 \\ -x_2\|x(t)\|^2 \end{bmatrix} \right),$$
$$x(t_0) = x_0, \quad t \geq t_0, \tag{5.97}$$

which is in the same form as (5.40) with $n = 2$, $m = 2$, $t_0 = 0$, $A = \begin{bmatrix} 1 & -1 \\ 1 & 1 \end{bmatrix}$ considered unknown, $B = I_2$, and $E_x f = \left[-x_1\|x\|^2, -x_2\|x\|^2\right]^{\mathrm{T}}$ considered unknown. Furthermore, we consider the reference model (5.41) with $A_r = \begin{bmatrix} -2 & -0.1 \\ 0.1 & -2 \end{bmatrix}$, $B_r = \begin{bmatrix} 2 & 0.1 \\ -0.1 & 2 \end{bmatrix}$, and

$$r(t) = \begin{bmatrix} 2\cos(0.1\pi t) \\ 4\sin(0.1\pi t) \end{bmatrix}, \quad t \geq 0; \tag{5.98}$$

it is easy to show that the matching conditions are verified.

As discussed in Remark 5.10, to prevent chattering, the discontinuous adaptive laws (5.81) are approximated by

$$\dot{\hat{\alpha}}_N(t) = \begin{cases} -\Gamma_\alpha x_N(t) e_N^{\mathrm{T}}(t) P B & \|e_N(t)\| > d + \varepsilon, \\ -\frac{\|e_N(t)\| - d}{\varepsilon} \Gamma_\alpha x_N(t) e_N^{\mathrm{T}}(t) P B & d < \|e_N(t)\| \leq d + \varepsilon, \\ 0 & \|e_N(t)\| \leq d, \end{cases} \tag{5.99a}$$

$$\dot{\hat{\beta}}_N(t) = \begin{cases} -\Gamma_\beta r(t) e_N^{\mathrm{T}}(t) P B & \|e_N(t)\| > d + \varepsilon, \\ -\frac{\|e_N(t)\| - d}{\varepsilon} \Gamma_\beta r(t) e_N^{\mathrm{T}}(t) P B & d < \|e_N(t)\| \leq d + \varepsilon, \\ 0 & \|e_N(t)\| \leq d, \end{cases} \tag{5.99b}$$

$$\dot{\hat{f}}_N(t) = \begin{cases} \Gamma_f \boldsymbol{\Pi}_N \mathcal{K}_{x_N(t)} B^{\mathrm{T}} P e_N(t) & \|e_N(t)\| > d + \varepsilon, \\ \frac{\|e_N(t)\| - d}{\varepsilon} \Gamma_f \boldsymbol{\Pi}_N \mathcal{K}_{x_N(t)} B^{\mathrm{T}} P e_N(t) & d < \|e_N(t)\| \leq d + \varepsilon, \\ 0 & \|e_N(t)\| \leq d, \end{cases} \tag{5.99c}$$

where $\varepsilon > 0$ captures a user-defined buffer zone; we set $\varepsilon = 0.1\,d$. Furthermore, in light of (5.91), we express (5.99c) as

$$\dot{\hat{\theta}}_N^l(t) = \begin{cases} \gamma_f^l \mathbb{K}_N^{-1} \begin{bmatrix} \mathfrak{K}(\xi_{N,1}, x_N(t)) \\ \vdots \\ \mathfrak{K}(\xi_{N,N}, x_N(t)) \end{bmatrix} [B^{\mathrm{T}} P e_N(t)]_l, & \|e_N(t)\| > d + \varepsilon, \\ \frac{\|e_N(t)\| - d}{\varepsilon} \gamma_f^l \mathbb{K}_N^{-1} \begin{bmatrix} \mathfrak{K}(\xi_{N,1}, x_N(t)) \\ \vdots \\ \mathfrak{K}(\xi_{N,N}, x_N(t)) \end{bmatrix} [B^{\mathrm{T}} P e_N(t)]_l, & d < \|e_N(t)\| \leq d + \varepsilon, \\ 0, & \|e_N(t)\| \leq d, \end{cases}$$
$$\text{for all } l \in \{1, \ldots, m\} \text{ and for all } t \geq t_0. \tag{5.100}$$

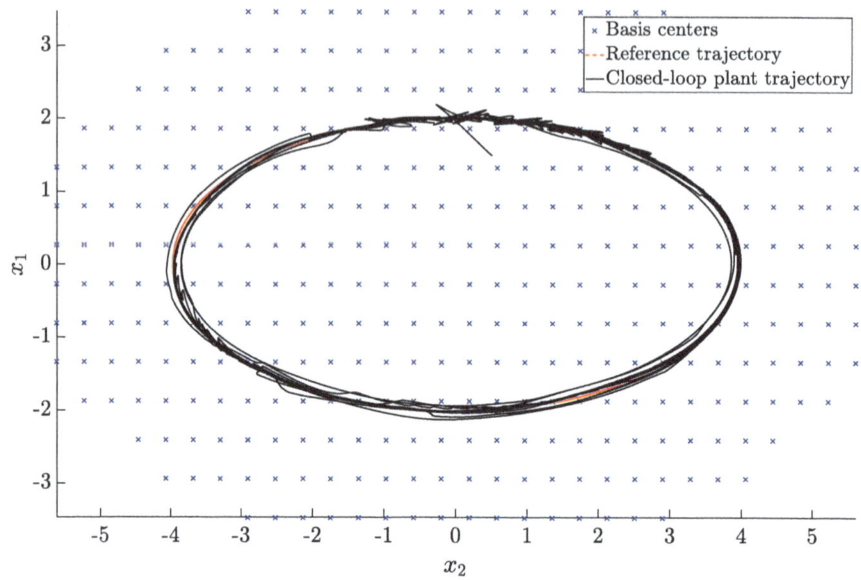

Fig. 5.5 Closed-loop plant trajectory, reference trajectory, and distribution of the centers where the kernel function have been evaluated. These results have been obtained by applying the dead-zone modification of MRAC and a 3/2 Sobolev–Matérn kernel

Figure 5.5 shows the closed-loop plant trajectory, the reference trajectory, and the centers of the kernels used. Kernel centers are located at the vertices of squares forming covering homogeneously a set Ω that contains the reference trajectory; the same center distribution is employed for all the following examples. The problem of identifying the centers is discussed in detail in Chap. 6.

The dead-zone width is defined as

$$d = \frac{2\|PB\|}{\lambda_{\min}(Q)} \sup_{\eta\in\Omega} \mathcal{P}(\eta) R_{\text{dead}}, \tag{5.101}$$

where $R_{\text{dead}} = 2$. Furthermore, we let $\Gamma_\alpha = 10^2 \cdot I_2$, $\Gamma_\beta = 10^2 \cdot I_2$, $\Gamma_f = 10^4 \cdot I_2$, and $Q = I_2$. Figure 5.6 shows the components of the control input as functions of time, and the MATLAB® codes generating these results can be found in [21].

To show the choice of the kernel function on the control system performance, the same simulations have been performed by applying 7 kernel functions of various continuity from 3 different families. The first family of kernels are specific types of Sobolev–Matérn kernel (see (3.23) and Table 3.1). These kernels can also be written as

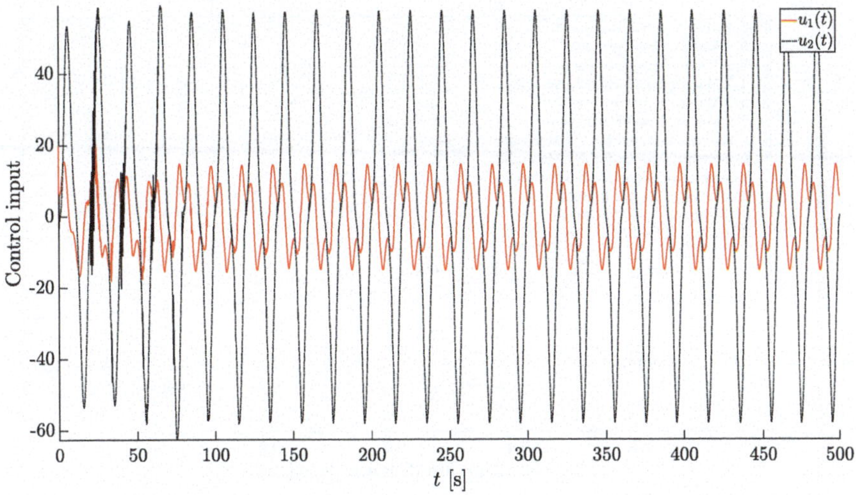

Fig. 5.6 Adaptive control input to generate the closed-loop trajectory shown in Fig. 5.5

$$\mathfrak{K}_{p+1/2}(d) = \exp\left(-\frac{\sqrt{2p+1}d}{l}\right)\frac{p!}{(2p)!}\sum_{i=0}^{p}\frac{(p+i)!}{i!(p-i)!}\left(\frac{2\sqrt{2p+1}d}{l}\right)^{p-i}, \tag{5.102}$$

where $d = \|x - y\|$, and $p \in \mathbb{N}$ denotes the power of the polynomial. The polynomial power index p corresponds to a Sobolev space of smoothness s defined by the Sobolev–Matérn kernel via half integers, with $s = p + 1/2$. In the proposed simulations, we let $p = 1$, 2, and 3, which induce the $s = 3/2$, $5/2$, and $7/2$ Sobolev–Matérn kernels, respectively. Note that the larger the index, the smoother the corresponding Sobolev space.

The second family of reproducing kernels are the Wendland kernels (see (3.26) and Table 3.2). In particular, Wendland kernels that correspond to 3 different smoothness parameters are chosen. For kernels having smoothness $C^2(\mathbb{R}^n)$, $C^4(\mathbb{R}^n)$, and $C^6(\mathbb{R}^N)$, the corresponding kernel functions for $n \leq 3$ ([15] or in Table 3.2) are defined as

$$\mathfrak{K}_2(r) = (1-r)_+^4(4r+1), \tag{5.103a}$$

$$\mathfrak{K}_4(r) = \frac{1}{3}(1-r)_+^6(35r^2+18r+3), \tag{5.103b}$$

$$\mathfrak{K}_6(r) = (1-r)_+^8(32r^3+25r^2+8r+1), \tag{5.103c}$$

respectively, where $r = \frac{\|x-y\|}{3l}$, and $l > 0$ is user-defined.

The last type of reproducing kernel studied in this section is the Gaussian kernel (see (3.23) and Table 3.1) given by

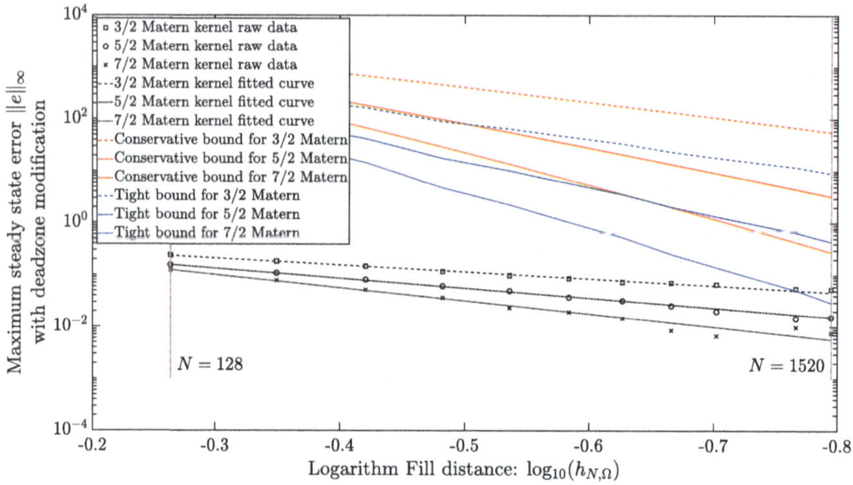

Fig. 5.7 Maximum norm of the steady-state error against the fill distance between basis centers for Sololev–Matérn kernels employing the dead-zone modification of MRAC. The blue lines show the tighter upper bounds captured by (5.84a). The red lines show the tighter upper bounds captured by (5.84b)

$$\mathfrak{K}_{\exp}(x, y) = \exp\left(-\frac{\|x-y\|^2}{2l^2}\right), \quad \text{for all } (x, y) \in \mathbb{X} \times \mathbb{X}. \tag{5.104}$$

The parameter l in (5.102) and (5.104) is set as $l = 1$.

We expect that the fidelity of the results will depend on the fill distance of the centers Ξ_N in Ω. Specifically, the larger the number N of centers in Ξ_N, the denser the kernel centers, and, consequently the smaller the fill distance $h_{\Xi_N,\Omega}$. Therefore, for each choice of kernel functions, we examine the steady-state error as a function of the fill distance.

Figures 5.7, 5.8, and 5.9 show that the maximum steady-state tracking error decreases as the fill distance between points decreases. It is noteworthy that in Fig. 5.7 the slope of the tracking error for the Sobolev–Matérn kernel is such that

$$\|e_N\|_\infty \lesssim (s-1)\log h_{\Xi_N,\Omega}, \tag{5.105}$$

which is the analytical upper bound for the Sobolev space, see [22]. Also, the conservative and tight bounds plotted in the figures correspond to the expressions

$$\text{conservative bound} = \frac{2}{\sigma\lambda_{\min}(Q)}\|PB\| \sup_{\eta\in\Omega}\mathcal{P}_N(\eta)R,$$

$$\text{tight bound} = \frac{2}{\sigma\lambda_{\min}(Q)}\|PB\| \sup_{\eta\in\Omega}\mathcal{P}_N(\eta)\epsilon_N,$$

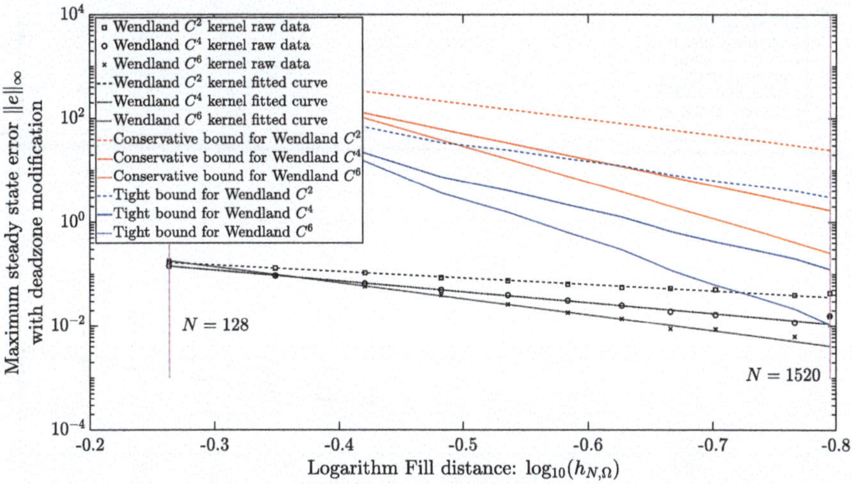

Fig. 5.8 Maximum norm of the steady-state error against the fill distance between basis centers for Wendland kernels employing the dead-zone-modification of MRAC. The blue lines show the tighter upper bounds captured by (5.84a). The red lines show the tighter upper bounds captured by (5.84b)

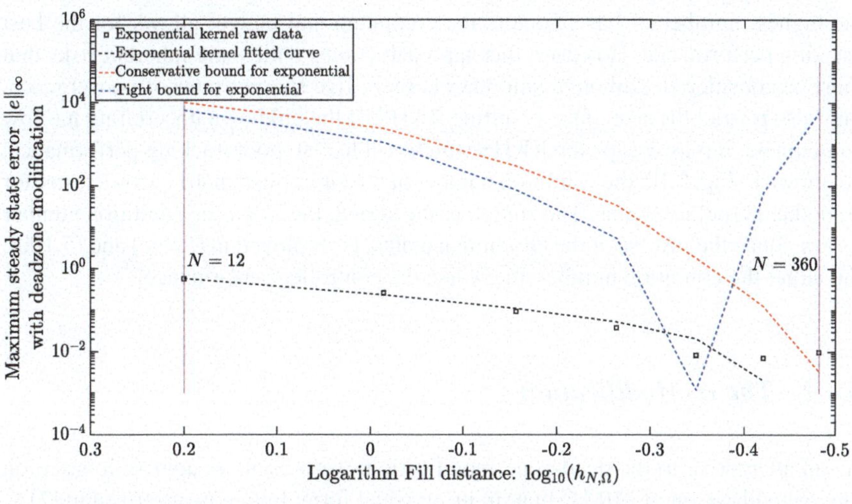

Fig. 5.9 Maximum norm of the steady-state error against the fill distance between basis centers for exponential kernels employing the dead-zone modification of MRAC. The blue lines show the tighter upper bounds captured by (5.84a). The red lines show the tighter upper bounds captured by (5.84b)

where $\epsilon_N \triangleq \|(I - \Pi_N)f\|_{\mathcal{H}}$, which appear in the analysis of the dead-zone method. It is apparent how smoother kernel functions yield steeper error bounds, and, hence, steeper descent in maximum steady-state tracking error as the fill distance decreases.

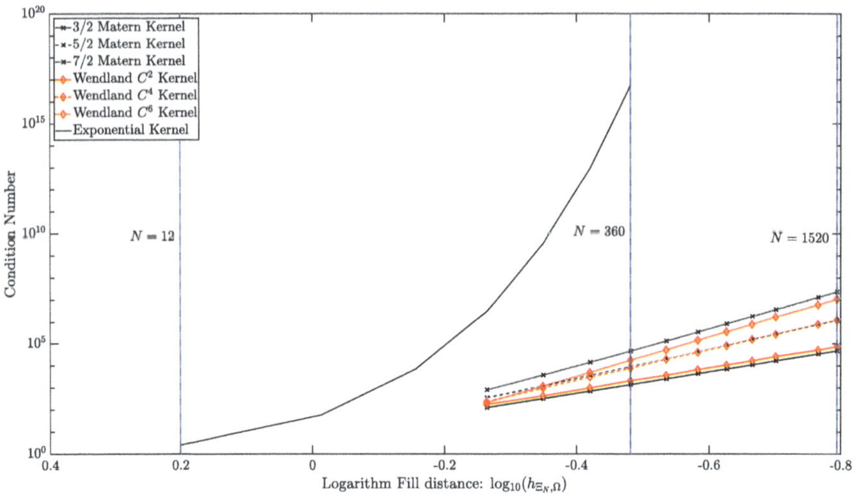

Fig. 5.10 Condition number of the Grammian matrix against the fill distance between basis centers

One may be tempted to use kernel functions with the greatest smoothness and the highest number of basis centers the computational budget allows for the best tracking performance. However, this approach comes with some inherent risks that must be considered. Although smoother kernels give sharper rates of convergence, they also restrict the size of the resulting RKHS. If the unknown uncertainty has low smoothness, it may escape the RKHS which will lead to poor tracking performance. As shown in Fig. 5.10, the condition number of the Grammian matrix grows faster for smoother kernel functions. The smoother the kernel, the faster the condition number grows. Since the inverse of the Grammian matrix is employed in (5.99c) and (5.100), the larger the condition number, the worse the controller performance.

5.4.2 *The σ-Modification*

As an alternative to the dead-zone modification, we present an approach based on the σ-modification of MRAC that, in its classical formulation, is discussed in [7].

5.4.2.1 Theoretical Formulation

As in the previous section, assume that the native space $\boldsymbol{\mathcal{H}}$ of vector-valued functions is given by the Cartesian product $\boldsymbol{\mathcal{H}} = \mathcal{H}^m$ of a native space $\mathcal{H}$ of scalar-valued functions. Furthermore, consider the trajectory tracking error dynamics (5.77) and the adaptive laws

$$\dot{\hat{\alpha}}_N(t) = -\Gamma_\alpha x_N(t) e_N^{\mathrm{T}}(t) PB - \sigma\hat{\alpha}_N(t), \quad \hat{\alpha}_N(t_0) = \hat{\alpha}_{N,0}, \quad t \geq t_0, \tag{5.106a}$$

$$\dot{\hat{\beta}}_N(t) = -\Gamma_\beta r(t) e_N^{\mathrm{T}}(t) PB - \sigma\hat{\beta}_N(t), \quad \hat{\beta}_N(t_0) = \hat{\beta}_{N,0}, \tag{5.106b}$$

$$\dot{\hat{f}}_N(t) = \Gamma_N \mathbf{\Pi}_N \mathcal{K}_{x_N(t)} B^{\mathrm{T}} P e_N(t) - \sigma \hat{f}_N(t), \quad \hat{f}_N(t_0) = \hat{f}_{N,0}, \tag{5.106c}$$

where $\Gamma_\alpha \in \mathbb{R}^{n\times n}$, $\Gamma_\beta \in \mathbb{R}^{m\times m}$, and $\Gamma_N \in \mathbb{R}^{m\times m}$ are user-defined, symmetric, and positive-definite adaptive rate matrices, P denotes the symmetric, positive-definite solution of the Lyapunov equation (5.45) for the some symmetric, positive-definite user-defined matrix $Q \in \mathbb{R}^{n\times n}$, and $0 < \sigma < \dfrac{\lambda_{\min}(Q)}{2\lambda_{\max}(P)}$, where $\lambda_{\max}(\cdot)$ denotes the maximum eigenvalue of its argument.

Theorem 5.7 *Consider the closed-loop trajectory tracking error dynamics given by (5.58), its finite-dimensional counterpart given by (5.77), and the adaptive laws (5.106). If the assumptions of Theorems 5.3 and 5.4 are verified, $f \in C_{N,\epsilon,R}$, and the kernel $\mathcal{K}(\cdot,\cdot)$ that induces the native space $\mathcal{H}$ of real-valued functions over $\mathbb{X}$ is uniformly bounded on the diagonal, then for any bounded subset $\Omega \supseteq \bigcup_{\tau\geq t_0>0} x(\tau)$, there exists $T \in (t_0, \infty)$ such that*

$$\begin{aligned} \|x_N(t) - x_r(t)\|_{\mathbb{X}}^2 \leq{}& \frac{2}{\sigma\lambda_{\min}(Q)} \|PB\|^2 \epsilon_N^2 \sup_{\eta\in\Omega} \mathcal{P}_N^2(\eta) \\ &+ \left\langle \alpha, \Gamma_\alpha^{-1}\alpha \right\rangle_{\mathrm{tr}} + \left\langle \beta, \Gamma_\beta^{-1}\beta \right\rangle_{\mathrm{tr}} + \left\langle f, \Gamma_f^{-1} f \right\rangle_{\mathcal{H}} \end{aligned} \tag{5.107}$$

for all $t \geq T$, where $\epsilon_N \triangleq \|(I - \mathbf{\Pi}_N) f\|_{\mathcal{H}}$.

Proof Consider the Lyapunov function candidate (5.82). By proceeding as in (5.83), along the trajectories of (5.77) and (5.106) it holds that

$$\begin{aligned} \dot{V}(t) ={}& -\langle Q e_N(t), e_N(t)\rangle_{\mathbb{X}} \\ &+ 2\left\langle (I - \mathbf{\Pi}_N)\mathcal{K}_{x_N(t)} B^{\mathrm{T}} P e_N(t), \tilde{f}_N(t,\cdot)\right\rangle_{\mathcal{H}} \\ &+ 2\sigma\left\langle \tilde{\alpha}_N(t), \Gamma_\alpha^{-1}\hat{\alpha}_N\right\rangle_{\mathrm{tr}} + 2\sigma\left\langle \tilde{\beta}_N(t), \Gamma_\beta^{-1}\hat{\beta}_N(t)\right\rangle_{\mathrm{tr}} \\ &+ 2\sigma\left\langle \tilde{f}(t,\cdot), \Gamma_N^{-1}\hat{f}_N(t,\cdot)\right\rangle_{\mathcal{H}}, \quad t \geq t_0, \end{aligned} \tag{5.108}$$

where $\dot{V}(\cdot)$ represents $\dot{V}(e_N(\cdot), \tilde{\alpha}_N(\cdot), \tilde{\beta}_N(\cdot), \tilde{f}_N(t,\cdot))$ for notational brevity, and $\langle\cdot,\cdot\rangle_{\mathrm{tr}}$ is defined in (2.26).

Next, for brevity, let $\langle\cdot,\cdot\rangle_G$ denote any weighted inner product so that, for example, $\langle\tilde{\alpha}_N, \hat{\alpha}_N\rangle_G = \left\langle\tilde{\alpha}_N, \Gamma_\alpha^{-1}\hat{\alpha}_N\right\rangle_{\mathrm{tr}}$. Furthermore, recalling that the inner product is symmetric (property (RI4) in Definition 2.46) and bilinear (properties (RI4) and (RI5) in Definition 2.46) over the field of real numbers, we obtain that, for all $t \geq t_0$,

$$\begin{aligned}\langle \tilde{\alpha}_N, \tilde{\alpha}_N \rangle_G &= \langle \alpha - \hat{\alpha}_N, \alpha \rangle_G - \langle \alpha - \hat{\alpha}_N, \hat{\alpha}_N \rangle_G \\ &= \langle \alpha, \alpha \rangle_G - 2\langle \alpha, \hat{\alpha}_N \rangle_G + \langle \hat{\alpha}_N, \hat{\alpha}_N \rangle_G\end{aligned}$$

which yields

$$2\langle \alpha, \hat{\alpha}_N \rangle_G = \langle \alpha, \alpha \rangle_G + \langle \hat{\alpha}_N, \hat{\alpha}_N \rangle_G - \langle \tilde{\alpha}_N, \tilde{\alpha}_N \rangle_G. \tag{5.109}$$

Subtracting $2\langle \hat{\alpha}_N, \hat{\alpha}_N \rangle_G$ from both sides of (5.109) yields

$$\begin{aligned}2\langle \tilde{\alpha}_N(t), \hat{\alpha}_N(t) \rangle_G = \langle \alpha, \alpha \rangle_G - \langle \tilde{\alpha}_N(t), \tilde{\alpha}_N(t) \rangle_G \\ - \langle \hat{\alpha}_N, \hat{\alpha}_N \rangle_G,\end{aligned} \tag{5.110}$$

and similar properties can be inferred for $\langle \tilde{\beta}_N(t), \hat{\beta}_N(t) \rangle_G$ and $\langle \tilde{f}_N(t), \hat{f}_N(t) \rangle_G$. Therefore, it follows from (5.108) that, for all $t \geq t_0$,

$$\begin{aligned}\dot{V}(t) = &-\langle Qe_N(t), e_N(t) \rangle_{\mathbb{X}} \\ &+ 2\left\langle (I - \mathbf{\Pi}_N) \mathcal{K}_{x_N(t)} B^{\mathrm{T}} P e_N(t), \tilde{f}_N(t, \cdot) \right\rangle_{\mathcal{H}} \\ &- \sigma \langle \tilde{\alpha}_N(t), \tilde{\alpha}_N(t) \rangle_G + \sigma \langle \alpha, \alpha \rangle_G - \sigma \langle \tilde{\beta}_N(t), \tilde{\beta}_N(t) \rangle_G \\ &+ \sigma \langle \beta, \beta \rangle_G - \sigma \langle \tilde{f}_N(t), \tilde{f}_N(t) \rangle_G + \sigma \langle f, f \rangle_G \\ &- \sigma \langle \hat{\alpha}_N, \hat{\alpha}_N \rangle_G - \sigma \langle \hat{\beta}_N, \hat{\beta}_N \rangle_G - \sigma \langle \hat{f}_N, \hat{f}_N \rangle_G.\end{aligned} \tag{5.111}$$

As the terms $-\sigma \langle \hat{\alpha}_N, \hat{\alpha}_N \rangle_G$, $-\sigma \langle \hat{\beta}_N, \hat{\beta}_N \rangle_G$, and $-\sigma \langle \hat{f}_N, \hat{f}_N \rangle_G$ are nonpositive, it follows from (5.111) that

$$\begin{aligned}\dot{V}(t) \leq &- \langle Qe_N(t), e_N(t) \rangle_{\mathbb{X}} \\ &+ 2\left\langle (I - \mathbf{\Pi}_N) \mathcal{K}_{x_N(t)} B^{\mathrm{T}} P e_N(t), \tilde{f}_N(t, \cdot) \right\rangle_{\mathcal{H}} \\ &- \sigma \langle \tilde{\alpha}_N(t), \tilde{\alpha}_N(t) \rangle_G + \sigma \langle \alpha, \alpha \rangle_G - \sigma \langle \tilde{\beta}_N(t), \tilde{\beta}_N(t) \rangle_G \\ &+ \sigma \langle \beta, \beta \rangle_G - \sigma \langle \tilde{f}_N(t), \tilde{f}_N(t) \rangle_G + \sigma \langle f, f \rangle_G.\end{aligned} \tag{5.112}$$

Adding and subtracting $\sigma \langle P e_N(t), e_N(t) \rangle_{\mathbb{X}}$, $t \geq t_0$, from (5.112) yields

$$\begin{aligned}\dot{V}(t) \leq &- \langle Qe_N(t), e_N(t) \rangle_{\mathbb{X}} \\ &+ 2\left\langle (I - \mathbf{\Pi}_N) \mathcal{K}_{x_N(t)} B^{\mathrm{T}} P e_N(t), \tilde{f}_N(t, \cdot) \right\rangle_{\mathcal{H}} \\ &+ \sigma \langle P e_N(t), e_N(t) \rangle_{\mathbb{X}} + \sigma \langle \alpha, \alpha \rangle_G + \sigma \langle \beta, \beta \rangle_G \\ &+ \sigma \langle f, f \rangle_G - \sigma V(t).\end{aligned} \tag{5.113}$$

Now, note that

$$\begin{aligned}
&2\left\langle (I - \Pi_N)\,\mathcal{K}_{x_N(t)} B^{\mathrm{T}} P e_N(t),\, \tilde{f}_N(t,\cdot)\right\rangle_{\mathcal{H}} \\
&= 2\left\langle B^{\mathrm{T}} P e_N(t),\, E_{x_N(t)}\,(I - \Pi_N)\,\tilde{f}_N(t,\cdot)\right\rangle_{\mathbb{U}} \\
&= 2\left\langle e_N(t),\, P B E_{x_N(t)}\,(I - \Pi_N)\,(f(\cdot) - \hat{f}_N(t,\cdot))\right\rangle_{\mathbb{R}}
\end{aligned} \tag{5.114}$$

for all $t \geq t_0$. Furthermore, since $\hat{f}_N(t,\cdot) \in \mathcal{H}_{d(N)}$, it holds that

$$E_{x_N(t)}\,(I - \Pi_N)\,\hat{f}_N(t,\cdot) = 0, \quad \text{for all } t \geq t_0. \tag{5.115}$$

Therefore, it follows from (5.114) and the Schwarz inequality (2.22a) that

$$\begin{aligned}
&-\langle Q e_N(t), e_N(t)\rangle_{\mathbb{X}} + 2\left\langle (I - \Pi_N)\,\mathcal{K}_{x_N(t)} B^{\mathrm{T}} P e_N(t),\, \tilde{f}_N(t,\cdot)\right\rangle_{\mathcal{H}} \\
&= -\langle Q e_N(t), e_N(t)\rangle_{\mathbb{X}} + 2\left\langle e_N(t),\, P B E_{x_N(t)}\,(I - \Pi_N)\, f(\cdot)\right\rangle_{\mathbb{R}} \\
&\leq -\lambda_{\min}(Q)\|e_N(t)\|_{\mathbb{X}}^2 + 2\|e_N(t)\|_{\mathbb{X}}\|P B E_{x_N(t)}\,(I - \Pi_N)\, f(\cdot)\|_{\mathbb{X}} \\
&\leq -\frac{1}{2}\lambda_{\min}(Q)\|e_N(t)\|_{\mathbb{X}}^2 - 2\lambda_{\min}(Q)\left(\frac{1}{4}\|e_N(t)\|_{\mathbb{X}}^2 \right.\\
&\quad \left. - \frac{\|e_N(t)\|_{\mathbb{R}^n}}{\lambda_{\min}(Q)}\left\|P B E_{x_N(t)}\,(I - \Pi_N)\, f(\cdot)\right\|_{\mathbb{R}^n}\right).
\end{aligned} \tag{5.116}$$

Now, for each $t \geq t_0$, let $\nu(t) \triangleq \lambda_{\min}^{-1}(Q)\left\|P B E_{x_N(t)}\,(I - \Pi_N)\, f(\cdot)\right\|_{\mathbb{R}^n}$, and note that

$$\begin{aligned}
&\frac{1}{4}\|e_N(t)\|_{\mathbb{R}^n}{}^2 - \frac{\|e_N(t)\|_{\mathbb{R}^n}}{\lambda_{\min}(Q)}\left\|P B E_{x_N(t)}\,(I - \Pi_N)\, f(\cdot)\right\|_{\mathbb{R}^n} \\
&= \frac{1}{4}\|e_N(t)\|_{\mathbb{R}^n}{}^2 - \nu(t)\|e_N(t)\|_{\mathbb{R}^n} \\
&= \left(\frac{1}{2}\|e_N(t)\|_{\mathbb{R}^n} - \nu(t)\right)^2 - \nu^2(t).
\end{aligned} \tag{5.117}$$

Thus, substituting (5.117) and (5.116) into (5.113) yields

$$\begin{aligned}
\dot{V}(t) \leq{}& -\frac{1}{2}\lambda_{\min}(Q)\,\|e_N(t)\|_{\mathbb{R}^n}{}^2 \\
&- 2\lambda_{\min}(Q)\left[\left(\frac{1}{2}\|e_N(t)\|_{\mathbb{X}} - \nu(t)\right)^2 - \nu^2(t)\right] \\
&+ \sigma\langle \alpha, \alpha\rangle_G + \sigma\langle \beta, \beta\rangle_G + \sigma\langle f, f\rangle_G - \sigma V(t) \\
&+ \sigma\langle P e_N(t), e_N(t)\rangle_{\mathbb{X}}, \quad t \geq t_0.
\end{aligned} \tag{5.118}$$

Since $\left(\frac{1}{2}\|e_N(t)\|_{\mathbb{X}} - \nu(t)\right)^2 \geq 0$ for all $t \geq t_0$, we obtain that

$$
\begin{aligned}
\dot{V}(t) \leq & -\frac{1}{2}\lambda_{\min}(Q)\,\|e_N(t)\|_{\mathbb{R}^n}{}^2 \\
& + 2\lambda_{\min}^{-1}(Q)\left\|PBE_{x_N(t)}(I-\boldsymbol{\Pi}_N)f(\cdot)\right\|_{\mathbb{R}^n}{}^2 \\
& + \sigma\langle\alpha,\alpha\rangle_G + \sigma\langle\beta,\beta\rangle_G \\
& + \sigma\langle f,f\rangle_G - \sigma V(t) + \sigma\langle Pe_N(t), e_N(t)\rangle_{\mathbb{X}} \\
\leq & -\sigma V(t) - \left(\frac{1}{2}\lambda_{\min}(Q) - \sigma\lambda_{\max}(P)\right)\|e_N(t)\|_{\mathbb{R}^n}{}^2 \\
& + \frac{2}{\lambda_{\min}(Q)}\left\|PBE_{x_N(t)}(I-\boldsymbol{\Pi}_N)f(\cdot)\right\|_{\mathbb{R}^n}{}^2 \\
& + \sigma\left(\langle\alpha,\alpha\rangle_G + \langle\beta,\beta\rangle_G + \langle f,f\rangle_G\right).
\end{aligned}
$$

By assumption, σ is such that $\frac{1}{2}\lambda_{\min}(Q) - \sigma\lambda_{\max}(P) > 0$, and, hence, we obtain that

$$
\begin{aligned}
\dot{V}(t) \leq & -\sigma V(t) + \frac{2}{\lambda_{\min}(Q)}\left\|PBE_{x_N(t)}(I-\boldsymbol{\Pi}_N)f(\cdot)\right\|_{\mathbb{R}^n}{}^2 \\
& + \sigma\left(\langle\alpha,\alpha\rangle_G + \langle\beta,\beta\rangle_G + \langle f,f\rangle_G\right).
\end{aligned}
$$

Next, initially set $\Omega = \mathbb{X}$. It follows from (3.157) and (5.2) that

$$
\|E_{x_N(t)}(I-\boldsymbol{\Pi}_N)f(\cdot)\|_{\mathbb{U}} \leq \sup_{\eta\in\Omega}\mathcal{P}_N(\eta)\epsilon_N < \infty,
$$

and, hence,

$$
\begin{aligned}
\dot{V}(t) \leq & -\sigma V(t) + \frac{2}{\lambda_{\min}(Q)}\|PB\|^2\epsilon_N^2\sup_{\eta\in\Omega}\mathcal{P}_N^2(\eta) \\
& + \sigma\left(\langle\alpha,\alpha\rangle_G + \langle\beta,\beta\rangle_G + \langle f,f\rangle_G\right), \quad \text{for all } t \geq t_0; \qquad (5.119)
\end{aligned}
$$

note that since $\mathcal{K}(\cdot,\cdot)$ is uniformly bounded on the diagonal (see Sect. 3.8.1), the power function $\mathcal{P}_N$ is also uniformly bounded over $\mathbb{X}$. Now, let

$$
D \triangleq \frac{2}{\lambda_{\min}(Q)}\|PB\|^2\sup_{\eta\in\Omega}\mathcal{P}_N^2(\eta)\epsilon_N^2 + \sigma\left(\langle\alpha,\alpha\rangle_G + \langle\beta,\beta\rangle_G + \langle f,f\rangle_G\right). \quad (5.120)
$$

It follows from (5.119) that

$$
V(t) \leq c_1 e^{-\sigma t} + \frac{D}{\sigma}, \quad \text{for all } t \geq t_0, \qquad (5.121)
$$

where $c_1 \in \mathbb{R}$ is a coefficient that depends on the initial condition of the system. Therefore, there exists $T \in (t_0, \infty)$ such that $\left(e_N(T), \tilde{\alpha}_N(T), \tilde{\beta}_N(T), \tilde{f}_N(T)\right) \in$

S_0, where

$$S_0 \triangleq \left\{ \left(e_N, \tilde{\alpha}_N, \tilde{\beta}_N, \tilde{f}_N\right) \in \mathbb{Z}_N | V\left(e_N, \tilde{\alpha}_N, \tilde{\beta}_N, \tilde{f}_N\right) \leq \frac{D}{\sigma} \right\}, \tag{5.122}$$

which implies (5.107). The proposed argument can be iterated for smaller sets $\Omega \subset \mathbb{X}$ containing the closed-loop trajectory to attain tighter bounds on the closed-loop system's uniform ultimate bounded given by (5.107). ■

The upper bound on the trajectory tracking error performance that is provided in (5.107) is relatively tight since it involves the unknown, ideal parameters α and β and the functional uncertainty f. These values can be replaced in (5.107) with user-defined estimates, and, hence, the a priori estimates on the trajectory tracking error performance will reflect the user's ability to estimate uncertainties.

Theorem 5.7 can be extended to the case whereby the native space $\mathcal{H}$ is induced by a generic operator-valued kernel $\mathcal{K}$. This task is left to the reader, who should leverage the same approach that led from Theorems 5.5 to 5.6. Furthermore, the same considerations as in Sect. 5.4.1.3 can be made to deduce a coordinate implementation of the adaptive law (5.106c); these arguments are omitted for brevity.

5.4.2.2 Numerical Example

In this section, we analyze the same numerical example as in Sect. 5.4.1.5, but employing the adaptive laws (5.106); the MATLAB® codes generating these results can be found in [21]. Figures 5.11, 5.11 and 5.12 show the controller's performance in terms of steady-state trajectory tracking error employing the Sololev–Matérn, Wendland, and exponential kernels, respectively. In each case, the controller's performance is compared with the theoretical bound provided by (5.107).

5.4.3 Convex Projection Methods

Thus far, we have discussed how two common robust modifications of real parametric adaptive control theory, namely the dead-zone and σ-modification, can be lifted to the setting when we want to accommodate nonparametric, vector-valued functional uncertainty classes. In this section, we discuss how the convex projection operator applied to the adaptive gains can be employed in the setting of vector-valued native spaces.

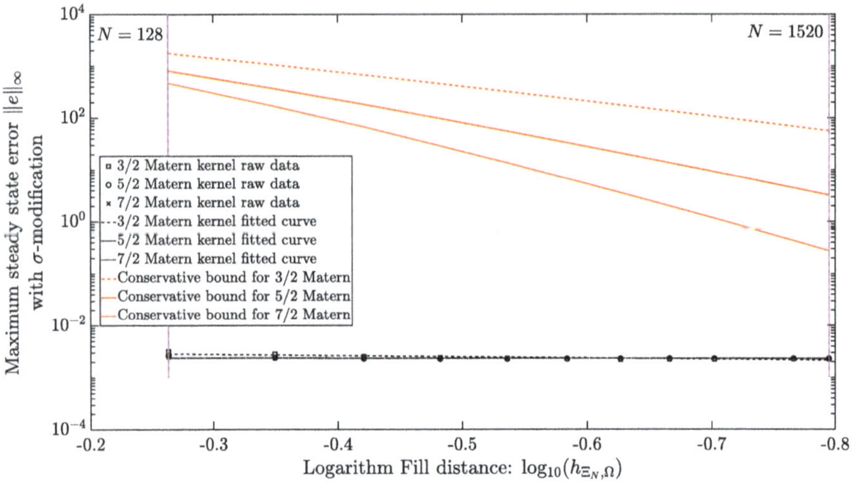

Fig. 5.11 Maximum norm of the steady-state error against the fill distance between basis centers for Sololev–Matérn kernels employing the σ-modification of MRAC. The red lines show the upper bounds captured by (5.107)

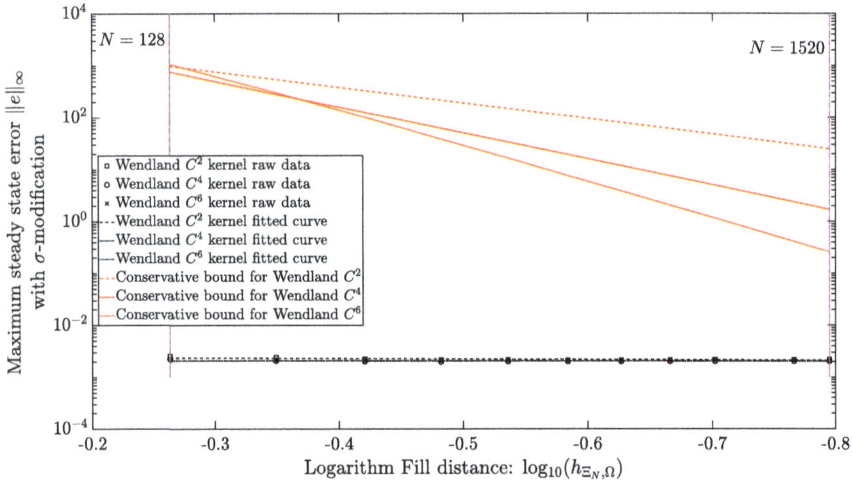

Fig. 5.12 Maximum norm of the steady-state error against the fill distance between basis centers for Wendland kernels employing the σ-modification of MRAC. The red lines show the upper bounds captured by (5.107)

5.4.3.1 Theoretical Formulation

To steer the trajectories of (5.40) toward the trajectories of (5.41), in this section, we employ the control input

$$\begin{aligned} u_N(t) =& \hat{\alpha}_N^{\mathrm{T}}(t)x_N(t) + \hat{\beta}_N^{\mathrm{T}}(t)r(t) \\ & - E_{x_N(t)}\left(\hat{f}_N(t,\cdot) + c_N(\cdot, e_N(t))\right), \quad t \geq t_0, \end{aligned} \tag{5.123}$$

where the *matrix adaptive gains* $\hat{\alpha} : [t_0, \infty) \to \mathbb{R}^{n\times m}$ and $\hat{\beta} : [t_0, \infty) \to \mathbb{R}^{m\times m}$ verify the *adaptive laws*

$$\dot{\hat{\alpha}}_N(t) = \operatorname{proj}(\hat{\alpha}_N(t), -\Gamma_\alpha x_N(t)e_N^{\mathrm{T}}(t)PB), \quad \hat{\alpha}(t_0) = \hat{\alpha}_0, \tag{5.124}$$

$$\dot{\hat{\beta}}_N(t) = \operatorname{proj}(\hat{\beta}_N(t), -\Gamma_\beta r(t)e_N^{\mathrm{T}}(t)PB), \quad \hat{\beta}(t_0) = \hat{\beta}_0, \tag{5.125}$$

the *adaptive rate matrices* $\Gamma_\alpha \in \mathbb{R}^{n\times n}$ and $\Gamma_\alpha \in \mathbb{R}^{m\times m}$ are symmetric and positive-definite, $P \in \mathbb{R}^{n\times n}$ denotes the symmetric, positive-definite solution of the *Lyapunov equation* (5.45), the *functional adaptive gain* $\hat{f} : [t_0, \infty) \to \mathcal{H}_{d(N)}$ verifies the *adaptive law*

$$\frac{\partial \hat{f}_N(t,\cdot)}{\partial t} = -\mathfrak{Proj}\left(\hat{f}_N(t,\cdot), \mathbf{\Pi}_N\Gamma_f E^*_{x_N(t)}B^{\mathrm{T}}Pe_N(t)\right), \quad \hat{f}(t_0,\cdot) = \hat{f}_0, \tag{5.126}$$

the *adaptive rate matrix* $\Gamma_f \in \mathbb{R}^{m\times m}$ is symmetric and positive-definite, $x_N : [t_0, \infty) \to \mathbb{X}$ denotes the solution of (5.40) with control input (5.123), $e_N(t) \triangleq x_N(t) - x_r(t)$, and the *compensator* $c_N : \mathbb{X} \times \mathbb{R}^n \to \mathbb{U}$ is user-defined; in Sect. 5.5.1 below, a specific form for $c_N(\cdot,\cdot)$ is provided. Note that (5.126) is a partial differential equation taking values in $\mathcal{H}$.

Let $\mathcal{D} \subseteq \mathbb{R}^m$ be convex and let $h : \mathcal{D} \to \mathbb{R}$ be convex and such that $\inf_{\theta\in\mathcal{D}} h(\theta) < 0$. The *convex vector projection operator over* $\mathbb{R}^n$ is defined as

$$\operatorname{proj}(\theta, \theta_d) \triangleq \begin{cases} \mathfrak{P}(\theta, \theta_d) \text{ if } h(\theta) > 0 \text{ and } \langle h'(\theta), \theta_d\rangle_{\mathbb{R}^n} > 0, \\ \theta_d \qquad\qquad \text{otherwise,} \end{cases} \tag{5.127}$$

where

$$\mathfrak{P}(\theta, \theta_d) \triangleq \theta_d - h(\theta)\frac{\left(\frac{\partial h(\theta)}{\partial \theta}\right)^{\mathrm{T}}\left(\frac{\partial h(\theta)}{\partial \theta}\right)}{\left(\frac{\partial h(\theta)}{\partial \theta}\right)\left(\frac{\partial h(\theta)}{\partial \theta}\right)^{\mathrm{T}}}\theta_d \tag{5.128}$$

for all $(\theta, \theta_d) \in \mathcal{D} \times \mathbb{R}^m$. Applying the vector projection operator to each column of matrices $\Theta, \Theta_d \in \mathbb{R}^{m\times N}$, we obtain the continuous matrix projection operator employed in (5.124) and (5.125). We employ the same symbol for both operators with minimal loss of clarity.

To introduce the convex projection operator on a Hilbert space, let $\mathcal{D} \subseteq \mathcal{H}$ be convex, and let $h : \mathcal{D} \to \mathbb{R}$ be convex, Fréchet differentiable, and such that $\inf_{f\in\mathcal{D}} h(f) < 0$. The *Fréchet derivative of* $h \in \mathcal{H}$ at $f \in \mathcal{H}$ is denoted by $Dh(f)$. The *convex projection operator over* $\mathcal{H}$, Proj : $\mathcal{D} \times \mathcal{H} \to \mathcal{H}$ is given by

$$\mathfrak{Proj}(f, f_d) \triangleq \begin{cases} \mathfrak{P}(f, f_d) & \text{if } h(f) > 0, \\ & \text{and } \langle Dh(f), f_d\rangle_{\mathcal{H}} > 0, \\ f_d & \text{otherwise}, \end{cases} \tag{5.129}$$

where

$$\mathfrak{P}(f, f_d) \triangleq \left[I - h(f)\left\langle \cdot, \frac{Dh(f)}{\|Dh(f)\|_{\mathcal{H}}} \right\rangle_{\mathcal{H}} \frac{Dh(f)}{\|Dh(f)\|_{\mathcal{H}}} \right] f_d \tag{5.130}$$

for all $(f, f_d) \in \mathcal{D} \times \mathcal{H}$. We purposely employed the same symbol in (5.128) and (5.130) to remark on their analogies.

The convex projection operator (5.127) is such that if $\hat{\alpha}_{i,N}(t) \in \overline{\Omega}_0$ for some $t \in [t_0, \infty)$, where

$$\overline{\Omega}_\delta \triangleq \{f \in \mathcal{D} : h(f) < \delta\}, \quad \text{for all } \delta \in (-\infty, 1], \tag{5.131}$$

then the classical adaptive law of MRAC for $\dot{\hat{\alpha}}_{i,N}(t)$, which is given by (5.44a), is not modified; here, $\hat{\alpha}_{i,N}(t)$ denotes the ith component of $\hat{\alpha}_N(t)$, $i \in \{1, \dots, m\}$. Alternatively, if $\hat{\alpha}_{i,N}(t) \in \overline{\Omega}_1 \setminus \mathring{\Omega}_0$ for some $t \in [t_0, \infty)$, and its gradient points toward $\partial\Omega_1$, then the magnitude of the component of $\dot{\hat{\alpha}}_{i,N}(t)$ locally normal to $\partial\Omega_\delta$ is reduced, with $\delta \in (0, 1]$ such that $\hat{\alpha}_{i,N}(t) = \delta$; here, $\mathring{\Omega}_0$ denotes the interior of $\overline{\Omega}_0$ (see Definition 2.9). This modification of the adaptive law for $\dot{\hat{\alpha}}_{i,N}(\cdot)$, which is captured by (5.128), is a continuous function of $\delta \in [0, 1]$. Finally, the convex projection operator (5.127) assures that $\hat{\alpha}_{i,N}(t)$ is always contained in the user-defined convex set $\overline{\Omega}_1$. An identical mechanism applies to the adaptive gain $\hat{\beta}_N(t)$, $t \in [t_0, \infty)$. Employing similar concepts, the convex projection operator (5.129) assures that the adaptive gain $\hat{f}_N(t, \cdot)$ is always contained in the user-defined convex set $\{f \in \mathcal{D} : h(f) \leq 1\}$. This result is obtained by not modifying $\dot{\hat{f}}_N(t, \cdot)$ whenever $\hat{f}_N(t, \cdot) \in \{f \in \mathcal{D} : h(f) \leq 0\}$ and gradually modifying $\dot{\hat{f}}_N(t, \cdot)$ whenever $\hat{f}_N(t, \cdot) \in \{f \in \mathcal{D} : h(f) \in (0, 1]\}$ and its gradient points toward the boundary of the constraint set.

Finally, for the statement of the key theorem of this section, we need to introduce the following result.

Lemma 5.1 *Consider the convex projection operator* $\mathfrak{Proj} : \mathcal{H} \times \mathcal{H} \to \mathcal{H}$. *Then,*

$$\langle f_b - f_i, \mathfrak{Proj}(f_b, f_d) - f_d\rangle_{\mathcal{H}} \leq 0 \tag{5.132}$$

for all $f_i \in \mathring{\Omega}_\delta$, $f_b \in \partial\Omega_\delta$, *and* $f_d \in \mathcal{H}$.

Proof This result follows by transposing the arguments of Lemma 11.3 in [4] from finite-dimensional matrix spaces into infinite-dimensional Hilbert spaces, and is omitted for brevity. ∎

Theorem 5.8 *Let $\mathcal{H} \triangleq \mathcal{H}^m$, and assume that the DPS given by (5.40) with control input (5.123), the reference model (5.41), and the adaptive laws (5.124)–(5.126) have complete solutions on $[t_0, \infty)$. If the compensator $c_N(\cdot, \cdot)$ is such that $\left\langle E_{x_N(t)} c_N(\cdot, e_N(t)), B^{\mathrm{T}} P e_N(t) \right\rangle_{\mathbb{U}} \le 0$ for all $t \ge t_0$, then the trajectories of the DPS are uniformly bounded in the set $\mathbb{Z}_N$ given by (5.78). Additionally, for any arbitrarily small constant $\eta > 0$, there exists $T \in (t_0, \infty)$ such that $e_N(t) \in \mathcal{B}_{\mathcal{S}}(0)$ for all $t \ge T$, where*

$$\mathcal{S} \triangleq (1+\eta) \frac{2\|B^{\mathrm{T}} P\|}{\lambda_{\min}(Q)} \overline{\epsilon}_N \sup_{\xi \in \Omega} \mathcal{P}_N(\xi) \tag{5.133}$$

denotes the radius of the open ball $\mathcal{B}_{\mathcal{S}}(0)$ centered at the origin, $\overline{\epsilon}_N \triangleq \sup_{x \in \Omega} \|E_x(I - \mathbf{\Pi}_N f))\|_{\mathcal{H}}$, and $\Omega \subset \mathbb{X}$ is compact and such that $x_N(t) \in \Omega$ for all $t \ge t_0$.

Proof Consider the Lyapunov function candidate (5.82). Taking the derivative of (5.82) along the trajectories of (5.40) with control input (5.123) and (5.124)–(5.126), we obtain that, for all $t \ge t_0$,

$$\begin{aligned}
\dot{V}(t) = &\underbrace{- \langle e_N(t), Q e_N(t) \rangle_{\mathbb{R}^n}}_{\text{Term 0}} \\
&\underbrace{-2 \left\langle \tilde{\alpha}_N(t), x_N(t) e_N^{\mathrm{T}}(t) P B + \Gamma_\alpha^{-1} \dot{\hat{\alpha}}_N(t) \right\rangle_{\mathrm{tr}}}_{\text{Term 1}} \\
&\underbrace{-2 \left\langle \tilde{\beta}_N(t), r(t) e_N^{\mathrm{T}}(t) P B + \Gamma_\beta^{-1} \dot{\hat{\beta}}_N(t) \right\rangle_{\mathrm{tr}}}_{\text{Term 2}} \\
&\underbrace{+2 \left\langle \tilde{f}_N(t, \cdot), \mathbf{\Pi}_N E_{x_N(t)}^* B^{\mathrm{T}} P e_N(t) + \Gamma_f^{-1} \dot{\hat{f}}_N(t, \cdot) \right\rangle_{\mathcal{H}}}_{\text{Term 3}} \\
&+ \underbrace{2 \left\langle E_{x_N(t)} (I - \mathbf{\Pi}_N) f, B^{\mathrm{T}} P e_N(t) \right\rangle_{\mathbb{U}}}_{\text{Term 4}} \\
&+ \underbrace{2 \left\langle E_{x_N(t)} c_N(\cdot, e_N(t)), B^{\mathrm{T}} P e_N(t) \right\rangle_{\mathbb{U}}}_{\text{Term 5}}
\end{aligned} \tag{5.134}$$

with $V(t) = V(e_N(t), \tilde{\alpha}_N(t), \tilde{\beta}_N(t), \tilde{f}_N(t, \cdot))$ for brevity.

Applying Lemma 11.3 of [4], we deduce that Terms 1 and 2 are nonpositive for all $t \ge t_0$. Furthermore, for all $t \ge t_0$, it holds that

$$\text{Term 3} = 2 \left\langle \mathbf{\Pi}_N f - \hat{f}_N(t, \cdot), \mathbf{\Pi}_N E_{x_N(t)}^* B^{\mathrm{T}} P e_N(t) - \mathfrak{Proj}(\hat{f}_N(t), \mathbf{\Pi}_N \Gamma_f E_{x_N(t)}^* B^{\mathrm{T}} P e(t)) \right\rangle_{\mathcal{H}}.$$

Now, let $\delta \in (-\infty, 1]$ and $\Theta_\delta \triangleq \{f \in \mathcal{D} : \|\mathbf{\Pi}_N f\|_{\mathcal{H}} < \delta\}$. If $f \in \Theta_\delta$, then $\mathbf{\Pi}_N f \in \Theta_\delta$ from the definition of Θ_δ. Thus, it follows from Lemma 5.1 that Term 3 in (5.134) is nonpositive.

By assumption, Term 5 is nonpositive. Finally, recalling that $\mathcal{H}_{d(N)}$ can be expressed as the Cartesian product of scalar-valued RKHS, it follows from (3.157) that

$$\begin{aligned}
\text{Terms0\&4} &= -\langle e_N(t), Qe_N(t)\rangle_{\mathbb{R}^n} \\
&\quad + 2\left\langle \tilde{f}_N(t,\cdot), (I - \mathbf{\Pi}_N)E^*_{x_N(t)} B^{\mathrm{T}} P e_N(t)\right\rangle_{\mathcal{H}} \\
&= -\langle e_N(t), Qe_N(t)\rangle_{\mathbb{R}^n} + 2\left\langle E_{x_N(t)}(I - \mathbf{\Pi}_N)f, B^{\mathrm{T}} P e_N(t)\right\rangle_{\mathcal{H}} \\
&\leq -\lambda_{\min}(Q)\|e_N(t)\| \left(\|e_N(t)\| - \frac{2\|B^{\mathrm{T}} P\|}{\lambda_{\min}(Q)} \|\epsilon_N\|_{\mathcal{H}} \sup_{x \in \Omega} \mathcal{P}_N(x) \right)
\end{aligned} \tag{5.135}$$

for all $t \geq t_0$, where $\Omega \subset \mathbb{X}$ is a compact set such that $x(t) \in \Omega$ at all times. Thus, the result is attained. ■

Similar to classical MRAC systems with adaptive laws modified by the convex projection operator [4, Chap. 10], the proposed formulation has the advantage of imposing user-defined bounds on the adaptive gains by designing the convex set $\mathcal{D}$ and the convex function h. However, a disadvantage of the proposed approach is that the uncertainty $f \in \mathcal{H}$ must lie in the set $\{f \in \mathcal{D} : h(f) < 0\}$. In general, this assumption is verified by designing this set sufficiently large.

5.4.3.2 Numerical Example

In this section, we analyze the same numerical example as in Sects. 5.4.1.5 and 5.4.2.2, but employing the adaptive laws (5.124)–(5.126); the MATLAB® codes generating these results can be found in [21]. Figures 5.13, 5.14 and 5.15 show the controller's performance in terms of steady-state trajectory tracking error employing the Sololev–Matérn, Wendland, and exponential kernels, respectively.

All simulations employ the same constraint function

$$h(\theta) = \frac{(1+\varepsilon)\|\theta\|^2 - \theta_{\max}^2}{\varepsilon\theta_{\max}^2}, \quad \text{for all } \theta \in \mathbb{R}^m, \tag{5.136}$$

where $\varepsilon > 0$ and $\theta_{\max} > 0$ are user-defined. Applying (5.136), we deduce that

$$\overline{\Omega}_0 = \left\{\theta \in \mathbb{R}^m : \|\theta\| \leq \frac{\theta_{\max}}{\sqrt{1+\varepsilon}}\right\} \tag{5.137}$$

and

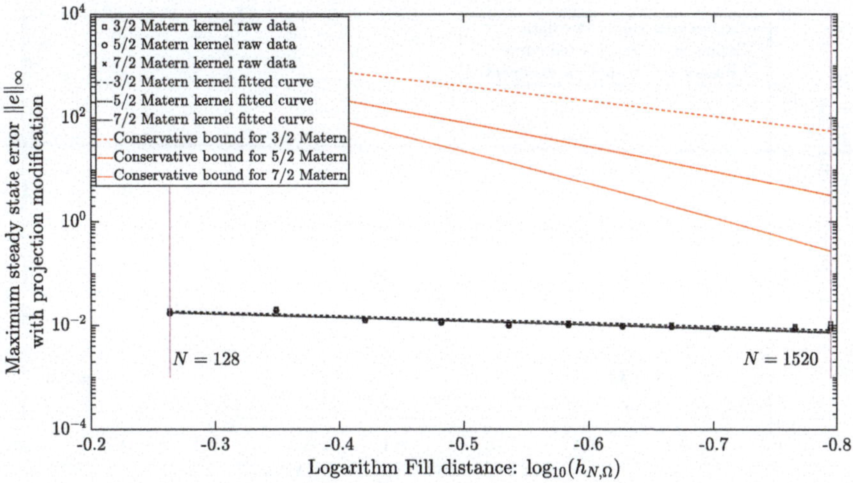

Fig. 5.13 Maximum norm of the steady-state error against the fill distance between basis centers for Sololev–Matérn kernels employing the projection operators to constrain the adaptive gains. The red lines show the upper bounds captured by (5.133)

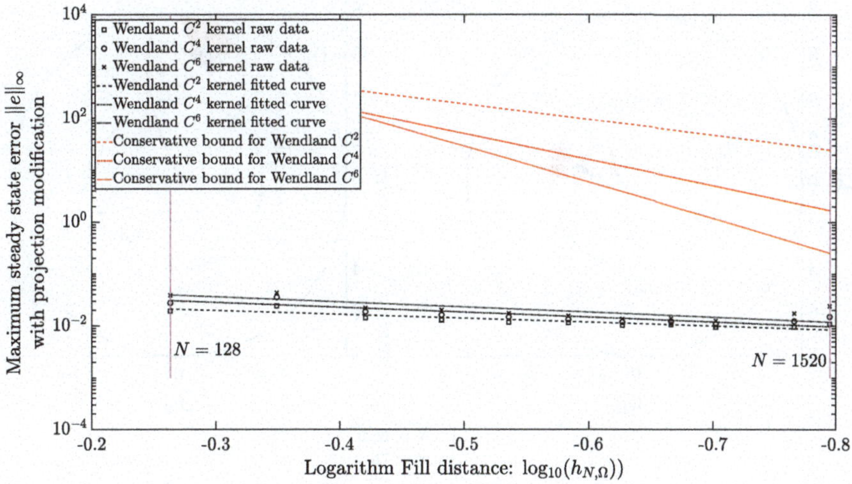

Fig. 5.14 Maximum norm of the steady-state error against the fill distance between basis centers for Wendland kernels employing the projection operators to constrain the adaptive gains. The red lines show the upper bounds captured by (5.133)

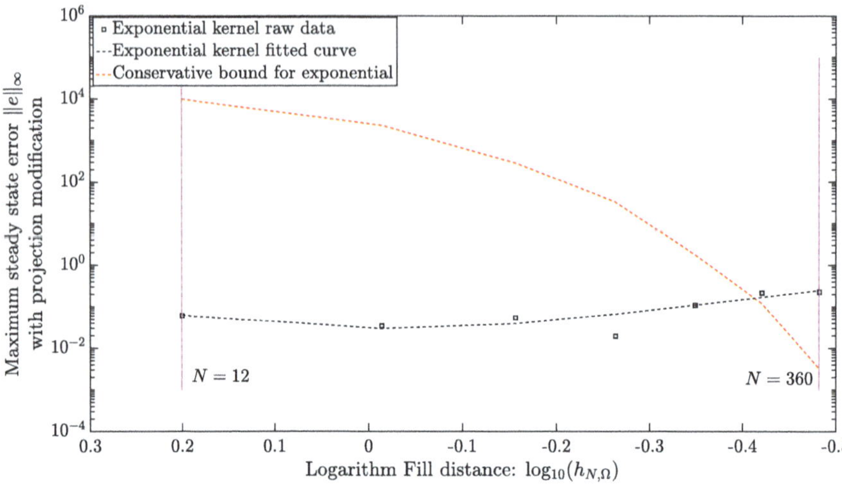

Fig. 5.15 Maximum norm of the steady-state error against the fill distance between basis centers for exponential kernels employing the projection operators to constrain the adaptive gains. The red lines show the upper bounds captured by (5.133)

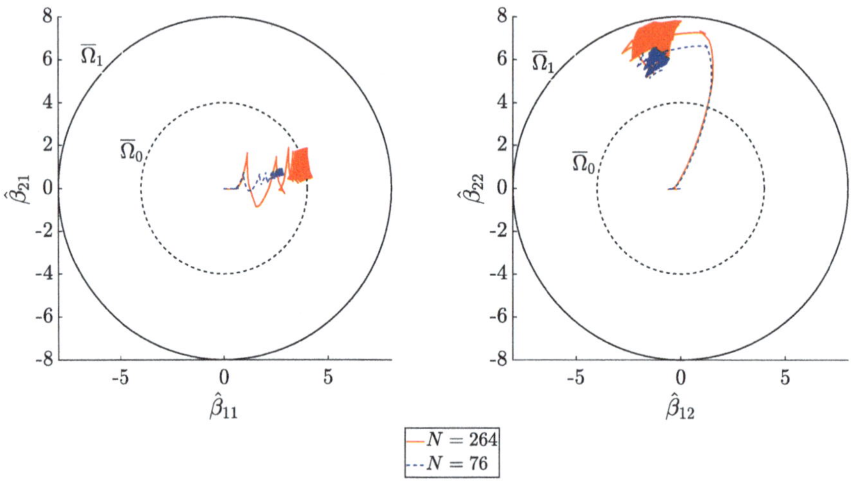

Fig. 5.16 The left figure shows the phase portrait of the first column of the adaptive gain matrix $\hat{\beta}_N(\cdot)$ and the right figure shows the phase portrait of the second column of this adaptive gain matrix for different values of N. Both figures show the sets $\overline{\Omega}_0$ and $\overline{\Omega}_1$

$$\overline{\Omega}_1 = \left\{\theta \in \mathbb{R}^m : \|\theta\| \leq \theta_{\max}\right\}. \tag{5.138}$$

For the proposed simulations, we set $\theta_{\max} = 8$ and $\varepsilon = 4$. Figure 5.16 shows the phase portraits of the columns of the adaptive gain matrix $\hat{\beta}_N$ and the sets $\overline{\Omega}_0$ and $\overline{\Omega}_1$ for different values of N. Neither phase portrait escapes the constraint set $\overline{\Omega}_1$.

5.5 Variable Structure Methods

In this section, we build upon the convex projection methods presented in Sect. 5.4.3 and discuss how the adaptive control framework discussed thus far can be further improved by introducing a variable structure approach. The class of variable structure controllers is very broad, and its most popular member is sliding mode control. According to this control design philosophy, the closed-loop trajectory is steered in finite time toward a smooth surface designed by the user so that, once on this surface, the closed-loop trajectory converges to the desired state.

As anticipated in Chap. 4, a careful analysis of the proposed results would require using Filippov or Krasovskii solutions. However, for brevity, and consistent with a large portion of the literature on nonlinear robust control, we consider Carathéodory solutions, assuming that they are well-defined.

5.5.1 *Error Bounding Control*

The first kind of variable structure control system we present in this book modifies the control law (5.74) by adding a *bounding control term*. This approach extends a technique presented in [3, Chaps. 4 and 6] to systems in the same form as (5.40), whose functional uncertainties lie in a native space.

To begin, it follows from (3.154) with $\alpha = e_i$, where e_i denotes the ith canonical basis in $\mathbb{R}^m$ (see Example 2.1), that, for any $f \in \mathcal{H}$,

$$\begin{aligned} |\langle E_x(I - \mathbf{\Pi}_N)f, e_i\rangle_{\mathbb{R}^n}| &= |E_x((I - \mathbf{\Pi}_N)f)_i| \\ &\leq \mathcal{P}_N^{e_i}(x)\|(I - \mathbf{\Pi}_N)f\|_{\mathcal{H}} \\ &\leq \mathcal{P}_N^{e_i}(x)\|f\|_{\mathcal{H}}, \quad \text{for all } x \in \mathbb{X}. \end{aligned} \tag{5.139}$$

Furthermore, it follows from (3.153) (see also [23]) that

$$|E_x((I - \mathbf{\Pi}_N)f)_i| \leq \sqrt{\Delta\mathcal{K}_{N,ii}(x,x)}\|f\|, \quad \text{for all } x \in \mathbb{X} \tag{5.140}$$

for each $i = 1, \ldots, m$, where

$$\Delta\mathcal{K}_N(x,x) \triangleq \left[\Delta\mathcal{K}_{N,11}(x,x),\ldots,\Delta\mathcal{K}_{N,mm}(x,x)\right]^{\mathrm{T}} \in \mathbb{R}^m, \tag{5.141}$$

$$\Delta\mathcal{K}_{N,ii}(x,x) \triangleq \mathcal{K}_{ii}(x,x) - \mathcal{K}_{N,ii}(x,x). \tag{5.142}$$

We design the *error bounding control term* $v_N : \mathbb{X} \times \mathbb{R}^n \to \mathbb{U}$, whose ith component is given by

$$v_{N,i}(x,e) \triangleq -\mathrm{sign}(B^{\mathrm{T}}Pe)_i \Delta\mathcal{K}_{N,ii}(x,x)\|f\|_{\mathcal{H}} \quad \text{for } i = 1,\ldots,m, \tag{5.143}$$

where the *signum function* $\mathrm{sign} : \mathbb{R} \to \{-1, 0, 1\}$ is such that $\mathrm{sign}(\alpha) = -1$ for $\alpha < 0$, $\mathrm{sign}(\alpha) = 0$ for $\alpha = 0$, and $\mathrm{sign}(\alpha) = 1$ for $\alpha > 0$. Finally, the adaptive control input is given by

$$u_N(t) = \hat{\alpha}_N^{\mathrm{T}}(t)x_N(t) + \hat{\beta}_N^{\mathrm{T}}(t)r(t) - E_{x_N(t)}\hat{f}_N(t,\cdot) + v_N(x_N(t), e_N(t)), \quad t \geq t_0. \tag{5.144}$$

We can now state the main result of this section.

Theorem 5.9 *Let $\mathcal{H} \triangleq \mathcal{H}^m$ and consider the DPS given by (5.40) with control input (5.144) and adaptive laws (5.124)–(5.126). Then,*

$$\lim_{t\to\infty} \|x_N(t) - x_r(t)\|_{\mathbb{X}} = 0 \tag{5.145}$$

uniformly in $t_0 \geq 0$.

Proof Following the same set of steps in the proof of Theorem 5.8, except for the analysis of Term 5, the derivative of the Lyapunov function candidate (5.82) along the trajectories of (5.40) with control input (5.144) and (5.124)–(5.126) is given by

$$\begin{aligned}\dot{V}(t) \leq& -e_N^{\mathrm{T}}(t)Qe_N(t) \\ &+ 2\left\langle E_{x_N(t)}(I - \mathbf{\Pi}_N)f + v_N(x_N(t), e_N(t)), B^{\mathrm{T}}Pe_N(t)\right\rangle_{\mathbb{R}}, \quad t \geq t_0.\end{aligned} \tag{5.146}$$

Now, note that, by virtue of the entry-wise bound in (5.140),

$$\begin{aligned}&\left\langle E_{x_N(t)}(I - \mathbf{\Pi}_N)f + v_N(x_N(t), e_N(t)), B^{\mathrm{T}}Pe_N(t)\right\rangle_{\mathbb{R}} \\ &\quad= \sum_{i=1}^{m} (B_P^{\mathrm{T}}e_N(t))_i \left(E_{x_N(t)}(I - \mathbf{\Pi}_N)f)_i + (v_N(x_N(t), e_N(t))_i\right) \\ &\quad= \sum_{i=1}^{m} \left|(B^{\mathrm{T}}Pe_N(t))_i\right| \left(\mathrm{sign}(B^{\mathrm{T}}Pe_N(t))_i E_{x_N(t)}(I - \mathbf{\Pi}_N)f)_i - \sqrt{\Delta\mathcal{K}_{N,ii}}\|f\|_{\mathcal{H}}\right) \\ &\quad\leq 0, \quad t \geq t_0.\end{aligned}$$

Thus, $\dot{V}(t) \leq 0$ for all $t \geq t_0$, and uniform Lyapunov stability of the system given by (5.40) with control input (5.144) and (5.124)–(5.126) is proven. Finally, (5.145) follows from Barbalat's lemma (Lemma 4.1). ∎

It is worthwhile mentioning that, with the control input (5.144), the closed-loop plant trajectory may suffer from chattering, that is, from high-frequency oscillations due to the sudden variations induced by the signum function in (5.143). This problem is usually solved by employing higher-order variable structure control systems, which, substantially, embed the signum function in some derivative of the control input with respect to time, and, hence, remove or, at least, attenuate, the chattering effect. The expansion of the results in Theorem 5.9 to higher-order variable structure systems is left to the interested reader as a research problem.

5.5.2 Adaptive Error Bounding Control

To implement the error bounding control term (5.143), knowledge of a (conservative) upper bound on $\|f\|_{\mathcal{H}}$ is required, which is typical of most robust control techniques. In this section, we discuss how the adaptive upper bounding method as described in [3, Chaps. 4, 6 and 7] can be modified to take advantage of the RKHS framework. We emphasize that the approach described in this section leverages the properties of the power function $\mathcal{P}_N^{\alpha}$; for details, see Section 3.8.1. The overall strategy is quite general and works in a wide variety of choices of the operator-valued native space $\mathcal{H}$.

We begin by modifying the upper bounding control v_N, whose ith component is defined in (5.143), as

$$v_{N,i}(x, e, \lambda) \triangleq -\text{sign}\left(B^{\mathrm{T}} P e\right)_i \sqrt{\Delta\mathcal{K}_{N,ii}(x, x)}\lambda \quad \text{for all } (x, e, \lambda) \in \mathbb{X} \times \mathbb{R}^n \times \mathbb{R}, \tag{5.147}$$

so that $v_N \triangleq \left[v_{N,1}, \ldots, v_{N,m}\right]^{\mathrm{T}}$. In Sect. 5.5.1, the function v_N takes two arguments, whereas in this section, v_N takes three arguments; this minimal abuse of notation aims to highlight similarities between the error bounding and the adaptive error bounding techniques. The two cases are related since

$$v_N(x, e) \triangleq v_N(x, e, \|f\|_{\mathcal{H}}) \quad \text{for all } (x, e, f) \in \mathbb{X} \times \mathbb{R}^n \times \mathcal{H}. \tag{5.148}$$

The scalar-valued adaptive gain λ in (5.147) is chosen as the solution $\hat{\lambda}_N : [t_0, \infty) \to \mathbb{R}$ of the adaptive law

$$\dot{\hat{\lambda}}_N(t) = -\Gamma_\lambda \Delta\mathcal{K}_N^{\mathrm{T}}(x_N(t), x_N(t)) B^{\mathrm{T}} P e_N(t), \quad \hat{\lambda}_N(t_0) = \hat{\lambda}_{N,0}, \quad t \geq t_0, \tag{5.149}$$

where $P \in \mathbb{R}^{n\times n}$ denotes the positive-definite solution of the algebraic Lyapunov equation (5.45) and $\Delta\mathcal{K}_N$ is given by (5.141). The adaptive gain $\hat{\lambda}_N$ can be interpreted as the counterpart of $\|f\|_{\mathcal{H}}$. In this book, we are employing direct adaptive control approaches, and, hence, we do not aim to design adaptive systems such that the adaptive gains converge to their unknown counterparts. The extension of the RKHS framework presented in this book to indirect adaptive control techniques is left to the reader as a research problem.

The adaptive control input considered in this section is given by

$$u_N(t) = \hat{\alpha}^{\mathrm{T}}(t)x(t) + \hat{\beta}^{\mathrm{T}}(t)r(t) - \hat{f}_N(t,\cdot) + v_N(x_N(t), e_N(t), \hat{\lambda}(t)), \quad t \geq t_0. \tag{5.150}$$

It is worthwhile remarking on the similarities between (5.150) and (5.144).

Theorem 5.10 *Let $\mathcal{H} \triangleq \mathcal{H}^m$ and consider the DPS given by (5.40) with control input (5.150) and adaptive laws (5.124)–(5.126) and (5.149). Then,*

$$\lim_{t\to\infty} \|x_N(t) - x_r(t)\|_{\mathbb{X}} = 0 \tag{5.151}$$

uniformly in $t_0 \geq 0$.

Proof To prove this result, we proceed as in the proof of Theorem 5.9. Consider the Lyapunov function candidate

$$\mathcal{V}(e_N, \tilde{\alpha}_N, \tilde{\beta}_N, \tilde{f}_N, \tilde{\lambda}_N) \triangleq V(e_N, \tilde{\alpha}_N, \tilde{\beta}_N, \tilde{f}_N) + \tilde{\lambda}^2,$$
$$(e_N, \tilde{\alpha}_N, \tilde{\beta}_N, \tilde{f}_N, \tilde{\lambda}_N) \in \mathbb{Z}_N \times \mathbb{R}, \tag{5.152}$$

where $V(\cdot)$ is given by (5.82) and $\tilde{\lambda}(t) \triangleq \|f\|_{\mathcal{H}} - \hat{\lambda}(t), t \geq t_0$. Taking the derivatives of $\mathcal{V}(\cdot)$ along the trajectories of (5.40) with control input (5.150) and adaptive laws (5.124)–(5.126) and (5.149), we obtain that

$$\begin{aligned}
\dot{\mathcal{V}}(t) = &- \langle Qe_N(T), e_N(t)\rangle_{\mathbb{R}^n} \\
&+ 2\Big\langle B^{\mathrm{T}}Pe_N(t), E_{x_N(t)}(I - \mathbf{\Pi}_N)f + v_N(x_N(t), e_N(t), \hat{\lambda}_N(t)\Big\rangle_{\mathbb{R}} \\
&+ 2\Big\langle -\dot{\hat{\lambda}}_N(t), \tilde{\lambda}_N(t)\Big\rangle_{\mathbb{R}} \\
= &-e_N^{\mathrm{T}}(t)Qe_N(t) \\
&+ 2\langle B^{\mathrm{T}}Pe_N(t), E_{x_N(t)}(I - \mathbf{\Pi}_N)f + v_N(x_N(t), e_N(t), \|f\|_{\mathcal{H}})\rangle_{\mathbb{R}} \\
&+ 2\Big\langle -\Delta\mathcal{K}_N^{\mathrm{T}}(x_N(t), x_N(t))B^{\mathrm{T}}Pe_N(t) - \dot{\hat{\lambda}}_N(t), \tilde{\lambda}_N(t)\Big\rangle_{\mathbb{R}} \\
= &- \langle Qe_N(T), e_N(t)\rangle_{\mathbb{R}^n}, \quad t \geq t_0,
\end{aligned}$$

where $\mathcal{V}(t) = \mathcal{V}(e_N(t), \tilde{\alpha}_N(t), \tilde{\beta}_N(t), \tilde{f}_N(t,\cdot), \tilde{\lambda}_N(t))$ for notational brevity. Thus, uniform Lyapunov stability of (5.40) with control input (5.150) and adaptive laws

(5.124)–(5.126) and (5.149) is proven. Finally, (5.151) follows from Barbalat's lemma (Lemma 4.1). ■

5.5.3 Numerical Example

In this section, we analyze the same numerical example as in Sects. 5.4.1.5, 5.4.2.2, and 5.4.3.2 but employing the control input (5.150) and a modification of the adaptive laws (5.124)–(5.126) and (5.149) aimed at avoiding chattering due to the discontinuities induced by the signum function in (5.147); the MATLAB® codes generating these results can be found in [21]. Specifically, we let

$$v_{N,i} = \begin{cases} -\text{sign}((B^{\mathrm{T}} P e_N(t))_i)\sqrt{\Delta\mathfrak{K}_i(x_N(t), x_N(t))}\hat{\lambda}_N(t), \\ \qquad\qquad \|B^{\mathrm{T}} P e_N(t)\| \geq \varepsilon, \\ -\frac{[B^{\mathrm{T}} P e_N(t)]_i}{\varepsilon}\sqrt{\Delta\mathfrak{K}_i(x_N(t), x_N(t))}\hat{\lambda}_N(t), \\ \qquad\qquad \|B^{\mathrm{T}} P e_N(t)\| < \varepsilon, \end{cases} \tag{5.153}$$

and

$$\dot{\hat{\lambda}}_N(t) = \begin{cases} -\Gamma_\lambda \sum_{i=1}^m \sqrt{\Delta\mathfrak{K}_i(x_N(t), x_N(t))}\left|(B^{\mathrm{T}} P e_N(t))_i\right|, \\ \qquad\qquad \|e_N(t)\| \geq d + \varepsilon, \\ -\frac{\|e_N(t)\| - d}{\varepsilon}\Gamma_\lambda \sum_{i=1}^m \sqrt{\Delta\mathfrak{K}_i(x_N(t), x_N(t))}\left|(B^{\mathrm{T}} P e_N(t))_i\right|, \\ \qquad\qquad d \leq \|e_N(t)\| \leq d + \varepsilon, \\ 0, \quad \|e_N(t)\| \leq d \text{ or } \hat{\lambda}_N(t) \leq 0.1 \cdot d, \end{cases} \tag{5.154}$$

where $\Delta\mathfrak{K}_i(x_N, x_N) \triangleq \mathfrak{K}_i(x_N, x_N) - \mathfrak{K}_{N,i}(x_N, x_N)$, and $\varepsilon > 0$ is user-defined. Figures 5.17 and 5.18 show the controller's performance in terms of steady-state trajectory tracking error employing the Sololev–Matérn and Wendland kernels, respectively.

5.6 Concluding Remarks

This chapter presented systematically how the theory of native spaces can be employed to capture functional uncertainties and, hence, to design robust and adaptive control systems for nonlinear dynamical models. Several architectures have been considered, ranging from an extension of a cornerstone of control systems design, namely robust control Lyapunov functions, to MRAC systems and variable structure controllers.

In addition to the relevance of these results *per se*, which lies in freeing the user from computing or designing parameterization of the functional uncertainties, the results presented in this book provide multiple examples of how, virtually, the theory

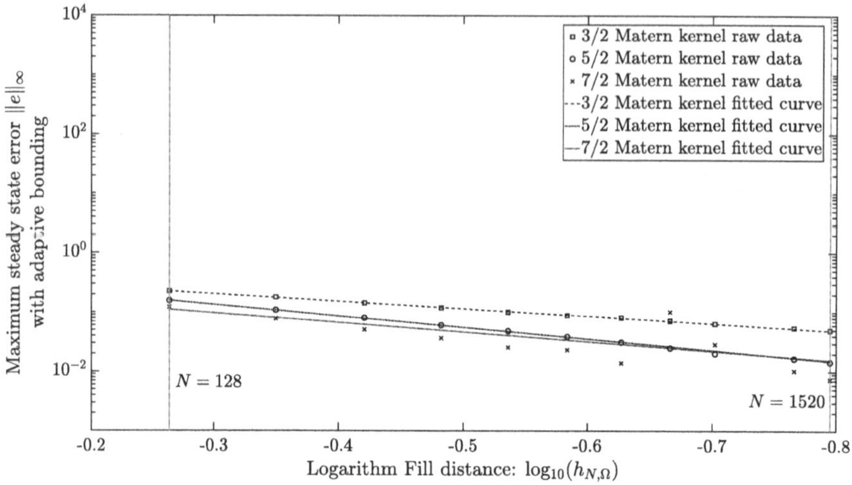

Fig. 5.17 Maximum norm of the steady-state error against the fill distance between basis centers for Sobolev–Matérn kernels employing the adaptive error bounding control method

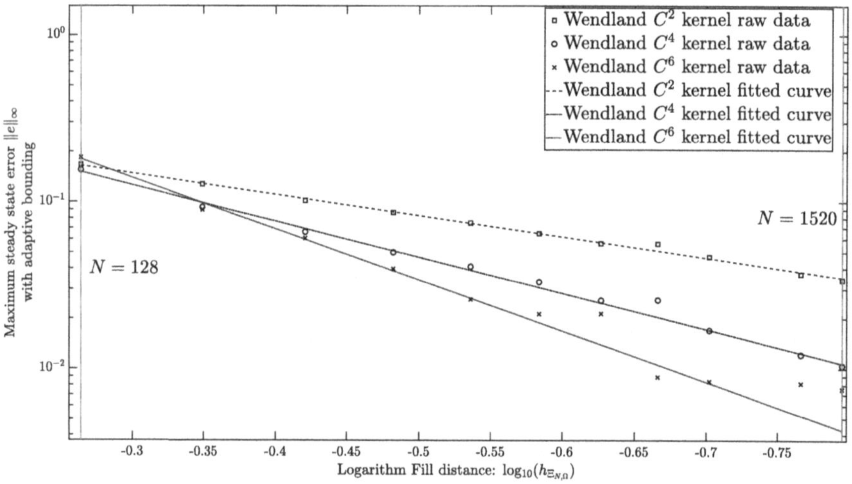

Fig. 5.18 Maximum norm of the steady-state error against the fill distance between basis centers for Wendland kernels employing the adaptive error bounding control method

of RKHSs can be used to extend any existing control system architecture. Finally, throughout this chapter, we suggested several research questions that may further advance the state-of-the-art in control systems design employing RKHSs.

5.7 Homework Problems

For an improved comprehension of the material exposed in this chapter, we recommend the following problems.

Exercise 5.1 Consider the vector-valued RKHS $\mathcal{H}(\mathbb{R}^n, \mathbb{R}^m) = \mathcal{H}^m(\mathbb{R}^n, \mathbb{R})$, where $\mathcal{H}$ is generated by the radial functions in Table 3.1. For each kind of RKHS, provide examples of functions in the uncertainty classes (5.1) and (5.2). △

Exercise 5.2 The equations of motion of a planar pendulum of mass $m > 0$ and length $L > 0$ are given by

$$\begin{bmatrix}\dot{\theta}(t)\\ \ddot{\theta}(t)\end{bmatrix} = \begin{bmatrix}0 & 1\\ -c_1 & -c_2\end{bmatrix}\begin{bmatrix}\theta(t)\\ \dot{\theta}(t)\end{bmatrix} + \begin{bmatrix}0\\ 1\end{bmatrix}\left[u(t) - \frac{g}{L}\sin\theta(t) + E_{\Theta(t)}f\right],$$

$$\begin{bmatrix}\theta(0)\\ \dot{\theta}(0)\end{bmatrix} = \begin{bmatrix}\theta_0\\ \theta_{\mathrm{d},0}\end{bmatrix}, \quad t \geq 0, \tag{5.155}$$

where $c_1, c_2 \in \mathbb{R}$ are unknown, $g > 0$ denotes the gravitational acceleration, and $\Theta \triangleq [\theta, \dot{\theta}]^{\mathrm{T}}$. Design an RCLF, and simulate the performance of the associated controller assuming that $f \in \mathcal{H}(\mathbb{R}, \mathbb{R})$ and $\mathcal{H}$ underlies a Gaussian, multi-quadratic, inverse multi-quadratic, or a Wendland kernel function. Provide multiple examples of uncertainties f and discuss the controller's performance in the case $f \in \mathcal{H}$ and $f \notin \mathcal{H}$. Modify the computer codes provided in [12]. △

Exercise 5.3 Consider the dynamical system given by (5.155). Design a backstepping controller, and simulate the performance of this controller assuming that $f \in \mathcal{H}(\mathbb{R}, \mathbb{R})$ and $\mathcal{H}$ underlies a Gaussian, multi-quadratic, inverse multi-quadratic, or Wendland kernel function. Provide multiple examples of uncertainties f and discuss the controller's performance in the case $f \in \mathcal{H}$ and $f \notin \mathcal{H}$. Modify the computer codes provided in [12]. △

Exercise 5.4 Consider the dynamical system given by (5.155). Design an MRAC system based on the dead-zone modification, and simulate the performance of this controller assuming that $f \in \mathcal{H}(\mathbb{R}, \mathbb{R})$ and $\mathcal{H}$ underlies a Gaussian, multi-quadratic, inverse multi-quadratic, or Wendland kernel function. Design the reference model so that

$$\lim_{t\to\infty} |\theta_{\mathrm{ref}}(t) - \theta_{\mathrm{user}}(t)| = 0, \tag{5.156a}$$

$$\lim_{t\to\infty} |\dot{\theta}_{\mathrm{ref}}(t) - \dot{\theta}_{\mathrm{user}}(t)| = 0, \tag{5.156b}$$

where $\left[\theta_{\mathrm{ref}}(t), \dot{\theta}_{\mathrm{ref}}(t)\right]^{\mathrm{T}}$ denotes the state of the reference model and $\theta_{\mathrm{user}} : [t_0, \infty) \to \mathbb{R}$ denotes a user-defined, bounded, piecewise twice continuously differentiable signal. Provide multiple examples of uncertainties f and discuss the controller's performance in the case $f \in \mathcal{H}$ and $f \notin \mathcal{H}$. Modify the computer codes provided in [21]. △

Exercise 5.5 Consider the dynamical system given by (5.155). Design an MRAC system based on the σ-zone modification, and simulate the performance of this controller assuming that $f \in \mathcal{H}(\mathbb{R}, \mathbb{R})$ and $\mathcal{H}$ underlies a Gaussian, multi-quadratic, inverse multi-quadratic, or Wendland kernel function. Design the reference model so that the design objective (5.156) is met. Provide multiple examples of uncertainties f and discuss the controller's performance in the case $f \in \mathcal{H}$ and $f \notin \mathcal{H}$. Modify the computer codes provided in [21]. △

Exercise 5.6 Consider the dynamical system given by (5.155). Design an MRAC system robustified by the use of the convex projection operator on the adaptive gains, and simulate the performance of this controller assuming that $f \in \mathcal{H}(\mathbb{R}, \mathbb{R})$ and $\mathcal{H}$ underlies a Gaussian, multi-quadratic, inverse multi-quadratic, or Wendland kernel function. Design the reference model so that the design objective (5.156) is met. Provide multiple examples of uncertainties f and discuss the controller's performance in the case $f \in \mathcal{H}$ and $f \notin \mathcal{H}$. Modify the computer codes provided in [21]. △

Exercise 5.7 Consider the dynamical system given by (5.155). Design an error bounding control method, and simulate the performance of this controller assuming that $f \in \mathcal{H}(\mathbb{R}, \mathbb{R})$ and $\mathcal{H}$ underlies a Gaussian, multi-quadratic, inverse multi-quadratic, or Wendland kernel function. Design the reference model so that the design objective (5.156) is met. Provide multiple examples of uncertainties f and discuss the controller's performance in the case $f \in \mathcal{H}$ and $f \notin \mathcal{H}$. Modify the computer codes provided in [21]. △

Exercise 5.8 Consider the dynamical system given by (5.155). Design an adaptive control system based on the adaptive error bounding control method, and simulate the performance of this controller assuming that $f \in \mathcal{H}(\mathbb{R}, \mathbb{R})$ and $\mathcal{H}$ underlies a Gaussian, multi-quadratic, inverse multi-quadratic, or Wendland kernel function. Design the reference model so that the design objective (5.156) is met. Provide multiple examples of uncertainties f and discuss the controller's performance in the case $f \in \mathcal{H}$ and $f \notin \mathcal{H}$. Modify the computer codes provided in [21]. △

References

1. Kurdila, A.J., L'Afflitto, A., Burns, J.A., Wang, H.: Nonparametric adaptive control in native spaces: a DPS framework (Part I). Ann. Rev. Control **58**, 100969 (2024)
2. Kurdila, A.J., L'Afflitto, A., Burns, J.A., Wang, H.: Nonparametric adaptive control in native spaces: finite-dimensional implementations (Part II). Ann. Rev. Control **58**, 100969 (2024)
3. Farrell, J.A., Polycarpou, M.M.: Adaptive Approximation Based Control: Unifying Neural, Fuzzy and Traditional Adaptive Approximation Approaches, vol. 48. Wiley, Danvers, MA (2006)
4. Lavretsky, E., Wise, K.: Robust and Adaptive Control: With Aerospace Applications. Springer, London, UK (2013)
5. Narendra, K.S., Annaswamy, A.M.: Stable Adaptive Systems. Dover, New York, NY (1989)

6. Sastry, S., Bodson, M.: Adaptive Control: Stability, Convergence and Robustness. Courier Corporation, New York, NY (2011)
7. Ioannou, P.A., Sun, J.: Robust Adaptive Control. Courier Corporation, Mineola, NY (2012)
8. Krstic, M., Kanellakopoulos, I., Kokotovic, P.: Nonlinear and Adaptive Control Design. Wiley, New York, NY (1995)
9. Tao, G.: Adaptive Control Design and Analysis, vol. 37. Wiley, Danvers, MA (2003)
10. Taylor, A.J., Ames, A.D.: Adaptive safety with control barrier functions. In: American Control Conference, pp. 1399–1405. IEEE, Denver, CO (2020)
11. Freeman, R., Kokotovic, P.V.: Robust Nonlinear Control Design: State-Space and Lyapunov Techniques. Springer, New York, NY (2008)
12. Wang, H., L'Afflitto, A., Kurdila, A.J.: Computer codes to implement control systems based on RKHS-embedded robust control Lyapunov functions (2024). https://lafflitto.com/Codes/Springer_RKHS_Book_RCLF_Codes.zip
13. Farkas, B., Wegner, S.-A.: Variations on barbălat's Lemma. Am. Math. Month. **123**(8), 825–830 (2016)
14. Demetriou, M.A.: Functional estimation of perturbed positive real infinite dimensional systems using adaptive compensators. In: American Control Conference, pp. 1582–1587. IEEE, Denver, CO (2020)
15. Wendland, H.: Scattered Data Approximation, vol. 17. Cambridge University Press, Cambridge, UK (2004)
16. Adams, R.A., Fournier, J.J.: Sobolev Spaces. Elsevier, London, UK (2003)
17. Pazy, A.: Semigroups of linear operators and applications to partial differential equations, vol. 44. Springer, New York, NY (2012)
18. Khalil, H.K., Grizzle, J.W.: Nonlinear Systems, vol. 3. Prentice Hall, Upper Saddle River, NJ (2002)
19. Peterson, B., Narendra, K.: Bounded error adaptive control. IEEE Trans. Autom. Control **27**(6), 1161–1168 (1982)
20. Kütük, Y.: Activation Functions: Activation Functions in Deep Learning with LaTeX Applications. Peter Lang GmbH, Internationaler Verlag der Wissenschaften, Berlin, Germany (2022)
21. Wang, H., L'Afflitto, A., Kurdila, A.J.: Computer codes to implement RKHS-embedded model reference adaptive control systems (2024). https://lafflitto.com/Codes/Springer_RKHS_Book_MRAC_Codes.zip
22. Fuselier, E., Wright, G.B.: Scattered data interpolation on embedded submanifolds with restricted positive definite kernels: Sobolev error estimates. SIAM J. Numer. Anal. **50**(3), 1753–1776 (2012)
23. Wittwar, D., Santin, G., Haasdonk, B.: Interpolation with uncoupled separable matrix-valued kernels. Dolomites Res. Notes Approx. **11**, 23–39 (2018)

Chapter 6
Data-Driven Methods and Adaptive Control: Deterministic Analysis

This chapter discusses how modern principles of statistical and machine learning theory, data-driven models of dynamics, and approximation theory can be merged into an RKHS embedding approach and produce a general theory of *deterministic*, nonparametric learning, and adaptive control. Chapter 5 introduced numerous nonparametric adaptive control methods, which have been deduced by extending classical parametric adaptive control theory to an RKHS embedding approach. The adaptive control systems in Chap. 5 were devised through a two-part process. We first defined a limiting (i.e., ideal) model reference adaptive control (MRAC) system for a nonparametric limiting distributed parameter system (DPS). Subsequently, we constructed multiple alternative finite-dimensional approximations of this DPS. These approximations assumed that the number and location of centers were given a priori by an "oracle" and remained constant at all times; in learning theory, *oracles* provide information that is not readily available in practice. In this chapter, we discuss how these assumptions can be lifted within a deterministic framework rooted in data-driven methods. A probabilistic framework involving adaptive control methods and Gaussian processes is deferred until Chap. 7.

Various authors have addressed aspects of the general problem of deriving adaptive controllers that reduce the reliance on oracles. Within the context of deterministic approaches, we recall [1–17]. Perusing these works, we observe that, despite substantial progress, deterministic approaches have not yet been synthesized into a general self-contained theory that gives simple and explicit closed-loop performance bounds for a wide variety of approximation or learning-theoretic hypotheses. The results in this chapter make a step forward in this direction while building on classical results in adaptive parametric control theory [18–24].

This chapter is organized as follows. Section 6.1 recalls some of the notation already presented in this book and needed to present the key arguments of this chapter. Section 6.2 presents three fundamental architectures in the context of data-driven control systems design, namely the learn-then-control (LTC) architecture, the

A. J. Kurdila et al., *Data-Driven, Nonparametric, Adaptive Control Theory*,
Lecture Notes in Control and Information Sciences 495,
https://doi.org/10.1007/978-3-031-78003-5_6

switched learn-and-control (SLC) architecture, and the concurrent learn-and-control (CLC) architecture. Section 6.3 introduces both linear and nonlinear approximation schemes and discusses their relations with LTC, SLC, and CLC control architectures. Thus, Sect. 6.4 presents multiple LTC methods, and Sects. 6.5 and 6.6 discuss specific examples of SLC and CLC methods. Finally, Sect. 6.7 outlines concluding remarks on learning-based control in a deterministic setting, and Sect. 6.8 recommends homework problems.

6.1 Notation, Symbols, and Terminology

This chapter makes use of many symbols and notational conventions introduced throughout this book. In this section, we recall some of the most relevant symbols and terminology for the reader's convenience.

We denote a generic Banach space by $\mathbb{B}$ and a generic Hilbert space by $\mathbb{H}$, which are referred to as the *universe space* and *hypothesis space*, respectively. It should be noted that, in general, the hypothesis space may be a generic Banach space; see Fig. 6.1 for a graphical representation.

As discussed in Sect. 5.3, we derive control systems captured by DPSs evolving on a Hilbert space. For this reason, as discussed in Sect. 5.4, we employ approximations in the hypothesis space. Specifically, we let

$$\Phi_{d(N)} \triangleq \left[\phi_1, \ldots, \phi_{d(N)}\right]^{\mathrm{T}} \tag{6.1}$$

denote the basis of a finite-dimensional space

$$\mathbb{H}_{d(N)} \triangleq \operatorname{span}\left\{\phi_i : \phi_i \in \Phi_{d(N)}, 1 \leq i \leq d(N)\right\}, \tag{6.2}$$

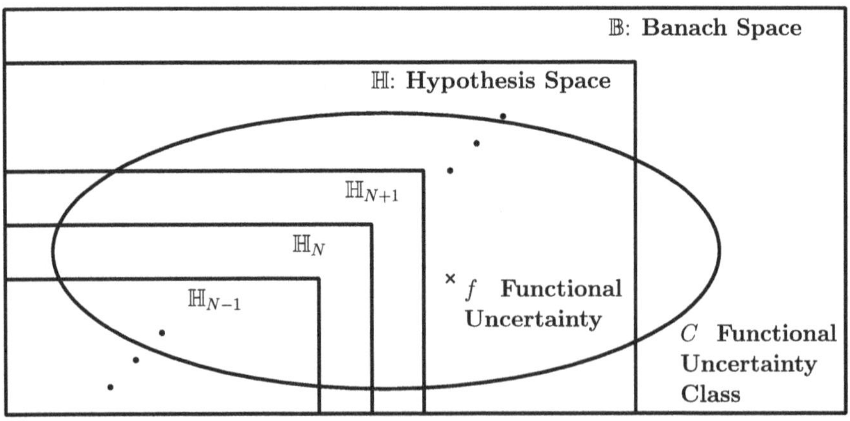

Fig. 6.1 Hypothesis space $\mathbb{H}$, uncertainty class C, and approximation spaces $\mathbb{H}_N$ with $N \in \mathbb{N}$

where $d(N) \triangleq mN$ denotes the dimension of the space of approximants and N denotes the number of centers in the set Ξ_N. Bases are sorted so that

$$\mathbb{H}_1 \hookrightarrow \ldots \hookrightarrow \mathbb{H}_{d(N)} \subset \ldots \subset \mathbb{H} \hookrightarrow \mathbb{B}, \tag{6.3}$$

as depicted in Fig. 6.1. In Chap. 5, we chose the hypothesis space to be an RKHS, that is, we set $\mathbb{H} = \mathcal{H}$, and $\mathbb{H}_{d(N)} = \mathcal{H}_{d(N)}$; see (3.148).

Remark 6.1 (Comment on Notation) While discussing classical parametric methods to characterize uncertainties, in (5.7), we employed the symbol Φ_N to denote the regressor vector parameterizing an approximation of the hypothesis space. In this chapter, we write $\Phi_{d(N)}$ to emphasize the fact that, in the case of vector-valued uncertainties, (6.1) captures approximations of the hypothesis space using kernel functions.

Remark 6.2 (Comment on Notation) In the case of scalar-valued RKHSs, we write $\mathcal{H}_N$ instead of $\mathcal{H}_{d(N)}$ to remark on the fact that, since $m = 1$, $d(N) = N$. For consistency, if $m = 1$, then we write Φ_N to denote parameterizations of the hypothesis space employing kernel functions.

In this chapter, we assume that the reproducing kernels $\mathfrak{K}$ or $\mathcal{K}$, which generate scalar-valued RKHSs and vector-valued RKHSs, respectively, are of strictly positive type (see Definitions 3.5 and 3.16). Furthermore, we consider restrictions of scalar-valued kernels $\mathfrak{K} : \mathbb{X} \times \mathbb{X} \to \mathbb{R}$ and operator-valued kernels $\mathcal{K} : \mathbb{X} \times \mathbb{X} \to \mathcal{L}(\mathbb{U})$ to $\Omega \times \Omega$ for some $\Omega \subset \mathbb{X} \triangleq \mathbb{R}^n$ to define native spaces containing functions over the subset $\Omega \subseteq \mathbb{X}$; for details, see Sects. 3.2, 3.5.3, 3.6.2, and 3.8.2. We therefore often use without further comment the same symbol for the reproducing kernel, $\mathfrak{K}$ or $\mathcal{K}$, for spaces of functions defined on $\mathbb{X}$ or for one of its subsets $\Omega \subseteq \mathbb{X}$. This convention is followed in standard treatises on RKHS; see, for example, [25].

6.2 Control Systems Design and Data-Driven Methods

The general strategy in control systems design based on learning theory in general and data-driven methods in particular is to let a learning policy choose a Hilbert space $\mathbb{H}_N$ that approximates the hypothesis space $\mathbb{H}$. Thus, a feedback control system steers the plant trajectory toward the reference trajectory despite functional and unmatched uncertainties. Chapter 5 was devoted to the design of this feedback control system by investigating control Lyapunov functions suitable for backstepping control, MRAC, and some adaptive variable structure methods, while assuming that the approximating space was provided by an oracle, or, at least, without discussing how this space was selected or designed. This section discusses three approaches to identifying approximating spaces. These approaches are LTC, SLC, and CLC, which integrate data-driven methods and control systems design.

6.2.1 The Learn-Then-Control Architecture

The LTC architecture is the most direct data-driven control system design approach as it consists of a single learning phase followed by an adaptive control phase. Figure 6.2 provides a schematic representation of an LTC architecture. Given some prior information on the plant model to control, the learning policy fixes the number of centers N, and, hence, the dimension of the approximation space $d(N) \triangleq mN$ and a parameterization of this space $\Phi_{d(N)}$. As shown in Fig. 6.3, the learning stage spans over the time interval $[t_0, \tau_1]$, and both N and $\Phi_{d(N)}$ may vary once or multiple times over this stage. Once the learning stage is completed, N and $\Phi_{d(N)}$ are fixed and the control stage starts employing classic, parametric methods, such as those in [18, 19], or nonparametric methods, such as those discussed in Chap. 5.

In LTC, the goal of the learning phase is not estimation of the functional uncertainty *per se*. It is not required that an *accurate* estimate of the functional uncertainty is generated during the learning phase. This approach is in stark contrast to the state-of-the-art of standard methods of (offline) statistical or machine learning theory, which are used to generate approximations of unknown functions; this topic is addressed in detail in Chap. 7. If such an accurate estimate is generated, either offline or online, then it certainly can be used to advantage. However, constructing an offline estimate can be substantially more time-consuming than generating candidate bases that define the subspace $\mathbb{H}_{d(N)}$ used to build feedback controllers u_N.

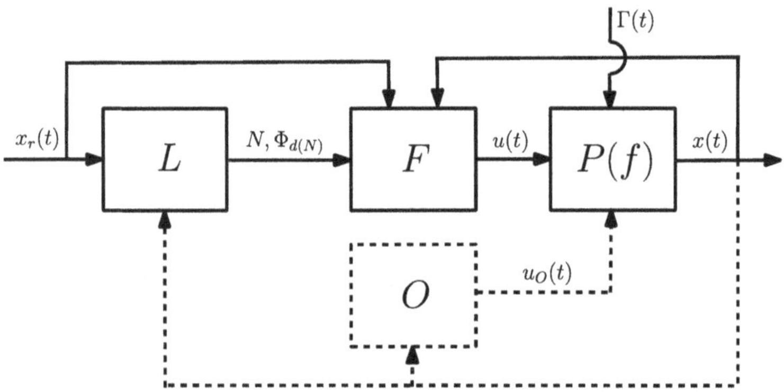

Fig. 6.2 Schematic representation of the LTC architecture. A learning policy L decides a priori the number of centers N and a parametrization of the approximating hypothesis space $\Phi_{d(N)}$. Fixed the finite-dimensional space $\mathbb{H}_{d(N)}$ that approximates the hypothesis space $\mathbb{H}$, the feedback control system F produces a feedback control law u that steers the plant trajectory x of the plant P toward the reference trajectory x_r despite the functional uncertainty f and the unmatched uncertainty Γ. The learning policy can be trained either offline or online by data-driven methods. Offline training is performed considering the reference signal to track. Online training employs some control system O that generates a control input u_O, and the plant's state x is employed to determine L. The control system O can be an open-loop control law or the nominal controller, that is, the controller we would apply if f and Γ were known. Once the learning policy is fixed, it cannot be changed; see Fig. 6.3

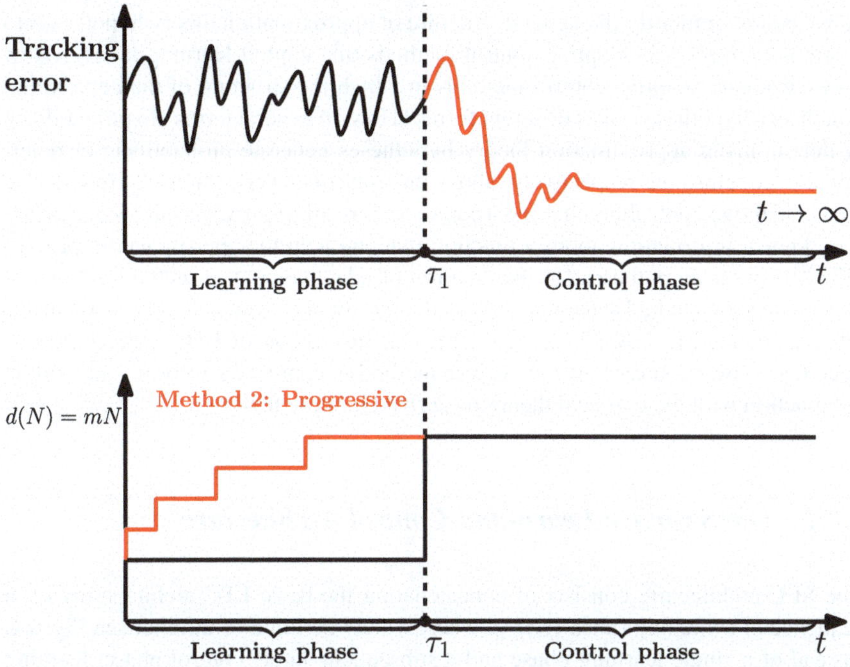

Fig. 6.3 Graphical representation of the tracking error and the dimensions of the approximating space $\mathbb{H}_{d(N)}$ in the learn-then-control (LTC) architecture. The parameter N that characterizes the number of centers can vary only once at the end of the learning phase, which is typical of linear approximation schemes (see Sect. 6.3), or throughout the learning phase (Method 2), which is typical of nonlinear approximation schemes

If the learning phase is performed online in an LTC architecture, then, initially, a nominal controller that is expressed in terms of a finite-dimensional space $\mathbb{H}_{d(N_0)}$, where $N_0 \in \mathbb{N}$ is smaller than the expected value of N, can be employed. Then, given some control input, the plant trajectory can be observed to refine the choice of N and $\mathbb{H}_{d(N)}$. In the online training stage, the control input does not need to be the adaptive control system employed to steer the plant trajectory x toward x_r. An open-loop control law or the *nominal* control law, that is, the control law we would use if the plant was not affected by any uncertainty, can be used.

Control methods based on parametric adaptive control, such as those in [18–23, 26], can be interpreted as examples of LTC architectures. In these texts, the finite-dimensional spaces $\mathbb{H}_N$ are assumed to be given offline by an oracle before the control strategy is applied, and the dimension $d(N)$ of the approximating hypothesis space is constant while the feedback controller is operating.

Our discussion of LTC architectures aims to provide an initial assessment of how modern data-driven methods can inform control designers. This discussion sets the stage for more advanced analysis. When we assess methods from learning theory for their potential use in adaptive control theory, we must consider questions such as how do we define or modify the general structure of approximation theory hypotheses to create *nonparametric* adaptive control methods that exploit learning theory? How do we evaluate adaptive control methods in which the structure of the approximation theory hypotheses takes different forms? How, in general terms, do probabilistic or deterministic approximation theory hypotheses generate probabilistic or deterministic conclusions, respectively, about the controller performance? How do we create adaptive controllers that incorporate notions of *approximation theory priors* (i.e., known information about a function that can help the approximation process [27–30]) in our hypotheses for the creation of adaptive control methods? None of these topics are studied systematically in the family of classical deterministic methods documented in [18–20, 22, 23, 26]. Our discussion of LTC architectures in Sect. 6.4 below captures how data-driven methods can precede, inform, and work in conjunction with the classical theorems derived in these texts.

6.2.2 The Switched Learn-and-Control Architecture

The SLC architecture consists of concatenating the basic LTC architectures for a sequence of switching times $\{\tau_i\}_{i=0}^{\bar{n}}$, where $\bar{n}$ may be finite or infinite; see Fig. 6.4. Instead of a single learning phase and a subsequent single control phase, learning and control epochs alternate. As shown in Fig. 6.5, a critical feature of this control architecture is that the dimension of the subspace representing uncertainties changes across the learning epochs according to data collected on the functional uncertainty or the plant's response to user-defined control inputs, which are not necessarily the adaptive control inputs used during the control epochs.

Although SLC systems are obtained as rather simple modifications of LTC systems, they provide a much broader collection of algorithms. The switching times can be fixed a priori or can be defined adaptively (see Sect. 6.3 below). If there is a finite number of switches, then the stability, convergence, and ultimate behavior of the SLC controller during its *final epoch* can be studied as for the LTC control schemes.

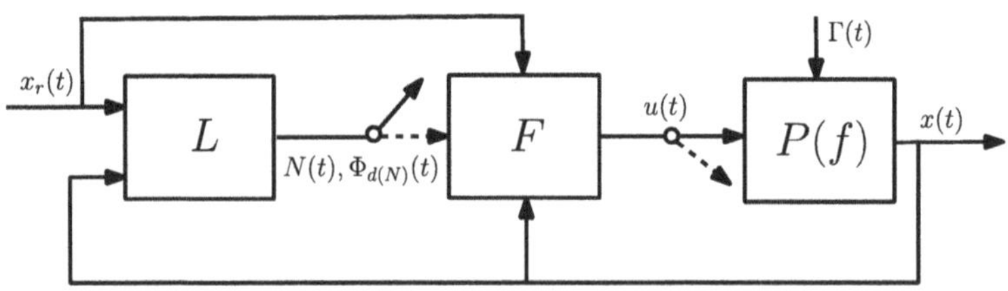

Fig. 6.4 Schematic representation of the switched learn-then-control (SLC) architecture. A learning policy L decides the number of centers N and a parametrization of this space $\Phi_{d(N)}$ according to the ability of the plant trajectory x to follow the reference trajectory x_r despite the functional uncertainty f and the unmatched uncertainty Γ affecting the plant P. The learning policy and the feedback control system F operate alternatively; see Fig. 6.5

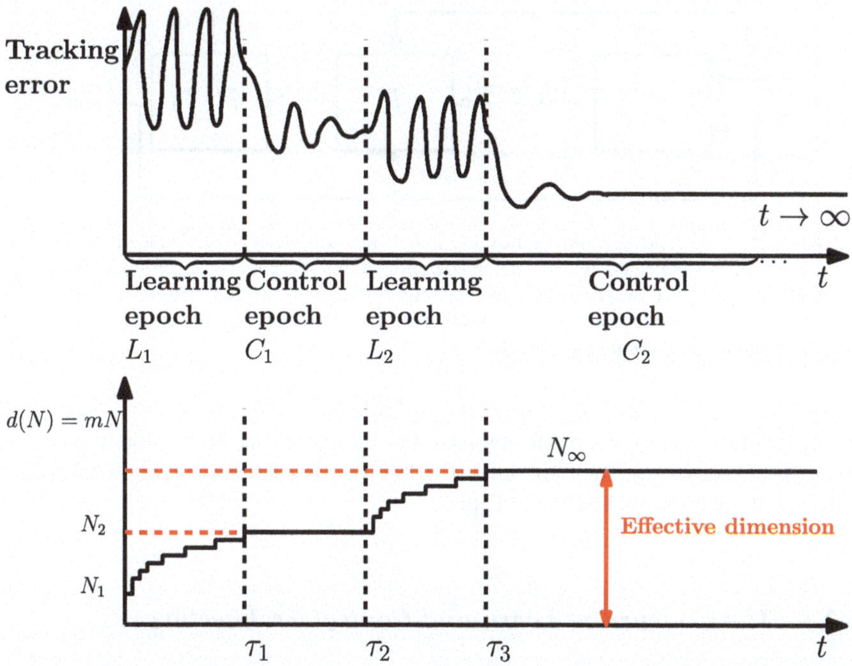

Fig. 6.5 Graphical representation of the tracking error and the dimensions of the approximating space $\mathbb{H}_N$ in the switched learn-and-control (SLC) architecture. The parameter N that characterizes the Hilbert space $\mathbb{H}_{d(N)}$ approximating the hypothesis space $\mathbb{H}$ varies over each learning epoch. Once an approximating hypothesis space is fixed at the end of a learning epoch, the control system steers the plant trajectory until a new learning epoch starts

However, a careful assessment of the *transient behavior* must address the delicate interplay of initial conditions and the change in dimension over finite-time intervals. Furthermore, infinitely many switches over finite-time horizons should be avoided. If infinitely many switches between learn-and-control stages occur over infinite time horizons, then stability and convergence of the overall system must be guaranteed. Finally, LTC approaches must deal with switched coordinate realizations with time-varying coefficients that evolve in Euclidean spaces of varying (usually increasing) dimensions.

While the theory of robust parametric adaptive control is routinely formulated to account for perturbations due to external disturbances and uncertain real parameters, the proposed adaptive control approach allows assessing the effect of perturbations due to approximation errors and the stability of computational processes, such as the sensitivity of numerical matrix factorizations and inversion for noisy inputs; these effects are usually ignored in classical parametric adaptive methods, owing to the typically low dimension of uncertainty approximations. In an SLC context, the effect of the error in approximating infinite-dimensional disturbances and numerical errors becomes more preponderant and induces a substantial perturbation source depending

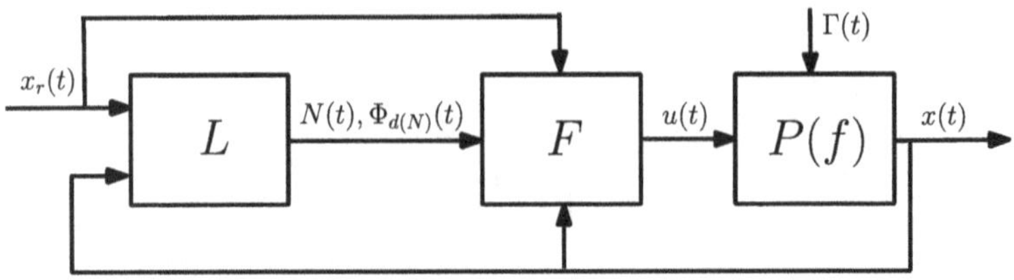

Fig. 6.6 Schematic representation of the CLC architecture. Similarly to SLC, a learning policy L decides the number of centers N and a parametrization of this space $\Phi_{d(N)}$ according to the ability of the plant trajectory x to follow the reference trajectory x_r despite the functional uncertainty f and the unmatched uncertainty Γ affecting the plant P. The learning policy and the feedback control system F operate concurrently; see Fig. 6.7

on the function spaces and bases involved. For these reasons, being able to quantify these effects is essential in nonparametric adaptive control theory. Section 6.5 below provides an example of SLC architecture.

6.2.3 *The Concurrent Learn-and-Control Architectures*

The CLC architecture is an evolution of SLC in which the discrete process, which builds candidates for basis enhancement or pruning, evolves concurrently with the continuous-time evolution of the closed-loop system; see Fig. 6.6. As in the SLC architecture, the dimension of the approximating state varies with the addition or deletion of basis functions. However, in the CLC architecture, the learning process and control process are not constrained to disjoint time intervals; see Fig. 6.7. In this algorithm, the adaptive control task and learning task for basis enrichment proceed concurrently. Section 6.6 below provides an example of CLC architecture.

6.3 Linear and Nonlinear Approximation Schemes

A key task of learning policies (L in Figs. 6.2, 6.4, and 6.6) is to identify an approximation scheme, that is, the dimension $d(N)$ of the approximants, the bases $\{\Phi_{d(N)}\}_{N\in\mathbb{N}}$, and, ultimately, the family $\{\mathbb{H}_{d(N)}\}_{N\in\mathbb{N}}$ of spaces approximating the hypothesis space $\mathbb{H}$. This section presents linear and nonlinear schemes to define approximations of the space in which functional uncertainties lie.

Recall the plant model

$$\dot{x}_N(t) = Ax_N(t) + B(u(t) + E_{x_N(t)}f) + \Gamma(t), \quad x(t_0) = x_0, \quad t \geq t_0 \tag{6.4}$$

discussed extensively in Chap. 5, where $x_N(t) \in \Omega \subseteq \mathbb{X} \triangleq \mathbb{R}^n$, $u(t) \in \mathbb{U} \triangleq \mathbb{R}^m$, $A \in \mathcal{L}(\mathbb{X}) \equiv \mathbb{R}^{n\times n}$ is unknown, $B \in \mathcal{L}(\mathbb{U}, \mathbb{X}) \equiv \mathbb{R}^{n\times m}$, and $\Gamma : [t_0, \infty) \to \mathbb{X}$ is such

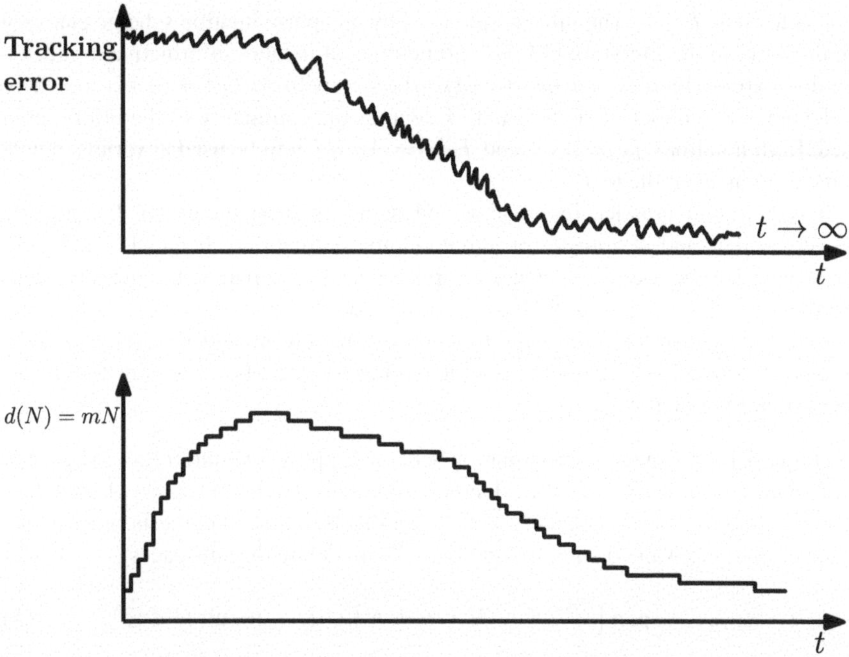

Fig. 6.7 Graphical representation of the tracking error and the dimensions of the approximating space $\mathbb{H}_N$ in the concurrent learn-and-control (CLC) architecture. The learning and control epochs are concurrent to one another

that $\|\Gamma(t)\|_{\mathbb{R}^n} \leq \Gamma_{\max}$; see (5.40), (5.92), and the discussion in Sect. 5.4.1.4. As anticipated in Sect. 6.1, in general, $f \in \mathbb{B}$. In the case of scalar-valued uncertainties, $\mathbb{B}$ may be the space of continuous functions $C^0(\Omega)$, the space of m-times continuously differentiable functions $C^m(\Omega)$ with $m \in \mathbb{N}$, the space of Lebesgue spaces $L^p(\Omega)$ with $p \in \mathbb{N}$, or the space r-times weakly differentiable Sobolev spaces $W^r(\Omega)$ over a subset $\Omega \subseteq \mathbb{R}^n$. In this chapter, we choose $\mathbb{B} \triangleq L^\infty(\Omega, \mathbb{U})$ or $\mathbb{B} \triangleq C_b(\Omega, \mathbb{U})$. To leverage the properties of Hilbert spaces, in most of the cases analyzed in this book, we assume that $f \in \mathbb{H} \hookrightarrow \mathbb{B}$ (see Fig. 6.1). Indeed, in Chap. 5, we chose a special class of hypothesis spaces, namely, RKHSs, by letting $\mathbb{H} = \mathcal{H}$. Special emphasis was given to the nonparametric adaptive control systems design problem, that is, to case whereby $\mathbb{H}_{d(N)} = \mathcal{H}_{d(N)}$ for all $N \in \mathbb{N}$ and $f \in \mathcal{H} \setminus \bigcup_{N\in\mathbb{N}} \mathcal{H}_{d(N)}$. If $f \in \mathbb{B} \setminus \mathcal{H}$, then we should address the issue of "escaping the native space" [31, 32]. This is an advanced topic in the study of approximations in native spaces that is the subject of much recent interest. With one notable exception in Chap. 7, in this book, we do not discuss the problem of escaping the native space, which we leave as a general topic for future research.

An *approximation method* or *scheme* is a family of algorithms $\{A_N\}_{N\in\mathbb{N}}$, with $A_N : \mathbb{H} \to \mathbb{H}_{d(N)}$, that uses some data about f, such as measurements or estimates of f at user-defined locations, to assign an approximation $f_N \triangleq A_N f \in \mathbb{H}_{d(N)}$ for

$N \in \mathbb{N}$ to each $f \in \mathbb{B}$. The information used by an approximation scheme can vary significantly in the literature [33–36]. Sometimes, it is more informative to index a family of approximation schemes by $\{A_{N,M}\}_{N,M \in \mathbb{N}}$ with $A_{N,M} : \mathbb{B} \to \mathbb{H}_{d(N)}$, where N denotes the number of centers and M denotes the cardinality of the information used. Both notations $f_N = A_N f$ and $f_{N,M} = A_{N,M} f$ will be used at various points in the remainder of the text.

It is common practice in learning and approximation theory to discriminate between *linear* and *nonlinear* (or *adaptive*) approximation schemes [27, 33, 35]. These will also be useful to categorize approaches of nonparametric adaptive control theory. If $\{\Phi_{d(N)}\}_{N \in \mathbb{N}}$ and $\{\mathbb{H}_{d(N)}\}_{N \in \mathbb{N}}$ are selected a priori, that is, before any estimates are constructed, then the function estimation method is a *linear method of approximation* [34]. The next example employs Fourier bases to construct linear approximation schemes.

Example 6.1 (A Linear Approximation Scheme Employing Fourier Bases) Let $\Omega \triangleq [0, \ell]$ with $\ell > 0$, let $\mathbb{B} \triangleq L^2(\Omega)$, that is, let the universe be the space of complex-valued functions over Ω, and let $\mathbb{H} \triangleq C^\infty(\Omega)$, that is, let the hypothesis space be the space of complex-valued smooth functions. Then, define the subspaces

$$\mathbb{H}_{2N+1} \triangleq \text{span}\left\{e^{\jmath n\pi x/\ell}, -N \leq n \leq N\right\} \subset \mathbb{H}, \quad \text{for all } N \in \mathbb{Z}^+. \tag{6.5}$$

We choose the approximation algorithm $A_N : \mathbb{B} \to \mathbb{H}_{2N+1}$ to be the truncated complex Fourier series

$$f_N \triangleq A_N f = \sum_{n=-N}^{N} C_n e^{n\pi x\jmath/\ell}, \tag{6.6}$$

where

$$C_n \triangleq \frac{1}{\ell}\int_0^\ell e^{-n\pi x\jmath/\ell} f(x)\mathrm{d}x. \tag{6.7}$$

The linear functionals of f in (6.6) define the "standard information about f" used to build an approximation. Note that the basis $\left\{e^{\jmath n\pi x/\ell}\right\}_{n=-N}^{N}$ that defines $\mathbb{H}_{2N+1}$ depends neither on f nor on the standard information about f, in this canonical method of approximating $f \in L^2(\Omega)$. The family $\{A_N\}_{N \in \mathbb{N}}$ is, therefore, an example of a *linear approximation method or scheme.* △

Classical parametric adaptive control schemes in standard treatises, such as [18, 19], make approximation hypotheses that can be interpreted in terms of linear approximation schemes. In Example 6.1, the bases are the complex exponentials common to Fourier analysis. References [18, 19] propose different bases, such as sigmoidal functions, radial basis functions, polynomials, splines, finite element functions, and wavelets, that can be used to define Φ_N. However, these texts do not consider how data about the uncertainty f can used to choose adaptively the basis Φ_N as N increases.

Such a construction would entail the definition of a nonlinear approximation scheme, which is discussed next.

If the bases $\{\Phi_{d(N)}\}_{n\in\mathbb{N}}$ that define the family of subspaces $\{\mathbb{H}_{d(N)}\}_{N\in\mathbb{N}}$ are selected recursively, depending on evolving data about a specific unknown $f \in \mathbb{H}$ at hand, then the method is an example of a *nonlinear or adaptive approximation scheme* [33, 34]. In this case, both $\Phi_{d(N)}$ and $\mathbb{H}_{d(N)}$ are explicit functions of f. However, to simplify our notation, we ordinarily avoid expressing these functional dependencies explicitly. Systematic studies of nonlinear schemes have been performed in approximation and learning theory for the past few decades [27, 34–37].

The following example illustrates a typical nonlinear approximation scheme. This is an example of *deterministic greedy method* of learning theory and approximation, which we discuss later in Sect. 6.4.2.5. Significantly, this method relies on constructions in an RKHS. Later, we discuss how such greedy methods can play an important role in defining nonparametric adaptive control methods.

Example 6.2 (The $f \cdot \mathcal{P}$ Data-Driven Approximation Scheme) In this example, we present a nonlinear approximation scheme known as *$f \cdot \mathcal{P}$ approximation scheme*. The $f \cdot \mathcal{P}$ scheme, which is summarized in Algorithm 6.1, is a greedy data-driven method and belongs to the collection of general strategies outlined in [38, 39], which are specifically designed using the properties of RKHSs.

Let $\mathcal{H}$ denote a scalar-valued RKHS generated by the scalar-valued kernel function $\mathfrak{K} : \Omega \times \Omega \to \mathbb{R}$ over the set Ω. Initially, we choose $\xi_1 \in \arg\max_{\xi\in\Omega} |f(\xi)|$, set $\Xi_1 = \{\xi_1\}$, and define $\mathcal{H}_1 = \text{span}\{\mathfrak{K}_{\xi_1}\}$. We then proceed recursively. At the Nth step, the center set Ξ_N, and the space of approximants $\mathcal{H}_N$ are set, and the next center ξ_{N+1} is selected as a maximizer of the *pointwise error indicator at* $\xi \in \Omega\backslash\Xi_N$ given by

$$\text{Err}(\xi; f, \mathcal{H}_N) \triangleq \mathcal{P}_N(\xi)|E_\xi(I - \Pi_N)f|, \tag{6.8}$$

where $\Xi_N \triangleq \{\xi_1, \ldots, \xi_N\}$, $\mathcal{P}_N(x)$ denotes the power function of the subspace

$$\mathcal{H}_N \triangleq \text{span}\left\{\mathfrak{K}_{\xi_i} \triangleq \mathfrak{K}(\cdot, \xi) \mid \xi_i \in \Xi_N, 1 \le i \le N\right\} \subseteq \mathcal{H}, \tag{6.9}$$

and $\Pi_N : \mathcal{H} \to \mathcal{H}_N$ denotes the $\mathcal{H}$-orthogonal projection of $\mathcal{H}$ onto $\mathcal{H}_N$; see Definition 3.8, (3.74), Definition 2.51, and (3.55). It is worthwhile to recall that the error $(I - \Pi_N)f$ of best approximation of $f \in \mathcal{H}$ from $\mathcal{H}_N$ satisfies the pointwise error bound

$$|E_x(I - \Pi_N)f| \le \mathcal{P}_N(x)\|(I - \Pi_N)f\|_{\mathcal{H}}, \quad \text{for all } x \in \Omega, f \in \mathcal{H}; \tag{6.10}$$

see Theorem 3.14.

To apply this method, there are many solutions to the problem of finding vector-valued maximizers of a cost function, and hence, find the center ξ_{N+1} for any $N \in \mathbb{N}$. Similar greedy methods have been studied in [40, 41] for a variety of choices of $\mathbb{H} \hookrightarrow \mathbb{B}$, even when $\mathbb{H}$ is not an RKHS. △

Algorithm 6.1 $f \cdot \mathcal{P}$ data-driven method for center selection

$N \leftarrow 0$

$\xi_1 \leftarrow \arg\max_{\xi \in \Omega} |f(\xi)|$

for $N \in \mathbb{N}$ **do**

$\quad \xi_{N+1} \leftarrow \arg\max_{\xi \in \Omega \setminus \Xi_N} \underbrace{(\mathcal{P}_N(\xi)|E_\xi(I - \Pi_N)f|)}_{\mathrm{Err}(\xi; f, \mathcal{H}_N)}$

end for

Linear and nonlinear approximation schemes, such as those in Examples 6.1 and 6.2, can generate *adaptive control methods with and without oracles*. In the following sections, we apply approximation schemes to design LTC, SLC, and CLC methods.

6.4 Deterministic LTC Methods

This section presents an in-depth analysis of LTC methods already introduced in Sect. 6.2.1. Examples of LTC architectures include but are not limited to, the classical methods of parametric adaptive control in [18, 19] and the nonparametric adaptive control techniques presented in Chap. 5. We begin by characterizing explicit controller performance bounds in terms of a rate function that depends on the number of centers N used to represent the uncertainty. These bounds on performance hold for any choice of the learning algorithm $L : N \rightarrow \Phi_{d(N)}$ that prescribes the basis for the subspace $\mathcal{H}_{d(N)}$ used to construct the controller u_N. Successively, we review several popular candidates for the learning module L that defines the collection $\{\mathcal{H}_{d(N)}\}_{N \in \mathbb{N}}$. Finally, we provide some numerical examples.

6.4.1 *Performance Bounds for LTC Methods*

In the design of LTC systems, we are interested in designing control strategies that are robust over a family of plant models, and, to this goal, we need to characterize the controller's performance in terms of its ability to steer the closed-loop plant trajectory toward the reference trajectory. To appreciate this point, first we present the performance bound assessment problem in light of the classical, parametric approach. Successively, we present the performance bound assessment problem in light of a nonparametric approach by revisiting some of the results of Chap. 5.

6.4.1.1 Performance Bounds: A Parametric Viewpoint

If we fix N and the bases Φ_N that define $\mathbb{H}_N \triangleq \text{span}\{\Phi_N\} \subset \mathbb{H}$ and consider an offline training of the LTC policy, then Fig. 6.2 captures the approach proposed in [18, 19] and the plethora of results inspired by these references to capture functional uncertainties. In this context of *parametric* adaptive control theory, for the scalar-valued uncertainty f, it is assumed that the uniform approximation assumption is verified, that is, $\mathbb{B} \equiv \mathbb{H} \equiv L^\infty(\Omega)$ for some $\Omega \subseteq \mathbb{R}^n$ where the closed-loop plant trajectory is expected to lie, and that the functional uncertainty in (6.4) is well captured by $\Theta^T \Phi_N$ for some $\Theta \in \mathbb{R}^{N\times m}$ within some error margin $\epsilon_a \geq 0$. Thus, parametric adaptive control systems guarantee that

$$\|x_N(t) - x_r(t)\|_{\mathbb{X}} < \varepsilon, \quad \text{for all } t \geq T_{\max} \tag{6.11}$$

for some $\varepsilon > 0$ and for some $T_{\max} \geq t_0$; note that $\varepsilon \neq \epsilon_a$. In the absence of unmatched uncertainties, that is, if $\Gamma(t) \equiv 0$ in (6.4), the pair $(\varepsilon, T_{\max})$ depends on the adaptive laws employed and the offline uniform approximation error ϵ_a. In the presence of unmatched uncertainties, that is, if $\Gamma \not\equiv 0$ and $\|\Gamma\| \leq \gamma$, then the pair $(\varepsilon, T_{\max})$ also depends on γ.

For many parametric methods, the relationship between the offline approximation error ϵ_a and tracking error bound ε is not explicit. Also, many of the robust parametric adaptive control methods are formulated in terms of various notions of parametric uncertainty classes. For example, projection methods (see [19, Ch. 11]) guarantee that the adaptive gains $\hat{f}_N(t, \cdot)$ corresponding to the functional uncertainty f lie in the parametric uncertainty class

$$C_{\Phi_N, R} \triangleq \left\{ f = \Theta^T \Phi_N : \Theta \in \mathbb{R}^{N\times m}, \|\Theta\|_\infty \leq R \right\} \tag{6.12}$$

for some fixed constant $R > 0$; see (5.7). The intuition behind these methods is that the control system should lead the norm of the trajectory tracking error below a threshold ε compatible with the norm of the unmatched uncertainty and of the offline approximation error. However, parametric methods are bound to a *specific* $\mathbb{H}_N$. Historically, parametric adaptive control theory does not systematically prescribe methods for choosing N, Φ_N, and ϵ_a as part of the adaptive control algorithm, to ensure that some tracking error bound ε is achieved. An oracle provides this information. Also, parametric approaches do not systematically study the effect of the regularity of the uncertainty f on the controller performance.

6.4.1.2 Performance Bounds: A Nonparametric Viewpoint

The nonparametric viewpoint is more ambitious than its parametric counterpart. For an LTC nonparametric adaptive control method, we seek to characterize the rates at which the performance measure varies *as a function of* N over the set of admissible initial conditions and plants defined in terms of a functional uncertainty class contained in a generally infinite-dimensional hypothesis space. The mechanism by which N varies and the basis functions are updated is part of the definition of the adaptive control method. In nonparametric control, we want this performance description to hold for all the coordinate representations of the functional space $\mathcal{H}$, and not for a specific representation as it occurs in parametric control.

In the nonparametric framework, we consider the *nonparametric uncertainty class of radius R*

$$C_R \triangleq \{f \in \mathcal{H} : \|f\|_{\mathcal{H}} \leq R\} \subset \mathcal{H}, \tag{6.13}$$

and the *nonparametric uncertainty class of radius R and projection error less than* ϵ

$$C_{N,\epsilon,R} \triangleq \{f \in \mathcal{H} \mid \|(I - \Pi_N)f\|_{\mathcal{H}} \leq \epsilon, \|f\|_{\mathcal{H}} \leq R\} \subset \mathcal{H}; \tag{6.14}$$

see (5.1) and (5.2). Recall that

$$C_{\Phi_N,R} \subset C_{N,\epsilon,R} \subset C_R. \tag{6.15}$$

Thus, fixed ϵ and R, by choosing different hypothesis spaces $\mathcal{H}$, nonparametric approximation schemes allow studying how the performance of adaptive control methods varies as a function of the regularity of the uncertainty f.

To define a rate of performance of an LTC system as a function of N, first we introduce the *cost* or *error functional* $J : C([t_0, \infty), \mathbb{R}^+) \to \mathbb{R}^+$; the minimum of the cost functional is assumed to be zero. Thus, assuming the unmatched disturbance is identically equal to zero, we derive a *rate function* $\mathcal{R}_{\mathcal{H}} : \mathbb{N} \to \mathbb{R}^+$ such that

$$J(e_N) \lesssim R_{\mathcal{H}}(N) \tag{6.16}$$

for all admissible trajectories generated by initial conditions $e_{N,0}$ contained in some subset $\mathbb{E}_0 \subset \mathbb{R}^n$ and functional uncertainties $f \in C \subseteq \mathcal{H} \hookrightarrow \mathbb{B}$, where $e_N(t) \triangleq x_N(t) - x_r(t)$ for all $t \geq t_0$, x_r defines the reference trajectory, and x_N denotes the closed-loop plant trajectory obtained by applying the control input u_N; see (5.74), (5.123), and (5.144). The constant underlying (6.16), which is not shown, is independent of N and the initial condition $e_{N,0}$ over $\mathbb{E}_0 \subset \mathbb{R}^n$.

Deriving a nonparametric LTC adaptive controller consequently involves three primary steps:

1. choose a learning algorithm $L : N \to \Phi_{d(N)}$ that defines $\mathcal{H}_{d(N)} \subseteq \mathcal{H} \hookrightarrow \mathbb{B}$;
2. define the family of feedback controllers $\{u_N\}_{N\in\mathbb{N}}$ that use $\Phi_{d(N)}$ to build $\mathcal{H}_{d(N)}$;
3. derive the closed-loop performance bound of the form (6.16) for all $f \in C \subset B$.

A *nonparametric LTC adaptive control method* is consequently a rule for choosing both a sequence of subspaces $\{\mathcal{H}_{d(N)}\}_{N\in\mathbb{N}} \subseteq \mathcal{H}$ and a family of controllers $\{u_N\}_{N\in\mathbb{N}}$ where the definition of u_N depends on $\mathcal{H}_{d(N)}$. The assessment of the performance of the closed-loop control scheme varies with the data-driven algorithm L used during the learning phase to pick $\mathcal{H}_{d(N)}$.

6.4.1.3 Asymptotic Approximation-Theory Optimality

The analysis of the performance bound presented in Sect. 6.4.1.2 can be specialized in multiple ways. In this section, we propose an approach that leverages the tools provided by the RKHS setting to capture approximation errors.

Recall that the error in approximating the functional uncertainty $f \in \mathcal{H}$ by means of $\mathcal{H}_{d(N)}$ can be captured by $(I - \mathbf{\Pi}_N)$, where $\mathbf{\Pi}_N$ denotes the orthogonal projection operator from $\mathcal{H}$ onto $\mathcal{H}_{d(N)}$; see (5.75). Furthermore, revisiting the proofs of all theorems in Sect. 5.4, we observe that if $(I - \mathbf{\Pi}_N)f \equiv 0$ for all $f \in \mathcal{H}$, i.e., if $\mathcal{H} \equiv \mathcal{H}_{d(N)}$, then (6.11) can be attained for any $\varepsilon > 0$. In light of these considerations, we introduce the following definitions.

Definition 6.1 (*Asymptotically approximation optimal LTC scheme*) Consider the plant model (6.4) with $\Gamma(t) \equiv 0$ for all $t \geq t_0$, the cost functional $J : C([t_0, \infty), \mathbb{R}^+) \to \mathbb{R}^+$ such that $J(0) = 0$ and $J(e) > 0$ for all $e \in C([t_0, \infty), \mathbb{R}^+) \setminus \{0\}$, and the rate function $\mathcal{R}_{\mathcal{H}} : \mathbb{N} \to \mathbb{R}^+$. If

$$J(e_N) \lesssim \mathcal{R}_{\mathcal{H}}(N) \lesssim \|(I - \mathbf{\Pi}_N)f\|_{\mathcal{H}} \tag{6.17}$$

for all $f \in C_R$ and initial conditions $e_{N,0} \in \mathbb{E}_0 \subseteq \mathbb{R}^n$, then the LTC scheme is *asymptotically approximation optimal* (AAO) in C_R.

Definition 6.2 (ϵ*-asymptotically approximation optimal LTC scheme*) Consider the plant model (6.4) with $\Gamma(t) \equiv 0$ for all $t \geq t_0$, the cost functional $J : C([t_0, \infty), \mathbb{R}^+) \to \mathbb{R}^+$ such that $J(0) = 0$ and $J(e) > 0$ for all $e \in C([t_0, \infty), \mathbb{R}^+) \setminus \{0\}$, and the rate function $\mathcal{R}_{\mathcal{H}} : \mathbb{N} \to \mathbb{R}^+$. If

$$J(e_N) \lesssim \mathcal{R}_{\mathcal{H}}(N) \lesssim \epsilon \tag{6.18}$$

for all $f \in C_{N,\epsilon,R}$ and initial conditions $e_{N,0} \in \mathbb{E}_0 \subseteq \mathbb{R}^n$, then the LTC scheme is ϵ*-asymptotically approximation optimal* (ϵ-AAO) in $C_{N,\epsilon,R}$.

Remark 6.3 The left-hand sides of (6.17) and (6.18) refer to the *closed-loop controller performance*, that is, applying the chosen adaptive control scheme after the representation of the functional uncertainty space has been fixed. The right-hand sides of (6.17) and (6.18) refer to the *error of best approximation of the uncertainty*, that is, the best approximation error measured in the norm on $\mathcal{H}$ obtained while defining the learning policy L.

Definitions 6.1 and 6.2 hold for all the coordinate representation of $\{\mathcal{H}_{d(N)}\}_{N \in \mathbb{N}}$. Thus, Definitions 6.1 and 6.2 capture the fact that, in general, approximations of any uncertainty $f \in \mathcal{H}$ generally harm the controller's performance. Assuming a uniform distribution of the centers that define the RKHSs $\{\mathcal{H}_{d(N)}\}_{N\in\mathbb{N}}$ around the closed-loop trajectory x_N, the intuition is that the controller's performance approaches that of the limiting DPS for the closed-loop system as the dimension N increases or as the fill distance over the set of centers decreases. By construction, for every $\epsilon > 0$, every AAO LTC method in $C_{N,\epsilon,R}$ is ϵ-AAO.

6.4.1.4 Rate and Cost Functions for a Cartesian Product RKHS

In this section, we provide examples of rate functions under the assumption that the hypothesis space is a vector-valued RKHS defined as the m-times Cartesian product of a scalar-valued RKHSs, that is, $\boldsymbol{\mathcal{H}} \triangleq \mathcal{H}^m$; these results have been recently developed in [16, 17, 42–50]. It follows from (3.155) that

$$\|(I - \boldsymbol{\Pi}_N)f\|_{L^\infty(\Omega,\mathbb{R}^m)} \le C \sup_{\xi\in\Omega} \mathcal{P}_N(\xi)\|(I - \boldsymbol{\Pi}_N)f\|_{\mathcal{H}}, \tag{6.19}$$

where $\mathcal{P}_N$ denotes the power function of the scalar-valued RKHS $\mathcal{H}$ computed as in (3.74). Now, if $f \in C_{N,\epsilon,R} \subset C_R \subset \boldsymbol{\mathcal{H}}$, then

$$\|(I - \boldsymbol{\Pi}_N)f\|_{L^\infty(\Omega,\mathbb{R}^m)} \le (C\epsilon) \sup_{\xi\in\Omega} \mathcal{P}_N(\xi), \tag{6.20}$$

and if $f \in C_R \subset \boldsymbol{\mathcal{H}}$, then

$$\|(I - \boldsymbol{\Pi}_N)f\|_{L^\infty(\Omega,\mathbb{R}^m)} \le (CR) \sup_{\xi\in\Omega} \mathcal{P}_N(\xi). \tag{6.21}$$

Thus, in these cases, the rate function can be defined as

$$\mathcal{R}_{\mathcal{H}}(N) \triangleq \sup_{\xi\in\Omega\subset\mathbb{X}} \mathcal{P}_N(\xi). \tag{6.22}$$

As shown by (3.80), power functions can be expressed in terms of the fill distance so that

$$\|\mathcal{P}_N\|_{L^\infty(\Omega)} \le C\sqrt{\mathcal{N}(h_{\Xi_N,\Omega})} \tag{6.23}$$

where $\mathcal{N}$ is given by Table 3.3 and

$$h_{\Xi_N,\Omega} \triangleq \sup_{x\in\Omega} \sup_{\xi_i\in\Xi_N} \|x - \xi_i\|_{\mathbb{R}^n}. \tag{6.24}$$

Thus, (6.23) can serve as a rate function, which is determined by the choice of the kernel that induces $\mathcal{H}$, the choice of the N centers in Ξ_N that define $\mathcal{H}_N$, and the fill distance $h_{\Xi_N,\Omega}$. This is a key difference between the deterministic RKHS embedding method and other techniques motivated by learning theory. None of the

the deterministic parametric adaptive control papers, and none of the methods based on Gaussian processes discussed in Chap. 7 and form the majority of approximation methods, derive such simple closed-loop controller performance bounds in terms of fill distances.

In light of Remark 6.3 and the fact that all nonparametric adaptive control systems presented in Chap. 5 assure uniform ultimate boundedness of the trajectory tracking error, that is, guarantee that (6.11) is verified, a viable performance metric is given by

$$J(e_N) \triangleq \limsup_{t\to\infty} \|x_N(t) - x_r(t)\|_{\mathbb{X}}. \tag{6.25}$$

Indeed, (5.84), (5.107), (5.133), (5.145), and (5.151) show that

$$J(x_N) \triangleq \limsup_{t\to\infty} \|x_N(t) - x_r(t)\|_{\mathbb{X}} \leq C\mathcal{N}(h_{\Xi_N,\Omega})\|(I - \mathbf{\Pi}_N)f\|_{\mathcal{H}} \tag{6.26}$$

for the dead-zone modification of MRAC, the σ-modification of MRAC, MRAC with convex projection operators, error bounding methods, and adaptive error bounding methods, respectively, in a nonparametric setting.

6.4.1.5 Rate and Cost Functions for a Generic Vector-Valued RKHS

As discussed at length in Chap. 3, not all vector-valued RKHS can be deduced as the Cartesian product of scalar-valued RKHS. To construct rate and cost functions in this more general setting, recall that

$$\left|\langle E_x(I - \mathbf{\Pi}_N)f, \alpha\rangle_{\mathbb{U}}\right| \leq \underbrace{\sqrt{\|\mathcal{K}(x,x) - \mathcal{K}_N(x,x)\|}}_{\triangleq \bar{\mathcal{P}}_N(x)}\|\alpha\|_{\mathbb{U}}\|(I - \mathbf{\Pi}_N)f\|_{\mathcal{H}} \tag{6.27}$$

for all $f \in \mathcal{H}$; see (5.85). This inequality (see [51, Cor. 2.11]) implies that

$$\|E_x(I - \mathbf{\Pi})f\|_{\mathbb{U}} \leq \sup_{\xi\in\Omega} \bar{\mathcal{P}}_N(\xi)\|(I - \mathbf{\Pi}_N)f\|_{\mathcal{H}} \quad \text{for all } f \in \mathcal{H}, \tag{6.28}$$

for any $\Omega \subseteq \mathbb{X}$, and hence, it is reasonable to set

$$\mathcal{R}_{\mathcal{H}}(N) \triangleq \sup_{\xi\in\Omega} \bar{\mathcal{P}}_N(\xi). \tag{6.29}$$

In Chap. 5, the design of adaptive control systems in the presence of generic vector-valued RKHSs was explicitly addressed for the dead-zone modification of MRAC in Sect. 5.4.1.2. However, very similar arguments can be applied to the other adaptive control systems, and these arguments have been omitted for brevity. Thus, as for the case whereby $\mathcal{H} \triangleq \mathcal{H}^m$, it is reasonable to set the cost function

$$J(e_N) \triangleq \limsup_{t\to\infty} \|x_N(t) - x_r(t)\|_{\mathbb{X}} \lesssim \begin{cases} \mathcal{R}_{\mathcal{H}}(N)R & \text{if } f \in C_R, \\ \epsilon\mathcal{R}_{\mathcal{H}}(N) & \text{if } f \in C_{N,\epsilon,R}. \end{cases} \tag{6.30}$$

We should note an important open problem when considering vector-valued RKHSs defined in terms of a generic operator kernel. A specific advantage in using diagonal operator kernels, that is, RKHSs produced as the Cartesian product of scalar-valued RKHSs, is that we can subsequently derive upper bounds on the cost function for an LTC scheme in terms of the fill distance. This result has not yet been extended to generic RKHSs defined in terms of general non-diagonal operator-valued kernels, to the authors' knowledge. However, it is possible to recover bounds on the cost function as an application of the many zeros theorem (see Theorems 2.8 and 3.23).

Suppose that the hypotheses of Theorem 3.23 hold. The most critical of these assumptions is that

$$\mathcal{H} \triangleq \mathcal{H}(\Omega, \mathbb{R}^m) \hookrightarrow \mathbf{W}^{k+s,p}(\Omega), \tag{6.31}$$

which can be interpreted as a regularity condition on the vector-valued RKHS $\mathcal{H}$. The condition in (6.31) requires that each coordinate function f_i in $f = [f_1, \ldots, f_m]^{\mathrm{T}}$ of any $f \in \mathcal{H}$ is an element of the Sobolev space that contains scalar-valued functions, $f_i \in W^{k+s,p}(\Omega)$. Thus, under the assumptions of Theorem 3.23, there exist constants $C > 0$ and $h_0 > 0$ such that

$$\|f - \mathbf{\Pi}_N f\|_{\mathbf{W}^{r,q}(\Omega)} \leq C h_{\Omega_N,\Omega}^{k+s-r-d(1/p-1/q)_+} \|f\|_{\mathcal{H}}, \quad \text{for all } f \in \mathcal{H} \tag{6.32}$$

and for all Ξ_N such that $h_{\Xi_N,\Omega} > h_0$, where $\mathbf{W}^{r,q}(\Omega) \triangleq (W^{r,q}(\Omega))^m$ as discussed in Sect. 3.8.2.

If the index r is sufficiently large so that $W^{r,p}(\Omega) \hookrightarrow C_b(\Omega)$, then we obtain the Jackson inequality

$$\|f - \mathbf{\Pi}_N f\|_{(C_b(\Omega))^m} \lesssim h_{\Omega_N,\Omega}^{k+s-r-d(1/p-1/q)_+} \|f\|_{\mathcal{H}}, \quad \text{for all } f \in \mathcal{H} \tag{6.33}$$

for $\mathcal{H} \hookrightarrow \mathbf{W}^{r,q}(\Omega) \hookrightarrow (C_b(\Omega))^m$ and $\mathbb{B} \triangleq (C_b(\Omega))^m$. This Jackson inequality can be used to derive performance bounds in the same form as

$$J(x_N) \triangleq \limsup_{t\to\infty} \|x_N(t) - x_r(t)\|_{\mathbb{X}} \lesssim \begin{cases} \mathcal{R}_{\mathcal{H}}(N)R & \text{if } f \in C_R, \\ \epsilon\mathcal{R}_{\mathcal{H}}(N) & \text{if } f \in C_{N,\epsilon,R}, \end{cases}, \tag{6.34}$$

with $\mathcal{R}_{\mathcal{H}}(N) \triangleq h_{\Xi_N,\Omega}^{k+s-r}$.

6.4.1.6 Summary of Results

In light of the proposed tools to assess LTC architectures, the results of Chap. 5 can be summarized by Tables 6.1 and 6.2 for the generic operator-valued hypothesis

Table 6.1 Summary of the performance of adaptive control methods encountered in Chap. 5 for the functional uncertainty classes $C_{\Phi_N,R} \subset C_{N,\epsilon,R} \subset C_R \subset \mathcal{H}$. Control techniques are in the same order as they appear in Chap. 5. The first row shows the performance of classical MRAC as presented in [19, Ch. 9], which only applies to the uncertainty class $C_{\Phi_N,R}$, and assures asymptotic convergence of the trajectory tracking error to zero. Without robust modifications, classical MRAC does not apply to the broader uncertainty classes $C_{N,\epsilon,R}$ and C_R. The second row shows the performance of the generalized gradient law for the limiting DPS system; see Sect. 5.3.1. This ideal controller assures asymptotic convergence of the ideal trajectory tracking error $e(\cdot)$ to zero for any uncertainty class considered. The remainder of this table shows the uniform ultimate bounds on the trajectory tracking error $e_N(\cdot)$ obtained by approximating functional uncertainties on $\mathcal{H}_N$, that is, $\bar{e}_{N,\infty}$, which we distinguish from $\bar{e}_\infty$ attained by the ideal, limiting DPS. In this table, $C_0(R) \triangleq C_1 + C_2 R^2$, the function $\mathcal{F}(\epsilon) \triangleq \sqrt{C\epsilon^2 \|\mathcal{P}_N\|_{L^\infty} + C_0(R)}$ captures the sub-optimality of the σ-modification in the class $C_{N,\epsilon,R}$, and $\mathcal{F}_1(N) \triangleq \sqrt{C\|\mathcal{P}_N\|_{L^\infty} R^2 + C_0(R)}$ bounds the sub-optimality in C_R. The term $\bar{\mathcal{P}}_N$ is defined in (3.156) for approximating spaces $\mathcal{H}_{d(N)}$ contained in the hypothesis space $\mathcal{H}$ defined in terms of a generic, possibly non-diagonal, kernel $\mathcal{K}$

Method	Type	Uncertainty class		
		$C_{\Phi_N,R}$	$C_{N,\epsilon,R}$	C_R
Parametric MRAC [19, Ch. 9]	Classical	0	–	–
Nonparametric gradient law	Limiting DPS	$\bar{e}_\infty = 0$	$\bar{e}_\infty = 0$	$\bar{e}_\infty = 0$
dead-zone		–	$\epsilon\|\overline{\mathcal{P}}_N\|_{L^\infty}$	$\|\overline{\mathcal{P}}_N\|_{L^\infty} R$
σ-modification		$\sqrt{C_0(R)}$	$\leq \mathcal{F}(\epsilon)$	$\mathcal{F}_1(N)$
Projection		–	$\epsilon\|\overline{\mathcal{P}}_N\|_{L^\infty}$	$\|\overline{\mathcal{P}}_N\|_{L^\infty} R$
Bounding method	Ideal	0	0	0
	Approx.	0	$\epsilon\|\overline{\mathcal{P}}_N\|_{L^\infty}$	$\|\overline{\mathcal{P}}_N\|_{L^\infty} R$
Adaptive error bounding	Ideal	0	0	0
	Approx.	0	$\epsilon\|\overline{\mathcal{P}}_N\|_{L^\infty}$	$\|\overline{\mathcal{P}}_N\|_{L^\infty} R$

Table 6.2 Summary of the performance of native space embedding methods in terms of approximation theory sub-optimality

Method	Class	Performance
Parametric MRAC [19, Ch. 9]	$C_{\Phi_N,R}$	AAO, $\bar{e}_{N,\infty} = 0$
Dead-zone modification	$C_{N,\epsilon,R}$	AAO, $\bar{e}_{N,\infty} \leq O(h^r_{\Xi_N,\Omega})$
σ-modification	$C_{N,\epsilon,R}$	$\mathcal{F}(\epsilon)$-suboptimal
Bounding Control	$C_{N,\epsilon,R}$	AAO*, $\bar{e}_{N,\infty} = 0$
Adaptive Bounding Control	$C_{N,\epsilon,R}$	AAO*, $\bar{e}_{N,\infty} = 0$

space $\mathcal{H}$ and under the assumption that the unmatched uncertainty Γ is identically equal to zero. These tables generalize the case described in Table 2 and Table 3 of [52], which only study the performance of adaptive controllers for scalar-valued uncertainty $f \in \mathcal{H}$ in terms of the power function $\mathcal{P}_N$.

6.4.2 Data-Driven Methods to Define the Learning Policies

This section discusses the effect of center distribution in RKHS-embedded control methods and presents some data-driven methods to define the subspace learning policy L. We will show how to center distributions that provide satisfactory bounds on the closed-loop plant trajectory tracking error at lower computational costs, though with some limitations in terms of uncertainty classes, are designed using some methods based on the analysis of the controlled plant behavior or observations of the functional uncertainty. Thus, employing a terminology typical of data-driven methods, in this section, we will refer to centers as *samples*.

In Sect. 6.4.2.1, we characterize the tracking error in RKHS-embedded control problems while employing deterministically prescribed samples $\{\Xi_N\}_{N\in\mathbb{N}}$ dense in some parallelepiped $\Omega \subset \mathbb{R}^n$. In Sect. 6.4.2.2, we characterize the tracking error with deterministically prescribed samples $\{\Xi_N\}_{N\in\mathbb{N}}$ dense in $\Omega \subset \mathbb{R}^n$ and distributed over a limit cycle (see Definition 4.36) known a priori. In Sect. 6.4.2.3, we consider samples distributed over a limit cycle of the controlled plant (see Definition 4.35) not known a priori. In Sect. 6.4.2.4, we discuss the performance of RKHS-embedded control systems while identifying the uncertainty; notably, the analysis in Sects. 6.4.2.1–6.4.2.3 was irrespective of the user's ability to characterize the functional uncertainty, but relied on the RKHS properties to characterize the hypothesis space. Finally, in Sect. 6.4.2.5, we present deterministic greedy methods of learning theory, which generalize the results in Example 6.2.

6.4.2.1 Prescribed Samples over Parallelepipeds

In this section, we assume that centers are located on regular grids that cover the user-defined set $\Omega \subset \mathbb{X} \triangleq \mathbb{R}^n$; this center distribution has already been employed in Sect. 5.4.1.5. Overall, the analysis in this section shows how we can balance the computational complexity of the performance bounds presented in Sect. 6.4.1, that is, how much computational work is needed to achieve a specified level of accuracy [27, 35]. The idea is to balance the size of the uncertainty class, which should be as large as possible, and the size of the set that contains a trajectory of interest, which should be as small as possible to reduce the computational burden when evaluating the kernel functions at each center in this set; see Remark 5.11. Judiciously choosing the uncertainty class and the size of the set of interests, we can obtain computationally suitable performance bounds.

Assume that, applying a feedback control policy F (see Fig. 6.2), the closed-loop plant trajectory x_N lies in Ω at all times. To simplify the exposition of our arguments, let

$$\Omega \triangleq [a_1, b_1] \times \cdots \times [a_n, b_n] \tag{6.35}$$

for some constants $a_i, b_i \in \mathbb{R}$, $1 \leq i \leq n$. The operator L deterministically chooses nested subsets

$$\ldots \subset \Xi_N \subset \Xi_{N+1} \subset \ldots \subset \Omega, \tag{6.36}$$

which are dense in Ω, that is,

$$\bigcup_{N \in \mathbb{N}} \Xi_N \subset \Omega \subset \overline{\bigcup_{N \in \mathbb{N}} \Xi_N}. \tag{6.37}$$

Assuming that $\mathcal{H} = \mathcal{H}^m$ and that the power function associated with the RKHS can be bounded by a function of the fill distance, as it occurs for Sobolev–Matérn and Wendland kernels (see (3.80) and Table 3.3), define the *minimum radius of separation* as

$$\rho_{\Xi_N} \triangleq \frac{1}{2} \min_{\substack{x_i, y_i \in \Xi_N \\ x_i \neq y_i}} \|x_i - y_i\|_{\mathbb{X}}. \tag{6.38}$$

We interpret ρ_{Ξ_N} as the radius of the largest ball that can placed between the centers Ξ_N in the set Ω. Note that, per definition, $\rho_{\Xi_N} \leq h_{\Xi_N,\Omega}$. If there are two constants $c_1, c_2 > 0$ such that

$$c_1 \rho_{\Xi_N} \leq h_{\Xi_N,\Omega} \leq c_2 \rho_{\Xi_N}, \quad \text{for all } N \in \mathbb{N}, \tag{6.39}$$

then the family of dense centers $\{\Xi_N\}_{N \in \mathbb{N}}$ contained in Ω is *quasi-uniform* [31, 53]. For a quasi-uniform collection $\{\Xi_N\}_{N \in \mathbb{N}}$ over Ω in $\mathbb{X} \triangleq \mathbb{R}^n$, it holds that

$$h_{\Xi_N,\Omega} \approx N^{-1/n}. \tag{6.40}$$

Indeed, assume that Ω is a unit cube in $\mathbb{R}^n$. This cube in n dimensions has a volume $1 \approx N h^n_{\Xi_N,\Omega}$ where $h^n_{\Xi_N,\Omega}$ is the volume in $\mathbb{R}^n$ of a cube whose sides are $h_{\Xi,\Omega}$ long, and (6.40) the result is proven.

Thus, if the performance bound for the adaptive control method in Table 6.1 is such that

$$\sup_{\xi \in \Omega} \mathcal{P}_N(\xi) \lesssim h^s_{\Xi_N,\Omega} \tag{6.41}$$

for a smoothness index $s > 0$, then

$$\sup_{\xi \in \Omega} \mathcal{P}_N(\xi) \lesssim N^{-s/n} = \left(\frac{d(N)}{m}\right)^{-s/n}, \tag{6.42}$$

where $d(N) \triangleq mN$. The bound on the power function given by (6.42) is a function of the number of basis functions $d(N) \triangleq mN$, which, as discussed in the following, can be an indicator of the computational complexity of an RKHS-embedded control system. Both the Sobolev–Matérn and Wendland kernels in Table 3.3 define native spaces of this type for an appropriate definition of the smoothness parameter s, as discussed in Sect. 3.5. This implies an ultimate tracking error performance bound of the form

$$\limsup_{t\to\infty} \|x_N(t) - x_r(t)\|_{\mathbb{X}} \lesssim \epsilon \left(\frac{Nm}{m}\right)^{-s/n} \tag{6.43}$$

for all f in the functional uncertainty class $C_{N,\epsilon,R}$.

The *computational complexity* of an algorithm describes the amount of computational work needed to generate a desired goal with a specified accuracy [27, 35]. Suppose that we desire an ultimate tracking performance bounded by ϵ^*, that is,

$$\limsup_{t\to\infty} \|x_N(t) - x_r(t)\|_{\mathbb{X}} \lesssim \epsilon^* \tag{6.44}$$

for all f in the nonparametric uncertainty class C_R. To this goal, it follows from (6.43) that N must be such that

$$d(N) \triangleq Nm \gtrsim \frac{m}{(\epsilon^*)^{n/s}}. \tag{6.45}$$

This bound scales poorly with the dimension n of the state space. This is a classical example of *curse of dimensionality*, which plays such an important role in the development of modern algorithms in statistical and machine learning theory; we discuss this matter in further detail in Chap. 7. If $n \approx O(10^5)$, as it occurs, for example, in genomic studies, then the curse of dimensionality proves disastrous in practice. In adaptive control problems involving vehicle dynamics, $n \approx O(10)$ and the dire implications of such poor scaling as n increases may not be problematic.

The proposed scaling estimates can be misleading if taken out of context. There is always a trade-off in computational complexity arguments between the size of the class in which the functional uncertainty lies and the guarantee of approximation error. Better complexity estimates are typically obtained by reducing the size of the class over which the estimates are derived. In the next section, we show how we can obtain better scaling results using data about the dynamics at hand rather than reducing the space over which the complexity estimates hold.

As a concluding remark, the classical parametric adaptive control systems illustrated in [18, 19] and the conspicuous number of works inspired by them can be analyzed according to the proposed procedure when the centers fill a parallelepiped. Thus, these strategies can be interpreted as defining LTC control architectures driven by linear approximation schemes. In particular, in these references, an oracle provides N, Φ_N, and $\mathbb{H}_N$. Thus, the proposed analysis shows that if the oracle defines the subspaces by filling a parallelepiped containing the trajectory of interest, then we obtain a bound of the same form as (6.43).

6.4.2.2 Prescribed Samples over Limit Cycles Known A Priori

The use of generic, regular distributions of centers, such as regular grids over parallelepipeds, does not leverage any insight into a specific problem about where trajectories of interest may lie. In tracking control problems, such as MRAC systems design, reference trajectories are always bounded and often converge to some periodic motion as $t \to \infty$. Knowledge of the fact that the closed-loop plant trajectory must lie in some neighborhood of the reference trajectory can be exploited, for instance, to define a set Ξ that in some sense supports the "ultimate behavior" of the controlled plant. Thus, L can be defined to choose centers from this set Ξ and ensure good local approximations of the uncertainty for the ultimate behavior of the adaptive controller. The intuition is that improved performance of the control scheme should be attained with better local estimates of the uncertainty as the closed-loop system's transient behavior has decayed. In this section, we study how this insight into the closed-loop plant dynamics and the reference model can be leveraged to design the learning policy L and improve the performance bounds.

To illustrate this point, we assume that the reference system converges to the limit cycle $\Xi \subset \Omega$ known a priori; see Definition 4.36. The learning policy L introduces scattered centers $\Xi_N \subseteq \Xi \subset \Omega \subset \mathbb{R}^n$ that are dense in Ξ, in the sense that

$$\bigcup_{N\in\mathbb{N}} \Xi_N \subset \Xi \subset \overline{\bigcup_{N\in\mathbb{N}} \Xi_N}, \tag{6.46}$$

and define $\mathcal{H}_{d(N)} \triangleq \text{span}\{\mathcal{K}_{\xi_i}\alpha \ : \ \xi_i \in \Xi_N, \alpha \in \mathbb{U}\}$. Now, define the closed subspaces

$$\begin{aligned} \mathcal{H}_\Xi &\triangleq \overline{\text{span}\left\{\mathcal{K}_\xi\alpha : \xi \in \Xi, \alpha \in \mathbb{U}\right\}} \\ &= \overline{\text{span}\left\{\mathcal{K}_\xi e_j : \xi \in \Xi, 1 \le j \le m\right\}}, \end{aligned} \tag{6.47a}$$

$$\mathcal{H}_\Omega \triangleq \overline{\text{span}\{\mathcal{K}_\xi\alpha : \xi \in \Omega, \alpha \in \mathbb{U}\}}, \tag{6.47b}$$

where e_j denotes the jth canonical basis in $\mathbb{U} \triangleq \mathbb{R}^m$ and the closure is taken with respect to the inner product on $\mathcal{H}$; see (3.148). Some key properties of the subspaces $\mathcal{H}_\Xi$ and $\mathcal{H}_\Omega$ can be deduced from the discussion in Sect. 3.6.2. It follows from (6.46) and (6.47) that the data-driven policy defines

$$\mathcal{H}_1 \subset \ldots \subset \mathcal{H}_{d(N)} \subset \mathcal{H}_{d(N)+1} \subset \ldots \subset \mathcal{H}_\Xi \subset \mathcal{H}_\Omega \subset \mathcal{H}. \tag{6.48}$$

If $\boldsymbol{\Pi}_\Xi$ denotes the $\mathcal{H}$-orthogonal projection from $\mathcal{H}$ onto the closed subspaces $\mathcal{H}_\Xi$, then $\mathcal{H}_\Xi$ is also an RKHS with kernel

$$\mathcal{K}_{\Xi}(x, y) \triangleq \mathcal{K}^*_{\Xi,x}\mathcal{K}_{\Xi,y}, \tag{6.49}$$

for all $(x, y) \in \Omega$, where $\mathcal{K}_{\Xi} \triangleq \Pi_{\Xi}\mathcal{K}_x$, as discussed in Sect. 3.6.2. Hence, we attain the pointwise bound

$$\|E_x(I - \mathbf{\Pi}_N)f\|_{\mathbb{U}} \leq \sup_{\xi \in \Omega} \sqrt{\|\mathcal{K}_{\Xi}(\xi, \xi) - \mathcal{K}_{\Xi,N}(\xi, \xi)\|}\ \|(I - \mathbf{\Pi}_N)f\|_{\mathcal{H}}, \tag{6.50}$$

for all $x \in \Omega$ and $f \in \mathcal{H}_{\Xi}$; see (3.155) for a related result. Applying any of the adaptive control systems discussed in Chap. 5, we obtain the ultimate performance bound

$$\begin{aligned}\limsup_{t \to \infty} &\|x_N(t) - x_r(t)\|_{\mathbb{X}} \\ &\lesssim \sup_{\xi \in \Omega} \sqrt{\|\mathcal{K}_{\Xi}(\xi, \xi) - \mathcal{K}_{\Xi,N}(\xi, \xi)\|}\ \|(I - \mathbf{\Pi}_N)f\|_{\mathcal{H}} \quad \text{for all } f \in \mathcal{H}_{\Xi}.\end{aligned} \tag{6.51}$$

With these choices, we conclude that this LTC method is AAO for the uncertainty classes

$$\tilde{C}_R \triangleq \{f \in \mathcal{H}_{\Xi} \ :\ \|f\|_{\mathcal{H}} \leq R\} = C_R \cap \mathcal{H}_{\Xi}, \tag{6.52}$$

$$\tilde{C}_{N,\epsilon,R} \triangleq \{f \in \mathcal{H}_{\Xi} \ :\ \|(I - \Pi_N)f\|_{\mathcal{H}} \leq \epsilon, \|f\|_{\mathcal{H}} \leq R\} = C_{N,\epsilon,R} \cap \mathcal{H}_{\Xi}. \tag{6.53}$$

Remark 6.4 The set Ω is selected to contain the closed-loop trajectory of interest. The set Ξ is chosen based on the ultimate support of the closed-loop dynamics. The performance bound on the tracking error given by (6.51) requires taking the supremum over the set Ω that contains the trajectory of interest, which conceivably could be large. On the other hand, we suppose that the uncertainty f is contained in the smaller set $\mathcal{H}_{\Xi} \subset \mathcal{H}_{\Omega} \subset \mathcal{H}$ that is defined in terms of where samples ultimately concentrate.

At this point, it can be helpful to review the numerical examples presented in Sect. 3.5.4.

Example 6.3 In this example, we study how to derive scaling laws in the case $\Xi \subset \Omega \subset \mathbb{X} \triangleq \mathbb{R}$ is a compact, smooth, one-dimensional Riemannian manifold that is regularly embedded in $\mathbb{X} \triangleq \mathbb{R}^n$.

We initially choose a global reproducing kernel $\mathfrak{K} : \mathbb{X} \times \mathbb{X} \to \mathbb{R}$ that defines the RKHS $\mathcal{H}(\mathbb{R}^n)$ of scalar-valued functions defined over $\mathbb{X} \triangleq \mathbb{R}^n$. We select $\mathfrak{K}$ to be a Mercer kernel that is strictly positive, bounded uniformly on the diagonal by a constant $\bar{\mathfrak{K}} > 0$, and satisfies the *algebraic decay condition*

$$\hat{\mathfrak{K}}_0(\xi) \approx (1 + \|\xi\|_2^2)^{-s}, \quad \text{for some } s > n/2, \tag{6.54}$$

where $\mathfrak{K}_0(x) \triangleq \mathfrak{K}(0, x)$; see Definition 3.10. It is a common assumption to ensure that the global kernel defines a native space $\mathcal{H}(\mathbb{R}^n)$ equivalent to a Sobolev space, that is, $\mathcal{H}(\mathbb{R}^n) \approx W^{s,2}(\mathbb{R}^n)$. Good examples of this type of kernel include the Sobolev–Matérn or Wendland kernels introduced in Chap. 3.

As discussed in Sect. 3.2, we denote by $\mathcal{R}_\Xi$ the RKHS space that consists of restrictions of functions in $\mathcal{H}(\mathbb{R}^n)$, where the space of restrictions is equipped with the restricted kernel $r : \Xi \times \Xi \to \mathbb{R}$ such that

$$r(x, y) \triangleq \mathfrak{K}|_{\Xi\times\Xi}(x, y), \qquad \text{for all } x, y \in \Xi. \tag{6.55}$$

Following the analysis in Sect. 3.2, we denote by $T_\Xi : \mathcal{H}(\mathbb{R}^n) \to \mathcal{R}_\Xi$ the restriction operator, while $\mathcal{E}_\Xi : \mathcal{R}_\Xi \to \mathcal{H}(\mathbb{R}^n)$ denotes the canonical minimal norm linear extension operator. As discussed in Sect. 3.2, both $\mathcal{E}_\Xi$ and T_Ξ are bounded linear operators, and

$$\|\mathcal{E}_\Xi\| \leq 1, \tag{6.56}$$

$$\|T_\Xi\| \leq 1, \tag{6.57}$$

$$T_\Xi \mathcal{E}_\Xi g = g \quad \text{for all } g \in \mathcal{R}_\Xi. \tag{6.58}$$

Furthermore, as discussed in [54, Th. 5], choosing a kernel that satisfies the algebraic decay condition (6.54) implies that the space of restrictions $\mathcal{R}_\Xi$ is a Sobolev space of a reduced smoothness defined over the compact Riemannian manifold Ξ, that is,

$$\mathcal{R}_\Xi \approx W^{s-(n-1)/2,2}(\Xi). \tag{6.59}$$

It is possible to derive pointwise errors for any $f = \mathcal{E}_\Xi g \in \mathcal{H}(\mathbb{R}^n)$ that is generated as the image of a restricted function $g \in \mathcal{R}_\Xi$. We begin by noting that

$$\begin{aligned} |E_x(I - \Pi_N)f| &= |E_x(\mathcal{E}_\Xi - \Pi_N \mathcal{E}_\Xi g)| \\ &= \left|E_x(\mathcal{E}_\Xi - \mathcal{E}_\Xi \Pi_{\Xi,N} g)\right| \\ &= \left|E_x \mathcal{E}_\Xi (I - \Pi_{\Xi,N})g)\right| \\ &\leq \bar{\mathfrak{K}} \|(I - \Pi_{\Xi,N})g\|_{\mathcal{R}_\Xi}, \end{aligned} \tag{6.60}$$

where $\Pi_N : \mathcal{H}(\mathbb{R}^n) \to \mathcal{H}_N$ denotes the $\mathcal{H}$-orthogonal projection onto $\mathcal{H}_d(N) \triangleq \text{span}\{\mathfrak{K}_{\xi_1}, \ldots, \mathfrak{K}_{\xi_N}\} \subset \mathcal{H}(\mathbb{R}^n)$, and $\Pi_{\Xi,N}$ denotes the $\mathcal{R}_\Xi$-orthogonal projection onto $\mathcal{R}_N \triangleq \text{span}\{\mathfrak{r}_{\xi_1}, \ldots, \mathfrak{r}_{\xi_N}\} \subset \mathcal{R}_\Xi$. Now, applying the many zeros theorem (see Theorem 2.8) in the space of restrictions $\mathcal{R}_\Xi \approx W^{s-(n-1)/2,2}(\Xi)$, it is possible to prove that

$$|E_x(I - \Pi_N)f| \leq \bar{\mathfrak{K}} h_{\Xi_N,\Xi}^{q-\mu} \|g\|_{W^{q,2}(\Xi)}, \quad \text{for all } 0 < n/2 < \mu < q < s - (n-1)/2 \tag{6.61}$$

provided that the uncertainty $f \in \mathcal{E}_\Xi(\mathcal{R}_\Xi) \subset \mathcal{H}(\mathbb{R}^n)$ is the extension of a function $g \in \mathcal{R}_\Xi$. As pointed out in [54], the best rate of decay of the fill distance $h_{\Xi_N,\Xi}$ in (6.61) is capped at $s-(n-1)/2$. The restriction process reduces the effective smoothness of restrictions in $\mathcal{R}_\Xi \approx W^{s-(n-1)/2,2}(\Xi)$ over the manifold Ξ, compared to smoothness of the extended function in $\mathcal{H}(\mathbb{R}^n) \approx W^{s,2}(\mathbb{R}^n)$.

The utility of the bound in (6.61) is that, if we pick centers $\Xi_N \subset \mathbb{X}$ that are quasi-uniform in the limit cycle Ξ, then,

$$h_{\Xi_N,\Xi} \approx \frac{1}{N}, \tag{6.62}$$

since this embedded manifold is one dimensional. Thus, the bound (6.62) does not depend on the dimension of the state space n since the approximation process is performed for the discretization of a one-dimensional submanifold. Furthermore, the tightest performance bounds on the controller are given by

$$\limsup_{t\to\infty} \|x_N(t) - x_r(t)\|_{\mathbb{X}} \lesssim O\left(\frac{1}{N^{s-(n-1)/2}}\right) \tag{6.63}$$

with $s > n/2$. Notably, the nonparametric uncertainty class

$$\overline{C}_R \triangleq \{f \in \mathcal{E}_\Xi(\mathcal{R}_\Xi) \subset \mathcal{H}(\mathbb{R}^n) : \|f\|_{\mathcal{H}(\mathbb{R}^n)} \leq R\} \subset C_R \subset \mathcal{H}(\mathbb{R}^n) \tag{6.64}$$

is much smaller than the classes C_R contained in $\mathcal{H}(\mathbb{R}^n)$.

This general process uses information about the dynamics and control problem to devise controllers that alleviate some of the worst implications of the curse of dimensionality noted in the last section. On the other hand, the control scheme only generates AAO or ϵ-AAO controllers that are robust over much smaller, but still infinite dimensional and nonparametric uncertainty classes. △

6.4.2.3 Prescribed Samples over Limit Sets Unknown A Priori

The approaches in Sects. 6.4.2.1 and 6.4.2.2 are based on the assumption that Ω is known a priori, and it is possible to deterministically prescribe samples Ξ_N that “fill out” the set Ω, or a smaller set $\Xi \subset \Omega$, as the number of centers increases. In this section, we consider the case wherein the centers that determine the space of approximants are generated by a deterministic dynamical system. The general study of data-driven methods for deterministic, as well as stochastic, dynamical systems, has been an active area of research over the past several years [55]. The following example illustrates how a dynamically data-driven sample approach may arise naturally and be implemented.

Example 6.4 Consider the plant model (6.4) with $\Gamma(t) \equiv 0$ for all $t \geq t_0$. As discussed in Sect. 6.2.1, to establish an LTC policy, it is possible to use a nominal feedback controller $u_O : [t_0, \infty) \to \mathbb{U}$ during the learning phase; see Fig. 6.2. In this case, during the learning phase, the plant dynamics are given by

$$\dot{x}(t) = Ax(t) + B\left(u_O(t) + E_{x(t)} f\right), \quad x(t_0) = x_0, \quad t \geq t_0. \tag{6.65}$$

We assume that, while applying the nominal feedback control $u_O(\cdot)$, the plant trajectories of interest lie in some known set Ω despite the uncertainty f. Given the sampling times$\{t_k\}_{k\in\mathbb{N}} \subset [t_0, \infty)$, we consider the sampling of the plant trajectory $\{x_k\}_{k\in\mathbb{N}}$, where $x_k \triangleq x(t_k)$. We then construct a collection $\{\Xi_N\}_{N\in\mathbb{N}}$ of nested sets of centers such that

$$\Xi_N \triangleq \{x_{k_1}, \ldots, x_{k_N}\} \subset \{x_k\}_{k\in\mathbb{N}} \tag{6.66}$$

to enforce some useful conditions on the centers Ξ_N, such as a minimum separation radius as defined in (6.38). Thus, we set

$$\mathcal{H}_{d(N)} \triangleq \text{span}\left\{\mathcal{K}_{\xi_i} e_j \mid \xi_i \in \Xi_N, 1 \leq i \leq N, 1 \leq j \leq m\right\}, \tag{6.67}$$

where $\mathcal{K} : \mathbb{X} \times \mathbb{X} \to \mathbb{U}$ denotes the operator kernel that defines $\mathcal{H}$. Identical arguments apply to the case wherein u_O is a state-feedback controller, that is, $u_O : \mathbb{X} \to \mathbb{U}$. △

Example 6.4 provided a very intuitive LTC approach. However, without additional information about f and without a careful design of $u_O(\cdot)$, it may be impossible to define sets $\{\Xi_N\}_{N\in\mathbb{N}}$ that are dense in Ω. For instance, if the nominal feedback controller drives the closed-loop trajectory to a fixed point in Ω, then the samples along such a trajectory cannot become dense in all of Ω. To define subsets $\{\Xi_N\}_{N\in\mathbb{N}}$ that are dense in Ξ while applying the same strategy as in Example 6.4, given a control law u_O, we can leverage several results from the theory of dynamical systems in Chap. 4.

Consider the autonomous dynamical system

$$\dot{x}(t) = Ax(t) + B\left(u_O(x(t)) + E_{x(t)} f\right), \quad x(0) = x_0, \quad t \geq 0, \tag{6.68}$$

and suppose, as it is routinely done in adaptive control of ODEs, that this system is forward complete on $[0, \infty)$; see Definition 4.4. In this case, noting that (6.68) is a linear, time-invariant dynamical system with control input $u_O(x_N(t)) + E_{x_N(t)} f$ for all $t \geq 0$, its solution can be expressed as

$$x_N(t) = S(t)x_0 \quad \text{for all } t \geq 0, \tag{6.69}$$

where S denotes a strongly continuous semigroup; see Definition 4.31. Thus, define the positive orbit $\gamma^+(x_0)$ and the positive limit set $\omega^+(x_0)$ for the initial condition x_0 as in (4.44) and (4.45), respectively. Choosing $u_O(t) = u_O(x_N(t))$ in (6.65), Example 6.4 shows how to create centers that are subset of the positive orbit of (6.68) with initial condition x_0. The next theorem provides sufficient conditions for this set of centers to be a subset of the positive limit set of (6.68)

Theorem 6.1 *Consider the autonomous dynamical system (6.68) with precompact positive orbit $\gamma^+(x_0)$, and consider an LTC strategy such that $\bigcup_{N\in\mathbb{N}} \Xi_N \subseteq \gamma^+(x_0)$. If $\{\Xi_N\}_{N\in\mathbb{N}}$ is dense in Ξ, then $\Xi \subseteq \omega^+(x_0)$.*

Proof Since $\bigcup_{N\in\mathbb{N}} \Xi_N$ is dense in $\mathbb{X}$, for each $z \in \mathbb{X}$, there is a sequence $\{z_k\}_{k\in\mathbb{N}}$ with $z_k \in \bigcup_{N\in\mathbb{N}} \Xi_N$ such that

$$z = \lim_{k\to\infty} z_k = \lim_{k\to\infty} S(t_k)x_0 \tag{6.70}$$

for some monotonic sequence $\{t_k\}_{k\in\mathbb{N}}$ with $\lim_{k\to\infty} t_k = \infty$. Since $z_k = S(t_k)x_0 \in \gamma^+(x_0)$, it follows from Theorem 4.15 that $z \in \omega^+(x_0)$, and the result is proven. ■

This result is attractive because of its simplicity and its generality. Theorem 6.1 holds under the mild assumption that the solutions of (6.68) are forward complete and its positive orbit with initial condition x_0 is precompact. It is just an exercise in unpacking the definitions to show that that the largest set for which we can hope such a dynamics-driven sampling to be dense is the positive limit set $\omega^+(x_0)$. In practice, depending on how the sub-selection process is structured, the collection of samples may be dense in a set that is considerably smaller than $\omega^+(x_0)$.

Much care needs to be given to the problem of designing the control law $u_O : \mathbb{X} \to \mathbb{U}$ employed in (6.68) and, hence, in the definition of the LTC policy. Indeed, the resulting trajectory may have a highly irregular limit set. A canonical example of an irregular limit set is given by the Lorenz system

$$\dot{x}_1(t) = \sigma(x_2(t) - x_1(t)), \quad x_1(t_0) = x_{1,0}, \quad t \geq 0, \tag{6.71a}$$

$$\dot{x}_2(t) = x_1(t)\,(\rho - x_3(t)) - x_2(t), \quad x_2(t_0) = x_{2,0}, \tag{6.71b}$$

$$\dot{x}_3(t) = x_1(t)x_2(t) - \beta z_3(t), \quad x_3(t_0) = x_{3,0}, \tag{6.71c}$$

whose trajectories are shown in Fig. 6.8. The limit set of a Lorenz system is precompact, has an empty interior, and zero Lebesgue measure in $\mathbb{R}^3$ [56].

To quantify the effect of choosing the set of centers as a subset of the limit set of (6.68), assume that the conditions of Theorem 6.1 are verified. Thus, define the closed subspace

$$\mathcal{H}_{\omega^+(x_0)} \triangleq \overline{\left\{\mathcal{K}_\xi \alpha : \xi \in \omega^+(x_0), \alpha \in \mathbb{U}\right\}}, \tag{6.72}$$

and note that

$$\mathcal{H}_\Xi \subseteq \mathcal{H}_{\omega^+(x_0)} \subset \mathcal{H}_\Omega \subseteq \mathcal{H}, \tag{6.73}$$

where $\mathcal{H}_\Xi$ is given by (6.47a). Ideally, we would like performance estimates that hold for all the uncertainty in the largest subspace $\mathcal{H}_\Omega$, but the practical limits on

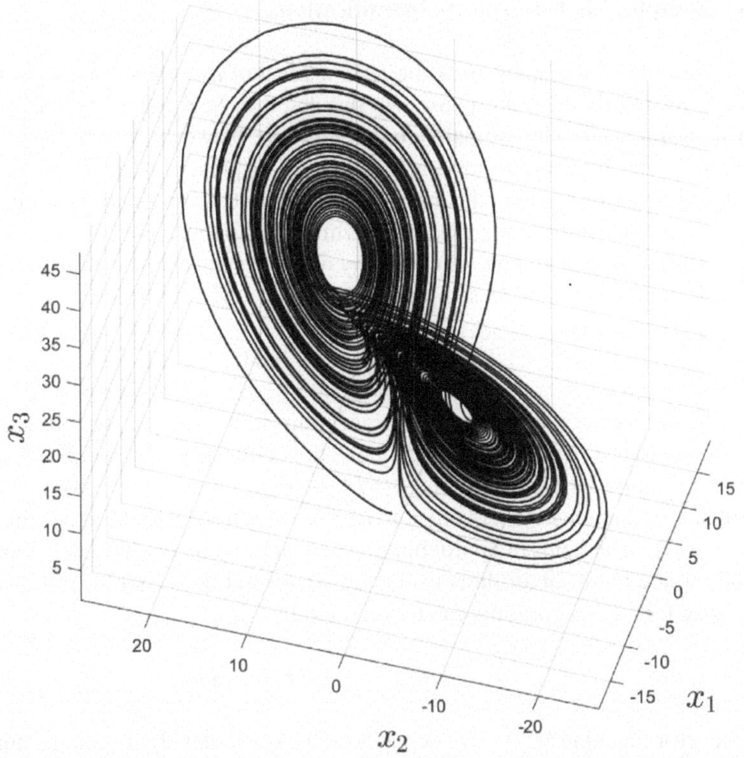

Fig. 6.8 Trajectory of a Lorenz system obtained by setting $\sigma = 10$, $\rho = 28$, and $\beta = 8/3$ in (6.71)

both sampling and approximation during the learning phase suggest that it is more reasonable to expect that they hold over the smaller space $\mathcal{H}_{\omega^+(x_0)}$.

Applying any of the adaptive control systems discussed in Chap. 5, we newly obtain the ultimate performance bound given by (6.50), and by proceeding as in Sect. 6.4.2.2, we conclude that the resulting adaptive control system is AAO for the uncertainty classes (6.52) and (6.53).

The proposed data-driven approach to LTC system design remains conceptually identical in the case the control system O employed during the learning phase (see Fig. 6.2) is state-feedback in discrete time. While Theorem 4.15 is stated for semi-groups in continuous time, similar results hold for discrete-time autonomous systems [57, 58]. Finally, (6.68) is an autonomous dynamical system. The proposed arguments can be extended to some types of plant models that, during the learning phase, are nonautonomous. We can use the analysis in [59], which derives a result similar to Theorem 4.15, but applies to nonautonomous systems that are asymptotically autonomous. However, ascertaining that a time-varying system is asymptotically time-invariant may be difficult.

6.4.2.4 Samples Via Uncertainty Identification

We have formulated the analysis of the LTC algorithm using a subspace selection operator L during the learning phase. As discussed in Sect. 6.2.1, the uncertainty $f \in \mathcal{H}$ does not need to be estimated to determine L. However, leveraging reliable estimates of f may be advantageous. In this section, we present a data-driven method based on deterministic learning theory to estimate f in (6.4), and show how knowing bounds on the error in estimating f helps determining ultimate bounds on the tracking error of a closed-loop plant controlled by any of the systems presented in Sect. 5.4.

We assume that we know M samples $\{(x_i, y_i)\}_{i=1}^M \subset \mathbb{X} \times \mathbb{U}$, where

$$y_i = f(x_i) + v_i \in \mathbb{U}, \quad \text{for all } i \in \{1, \ldots, M\}, \tag{6.74}$$

and $v_i \in \mathbb{U}$ denotes some deterministic uncertainty or error in the observation. Employing classical regression techniques, this collection of samples can be employed to define an algorithm $A_{N,M} : \mathcal{H} \to \mathcal{H}_{d(N)}$ such that $f_{N,M} \triangleq A_{N,M} f$ approximates f pointwise. There are a myriad of ways to realize the approximation algorithm $A_{N,M}$. The fields of approximation and learning theory have developed an enormous collection of algorithms over the past several decades [33, 60], and the performance for such algorithms can be captured by

$$\|f - f_{M,N}\|_{L^\infty(\mathbb{X},\mathbb{U})} \lesssim \epsilon(M, N; \mathcal{H}), \tag{6.75}$$

where the error bound $\epsilon(M, N; \mathcal{H})$ depends on the particular algorithm, the number of samples M, and the dimension of the space of approximants N. We are interested in understanding how the knowledge of the bound $\epsilon(M, N; \mathcal{H})$ can be used to infer related bounds on the performance of nonparametric adaptive control schemes. In this section, we only are concerned with deterministic identification methods. In Chap. 7, we show that deterministic methods can be considered special cases of statistical methods.

As shown for any of the approaches in Chap. 5, the trajectory tracking error dynamics are written in the form

$$\begin{aligned} \dot{e}_N(t) &= A_r e_N(t) - B\left(\tilde{\alpha}_N^{\mathrm{T}}(t) x_N(t) + \tilde{\beta}_N^{\mathrm{T}}(t) r(t) - E_{x_N(t)} \Pi_N \tilde{f}_N(t, \cdot)\right) \\ &\quad + B E_{x_N(t)}\left((I - \Pi_N)\tilde{f}_N(t, \cdot)\right), \quad e_N(t_0) = x_N(t_0) - x_r(t_0), \quad t \geq t_0; \end{aligned} \tag{6.76}$$

for example, see (5.77). Since

$$E_{x_N(t)} f = E_{x_N(t)} f_{N,M} + E_{x_N(t)} \tilde{f}_{N,M}, \tag{6.77}$$

where $\tilde{f}_{N,M} \triangleq f - f_{N,M}$, it follows from (6.76) that

$$\dot{e}_N(t) = A_r e_N(t) - B\left(\tilde{\alpha}_N^{\mathrm{T}}(t)x_N(t) + \tilde{\beta}_N^{\mathrm{T}}(t)r(t) - E_{x_N(t)}\tilde{f}_{N,M}(t,\cdot)\right)$$
$$+ BE_{x_N(t)}\tilde{f}_{N,M}, \quad e_N(t_0) = x_N(t_0) - x_r(t_0). \quad t \geq t_0. \tag{6.78}$$

The adaptive laws for $\hat{\alpha}(t)$, $\hat{\beta}(t)$, and $\hat{f}_N(t,\cdot)$ described in Sect. 5.4 can be employed. Now, to exploit the knowledge of the bound $\epsilon(M, N; \mathcal{H})$, we consistently replace $\tilde{f}_N$ with $\tilde{f}_{N,M}$ in any of the derivations in Sect. 5.4. For example, in the case of the dead-zone modification of MRAC, proceeding as in (5.83) we obtain

$$\begin{aligned}\dot{V}(t) &\leq -\langle Qe_N(t), e_N(t)\rangle_{\mathbb{X}} + 2\|PB\|\|e_N(t)\|_{\mathbb{X}}\|E_{x_N(t)}\tilde{f}_{M,N}\|_{\mathbb{U}}\\ &\leq -\lambda_{\min}(Q)\|e_N(t)\|_{\mathbb{X}}\left(\|e_N(t)\|_{\mathbb{X}} - \frac{2\|PB\|}{\lambda_{\min}(Q)}\|\tilde{f}_{N,M}\|_{L^\infty(\Omega,\mathbb{U})}\right)\\ &\leq -\lambda_{\min}(Q)\|e_N(t)\|_{\mathbb{X}}\left(\|e_N(t)\|_{\mathbb{X}} - \frac{2\|PB\|}{\lambda_{\min}(Q)}\epsilon(N, M; \mathcal{H})\right),\end{aligned} \tag{6.79}$$

whenever the trajectory is outside the dead-zone. Thus, we can choose our dead-zone width as

$$d = \frac{2\|PB\|}{\lambda_{\min}(Q)}\epsilon(M, N; \mathcal{H}). \tag{6.80}$$

This way, the dead-zone method studied in Sect. 5.4.1 guarantees that the ultimate bound on tracking error is $O(\epsilon(M, N; \mathcal{H}))$. Similar reasoning can be applied to other control techniques in Sect. 5.4, and are left to the reader as exercises.

It must be emphasized that the adaptive laws that define the adaptive feedback control scheme in this section are the same as those studied in Sect. 5.4. We only have modified our Lyapunov-based analysis of the same controllers to show how knowing the approximation error (6.75) impacts our estimates on the upper bounds of the tracking error.

6.4.2.5 Relaxed and Greedy Data-Driven Methods

Greedy methods, already mentioned in Example 6.2, are particularly popular in learning theory to construct approximations in Hilbert and Banach spaces [30, 41]. In this section, we present some greedy methods over generic Banach spaces, Hilbert spaces, and scalar-valued RKHSs. To present these results, recall that the approximating space $\mathbb{H}_i$ in (6.3) has dimension i for any $i \in \mathbb{N}$.

Greedy methods in a generic Banach space. Assume that the uncertainty class C is a compact subset of functions contained in the Banach space $\mathbb{B}$. To construct the bases $\{\Phi_N\}_{N\in\mathbb{N}}$ and subspaces $\{\mathbb{H}_N\}_{N\in\mathbb{N}}$, with $\mathbb{H}_N \subseteq \mathbb{B}$, choose the first basis function so that

$$\phi_1 \in \arg\max_{\phi \in C} \lambda \|\phi\|_{\mathbb{B}}, \tag{6.81}$$

where $\lambda \in (0, 1]$ denotes the *relaxation parameter*; if $\lambda = 1$, then the proposed method is a *greedy method*, and if $\lambda \in (0, 1)$, then the proposed method is a *relaxed greedy method*. Thus, we set $\mathbb{H}_1 = \text{span}\{\phi_1\}$, and we proceed recursively so that

$$\phi_{i+1} \in \arg\max_{\phi \in C} \lambda \text{dist}_{\mathbb{B}}(\phi, \mathbb{H}_i) \tag{6.82}$$

and $\mathbb{H}_{i+1} \triangleq \text{span}\{\phi_1, \ldots, \phi_{i+1}\}$ with $i \in \mathbb{N}$, where the distance between a vector $\phi \in \mathbb{B}$ and a set A is defined as in (2.73). This algorithm augments the current space $\mathbb{H}_i$ to obtain the next space $\mathbb{H}_{i+1}$ by adding a new basis function that is furthest from the current pool that defines $\mathbb{H}_i$.

From a practical standpoint, the optimization problem in (6.82) is often difficult, if not impossible, to solve. In many cases, the relaxation parameter makes this optimization problem tractable. However, tractable algorithms depend crucially on a judicious selection of the set C that contains the pool of possible bases. The authors in [30, 41] derived some powerful theorems that describe the performance, in terms of the worst-case approximation error, of families of methods to solve constrained nonlinear optimization problem captured by (6.82).

If $\mathbb{B}$ is a generic Banach space, then the ability of $\mathbb{H}_i$ to approximate C is quantified by the worst-case error

$$\sigma_i(C) \triangleq \sup_{f \in C} \text{dist}_{\mathbb{B}}(f, \mathbb{H}_i) = \sup_{f \in C} \inf_{\phi \in \mathbb{H}_i} \|f - \phi\|_{\mathbb{B}}, \tag{6.83}$$

which measures how well a specific subspace, namely, $\mathbb{H}_i$, approximates *all of the elements* in the compact set C. The best worst-case error over all linear approximating subspaces of dimension $i \in \mathbb{N}$ is captured by the *Kolmogorov width*

$$w_i \triangleq w_i(C, \mathbb{B}) \triangleq \inf_{\{S_i : \dim(S_i) = i\}} \sup_{f \in C} \text{dist}_{\mathbb{B}}(f, S_i), \tag{6.84}$$

where $\dim(\cdot)$ denotes the dimension of its argument. Note that there may not be any particular i-dimensional subspace $S_i^* \subseteq B$ that attains the infimum in (6.84). However, if such an S_i^* exists, then we can compare the performance of the greedily defined subspace $\mathbb{H}_i$ to that of the optimal subspace S_i^*.

Greedy methods in a generic Hilbert space. References [30, 41, 61] analyze how well the greedily defined subspace $\mathbb{H}_i$ performs in terms of the Kolmogorov width w_i in several settings with the universe space $\mathbb{B}$ being a generic Banach space. Such bounds can be complicated to quantify. If the universe space is a Hilbert space and $\mathbb{H} = \mathbb{B}$, then [41] shows that these error bounds can be simplified. This case is particularly important to the scope of this book since, in most problems, we choose $\mathbb{H}$ to be an RKHS, $\mathcal{H}$ or $\boldsymbol{\mathcal{H}}$. In this case, Corollary 3.3 of [41] shows weak greedy methods to generate subspaces $\mathbb{H}_i$ such that

$$\sigma_{2i}(C) \leq \frac{\sqrt{2}}{\lambda}\sqrt{w_i(C)}, \quad i \in \mathbb{N}. \tag{6.85}$$

This remarkable result implies that the greedy method generates a subspace of dimension $2i$, whose worst-case error over C is comparable to the worst-case error for the optimal i-dimensional subspace S_i^*, if it exists.

Greedy methods in a scalar-valued RKHS $\mathcal{H}$. The worst-case performance bound in (6.85) is attractive in its simplicity since it provides a rule of thumb regarding the worst-case performance of greedy data-driven methods in generic Hilbert spaces. However, (6.85) does not allow to develop nonparametric feedback controllers for practical problems. The problem is that the Kolmogorov width on the right-hand side of (6.85) is difficult to obtain in most cases. The authors in [27, 35, 62] discuss some cases in which the Kolmogorov width can be bounded in terms of known functions of N. If we assume that $\mathbb{B} = \mathbb{H}$ and the hypothesis space is a scalar-valued native space $\mathcal{H}$, then there are standard choices of C for which the Kolmogorov width is easy to compute in terms of known quantities.

Suppose that $\mathbb{H} = \mathcal{H}$ is defined in terms of the scalar-valued kernel function $\mathfrak{K} : \Omega \times \Omega \to \mathbb{R}$, with $\Omega \subseteq \mathbb{R}^n$ compact. In this case, we choose the pool of bases as $C \triangleq \{\mathfrak{K}_x \ : \ x \in \Omega\}$. Thus, we set $\{\mathcal{H}_i\}_{i\in\mathbb{N}}$ with $\mathcal{H}_i \triangleq \text{span}\{\mathfrak{K}_{\xi_1}, \ldots \mathfrak{K}_{\xi_i}\} = \text{span}\{\mathfrak{K}_\xi \ : \ \xi \in \Xi_i\}$, $\Xi_i \triangleq \{\xi_1, \ldots, \xi_i\}$, and reformulate the relaxed data-driven greedy method by recasting (6.82) as

$$\xi_{i+1} \in \arg\max_{\xi\in\Omega} \mathcal{P}_i(\xi). \tag{6.86}$$

Indeed, it follows from Theorem 3.12 that

$$\begin{aligned} \text{dist}_H(f, \mathcal{H}_i) &\triangleq \inf_{\phi\in\mathcal{H}_i} \|f - \phi\|_{\mathcal{H}} \equiv \|(I - \Pi_i)f\|_{\mathcal{H}} \\ &= \|(I - \Pi_i)\mathfrak{K}_x\|_{\mathcal{H}} \\ &= \mathcal{P}_i(x) \quad \text{for any } f \in C \end{aligned}$$

with $f = \mathfrak{K}_x$; Example 6.2 anticipated these arguments. In this case, the Kolmogorov width can be expressed as

$$w_i(C) = \inf_{\Omega_i\subset\Omega} \sup_{\xi\in\Omega} \mathcal{P}^2_{\Omega_i}(\xi), \tag{6.87}$$

where the outer infimum is taken over all subsets Ω_i that contain i centers in Ω [63]. In (6.87), we have used the more explicit notation $\mathcal{P}_{\Omega_i}$ to denote the power function of the subspace $\mathcal{H}_{\Omega_i} \triangleq \text{span}\{\mathfrak{K}_{\omega_j} \mid \omega_j \in \Omega_i, 1 \leq j \leq i\} \subset \mathcal{H}$, and it is important to distinguish the power function $\mathcal{P}_{\Omega_i}$ of a subspace $\mathcal{H}_{\Omega_i}$ from the power function $\mathcal{P}_i \triangleq \mathcal{P}_{\Xi_i}$ generated by greedy method for the space $\mathcal{H}_i$. Finally, it follows from (6.83), (6.85), and (6.87) that the worst-case error bounds is given by

Table 6.3 Error indicators deduced in [38] to compute centers through the data-driven method captured by (6.89)

Data-driven greedy method	Error Indicator $\text{Err}(\cdot;\, f, \mathcal{H}_i)$
$\mathcal{P}$	$\mathcal{P}_i(\cdot)$
$f \cdot \mathcal{P}$	$\lvert (f - \Pi_i f)(\cdot) \rvert \mathcal{P}_i(\cdot)$
f	$\lvert (f - \Pi_i f)(\cdot) \rvert$
$f/\mathcal{P}$	$\lvert (f - \Pi_i f)(\cdot) \rvert / \mathcal{P}_i(\cdot)$

$$\sigma_{2i}(C) \triangleq \sup_{f \in C} \inf_{\phi \in \mathcal{H}_i} \|f - \phi\|_{\mathcal{H}_i} \leq \frac{\sqrt{2}}{\lambda} \inf_{\Omega_i \subset \Omega} \sup_{\xi \in \Omega} \mathcal{P}_{\Omega_i}(\xi), \tag{6.88}$$

and (3.80) and Table 3.3 can be employed to quantify the power function in terms of the fill distance.

The authors in [38, 39, 64] describe greedy data-driven methods in an RKHS that are based on error indicators. Specifically, (6.86) is replaced with

$$\xi_{i+1} \in \arg\max_{\xi \in \Omega} \text{Err}(\xi;\, f, \mathcal{H}_i), \tag{6.89}$$

where Err denotes an error indicator function; see Table 6.3 for a list indicator functions derived from [38] for the $\mathcal{P}$, $f \cdot \mathcal{P}$, f, and $f/\mathcal{P}$ approaches. The $\mathcal{P}$, $f \cdot \mathcal{P}$, and $f/\mathcal{P}$ approaches require the calculation of the power function $\mathcal{P}_i(\cdot)$ over the domain Ω. The $\mathcal{P}$, $f \cdot \mathcal{P}$, and $f/\mathcal{P}$ methods also require the knowledge of the residual $|(f - \Pi_i f)(\cdot)|$ at each point in Ω. The $\mathcal{P}$-method generates a linear approximation scheme since the generated basis does not depend on particular information about f. The f, $f \cdot \mathcal{P}$, and $f/\mathcal{P}$ methods are examples of nonlinear approximation schemes, and the basis generated is tied to the residual that depends specifically on f.

The proposed greedy methods apply to scalar-valued RKHSs $\mathcal{H}$. The setting for vector-valued RKHSs induced by a general operator kernel $\mathcal{K}$ are not as mature.

6.4.3 Numerical Example

All the numerical examples discussed in Sects. 5.4 and 5.5 show applications of the LTC procedure placing samples over parallelepipeds or limit cycles provided by the reference trajectory, which is explicitly designed to produce a limit cycle known a priori. In this section, we present an example of LTC architecture designed leveraging the actor-critic method. This example is based on recent results presented in [65, 66]. For overviews of the state-of-the-art on the important topic of nonlinear adaptive control and reinforcement learning, see, for example, [67, 68].

For clarity of exposition, before we introduce the actor-critic method, first, we present one of its precursors known as *policy iteration* (PI) method [69]. Consider the continuous-time nonlinear dynamical system

$$\dot{x}(t) = a(x(t)) + g(x(t))u(t), \quad x(0) = x_0, \quad t \geq 0, \tag{6.90}$$

where $x : [0, \infty) \to \mathbb{X} \triangleq \mathbb{R}^n$, $a : \mathbb{X} \to \mathbb{R}^n$, $g : \mathbb{X} \to \mathbb{R}^{n \times m}$, and $u : [0, \infty) \to \mathbb{U} \triangleq \mathbb{R}^m$ represent the state of the system, the drift dynamics, the control input operator, and the control input, respectively. We assume that both $a(\cdot)$ and $g(\cdot)$ are known.

Remarkably, the plant models considered in Chap. 5 and so far in this chapter can be reduced to the same form as (6.90). For example, (6.4) with $\Gamma(t) \equiv 0$ and $t_0 = 0$ is in the same form as (6.90) with $a(x)$ replaced by $Ax + BE_x f$ and $g(x)$ replaced by B. The assumption whereby $a(\cdot)$ and $g(\cdot)$ are known can be interpreted as assuming that, in the learning phase, the plant model operates according to some *nominal* model. This assumption is not limiting, especially in the LTC setting. Indeed, in practice, as soon as a plant becomes operative, its dynamical models should be well known, and hence, using control systems based on the assumption that the nominal plant model is sufficiently accurate is a good design practice. Robust and adaptive control systems are primarily designed to account for off-nominal conditions that may occur throughout the life of the controlled plant. It would be a questionable design practice to assume that a robust or adaptive controller is the nominal controller of a plant. Besides the ontological aspects of a discussion on whether adaptive controllers should be considered the nominal controllers for a plant model, as noted in [19, 70], adaptive controllers tend to be aggressive, especially in the transient phase, and are best used to augment nominal control systems.

We choose a subset $\Omega \subset \mathbb{X} \triangleq \mathbb{R}^n$ of initial conditions, and we are interested in finding a stabilizing feedback controller $u(t) \triangleq \mu(x(t))$, $t \geq 0$, such that $\lim_{t\to\infty} x(t) = 0$, where $x(\cdot)$ denotes the controlled plant trajectory, and the quadratic cost function

$$J(x_0, u) \triangleq \int_0^\infty r(x(t), u(t))\mathrm{d}t \tag{6.91}$$

is minimized, where

$$r(x, u) \triangleq Q(x) + u^{\mathrm{T}} Ru, \quad (x, u) \in \mathbb{X} \times \mathbb{U}, \tag{6.92}$$

$Q : x \mapsto Q(x)$ is user-defined, symmetric, and nonnegative-definite, and $R \in \mathbb{R}^{m \times m}$ is symmetric and positive-definite.

According to the classical Hamilton–Jacobi–Bellman (HJB) theory, if there is a sufficiently smooth optimal *value function* $V^\star \in \mathbb{R}$ that satisfies the *HJB equation*

$$\begin{aligned}\mathcal{H}_{u^\star}(V^\star(x)) \triangleq -\frac{1}{4}\nabla V^{\star\mathrm{T}}(x)g(x)R^{-1}g^{\mathrm{T}}(x)\nabla V^\star(x)\\ +\nabla V^{\star\mathrm{T}}(x)a(x)+Q(x)=0,\quad V^\star(0)=0,\quad \text{for all } x\in\Omega,\end{aligned} \tag{6.93}$$

where $\nabla(\cdot)$ denotes the gradient operator, then the controller that minimizes (6.91) for any $x_0 \in \mathbb{X}$ is given by

$$u^\star(x) = -\frac{1}{2}R^{-1}g^{\mathrm{T}}(x)\nabla V^\star(x). \tag{6.94}$$

The state-feedback control law (6.94) also guarantees asymptotic stability of the closed-loop plant model [71, Ch. 8]. Over the past few decades, numerous direct and inverse methods have been studied to generate approximations of such an optimal controller. For a review of direct methods, see [67–69], and for a review of inverse methods, see [72–76].

According to the PI method, to approximate solutions of the HJB partial differential equation (PDE) (6.93), we define the differential operator

$$(Av)(x) \triangleq (a(x)+g(x)\mu^{\mathrm{T}}(x(t)))(\nabla v)(x),\quad \text{for all } x\in\mathbb{X}, \tag{6.95}$$

where $\mu : \mathbb{X} \to \mathbb{U}$ and $v : \mathbb{X} \to \mathbb{R}$ is sufficiently smooth. Thus, we define an offline recursion, where, given a fixed input feedback function $\mu_i \in \mathbb{U} \triangleq \mathbb{R}^m$, we solve for $V_{\mu_i} \in \mathbb{R}$ from the operator equation

$$(AV_{\mu_i})(x) = b(x) \triangleq -r(x,\mu_i(x)),\quad \text{for all } x\in\Omega. \tag{6.96}$$

After V_{μ_i} is determined from (6.96), we construct the updated control law

$$\mu_{i+1}(x) = -\frac{1}{2}R^{-1}g^{\mathrm{T}}(x)\nabla v_i(x),\quad \text{for all } x\in\mathbb{X}. \tag{6.97}$$

The convergence $V_{\mu_i} \to V^\star$ as $i \to \infty$ is studied in detail in [67–69] and the references therein.

The PI method motivated a large family of online methods referred to as *actor-critic methods*, which have a dominant role in online adaptive approaches based on reinforcement learning [67–69]. The actor-critic method replaces the discrete offline computational steps of the PI approach with a pair of coupled learning laws that evolve in continuous time [77]. The learning law in continuous time for the critic replaces the discrete step in (6.96), while the learning law in continuous time for the actor replaces the discrete step in (6.97). In this brief section, we only analyze the online critic learning law.

Proceeding as in [66], suppose that we have fixed the nominal feedback control μ, and we are interested in defining an online estimate $\hat{V}_\mu(t,\cdot)$ of the solution V_μ of (6.96) with V_{μ_i} replaced by V_μ. We define the value function error

$$\tilde{V}(t,\cdot) \triangleq V_\mu(\cdot) - \hat{V}_\mu(t,\cdot), \tag{6.98}$$

and we assume that $V_\mu, \hat{V}_\mu(t,\cdot) \in \mathcal{H}(\Omega)$, where $\mathcal{H}$ denotes a scalar-valued RKHS over the set Ω. The associated gradient descent learning law is given by

$$\frac{\partial}{\partial t}\tilde{V}(t,\cdot) = -\gamma A^* E^*_{x(t)}\left(y(t) - E_{x(t)} A \hat{V}_\mu(t,\cdot)\right), \quad \tilde{V}(0,\cdot) = V_0, \tag{6.99}$$

where $y(t) \triangleq E_{x(t)} A V_\mu$ and $\gamma > 0$ denotes the learning rate. This ideal gradient law (6.99) evolves in $\mathcal{H}(\Omega)$, and defines a distributed parameter system (DPS). This DPS can be viewed either as a PDE or as an ODE that takes values in a Hilbert space. Note the resemblance between the learning law (6.99) and the adaptive law (5.44c) for RKHS-embedded MRAC.

Similarly to (5.87), fixed the number of centers $N \in \mathbb{N}$, for each $t \in [0, \infty)$ approximations of $V_\mu(t,\cdot)$ are parameterized as

$$\hat{V}_N(t,\cdot) \triangleq \sum_{j=1}^{N} \hat{W}_j(t)\mathfrak{K}_{\xi_j}(\cdot), \tag{6.100}$$

where $\mathfrak{K}_\xi$ denotes the kernel function with center located at ξ. By proceeding as in Sect. 5.4.1.3, the coefficients $\left\{\hat{W}_j(t)\right\}_{j=1}^{N}$ can be computed as the solutions of

$$\mathbb{K}_N \dot{\hat{W}}(t) = -\gamma\left[\Phi^{\mathrm{T}}(x,\Xi_N)\psi(x)\psi(x)^{\mathrm{T}}\Phi(x,\Xi_N)\right]\hat{W}(t) + \gamma Y(t),$$
$$\hat{W}(0) = \hat{W}_0, \quad t \geq 0, \tag{6.101}$$

where $\hat{W}(t) \triangleq \left[\hat{W}_1(t), \ldots, \hat{W}_N(t)\right]^{\mathrm{T}}$, $\mathbb{K}_N \triangleq [\mathfrak{K}(\xi_i, \xi_j)] \in \mathbb{R}^{N\times N}$ denotes the Grammian matrix,

$$\psi(x) \triangleq a(x) + g(x)\mu(x), \tag{6.102a}$$

$$Y(t) \triangleq -\Phi(x(t),\Xi_N)^{\mathrm{T}}\psi(x(t))r(x(t),\mu(x(t))), \tag{6.102b}$$

$$\Phi(x,\Xi_N) \triangleq \begin{bmatrix} \frac{\partial\mathfrak{K}(x,\xi_1)}{\partial x_1} & \cdots & \frac{\partial\mathfrak{K}(x,\xi_N)}{\partial x_1} \\ \vdots & \ddots & \vdots \\ \frac{\partial\mathfrak{K}(x,\xi_1)}{\partial x_n} & \cdots & \frac{\partial\mathfrak{K}(x,\xi_N)}{\partial x_n} \end{bmatrix} \in \mathbb{R}^{n\times N}. \tag{6.102c}$$

To illustrate this actor-critic procedure, let $n = 2$, $m = 1$,

$$a(x) = \begin{bmatrix} -x_1 + x_2 \\ -0.5x_1 - 0.5x_2(1 - (\cos(2x_1) + 2)^2 \end{bmatrix}, \quad g(x) = \begin{bmatrix} 0 \\ \cos(2x_1) + 2 \end{bmatrix}, \tag{6.103}$$

$R = 1$, and $Q(x) = x^{\mathrm{T}} I_2 x$, where I_2 denotes the identity matrix in $\mathbb{R}^{2\times 2}$. In this case,

$$V^{\star}(x) \triangleq \frac{1}{2}x_1^2 + x_2^2, \quad x \in \mathbb{X} \triangleq \mathbb{R}^n. \tag{6.104}$$

The Gaussian and Sobolev–Matérn kernels are used for the learning (6.101). The learning rate γ is set to 10.

To outline an LTC architecture for the problem at hand, when we initialize the learning phase, we choose a group of $N = 9$ centers, as shown in Fig. 6.9. For simplicity, in this initial example, these centers are placed evenly over the square $\Omega = [-1, 1] \times [1, 1]$. Later, we compare the convergence of the value function estimates $\hat{V}_{\mu}(t, \cdot) \in \mathcal{H}$, and the associated controllers, as N is varied. We also choose the nominal feedback controller, which is intended to explore the state space during the learning phase, as

$$\mu_1(t, x) = -(\cos(2x_1) + 2)\, x_2 + (\sin(t)\mathrm{rem}(\mathrm{ceil}(t), 6)1.3)\sin(t)\cos(t) \tag{6.105}$$

for all $(t, x) \in [0, \infty) \times \mathbb{X}$, where $\mathrm{ref}(a, b)$ defines the remainder after division of a by b and $\mathrm{ceil}(a)$ rounds a to the nearest integer greater than or equal to a.

During the learning phase, the plant states evolve according to (6.90) with $u(t) = \mu_1(t, x(t))$, and the online estimates of $\hat{V}_{\mu}(t, \cdot) \in \mathcal{H}$ are computed by applying (6.100) and solving (6.101). After the learning phase has finished, the control phase begins. During the control phase, the system of governing equations is driven

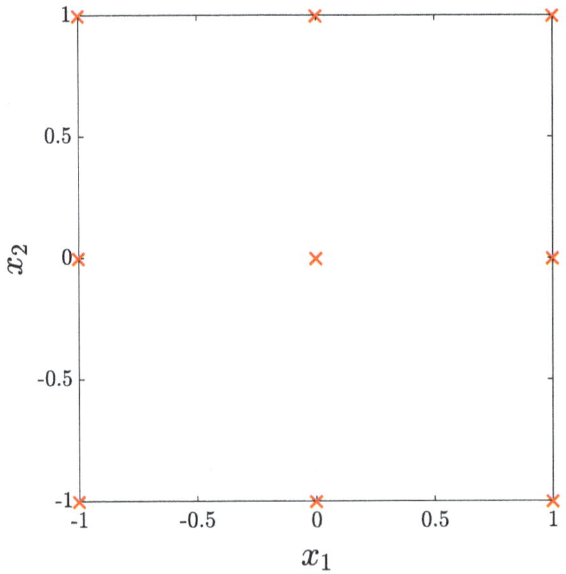

Fig. 6.9 LTC for the online critic, initial centers placement, $N = 9$

by the updated feedback controller μ_1 obtained from (6.97) with $i = 0$. This nominal controller can be augmented by any of the RKHS-embedded MRAC laws discussed in Chap. 5.

Employing a Gaussian kernel with hyperparameter $\alpha = 1$ in Table 3.1, the controlled plant trajectory is shown in Fig. 6.10, while the time history of the associated control input is illustrated in Fig. 6.12. Figures 6.11 and 6.13 depict the rates of convergence of the estimates $V_N(\tau_c, \cdot) \in \mathcal{H}$ to $V^\star$, and the associated rates of convergence of the controllers $u_N(\tau_c, \cdot)$, where $\tau_c > 0$ denotes the start time for the control phase starts, as functions of the fill distance $h_{\Xi_N,\Omega}$ of the samples Ξ_N in the set Ω. When employing the Sobolev–Matérn kernels, we chose the indices smoothness $s = 1.5, 2.5$. The straight lines in Figs. 6.11 and 6.13 show conservative theoretical bounds determined by either the power function or the many zeros theorem; see [66] for the details and a complete analysis.

For all the cases studied for this LTC architecture, the error analysis presented in Sect. 3.5 using the power function or the many zeros theorem yields sharp estimates of the performance of the online critic in terms of the dimension N of $\mathcal{H}_N$ used.

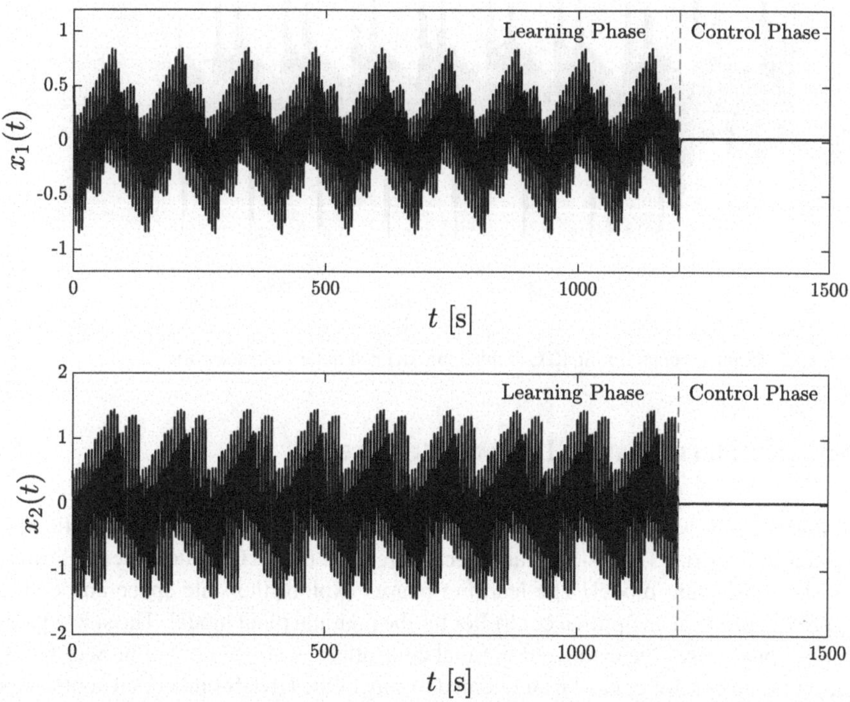

Fig. 6.10 States $x_1(t)$ and $x_2(t)$ of the controlled plant model applying an actor-critic-based LTC system

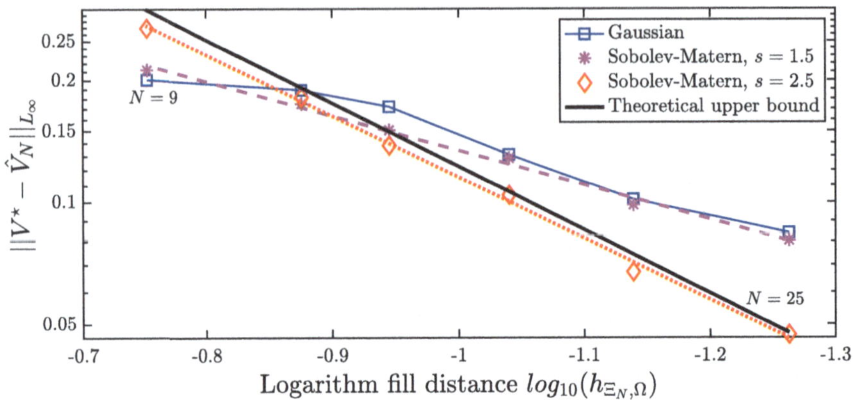

Fig. 6.11 Performance of the LTC architecture defined in terms of the online critic. Estimates of the approximate function $V_N(\tau_c, \cdot) \in \mathcal{H}$ are compared to $V^\star$ versus fill the distance

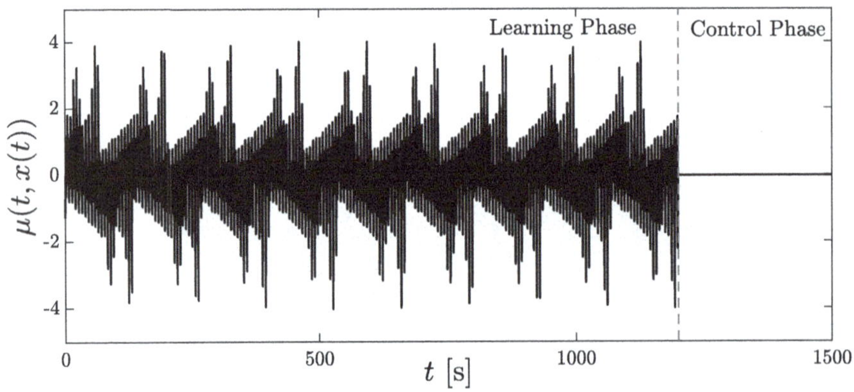

Fig. 6.12 Control inputs for the LTC architecture defined in terms of the critic

6.5 A Deterministic SLC Architecture

In this section, we construct a simple SLC architecture using the online critic discussed in Sect. 6.4.3. This SLC architecture repeats the LTC strategy several times for the same plant model. The learning phases explore the state space and define approximations of an optimal controller for the nominal plant model. The subsequent control phase uses the estimated optimal controller; as already noted in Sect. 6.4.3, this nominal controller can be augmented by any of the RKHS-embedded controllers presented in Chap. 5.

This SLC control requires the user to design the time horizon of each learning and control phase, the nominal controllers that are used during each exploration and learning phase, the dimensions $N(t)$ of the approximants used during each phase, and the subspaces $\mathcal{H}_{N(t)}$ used for building approximations during each phase. However, as

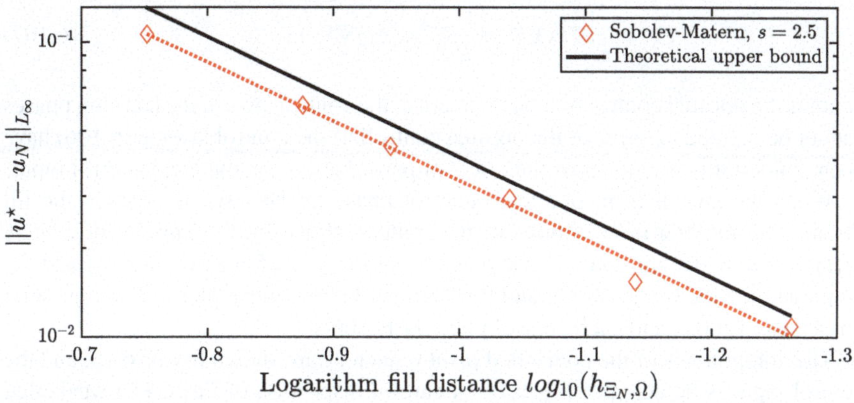

Fig. 6.13 Performance of the LTC architecture defined in terms of the online critic. The feedback control $u_N(\tau_c, \cdot)$ is compared to $u^\star$ versus fill distance

discussed in Sect. 6.2.2, concatenating infinitely many control laws, each of which is defined over a finite-time interval, may induce instability or other undesirable effects. This problem, which has not been addressed in the context of SLC systems design, is left for future research.

Three test cases are examined in this section. In the first two cases, the centers that define the kernel functions are user-defined. In particular, in the first test case, we illustrate how moving the location of the centers influences the performance of the feedback controllers used during the control phases. In the second test case, we show how reducing the fill distance between centers influences the performance of the feedback controllers used during the control phases. In the third test case, the location of the centers is defined automatically by the $\mathcal{P}$-greedy method discussed in Sect. 6.4.2.5.

6.5.1 Case Study 1

In the first study case, there are three learning and control phases. The control laws for the first, second, and third learning phases are

$$\mu_1(t,x) = -(\cos(2x_1+2)x_2 + (\sin(t)\mathrm{rem}(\mathrm{ceil}(t),6)1.3)\sin(t)\cos(t), \tag{6.106a}$$

$$\mu_2(t,x) = \mu^\star(x) + (\sin(t)\mathrm{rem}(\mathrm{ceil}(t),4)1.3)\sin(t)\cos(t), \tag{6.106b}$$

$$\mu_3(t,x) = \mu^\star(x) + (\sin(t)\mathrm{rem}(\mathrm{ceil}(t),2.5)1.3)\sin(t)\cos(t), \tag{6.106c}$$

respectively, for all $(t,x) \in [0,\infty) \times \mathbb{X}$, where

$$\mu^\star(x) \triangleq -\frac{1}{2}R^{-1}g^{\mathrm{T}}(x)\nabla V^\star(x) \tag{6.107}$$

denotes the optimal control policy. In practice, the control laws in the learning phases cannot be defined in terms of the optimal controller; the control laws in (6.106) have been defined this way to show the effect of perturbations on the ideal control input.

When we switch from one learn-control phase to the next, we reduce the fill distance by moving the locations of the centers. However, the number of centers $N(t) = 9$ remains constant. Figure 6.14 shows the sets of centers Ξ_1, Ξ_2, and Ξ_3 employed for the first, second, and third pair of learn-control phase. Each learning phase lasts 1200 s, and each control phase lasts 300 s.

The components of the controlled plant trajectory are shown in Fig. 6.15, and the control input is depicted in Fig. 6.16. A careful inspection of Fig. 6.15 reveals that the steady-state error at the end of each control phase is reduced after each learning process.

Figure 6.17 shows irregular positive limit sets during the learning phases. A topic of future research is extending the analysis in Sect. 6.4.2.3 to positive limit sets defined for collections of initial conditions $E_0 \subset \mathbb{X}$ and for time-varying feedback functions, such as those employed in this example during the learning phase.

6.5.2 Case Study 2

In the second study case, we define three learning and control phases. However, in contrast to Case Study 1, we reduce the fill distance by increasing the dimension $N(t)$ of the subspaces $\mathcal{H}_{N(t)}$ used to construct approximations of the ideal value

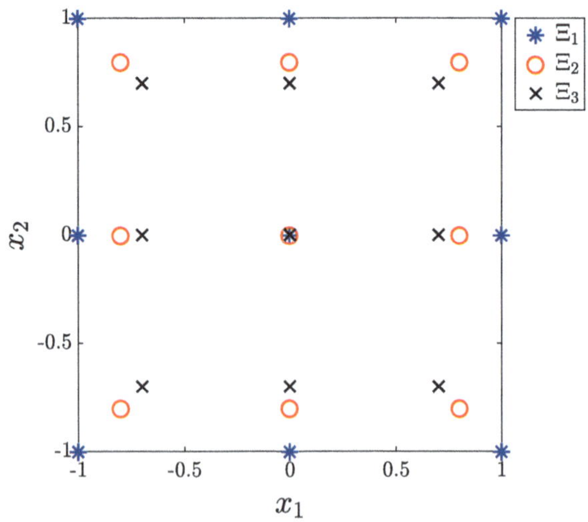

Fig. 6.14 SLC control architecture, Case Study 1. Location of centers for each of the three learning phases

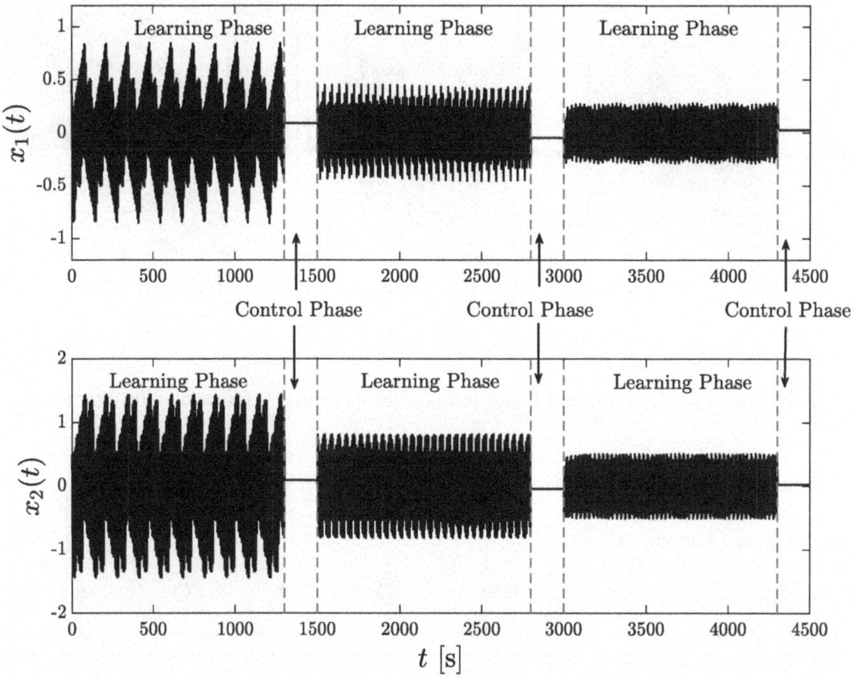

Fig. 6.15 SLC control architecture, Case Study 1. Plant states

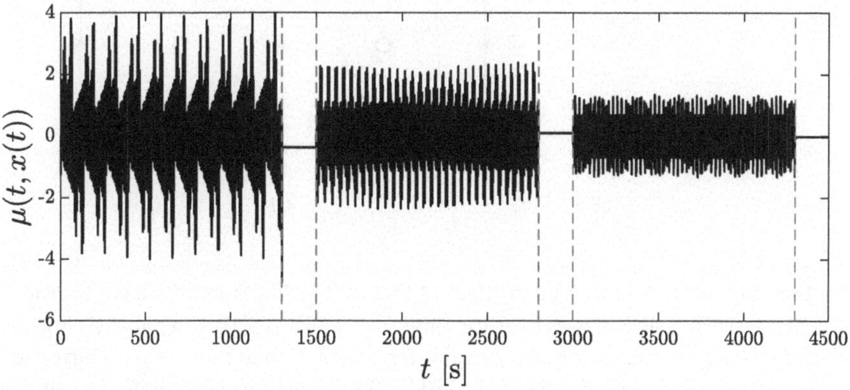

Fig. 6.16 SLC control architecture, Case Study 1. Control input

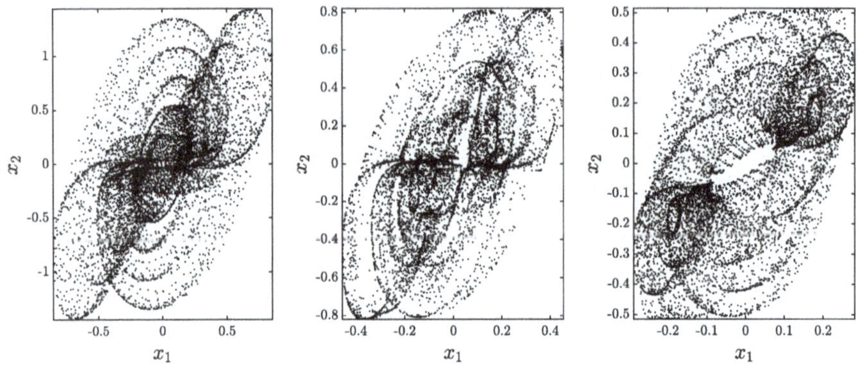

Fig. 6.17 SLC control architecture, Case Study 1. Positive limit sets for each of the three learning phases

Fig. 6.18 SLC architecture, Case Study 2. Locations of centers for each learning phase

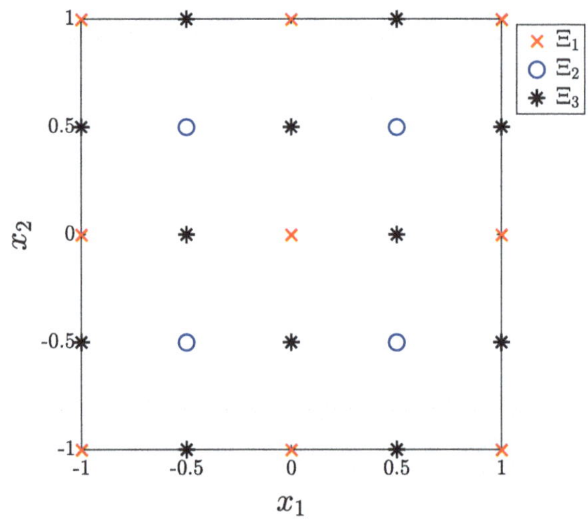

function and of the optimal controllers. As shown in Fig. 6.18, we have 3 groups of centers. During first phase, over the time interval [0, 1500] s, we use Ξ_1 for the centers, and choose the nominal controller for exploration to be $\mu = \mu_1$. During the second phase, over the time interval [1500, 3000] s, we expand the basis and use $N = 13$ centers in Ξ_2, and we pick the nominal controller $\mu = \mu_2$. Finally, during the third phase, over the time interval [3000, 4500] s, we set $N(t) = 25$ and use the centers in Ξ_3, and we pick the nominal controller $\mu = \mu_3$. As in Case Study 1, the learning stage lasts 1200 s, and the control stage that lasts 300 s. Figures 6.19 and 6.20 show the plant state and the control input for this case study, respectively.

This overall SLC control architecture also improves the steady-state performance at the end of each control phase, over the three control phases. As in Case Study 1, the positive limit sets shown in Fig. 6.21 are highly irregular.

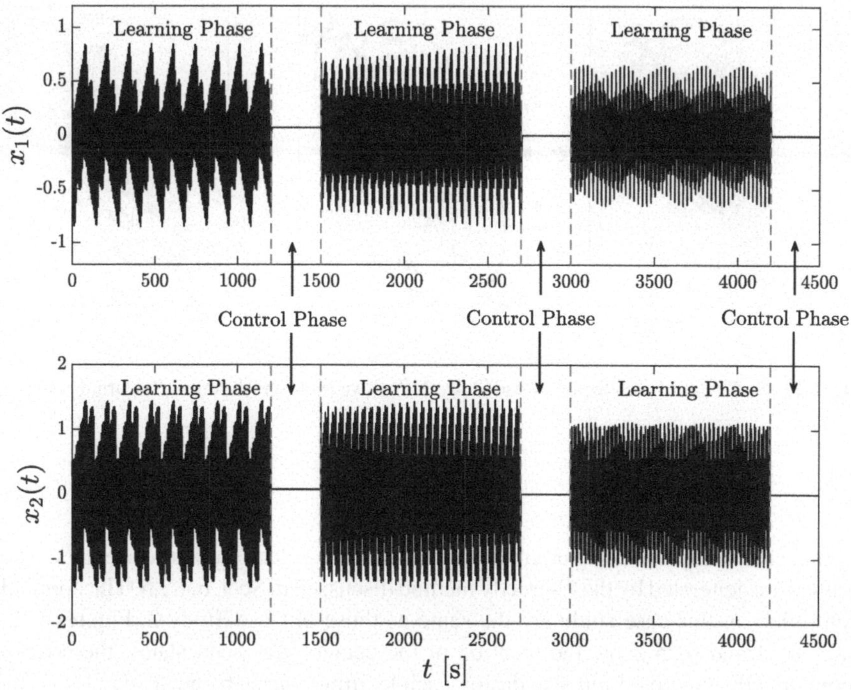

Fig. 6.19 SLC control architecture, Case Study 2. Plant states

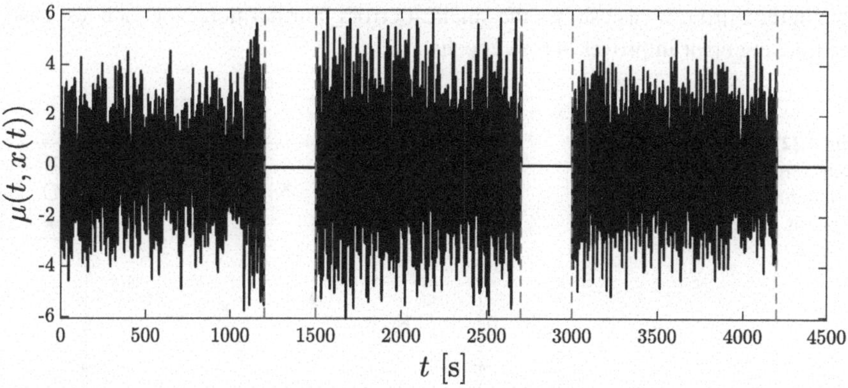

Fig. 6.20 SLC control architecture, Case Study 2. Control input

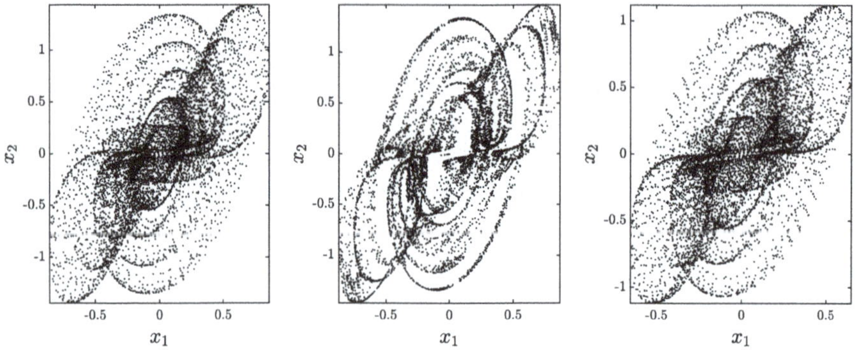

Fig. 6.21 SLC control architecture, Case Study 2. Positive limit sets during each learning phase

6.5.3 *Case Study 3*

In this case study, no user judgment is exercised to select the kernel centers. The centers are generated by the $\mathcal{P}$-greedy method discussed in Sect. 6.4.2.5. The nominal controllers in this case study are the same as those in Case Study 2. Figures 6.22, 6.23, 6.24 and 6.25 show the location of the centers, the plant states, the control input, and the positive limit sets during each learning phase, respectively. .

In conclusion, the performances of the proposed SLC method for each of the three test cases are compared in Table 6.4. It is apparent how, at the end of the each epoch, the third test case shows the smallest errors. Furthermore, for each test case, the tracking error improves across epochs.

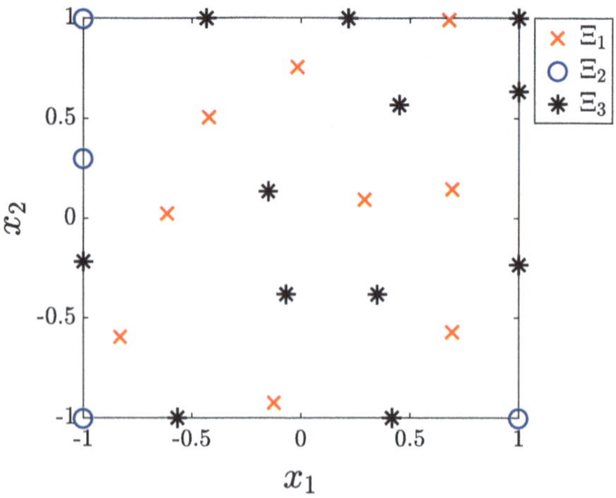

Fig. 6.22 SLC control architecture, Case Study 3. Center locations deduced by applying the $\mathcal{P}$-greedy method

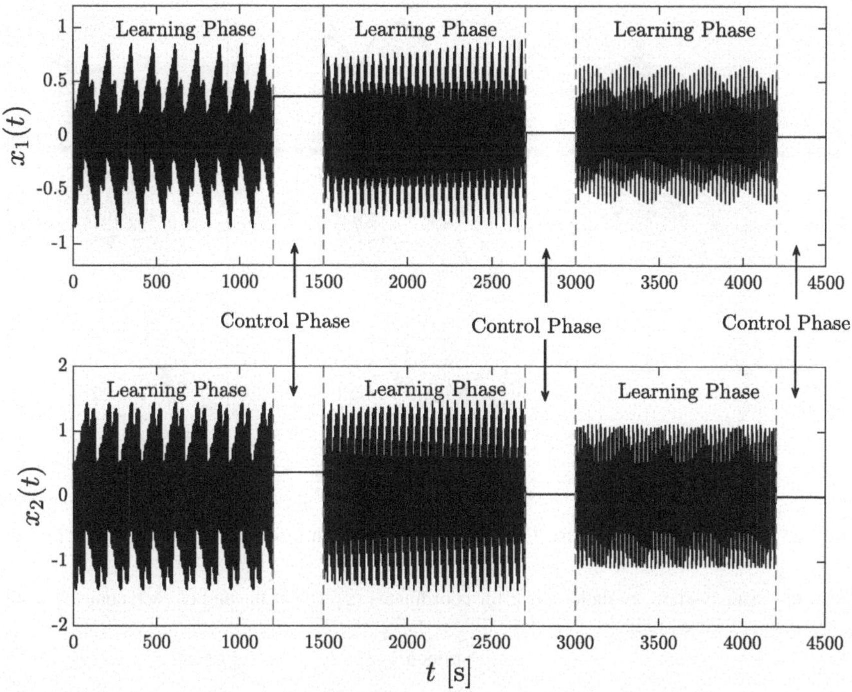

Fig. 6.23 SLC control architecture, Case Study 3. Plant states

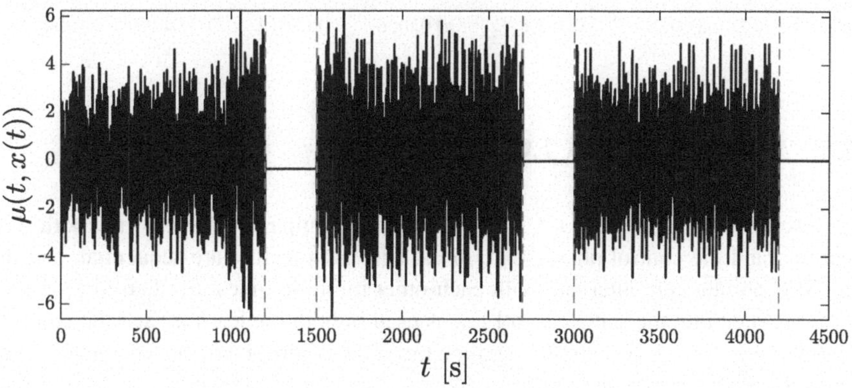

Fig. 6.24 SLC control architecture, Case Study 3. Control input

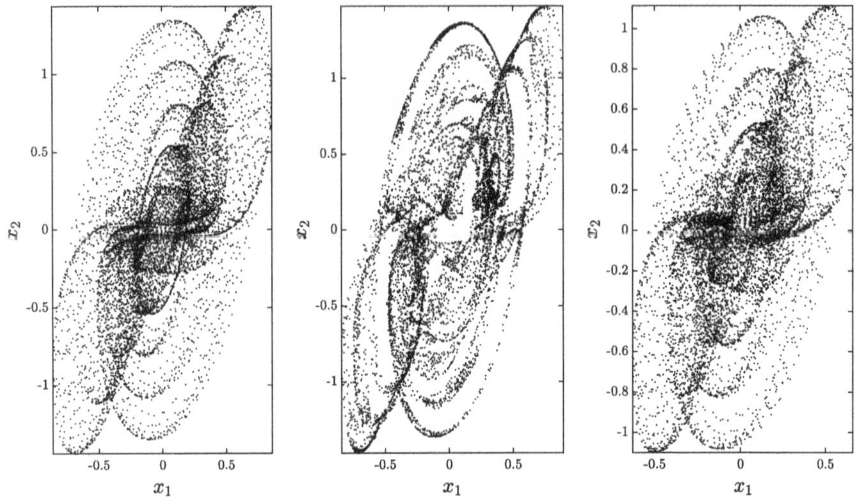

Fig. 6.25 SLC control architecture, Case Study 3. Positive limit sets during each learning phase

Table 6.4 Steady-state regulation error for coordinates (x_1, x_2) at the end of each control epoch. Comparison of Case Studies 1, 2, and 3

	Epoch 1		Epoch 2		Epoch 3	
	x_1	x_2	x_1	x_2	x_1	x_2
Case Study 1	0.08826	0.08843	−0.04742	−0.04740	0.02526	0.02530
Case Study 2	0.07141	0.07140	−0.0002571	−0.0002584	−0.00168	−0.001680
Case Study 3	0.3626	0.3623	0.03026	0.03014	−0.004380	−0.004380

6.6 A Deterministic CLC Architecture

As discussed in Sect. 6.2.3, a CLC architecture simultaneously performs both the learning and the control phases that, in the LTC architecture are sequential, and in the SLC architecture alternate with each other multiple times. Furthermore, in the CLC architecture, the same control law is used in both the learning and the control phases, whereas in LTC and SLC architectures, the control input does not need to be the same for these two phases.

In this section, we present a specific CLC architecture that applies to plant models in the same form as (6.4) with $\mathbb{U} = \mathbb{R}$ and, hence, uses hypothesis spaces that are scalar-valued RKHSs. However, in principle, the proposed arguments can be ported to the case of vector-valued RKHSs obtained as the Cartesian product of scalar-valued RKHSs. We leave these details as a subject for future research.

The proposed CLC architecture seeks to drive the plant trajectory to follow the *reference trajectory* $x_r : [t_0, \infty) \to \mathbb{X}$ such that

$$\dot{x}_r(t) = A_r x_r(t) + B_r r(t), \quad x_r(t_0) = x_{r,0}, \quad t \geq t_0, \tag{6.108}$$

where $r : [t_0, \infty) \to \mathbb{U}$ denotes the *reference command input*, which is user-defined, continuous, and such that $|r(t)| \leq r_{\max}$ for all $t \geq t_0$, $A_r \in \mathbb{R}^{n\times n}$ is Hurwitz, and $B_r \in \mathbb{R}^{n\times m}$. The matrices A_r, B_r verify the *matching conditions*

$$A_r = A + B\alpha^{\mathrm{T}}, \tag{6.109a}$$

$$B_r = B\beta^{\mathrm{T}} \tag{6.109b}$$

for ideal gains $\alpha \in \mathbb{R}^n$ and $\beta \in \mathbb{R}$; see (5.41) and (5.42).

Of all the data-driven schemes encountered thus far, this CLC method comes closest to the goal of eliminating the reliance on oracles to choose the subspaces used to construct approximations of matched functional uncertainties. This learning method can be interpreted as a combination of three techniques: an extension of the approach in [19, Ch. 11] to account for a varying dimension $N(t)$ of the approximating hypothesis space; the use of deterministic greedy methods summarized in Sect. 6.4.2.5; and an extension of the RKHS embedding strategy discussed in Chap. 5.

6.6.1 Overview

In the CLC architecture presented in this section, the dimension $N(t) \in \mathbb{Z}^+$ of the space $\mathcal{H}_{N(t)}$ approximating the hypothesis space $\mathcal{H}$ is defined as a function of time, which is constant over each interval $\mathcal{I}_m \triangleq [t_m, t_{m+1}) \subseteq [t_0, \infty)$, where $\{t_m\}_{m=0}^{\bar{m}}$ denote switching times defined by an asynchronous triggering mechanism based on the size of the tracking error, and $0 \leq \bar{m} \leq \infty$. During the interval $\mathcal{I}_m$, the dimension of the approximating hypothesis space is denoted by $N(m)$ to recall explicitly the current phase, and we define the associated approximating RKHS as

$$\mathcal{H}_{N(m)} \triangleq \operatorname{span}\left\{\mathfrak{K}_{\xi_i} : \xi_i \in \Xi_{N(m)}, 1 \leq i \leq N(m)\right\}, \quad \text{for all } m \in \{1, \ldots, \bar{m}\}, \tag{6.110}$$

where

$$\Xi_{N(m)} \triangleq \{\xi_i\}_{i=1}^{N(m)} \subset \Omega_m \tag{6.111}$$

denotes the family of centers selected recursively, and $\{\Omega_m\}_{m=0}^{\bar{m}} \subset \mathbb{X} \triangleq \mathbb{R}^n$ denotes a collection of user-defined compact sets.

The definition of the mth approximating hypothesis space in terms of the scattered basis provided by (6.110) is not used for actual implementations, but it remains useful to state theoretical results and complete proofs. For the sake of numerical computations and practical implementations, the mth approximating hypothesis space is computed as

$$\mathcal{H}_{N(m)} = \text{span}\left\{\boldsymbol{b}_1, \ldots, \boldsymbol{b}_{N(m)}\right\}, \tag{6.112}$$

where $\boldsymbol{b}_i \in \mathcal{H}$, $i \in \{1, \ldots, N(m)\}$, denotes the ith Newton basis function defined later in this section; for details, see on Newton bases [78, 79].

A critical feature of the proposed CLC is that, for each $m \in \{1, \ldots, \bar{m}\}$, the set of centers $\Xi_{N(m+1)}$ used over the time interval $\mathcal{I}_{m+1}$ to define the subspace $\mathcal{H}_{N(m+1)}$ is obtained by using a greedy update (see Sect. 6.4.2.5) of the basis used to define $\mathcal{H}_{N(m)}$, so that

$$\mathcal{H}_{N(m)} \subset \mathcal{H}_{N(m+1)} \subseteq \mathcal{H}. \tag{6.113}$$

This is achieved by recursively selecting the centers for a new basis by maximizing the power function. In practice, the event trigger occurs if the *tracking error* $e_N(t) \triangleq x_N(t) - x_r(t)$ of the closed-loop plant trajectory enters a neighborhood of the reference trajectory, and the diameter of this neighborhood is defined in terms of the approximation error that is associated with using $\mathcal{H}_{N(m)}$ during $\mathcal{I}_m$. In this expression, x_N stands for $x_{N(m)}$ and denotes the closed-loop plant state when applying a control input designed employing the approximating space $\mathcal{H}_{N(m)}$.

Any control law based on the assumption that the functional uncertainty lies in an RKHS includes an adaptive gain that corresponds to this nonlinear uncertainty. For instance, applying the control law (5.74) for the dead-zone modification and σ-modification of MRAC, (5.123) for the use of the projection operator in the MRAC framework, (5.144) for error bounding methods, and (5.150) for adaptive error bounding methods, this terms is given by $\hat{f}_{N(m)}(t, \cdot)$ for all $t \in [t_0, \infty)$. The adaptive laws underlying these adaptive gains may experience instantaneous variations across two consecutive time intervals. However, computing the adaptive gains $\hat{f}_{N(m)}(t, \cdot)$ as Carathéodory solutions of these piecewise continuous adaptive laws, the adaptive gains are absolutely continuous functions of time over $[t_0, \infty)$.

As shown in the following, the proposed CLC method guarantees that if $\bar{m} = \infty$ and $\lim_{m\to\infty} t_m = \infty$, then

$$\lim_{t\to\infty} \|x_N(t) - x_r(t)\|_{\mathbb{X}} = 0. \tag{6.114}$$

However, this is an ideal result. In practice, $\bar{m}$ is finite, and, in this case, the proposed CLC method guarantees an ultimate bound that scales like

$$\limsup_{t\to\infty} \|x_N(t) - x_r(t)\|_{\mathbb{X}} \lesssim \sup_{\xi\in\Omega} \mathcal{P}_{N(m)}(\xi), \tag{6.115}$$

where

$$\mathcal{P}_{N(m)}(x) \triangleq \sqrt{\mathfrak{K}(x,x) - \mathfrak{K}_{N(m)}(x,x)}, \quad \text{for all } x \in \mathbb{X} \tag{6.116}$$

denotes the power function of the subspace $\mathcal{H}_{N(m)} \subseteq \mathcal{H}$. Once again, it is worthwhile recalling that for some popular choices of kernels, the power function can be further bounded as a function of the fill distance exploiting, for instance, (3.80) and Table 3.3.

6.6.2 An Event-Driven Controller and Adaptive Law

During the time mth interval $\mathcal{I}_m$, the control input is defined in terms of two alternative control laws. A simpler control law is given by

$$u_{N(m)}(t) \triangleq \hat{\alpha}^{\mathrm{T}}(t)x_N(t) + \beta r(t) - \hat{f}_{N(m)}(t, x_N(t)), \quad \text{for all } t \in \mathcal{I}_m \text{ and } m \in \{0, \dots, \bar{m}\}. \tag{6.117}$$

As discussed in the following, the hypotheses to use this control law are rather restrictive in practice. Thus, as an alternative, we define a modification of the control law in [19, Ch. 11] given by

$$\begin{aligned} u_{N(m)}(t) \triangleq & \hat{\alpha}^{\mathrm{T}}(t)x_N(t) + (1-\mu(x_N(t)))\beta r(t) \\ & - \hat{f}_{N(m)}(t, x_N(t)) + \mu(x_N(t))u_{SL}(t) + v_{N(m)}(x_N(t), e_N(t)). \end{aligned} \tag{6.118}$$

This control law is applicable under more reasonable hypotheses, which are better suited for practical implementations.

The adaptive laws to update the gains $\hat{\alpha}(t)$ and $\hat{f}_{N(m)}(t,\cdot)$ are familiar from Chap. 5. In (6.117), the *adaptive gain matrix* $\hat{\alpha} : [t_0, \infty) \to \mathbb{R}^n$ is such that

$$\dot{\hat{\alpha}}(t) = -\Gamma_\alpha x_N(t)e_N^{\mathrm{T}}(t)PB, \quad t \in \mathcal{I}_m \quad m \in \{0, \dots, \bar{m}\}, \tag{6.119}$$

and, in (6.118), $\hat{\alpha} : [t_0, \infty) \to \mathbb{R}^n$ is such that

$$\dot{\hat{\alpha}}(t) = \begin{cases} -\Gamma_\alpha x_N(t)e_N^{\mathrm{T}}(t)PB & \text{if } \|P^{1/2}e_N(t)\| \le \epsilon, \\ 0 & \text{otherwise,} \end{cases} \quad t \in \mathcal{I}_m, \quad m \in \{0, \dots, \bar{m}\}, \tag{6.120}$$

where the *adaptive rate matrix* $\Gamma_\alpha \in \mathbb{R}^{n\times n}$ is user-defined, symmetric, and positive-definite, $e_N(t) \triangleq x_N(t) - x_r(t)$ denotes the *trajectory tracking error*, $\epsilon > \frac{2\|PB\|}{\lambda_{\min}(Q)}\epsilon_{N(m)}$ is user-defined,

$$\epsilon_{N(m)} \triangleq \sup_{x\in\Omega_m} \bar{e}_{N(m)}(x), \tag{6.121}$$

$$\bar{e}_{N(m)}(x) \triangleq \mathcal{P}_{N(m)}(x)\|f\|_{\mathcal{H}}, \quad x \in \Omega_m, \tag{6.122}$$

and $P \in \mathbb{R}^{n\times n}$ denotes the symmetric, positive-definite solution of the *algebraic Lyapunov equation*

$$A_r^{\mathrm{T}} P + P A_r = -Q \tag{6.123}$$

for $Q \in \mathbb{R}^{n\times n}$ symmetric, positive-definite, and user-defined. The initial condition $\hat{\alpha}(t_m)$ is defined to enforce continuity of $\hat{\alpha}(\cdot)$ across the consecutive intervals $\mathcal{I}_{m-1}$ and $\mathcal{I}_m$ for any $m \in \mathbb{N}$.

For the control laws in (6.117) and (6.118), the functional gains $\hat{f}_{N(m)}(\cdot,\cdot)$ verifies the *adaptive law*

$$\frac{\partial \hat{f}_{N(m)}(t,\cdot)}{\partial t} = \begin{cases} -\gamma_f \Pi_{N(m)} \mathfrak{K}(x_N(t),\cdot) e_N(t) P B & \text{if } \|P^{1/2} e_N(t)\| \le \epsilon, \\ 0 & \text{otherwise,} \end{cases} \tag{6.124}$$

for all $t \in \mathcal{I}_m$ and $m \in \{0,\ldots,\bar{m}\}$, where $\gamma_f > 0$ denotes the *adaptive rate*, $\Pi_{N(m)} : \mathcal{H} \to \mathcal{H}_{N(m)}$ denotes the *$\mathcal{H}$-orthogonal projection onto $\mathcal{H}_{N(m)}$*, $P^{1/2}$ is symmetric, positive-definite, and such that $P^{1/2}P^{1/2} = P$, and the initial condition $\hat{f}_{N(m)}(t_m,\cdot)$ is chosen so that $\hat{f}_{N(m-1)}(t_m,\cdot) = \hat{f}_{N(m)}(t_m,\cdot)$ to enforce continuity of the adaptive gain across intervals. To define (6.118), let

$$\mu(x) \triangleq \max\left\{0, \min\left\{1, \frac{\|x\| - \delta b}{(1-\delta)b}\right\}\right\}, \quad \text{for all } x \in \mathbb{X} \tag{6.125}$$

define the *blending function* (see [19, Ch. 11]), where $\delta \in (0,1)$ is user-defined,

$$b \triangleq \bar{e}_1 + \max_{t\ge t_0} \|x_r(t)\|_{\mathbb{X}}, \tag{6.126}$$

$$\bar{e}_1 \triangleq 2\sqrt{\frac{\lambda_{\max}(P)}{\lambda_{\min}(P)}\frac{\lambda_{\max}(P)}{\lambda_{\min}(Q)}}\left(\Gamma_{\max} + 2\sup_{\xi\in S_\epsilon}\|PB\|\mathcal{P}_{N(m)}(\xi)\|f\|_{\mathcal{H}}\right), \tag{6.127}$$

$\lambda_{\max}(\cdot)$ and $\lambda_{\min}(\cdot)$ denote the largest and the smallest eigenvalues of their arguments, respectively, and $S_\epsilon \triangleq \bigcup_{t\in[t_0,\infty)} \left\{x \in \mathbb{X} : \|B^{\mathrm{T}} P e_N(t)\| < \epsilon\right\}$. Furthermore, in (6.118), let

$$u_{SL}(t) \triangleq -K_r r_{\max} \operatorname{sign}\left(B^{\mathrm{T}} P e_N(t)\right), \quad \text{for all } t \in \mathcal{I}_m, \quad m \in \{0,\ldots,\bar{m}\} \tag{6.128}$$

denote the *state-limiting term*. We remark that $\mu(x) = 0$ for all $x \in \{x \in \mathbb{X} : \|x\| \le \delta b\}$, $\mu(x) = 1$ for all $x \in \{x \in \mathbb{X} : \|x\| \ge b\}$, and $\mu(\cdot)$ linearly increases from 0 to 1 with $\|x\|$ for all $x \in \{x \in \mathbb{X} : \|x\| \in [\delta b, b]\}$. As discussed in Theorem 6.2 below, the control law given by (6.117) assumes positive invariance of the triggering sets Ω_m. The control law (6.118) includes the smooth bounded control term

$$v_{N(m)}(\cdot, e_N(t)) \triangleq \begin{cases} \operatorname{sign}\left(B^{\mathrm{T}} P e_N(t)\right) \bar{e}_{N(m)}(\cdot) & \text{if } |B^{\mathrm{T}} P e_N(t)| \geq \epsilon, \\ \epsilon^{-1} B^{\mathrm{T}} P e_N(t) \bar{e}_{N(m)}(\cdot) & \text{otherwise} \end{cases} \tag{6.129}$$

for all $t \in \mathcal{I}_m$ and $m \in \{0, \ldots, \bar{m}\}$. Theorem 6.3 for the control law (6.118) does not require positive invariance of the triggering sets Ω_m.

6.6.3 A Data-Driven Basis Refinement Algorithm

In the proposed CLC architecture, the procedure to select $\mathcal{I}_m \triangleq [t_m, t_{m+1})$ is the following. Suppose that the last trigger event occurred at $t_m \geq t_0$ for some $m \in \{0, \ldots, \bar{m}\}$; if no trigger has yet occurred, then $m = 0$. The next trigger event depends on four user-defined parameters, namely the *holding time* $T_m > 0$, the *termination offset* $\Delta_m \in (T_m, \infty]$, the compact set $\Omega_m \subset \mathbb{X}$, and the *threshold*

$$c_m \in \left(0, (1-\sigma)\frac{\lambda_{\min}(Q)}{2}\|PB\|\|f\|_{\mathcal{H}}\right], \tag{6.130}$$

where $\sigma \in (0, 1)$. If, for some $T_{b,m} \in [0, \Delta_m)$ and all $t \in \left[t_m + T_{b,m}, t_m + T_{b,m} + T_m\right]$ it holds that

$$\sup_{\xi \in \Omega_m} \mathcal{P}_{N(m)}(\xi) \geq c_m \|e_N(t)\|, \tag{6.131}$$

then we say that the trajectory $t \mapsto x_N(t)$ of the closed-loop system satisfies the *next event trigger at* $T_{b,m}$ *over* Ω_m for the threshold c_m.

Whereas (6.131) can be easily verified a posteriori, verifying this condition in real time may be challenging. Indeed, because of the oscillatory nature of the controlled trajectory in the transient stage, the condition (6.131) might be briefly satisfied over an arbitrarily small interval. These problems can be avoided not by reducing the approximation error but by employing a holding time T_m sufficiently large in the definition of $T_{b,m}$. Larger values of T_m allow excluding the activation of (6.131) due to oscillatory behaviors of the transient dynamics but may delay the proposed triggering mechanism. Conversely, smaller values of T_m may lead to more aggressive responses, but may trigger unnecessary greedy updates. Finally, if the condition (6.131) is not satisfied for a sufficiently long time, as indicated in Fig. 6.26 by the initial time circled on the figure, then $T_{b,m}$ will be reset to the subsequent time when $c_m\|x_N(t)\|$ enters the bounded region and satisfies (6.131) for all $t > T_{b,m}$.

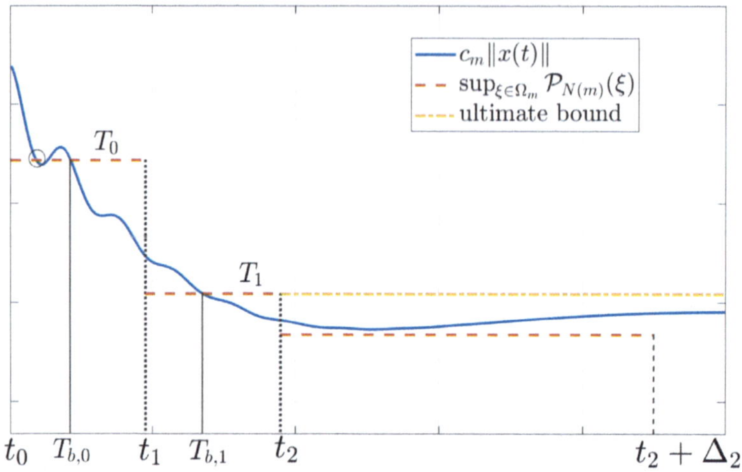

Fig. 6.26 An illustration of the timing of the trigger events. The trigger t_{m+1} occurs only after the tracking error enters the bounded region given by the red dashed line for a holding time T_m. Here the time to enter each bounded region $T_{b,0} = t_1 - T_0$ and $T_{b,1} = t_2 - T_1$ are given by the solid vertical black lines. Note that while Eq. 6.131 is briefly satisfied at the ellipse to the far left, the condition does not hold for sufficient time T_0, and thus the time $T_{b,0}$ is reset to the next occurrence where Eq. 6.131 is satisfied. After a trigger t_m occurs, represented by the vertical dashed lines, the threshold bound for the next trigger event is decreased. After some time Δ_m it may be the convergence cannot be achieved in a reasonable amount of time and thus the triggering events will no longer occur. Here we see two trigger events. For this illustration, adaptation turns off after time $t_2 + \Delta_2$ because the error is no longer decreasing sufficiently fast. In this case, the error bound follows the standard MRAC ultimate error bound given by the orange dashed line

The termination offset Δ_m specifies a time horizon to check for condition (6.131) to be verified. If $\Delta_m < \infty$, then, after the event t_m, the possible occurrence of the next trigger event is monitored over the interval $[t_m, t_m + \Delta_m]$. Alternatively, if $\Delta_m = \infty$, then, after the event t_m, the possible occurrence of the next trigger event is monitored over the interval $[t_m, \infty)$. For a visual representation of the timing of trigger events, refer to Fig. 6.26.

We set

$$t_{m+1} = t_m + T_{b,m} + T_m \tag{6.132}$$

and augment the subset of centers $\Xi_{N(m)}$ to form a new set of centers $\Xi_{N(m+1)}$. This set of centers is created using a greedy algorithm, whereby we recursively add L centers so that

$$\xi_{m+\ell} \in \underset{\xi\in\Omega_{m+1}\setminus\Xi_{N(m)+\ell-1}}{\arg\max} \mathcal{P}_{N(m)+\ell-1}(\xi), \quad \text{for each } \ell = 1, 2, \ldots, L. \tag{6.133}$$

See 6.4.2.5, and specifically the $\mathcal{P}$-greedy method, to build intuition about this recursive process. This process terminates when the trigger condition no longer holds, that is, when

$$\sup_{\xi \in \Omega_{m+1}} \mathcal{P}_{N(m+1)} \leq c_{m+1} \| e_N(t) \|, \tag{6.134}$$

where $N(m+1) \triangleq N(m) + L$.

6.6.4 Stability and Convergence Analysis

This section studies the stability and convergence of the greedy basis selection scheme summarized in Sect. 6.6.3. First, we address the effectiveness of the simpler control law (6.117), and then of the more realistic control law (6.118). For the study of the control law (6.117) with the basis refinement scheme discussed in Sect. 6.6.3, we assume that each user-defined compact set Ω_m, $m \in \{0, \ldots, \bar{m}\}$, is positively invariant with respect to the controlled plant dynamics $x(\cdot)$. The analysis of the subsequent control law in 6.118 dispenses with this assumption. The structure of this development, first assuming positive invariance and then eliminating this assumption, is a typical way to proceed in developing adaptive control laws; for additional details, see the discussions in the seminal book [18].

Theorem 6.2 *Consider the plant model (6.4) with $\mathbb{U} \triangleq \mathbb{R}$ and $\Gamma(t) \equiv 0$ for all $t \in [t_0, \infty)$, control law (6.117), and $f \in C_R$. Consider also the reference model (6.108) and the adaptive laws (6.119) and (6.124). Let the trigger condition in (6.131) define trigger times $\{t_m\}_{m=1}^{\bar{m}}$ and associated intervals $\{\mathcal{I}_m\}_{m=1}^{\bar{m}}$. Then, $e_N(\cdot)$, $\hat{\alpha}(\cdot)$, and $\hat{f}_{N(m)}(\cdot,\cdot)$ are uniformly bounded for all $m \in \{1, \ldots, \overline{m}\}$ and for all $t \in [t_0, \infty)$. Furthermore, if there is an infinite number of trigger times and $\lim_{m\to\infty} \left(T_{b,m} + T_m\right) = \infty$, then*

$$\lim_{t\to\infty} \|x_N(t) - x_r(t)\|_{\mathbb{X}} = 0. \tag{6.135}$$

Alternatively, if there is a finite number of trigger times $\overline{m} \in \mathbb{N}$, $T_{b,\overline{m}-1} + T_{\overline{m}-1} < \infty$, and $\Omega_{\overline{m}} \supseteq \bigcup_{t \geq t_{\overline{m}}} x_N(t)$, where $t_{\overline{m}} \geq t_0$ denotes the last trigger time, then the trajectory tracking error $e_N(\cdot)$ is uniformly ultimately bounded, that is, for any $\eta > 0$, there exists $T(\eta) > 0$ such

$$\|x_N(t) - x_r(t)\|_{\mathbb{X}} \leq (1+\eta) \frac{2\|B^{\mathrm{T}} P\|}{\lambda_{\min}^{1/2}(P)\lambda_{\min}(Q)} \sup_{\xi\in\Omega_m} \mathcal{P}_{N(m)}(\xi) \underbrace{\|(I - \Pi_{N(m)}) f\|_{\mathcal{H}}}_{\leq \|f\|_{\mathcal{H}}} \tag{6.136}$$

for all $t > T(\eta)$.

Proof This proof is divided into two parts. In the first part, we assume that there is an infinite number of trigger events on the interval $[t_0, \infty)$, and in the second part, we assume that there is a finite number of trigger events on the interval $[t_0, \infty)$. In both parts of this proof, we will consider the common Lyapunov function candidate

$$V\left(e_N, \tilde{\alpha}, \tilde{f}_{N(m)}\right) \triangleq e_N^{\mathrm{T}} P e_N + \operatorname{tr}\left(\tilde{\alpha}^{\mathrm{T}} \Gamma_\alpha \tilde{\alpha}_x\right) + \gamma_f^{-1} \langle \tilde{f}_{N(m)}, \tilde{f}_{N(m)} \rangle_{\mathcal{H}}, \tag{6.137}$$

for all $\left(e_N, \tilde{\alpha}, \tilde{f}_{N(m)}\right) \in \mathbb{R}^n \times \mathbb{R}^n \times \mathcal{H}$, $m \in \{0, \ldots, \bar{m}\}$, where $\tilde{\alpha}(t) \triangleq \hat{\alpha}(t) - \alpha$, α is unknown and verifies (6.109a), and $\tilde{f}_{N(m)}(t, \cdot) \triangleq \hat{f}_{N(m)}(t, \cdot) - f$.

Consider the case whereby there is an infinite number of trigger times and $\lim_{m \to \infty} \left(T_{b,m} + T_m\right) = \infty$. The time derivative of $V(\cdot, \cdot)$ along the trajectories of the plant model (6.4) with control law (6.117) and the adaptive laws (6.119) and (6.124) is given by

$$\begin{aligned}
&\dot{V}(e_N(t), \tilde{\alpha}(t), \tilde{f}_{N(m)}(t)) \\
&\quad = -e_N^{\mathrm{T}}(t) Q e_N(t) + 2\big\langle B^{\mathrm{T}} P e_N(t), E_{x_N(t)}(I - \Pi_{N(m)}) f \big\rangle_{\mathbb{R}} \\
&\qquad + 2\operatorname{tr}\left(\tilde{\alpha}^{\mathrm{T}}(t)\left[\Gamma_\alpha^{-1} \dot{\hat{\alpha}}(t) + x_N(t) e_N(t) P B\right]\right) \\
&\quad \leq -\lambda_{\min}(Q)\|e_N(t)\|^2 + 2\|PB\| \left|E_{x_N(t)}(I - \Pi_{N(m)}) f\right| \|e_N(t)\| \\
&\quad \leq -\lambda_{\min}(Q)\|e_N(t)\|^2 + 2\|PB\| \sup_{\xi \in \Omega_m} \mathcal{P}_{N(m)}(\xi) \|f\| \|e_N(t)\| \\
&\quad \leq -\sigma \lambda_{\min}(Q)\|e_N(t)\|^2, \quad \text{for all } t \in \mathcal{I}_m, \quad m \in \{0, \ldots, \bar{m}\},
\end{aligned} \tag{6.138}$$

since $\|f\|_{\mathcal{H}} \leq R$. Thus, convergence of $e_N(t)$ to zero as $t \to \infty$ follows by applying Barbalat's lemma.

Consider the case whereby there is a finite number of trigger times, which we denote by $\overline{m}$, and $T_{b,\overline{m}-1} + T_{\overline{m}-1}$ is finite, that is, consider the case whereby the trigger condition in (6.131) does not occur again after $t_{\overline{m}} < \infty$. Then, the time derivative of V along the trajectories of (6.4) with control law (6.117), (6.108), (6.119), and (6.124) is given by

$$\begin{aligned}
&\dot{V}\left(e_N(t), \tilde{\alpha}(t), \tilde{f}_{N(m)}(t)\right) \\
&\quad \leq -\lambda_{\min}(Q)\|e_N(t)\|^2 + 2\|PB\| \|f\|_{\mathcal{H}} \epsilon_{N(m)} \|e_N(t)\| \\
&\quad \leq -\frac{\lambda_{\min}(Q)}{\lambda_{\min}(P)} \|P^{1/2} e_N(t)\|^2 + \frac{2\|PB\|}{\lambda_{\min}(P)} \|f\|_{\mathcal{H}} \epsilon_{N(m)} \|P^{1/2} e_N(t)\| \\
&\quad \leq -\frac{\lambda_{\min}(Q)}{\lambda_{\min}(P)} \|P^{1/2} e_N(t)\| \left(\|P^{1/2} e_N(t)\| - \frac{2\|PB\|}{\lambda_{\min}(Q)} \epsilon_{N(m)}\right),
\end{aligned} \tag{6.139}$$

where $\epsilon_{N(m)}$ is given by (6.121).

To conclude the proof, we prove that the trajectory tracking error $e_N(\cdot)$ spends a finite-time outside the dead-zone $\mathcal{D} \triangleq \left\{x \in \mathbb{X} : \|P^{1/2} e_N\| \leq \epsilon\right\}$. Indeed, by proceeding in (6.138), we deduce that, for all $a_j \geq t_{\overline{m}}$, $j \in \mathbb{Z}^+$, such that $x_N(t) \notin \mathcal{D}$, $t \geq a_j$, there must exist $W \geq 0$ such that

$$\dot{V}\left(e_N(t), \tilde{\alpha}(t), \tilde{f}_{N(m)}(t)\right) \leq W, \tag{6.140}$$

which implies that there must exist some finite time $b_j > a_j$ such that $x_N(t) \in \mathcal{D}$, $t \geq b_j$. Now, for all $\left\{(a_j, b_j)\right\}_{j\in\mathbb{N}}$, it holds that

$$V\left(e_N(b_i), \tilde{\alpha}(b_i), \tilde{f}_{N(m)}(b_i)\right) - V\left(e_N(a_i), \tilde{\alpha}(a_i), \tilde{f}_{N(m)}(a_i)\right) \leq -W(b_i - a_i). \tag{6.141}$$

This means that the total time spent outside the dead zone is bounded by

$$\sum_{i\in\mathbb{N}}(b_i - a_i) \leq \frac{1}{W}\sum_{i\in\mathbb{N}}\Big[V\left(e_N(a_i), \tilde{\alpha}(a_i), \tilde{f}_{N(m)}(a_i)\right) - V\left(e_N(b_i), \tilde{\alpha}(b_i), \tilde{f}_{N(m)}(b_i)\right)\Big], \tag{6.142}$$

and, since $e_N(b_i) = e_N(a_{i+1})$, $\tilde{\alpha}(b_i) = \tilde{\alpha}(a_{i+1})$, and $\tilde{f}(b_i, \cdot) = \tilde{f}(a_{i+1}, \cdot)$ for any $i \in \mathbb{N}$, the summation on the right-hand side of (6.142) is bounded by $V\left(e_N(a_0), \tilde{\alpha}(a_0), \tilde{f}_{N(m)}(a_0)\right)$. Consequently, the trajectory tracking error $e_N(\cdot)$ spends a finite-time outside the dead-zone $\mathcal{D}$. ■

Theorem 6.2 provides sufficient conditions that prove the efficacy of the control law (6.117), the adaptive laws (6.119) and (6.124), and the data-driven basis refinement algorithm discussed in Sect. 6.6.3. The case whereby there is an infinite number of triggers and $m \to \infty$ is ideal and serves as a limiting case. The case whereby the number of trigger m is finite is observed in practice. This result resembles the case studied in [18, 19], wherein the regressor vector, that is, the basis functions that characterize the functional uncertainty f are fixed. In Theorem 6.2, instead, such a basis is progressively refined. Finally, Theorem 6.2 relies on the assumption whereby the user-defined compact sets $\{\Omega_m\}_{m=1}^{\bar{m}}$ contain the closed-loop plant trajectory $x_N(\cdot)$. However, in general, this assumption, which is already employed in [18], can only be verified a posteriori. The next result allows omitting this limiting assumption.

Theorem 6.3 *Consider the plant model (6.4) with* $\mathbb{U} = \mathbb{R}$*, control law (6.118), and* $f \in C_R$*, the reference model (6.108), and the adaptive laws (6.120) and (6.124). Let* $\Delta_m = \infty$*, and let the trigger condition in (6.131) define trigger times* $\{t_m\}_{m=0}^{\overline{m}}$ *and associated intervals* $\{\mathcal{I}_m\}_{m=0}^{\overline{m}}$*. Then,* $e_N(\cdot)$*,* $\hat{\alpha}(\cdot)$*, and* $\hat{f}_{N(m)}(\cdot,\cdot)$ *are uniformly bounded over the time interval* $[t_0,\infty)$ *for all* $m \in \{1,\ldots,\overline{m}\}$*. Furthermore, if there is a finite number of trigger times* $\overline{m} \in \mathbb{Z}^+$ *and if* $\Omega_{\overline{m}} \supseteq \bigcup_{t\geq t_{\overline{m}}} x_N(t)$*, where* $t_{\overline{m}} \geq t_0$ *denotes the last trigger time, then the trajectory tracking error dynamics* $e_N(\cdot)$ *is uniformly ultimately bounded, that is, there is a time* $T > t_0$ *such that*

$$\|e_N(t)\| \leq \bar{e}_1, \quad \text{for all } t > T, \tag{6.143}$$

where $\overline{e}_1$ *is given by (6.127).*

Proof With the control law (6.118), the controlled plant dynamics are captured by

$$\begin{aligned}\dot{x}_N(t) = Ax_N(t) + B\Big[&\hat{\alpha}^{\mathrm{T}}(t)x_N(t) + (1-\mu(x_N(t)))\beta r(t)\\ &+ E_{x_N(t)}\Big(f - \hat{f}_{N(m)}(t, x_N(t))\Big)\\ &+ \mu(x_N(t))u_{SL}(x_N(t)) + v_{N(m)}x_N(t), e_N(t))\Big]\\ &+\Gamma(t),\quad x_N(t_0) = x_0,\quad t\in\mathcal{I}_m,\quad m\in\{0,\ldots,\bar{m}\}.\end{aligned}\tag{6.144}$$

Thus, it follows from the matching conditions (6.109) that the trajectory tracking error dynamics are given by

$$\begin{aligned}\dot{e}_N(t) = A_r e_N(t) + B\Big[&\tilde{\alpha}^{\mathrm{T}}(t)x_N(t) + E_{x_N(t)}\Big(f - \hat{f}_{N(m)}(t,\cdot)\Big)\\ &+ \mu(x_N(t))\left(-\beta r_{\max}\operatorname{sign}(B^{\mathrm{T}}Pe_N(t)) - \beta r(t) + v_{N(m)}(x_N(t), e_N(t))\right)\Big]\\ &+\Gamma(t),\quad e_N(t_0) = x_0 - x_{r,0},\quad t\in\mathcal{I}_m,\quad m\in\{0,\ldots,\bar{m}\}.\end{aligned}\tag{6.145}$$

Next, consider the candidate Lyapunov function candidate (6.137). The time derivative of (6.137) along the trajectories of (6.144), (6.119), and (6.124) is given by

$$\begin{aligned}\dot{V}(t) &= -e_N^{\mathrm{T}}(t)Qe_N(t) + 2B^{\mathrm{T}}Pe_N(t)\Big[E_{x_N(t)}\left(f - \hat{f}_{N(m)}(t,\cdot)\right)\\
&\qquad + \mu(x_N(t))\left(-\beta r_{\max}\mathrm{sign}\left(B^{\mathrm{T}}Pe_N\right) - \beta r(t)\right) + \upsilon_{N(m)}(x_N(t), e_N(t))\Big]\\
&\quad + 2\Gamma^{\mathrm{T}}(t)Pe_N(t) + \left\langle \tilde{f}_{N(m)}(t,\cdot), 2\gamma_f^{-1}\frac{\partial \tilde{f}_{N(m)}(t,\cdot)}{\partial t}\right\rangle_{\mathcal{H}}\\
&\quad + 2\mathrm{tr}\left(\tilde{\alpha}^{\mathrm{T}}(t)\left[\Gamma_\alpha^{-1}\dot{\hat{\alpha}}(t) + x_N(t)e_N^{\mathrm{T}}(t)PB\right]\right)\\
&= -e_N^{\mathrm{T}}(t)Qe_N(t) + 2B^{\mathrm{T}}Pe_N(t)\Big[E_{x_N(t)}\left(I - \Pi_{N(m)}\right)f\\
&\qquad + \mu(x_N(t))\left(-\beta r_{\max}\mathrm{sign}\left(B^{\mathrm{T}}Pe_N\right) - \beta r(t)\right) + \upsilon_{N(m)}(x_N(t), e_N(t))\Big]\\
&\quad + 2\Gamma^{\mathrm{T}}(t)Pe_N(t) + \left\langle \tilde{f}_{N(m)}(t,\cdot), 2\Pi_{N(m)}E^*_{x_N(t)}B^{\mathrm{T}}Pe_N(t)\right\rangle_{\mathcal{H}}\\
&\quad + \left\langle \tilde{f}_{N(m)}(t,\cdot), 2\gamma_f^{-1}\frac{\partial \tilde{f}_{N(m)}(t,\cdot)}{\partial t}\right\rangle_{\mathcal{H}}, \quad t \in \mathcal{I}_m, \quad m \in \{0, \ldots, \bar{m}\}.\end{aligned} \tag{6.146}$$

Here, we write $V(t) = V\left(e_N(t), \tilde{\alpha}(t), \tilde{f}_{N(m)}(t,\cdot)\right)$ for shortness and E_x^* denotes the *adjoint* of the evaluation operator E_x. Thus, recalling that for symmetric kernels, such as admissible kernels, it holds that

$$(\mathfrak{K}(x_1, x_2))^* = (E_{x_1}E^*_{x_2})^* = (E_{x_2}, E^*_{x_1}) = \mathfrak{K}(x_2, x_1), \quad \text{for all } x_1, x_2 \in \mathbb{X}, \tag{6.147}$$

it follows from (6.124) that

$$\begin{aligned}\dot{V}(t) &= -e_N^{\mathrm{T}}(t)Qe_N(t) + 2B^{\mathrm{T}}Pe_N(t)\Big[E_{x_N(t)}\left(I - \Pi_{N(m)}\right)f\\
&\qquad + \mu(x_N(t))\left(-\beta r_{\max}\mathrm{sign}\left(B^{\mathrm{T}}Pe_N(t)\right) - \beta r(t)\right) + \upsilon_{N(m)}(x_N(t), e_N(t))\Big]\\
&\quad + 2\Gamma^{\mathrm{T}}(t)Pe_N(t), \qquad t \in \mathcal{I}_m, \quad m \in \{0, \ldots, \bar{m}\}.\end{aligned} \tag{6.148}$$

Next, we analyze (6.148) in four alternative cases. Firstly, we consider the case wherein $\|x_N(t)\| > b$ for some $t \geq t_0$, where b is given by (6.126). Secondly, we consider the case wherein $\|x_N(t)\| < \delta b$ for some $t \geq t_0$. Successively, we consider the case wherein $\|x_N(t)\| \in [\delta b, b]$ and $\left|B^{\mathrm{T}}Pe_N(t)\right| > \epsilon$ for some $t \geq t_0$. Finally, we consider the case wherein $\|x_N(t)\| \in [\delta b, b]$ and $\left|B^{\mathrm{T}}Pe_N(t)\right| \leq \epsilon$ for some $t \geq t_0$.

Case 1: If $\|x_N(t)\| \geq b$ for some $t \geq t_0$, where b is given by (6.126), then $\mu(x_N(t)) = 1$, $\left|B^{\mathrm{T}}Pe_N(t)\right| > \epsilon$, and it follows from (6.129) and (6.122) that $\upsilon_{N(m)}(x_N(t), e_N(t)) = -\mathrm{sign}\left(B^{\mathrm{T}}Pe_N(t)\right)\mathcal{P}_{N(m)}(x_N(t))\|f\|_{\mathcal{H}}$. Thus, it follows from (6.148) that

$$
\begin{aligned}
\dot{V}(t) &= -e_N^{\mathrm{T}}(t)Qe_N(t) + 2\Gamma^{\mathrm{T}}(t)Pe_N(t) + 2B^{\mathrm{T}}Pe_N(t)\Big[E_{x_N(t)}(I - \Pi_{N(m)})f \\
&\qquad - \beta r_{\max}\operatorname{sign}\left(B^{\mathrm{T}}Pe_N(t)\right) - \beta r(t) + \upsilon_{N(m)}(x_N(t), e_N(t))\Big] \\
&\leq -\lambda_{\min}(Q)\|e_N(t)\|^2 + 2\Gamma_{\max}\lambda_{\max}(P)\|e_N(t)\| \\
&\quad - 2|B^{\mathrm{T}}Pe_N(t)|\Big[\underbrace{\beta r_{\max} + \operatorname{sign}\left(B^{\mathrm{T}}Pe_N(t)\right)\beta r(t)}_{\text{Term 1}} \\
&\qquad \underbrace{+\,\mathcal{P}_{N(m)}(x_N(t))\|f\|_{\mathcal{H}} - \operatorname{sign}\left(B^{\mathrm{T}}Pe_N(t)\right)E_{x_N(t)}(I - \Pi_{N(m)})f}_{\text{Term 2}}\Big]
\end{aligned}
\tag{6.149}
$$

for all $t \in [t_0, \infty)$. Since Terms 1 and 2 in (6.149) are positive, we deduce that for all $t \in [t_0, \infty)$ we have

$$
\dot{V}(t) \leq -\lambda_{\min}(Q)\|e_N(t)\|^2 + 2\Gamma_{\max}\lambda_{\max}(P)\|e_N(t)\| \tag{6.150}
$$

$$
= -\lambda_{\min}(Q)\|e_N(t)\|\left(\|e_N(t)\| - 2\frac{\lambda_{\max}(P)}{\lambda_{\min}(Q)}\Gamma_{\max}\right), \tag{6.151}
$$

whose right-hand side is negative for all e_N outside of the compact set

$$
E_0 \triangleq \left\{e_N \in \mathbb{R}^n \;\middle|\; \|e_N\| \leq 2\frac{\lambda_{\max}(P)}{\lambda_{\min}(Q)}\Gamma_{\max}\right\}. \tag{6.152}
$$

Therefore, for any $\bar{e}_0 > \lambda_{\max}(P)\lambda_{\min}^{-1}(Q)\Gamma_{\max}$, there exists $T > t_0$ such that (6.143) is verified. Note that parameter drift for $\hat{\alpha}(\cdot)$ and $\hat{f}_{N(m)}(\cdot,\cdot)$ is averted by the switched form of the adaptive laws (6.120) and (6.124), which is known as dead-zone modification of the adaptive laws.

Case 2: If $\|x_N(t)\| \leq \delta b$ for some $t \geq t_0$, then $\mu(x_N(t)) = 0$. Thus, we consider two alternative scenarios, namely $\left|B^{\mathrm{T}}Pe_N(t)\right| \geq \epsilon$ and $B^{\mathrm{T}}Pe_N(t) < \epsilon$. If $\left|B^{\mathrm{T}}Pe_N(t)\right| \geq \epsilon$, then $\upsilon_{N(m)}(x_N(t), e_N(t)) = -\operatorname{sign}\left(B^{\mathrm{T}}Pe_N(t)\right)\mathcal{P}_{N(m)}(x_N(t))\|f\|_{\mathcal{H}}$, and it follows from (6.148) that

$$
\begin{aligned}
\dot{V}(t) &= -e_N^{\mathrm{T}}(t)Qe_N(t) + 2\Gamma^{\mathrm{T}}(t)Pe_N(t) + 2B^{\mathrm{T}}Pe_N(t)\Big(E_{x_N(t)}(I - \Pi_{N(m)})f \\
&\qquad - \operatorname{sign}\left(B^{\mathrm{T}}Pe_N(t)\right)\mathcal{P}_{N(m)}(x_N(t))\|f\|_{\mathcal{H}}\Big) \\
&\leq -\lambda_{\min}(Q)\|e_N(t)\|\left(\|e_N(t)\| - 2\frac{\lambda_{\max}(P)}{\lambda_{\min}(Q)}\Gamma_{\max}\right),
\end{aligned}
\tag{6.153}
$$

and (6.143) is verified by proceeding as in Case 1. Alternatively, if $\left|B^{\mathrm{T}}Pe_N(t)\right| < \epsilon$, then $\upsilon_{N(m)}(x_N(t), e_N(t)) = \epsilon^{-1}B^{\mathrm{T}}Pe_N(t)\mathcal{P}_{N(m)}(x_N(t))\|f\|_{\mathcal{H}}$. The bound

$$\left(\epsilon^{-1}|B^{\mathrm{T}}Pe_N|\mathcal{P}_{N(m)}(x)\|f\|_{\mathcal{H}} + \mathrm{sign}(B^{\mathrm{T}}Pe_N E_{x_N(t)}(I - \Pi_{N(m)})f\right)$$
$$\leq 2\sup_{\xi\in S_\epsilon}\mathcal{P}_{N(m)}(\xi)\|f\|_{\mathcal{H}} \tag{6.154}$$

is illustrated from Fig. 6.27. It follows then from (6.148) and (6.154) that

$$\begin{aligned}\dot{V}(t) &= -e_N^{\mathrm{T}}(t)Qe_N(t) + 2\Gamma^{\mathrm{T}}(t)Pe_N(t)\\ &\quad + 2B^{\mathrm{T}}Pe_N(t)\left(\epsilon^{-1}B^{\mathrm{T}}Pe_N(t)\mathcal{P}_{N(m)}(x_N(t))\|f\|_{\mathcal{H}} + E_{x_N(t)}(I - \Pi_{N(m)})f\right)\\ &\leq -\lambda_{\min}(Q)\|e_N(t)\|\Bigg[\|e_N(t)\|\\ &\quad - 2\frac{\lambda_{\max}(P)}{\lambda_{\min}(Q)}\left(\Gamma_{\max} + 2\|B^{\mathrm{T}}P\|\sup_{\xi\in S_\epsilon}\mathcal{P}_{N(m)}(\xi)\|f\|_{\mathcal{H}}\right)\Bigg].\end{aligned} \tag{6.155}$$

Therefore, the right-hand side of (6.155) is negative for all e_N outside of the compact set

$$E_1 \triangleq \left\{e_N \in \mathbb{R}^n \;\middle|\; \|e_N\| \leq 2\frac{\lambda_{\max}(P)}{\lambda_{\min}(Q)}\left(\Gamma_{\max} + \|B^{\mathrm{T}}P\|2\sup_{\xi\in S_\epsilon}\mathcal{P}_{N(m)}(\xi)\|f\|_{\mathcal{H}}\right)\right\}, \tag{6.156}$$

and the result follows as in Case 1.

Case 3: If $\|x_N(t)\| \in (\delta b, b)$ and $\left|B^{\mathrm{T}}Pe_N(t)\right| > \epsilon$ for some $t \geq t_0$, then $\mu(x_N(t)) \in (0,1)$ and $v_{N(m)} = -\mathrm{sign}\left(B^{\mathrm{T}}Pe_N(t)\right)\mathcal{P}_{N(m)}(x_N(t))\|f\|_{\mathcal{H}}$, and it follows from (6.148) and noting for this case $\mu(x_N(t)) \leq 1$ that

$$\begin{aligned}\dot{V}(t) &= -e_N^{\mathrm{T}}(t)Qe_N(t) + 2\Gamma^{\mathrm{T}}(t)Pe_N(t)\\ &\quad + 2B^{\mathrm{T}}Pe_N(t)\Big[\mu(x_N(t))\left[-\beta r_{\max}\mathrm{sign}\left(B^{\mathrm{T}}Pe_N(t)\right) - \beta r(t)\right]\\ &\quad - \mathrm{sign}\left(B^{\mathrm{T}}Pe_N(t)\right)\mathcal{P}_{N(m)}(x_N(t))\|f\|_{\mathcal{H}} + E_{x_N(t)}(I - \Pi_{N(m)})f\Big]\\ &\leq -\lambda_{\min}(Q)\|e_N(t)\|^2 + 2\Gamma_{\max}\lambda_{\max}(P)\|e_N(t)\|\\ &\quad - 2\|B^{\mathrm{T}}Pe_N(t)\|\Big[\mu(x_N(t))[\beta r_{\max} + \mathrm{sign}\left(B^{\mathrm{T}}Pe_N(t)\right)\beta r(t)]\\ &\quad + \mathcal{P}_{N(m)}(x_N(t))\|f\|_{\mathcal{H}} - \mathrm{sign}\left(B^{\mathrm{T}}Pe_N(t)\right)E_{x_N(t)}(I - \Pi_{N(m)})f\Big],\\ &\leq -\lambda_{\min}(Q)\|e_N(t)\|^2 + 2\Gamma_{\max}\lambda_{\max}(P)\|e_N(t)\|\\ &\quad - 2|B^{\mathrm{T}}Pe_N(t)|\Big[\beta r_{\max} + \mathrm{sign}\left(B^{\mathrm{T}}Pe_N(t)\right)\beta r(t)\\ &\quad + \mathcal{P}_{N(m)}(x_N(t))\|f\|_{\mathcal{H}} - \mathrm{sign}\left(B^{\mathrm{T}}Pe_N(t)\right)E_{x_N(t)}(I - \Pi_{N(m)})f\Big].\end{aligned} \tag{6.157}$$

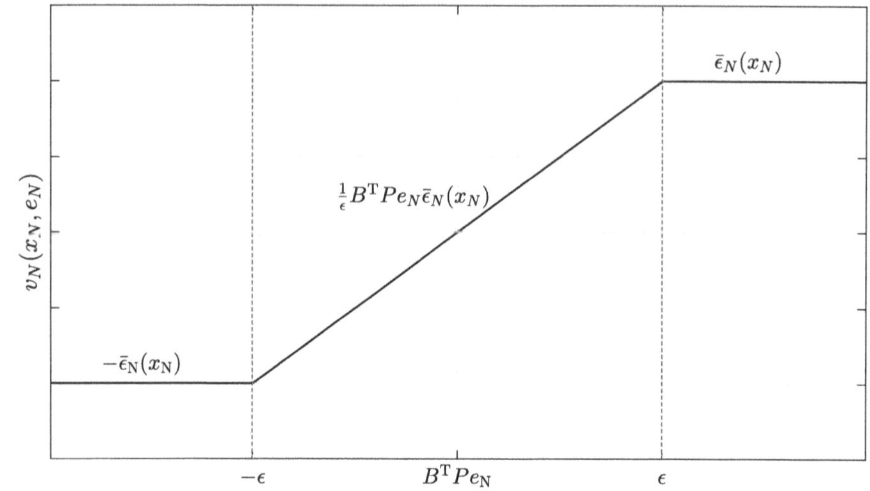

Fig. 6.27 A visualization of the bound provided by (6.154)

The result follows as in Case 1.

Case 4: If $\|x_N(t)\| \in (\delta b, b)$ and $|B^{\mathrm{T}}Pe_N(t)| < \epsilon$ for some $t \geq t_0$, then $\mu(x_N(t)) \in (0, 1)$, $v_{N(m)}(x_N(t), e_N(t)) = \epsilon^{-1}B^{\mathrm{T}}Pe_N(t)\mathcal{P}_{N(m)}(x_N(t))\|f\|_{\mathcal{H}}$, and it follows from (6.148) that

$$
\begin{aligned}
\dot{V}(t) &= -e_N^{\mathrm{T}}(t)Qe_N(t) + 2\Gamma^{\mathrm{T}}(t)Pe_N(t) \\
&\quad + 2B^{\mathrm{T}}Pe_N(t)\Big(\mu(x_N(t))[-\beta r_{\max}\operatorname{sign}(B^{\mathrm{T}}Pe_N(t)) - \beta r(t)] \\
&\quad + \epsilon^{-1}B^{\mathrm{T}}Pe_N(t)\mathcal{P}_{N(m)}(x_N(t))\|f\|_{\mathcal{H}} + E_{x_N(t)}(I - \Pi_{N(m)})f\Big) \\
&\leq -\lambda_{\min}(Q)\|e_N(t)\|\Big[\|e_N(t)\| - 2\frac{\lambda_{\max}(P)}{\lambda_{\min}(Q)} \\
&\qquad \cdot \Big(\Gamma_{\max} + \|B^{\mathrm{T}}P\|2\sup_{\xi\in S_\epsilon}\mathcal{P}_{N(m)}(\xi)\|f\|_{\mathcal{H}}\Big)\Big], \qquad (6.158)
\end{aligned}
$$

and the result follows as in the second part of Case 2. ■

6.6.5 Implementations Using Recursive Newton Bases

By proceeding as in Sect. 5.4.1.3, the adaptive law (6.124) leads to coordinate implementations where the coefficients evolve according to

$$\dot{\Theta}(t) = \begin{cases} -\gamma_f \mathbb{K}_N^{-1} \mathfrak{K}_{\Xi_N}(x(t)) B^{\mathrm{T}} P e_N(t) & \text{if } \|P^{1/2} e_N(t)\| \leq \epsilon, \\ 0 & \text{otherwise,} \end{cases} \tag{6.159}$$

where $\hat{\Theta}(t) \triangleq [\theta_1(t), \ldots, \theta_N(t)]^{\mathrm{T}}$, $\mathfrak{K}_{\Xi_N}(\cdot) \triangleq \left[\mathfrak{K}_{\xi_1}(\cdot), \ldots, \mathfrak{K}_{\xi_N}(\cdot)\right]^{\mathrm{T}}$ denotes the vector of basis functions, and $\mathbb{K}_N \triangleq \mathbb{K}(\Xi_N, \Xi_N) \triangleq [\mathfrak{K}(\xi_i, \xi_j)] \in \mathbb{R}^{N \times N}$ denotes the Grammian matrix.

As noted in [25, 78–80], the Grammian matrix $\mathbb{K}_N$ can become ill-conditioned for a number of reasons. For instance, the condition number of the Grammian typically increases as the smoothness of the kernel $\mathfrak{K}$ increases, and it increases with decreasing separation distance

$$q_{\Xi_N} = \min\left\{\|\xi_i - \xi_j\| \mid \xi_i \neq \xi_j, \xi_i, \xi_j \in \Xi_N\right\}; \tag{6.160}$$

see [25, 81]. Irregular distributions of centers in $\Xi_N \subset \Omega$ can also lead to larger condition numbers. Maintaining good conditioning of the Grammian as the number of centers increases can be challenging [80, 82]. With automated basis augmentation defined by selecting new centers, maintaining good conditioning of the Grammian, as it evolves when applying a greedy algorithm, can give rise to additional challenges [82]. In this section, we present an approach to mitigate this effect on the performance of the closed-loop system.

First, we consider a change of bases between the classical scattered basis defined in terms of the kernels $\mathfrak{K}_{\Xi_N} \triangleq \left[\mathfrak{K}_{x_1}, \ldots, \mathfrak{K}_{x_N}\right]^{\mathrm{T}}$ to a basis $\boldsymbol{b}_{\Xi_N} \triangleq [\boldsymbol{b}_1, \ldots, \boldsymbol{b}_N]^{\mathrm{T}}$ for $\mathcal{H}_N$ via the change of basis formula

$$\boldsymbol{b}_i(x) = \sum_{j=1}^{N} C_{ij} \mathfrak{K}_{\xi_j}(x) = \sum_{j=1}^{N} C_{ij} \mathfrak{K}(x, \xi_j), \quad \text{for all } i \in \{1, \ldots, N\}, \tag{6.161}$$

where $C = [C_{ij}]$ denotes a full-rank matrix. There is a multitude of possible bases to improve the numerical stability of realization of projection or interpolation operators onto $\mathcal{H}_N$ [78, 79, 83]. In this section, we focus on the *Newton basis* $\boldsymbol{b}_{\Xi_N} \triangleq [\boldsymbol{b}_i, \ldots, \boldsymbol{b}_N]^{\mathrm{T}}$ for $\mathcal{H}_N$.

Suppose for the moment that we have selected some (possibly quite large) set of centers Ξ_N. In this case, the Cholesky decomposition of the kernel matrix can be expressed as

$$\mathbb{K}_N = \mathbb{L}\mathbb{L}^{\mathrm{T}}, \tag{6.162}$$

where $\mathbb{L} \in \mathbb{R}^{N \times N}$ is a lower-triangular matrix. One way to define the Newton basis chooses the matrix $[C] \triangleq \mathbb{L}^{-1}$ so that

$$\boldsymbol{b}_j = \sum_{i=1}^{N} L_{ij}^{-1} \mathfrak{K}_{\xi_i}. \tag{6.163}$$

The Newton basis is an orthonormal basis for $\mathcal{H}_N$ as can be seen through calculation of the associated Grammian $\mathbb{B}_N \triangleq [(\boldsymbol{b}_i, \boldsymbol{b}_j)_{\mathcal{H}}]$, which becomes

$$\mathbb{B}_N \triangleq \mathbb{L}^{-1}\mathbb{K}_N\mathbb{L}^{-T} = \mathbb{L}^{-1}\mathbb{L}\mathbb{L}^{\mathrm{T}}\mathbb{L}^{-T} = I_N. \tag{6.164}$$

Employing the basis $\{\boldsymbol{b}_1, \ldots, \boldsymbol{b}_N\}$ for the subspace $\mathcal{H}_N$, the adaptive law (6.159) can be expressed equivalently as

$$\dot{\hat{\Theta}}(t) = \begin{cases} -\gamma_f \boldsymbol{b}_{\Xi_N}(x(t))B^{\mathrm{T}}Pe_N(t) & \text{if } \|P^{1/2}e_N(t)\| \leq \epsilon, \\ 0 & \text{otherwise,} \end{cases} \tag{6.165}$$

where now we have $\hat{f}_N(t,\cdot) \triangleq \sum_{i=1}^{N} \theta_i(t)\boldsymbol{b}_i$. Comparing the coordinate realizations (6.159) and (6.165), the latter does not explicitly include the inversion of the Grammian $\mathbb{K}_N$. As discussed in Chap. 8, this structural change can have important effects in ameliorating sharp gradient changes at switching times.

The definition of the Newton basis in (6.161) is used because of its simplicity. Since $\boldsymbol{b}_{\Xi_N}(\cdot) = \mathbb{L}^{-1}\mathfrak{K}_{\Xi_N}(\cdot)$, if the centers in Ξ_N are not distributed well and the condition of $\mathbb{K}_N$ is poor, then the condition of $\mathbb{L}$ will be also be poor. Thus, one may surmise that not much is gained by passing to Newton bases. However, an important feature of the Newton basis is that it is possible to implement its definition recursively by adaptively choosing centers, and the recursive center selection itself can have a beneficial effect on conditioning. This recursive process was described initially in [79], and can be interpreted as an application of the Gram-Schmidt orthogonalization procedure for bases selected recursively by the $\mathcal{P}$-greedy algorithm.

To define the Newton basis recursively, consider a very dense grid of isolated points $\Omega_M \subset \Omega$, which defines the pool of possible sites at which recursively to choose centers. Thus, let

$$\xi_1 \in \arg\max\{\mathfrak{K}(x,x)\ :\ x \in \Omega\} \tag{6.166}$$

and set $\Xi_1 = \{\xi_1\}$. Additionally, define the first Newton basis function as

$$\boldsymbol{b}_1(\cdot) \triangleq \frac{\mathfrak{K}(\xi_1,\cdot)}{\sqrt{\mathfrak{K}(\xi_1,\xi_1)}} \tag{6.167}$$

and

$$\mathcal{P}_1(\xi_1) = \sqrt{z_1 - w_1}, \tag{6.168}$$

where $z_1 \triangleq \mathfrak{K}(\xi_1,\xi_1)$ and $w_1 \triangleq \boldsymbol{b}_1^2(\xi_1)$.

The second kernel center is chosen as

$$\xi_2 \in \arg\max_{\xi\in\Omega_M\setminus\Xi_1}\mathcal{P}_1(\xi), \tag{6.169}$$

and we set $\Xi_2 = \{\xi_1, \xi_2\}$. Now, let

$$\boldsymbol{u}_2(\cdot) \triangleq \mathfrak{K}(\xi_2, \cdot) - \boldsymbol{b}_1(\cdot)\boldsymbol{b}_1(\xi_2), \tag{6.170}$$

and define the next Newton basis function as

$$\boldsymbol{b}_2(\cdot) \triangleq \frac{\boldsymbol{u}_2(\cdot)}{\|\boldsymbol{u}_2(\cdot)\|_{\mathcal{H}}}. \tag{6.171}$$

We then compute

$$\mathcal{P}_2(x_j) = \sqrt{z_2 - w_2} \tag{6.172}$$

with $z_2 \triangleq \mathfrak{K}(\xi_2, \xi_2)$ and $w_2 \triangleq \sum_{j=1}^{2} \boldsymbol{b}_j^2(\xi_2)$.

At the $(j+1)$th step, we let

$$\xi_{j+1} \in \arg\max_{\xi \in \Omega_M \setminus \Xi_j} \mathcal{P}_j(\xi), \tag{6.173}$$

where

$$\mathcal{P}_j(\xi) = \sqrt{z_j - w_j}, \tag{6.174}$$

$$z_j \triangleq \mathfrak{K}(x_j, x_j), \tag{6.175}$$

$$w_j \triangleq \sum_{k=1}^{j} \boldsymbol{b}_k^2(x_j). \tag{6.176}$$

Finally, the Newton basis function is calculated as

$$\boldsymbol{b}_j(\cdot) \triangleq \frac{\boldsymbol{u}_j(\cdot)}{\|\boldsymbol{u}_j(\cdot)\|_{\mathcal{H}}}, \tag{6.177}$$

where

$$\boldsymbol{u}_j(\cdot) \triangleq \mathfrak{K}(\xi_j, \cdot) - \sum_{i=1}^{j} \boldsymbol{b}_i(\cdot)\boldsymbol{b}_i(\xi_j). \tag{6.178}$$

This procedure can be iterated until some user-defined criterion is met. For instance, until $\max_{x_j \in \Omega_M} \mathcal{P}_n(x_j) \leq \epsilon$, where $\epsilon > 0$ is user-defined. It is apparent how, by using the proxy Ω_M in place of Ω, the problem of maximizing the power function, is addressed on a smaller set and, hence, is computationally more tractable. Otherwise, classical optimization approaches can be applied for the maximization of the power function over the original set Ω.

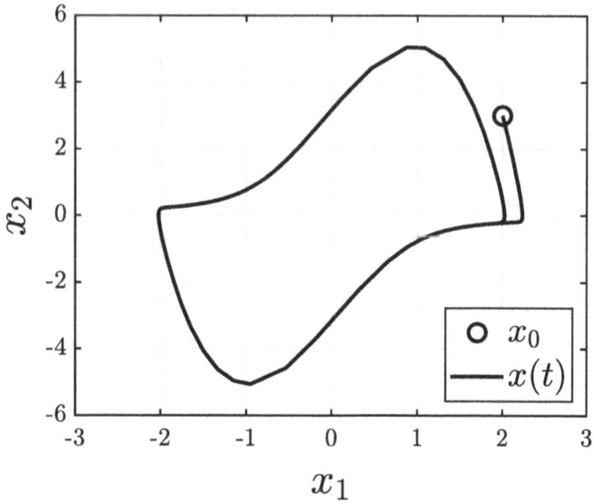

Fig. 6.28 Phase portrait of the Van der Pol oscillator without an external control input for initial conditions $x_0 = [2, 3]^{\mathrm{T}}$

6.6.6 *Simulation Results*

In this section, we give a brief survey of results from the paper [84]. MATLAB® codes to reproduce these results are available at [85]. The model problem is based on the Van der Pol oscillator with external forcing, whose dynamics are given by

$$\dot{x}_1(t) = x_2(t), \quad x_1(0) = x_{10}, \quad t \geq 0, \tag{6.179a}$$

$$\dot{x}_2(t) = -x_1(t) + \nu x_2(t) - \nu x_1^2(t) x_2(t) + u(t), \quad x_2(0) = x_{20}, \tag{6.179b}$$

where $\nu = 3$ for this example. The uncontrolled dynamics are shown in Fig. 6.28.

The reference trajectory $x_r(t) = \left[x_{r,1}(t), x_{r,2}(t)\right]^{\mathrm{T}}$ is generated by the linear system

$$\begin{bmatrix} \dot{x}_{r,1}(t) \\ \dot{x}_{r,2}(t) \end{bmatrix} = \underbrace{\begin{bmatrix} 0 & 1 \\ -1 & -1 \end{bmatrix}}_{A_r} \begin{bmatrix} x_{r,1}(t) \\ x_{r,2}(t) \end{bmatrix} + \underbrace{\begin{bmatrix} 0 \\ 1 \end{bmatrix}}_{B_r} r(t), \quad x_r(0) = x_{r,0}, \quad t \geq 0, \tag{6.180}$$

where

$$r(t) = x_2(t). \tag{6.181}$$

The reference trajectory follows a circular path with radius $R = \|x_r(t_0)\|_{\mathbb{R}^2}$. For this problem, we set $x_r(0) = [2, 2]^{\mathrm{T}}$.

To recast this problem in the standard form considered in this text, (6.179) can be equivalently expressed as

$$\dot{x}(t) = \underbrace{\begin{bmatrix} 0 & 1 \\ -1 & -1 \end{bmatrix}}_{A} x(t) + \underbrace{\begin{bmatrix} 0 \\ 1 \end{bmatrix}}_{B} (u(t) + f(x(t))), \quad x(0) = x_0, \quad t \geq 0, \tag{6.182}$$

where

$$f(x) = (1+\nu)x_2 - \nu x_1^2 x_2, \quad \text{for all } x \in \mathbb{R}^2 \tag{6.183}$$

denotes the matched functional uncertainty. This uncertainty is assumed to lie in the user-defined RKHS $\mathcal{H}(\Omega)$ of scalar-valued functions.

In this simple example, note that we have $A = A_r$ and $B = B_r$. Hence, the adaptive law underlying the CLC architecture only needs to account for the matched uncertainty (see Sect. 5.3), and the control law (6.118) employed for the proposed simulations reduces to

$$\begin{aligned} u_{N(m)}(t) &\triangleq x_N(t) + (1 - \mu(x_N(t)))r(t) \\ &\quad - \hat{f}_{N(m)}(t, x_N(t)) + \mu(x_N(t))u_{SL}(t) + v_{N(m)}(x_N(t), e_N(t)), \end{aligned} \tag{6.184}$$

where $\hat{f}_{N(m)}(t, \cdot)$ captures the $N(m)$-dimensional adaptive gain associated with f.

The initial condition used in all simulations is $x(0) = [2, 3]^{\mathrm{T}}$, and and the set of centers $\Xi_{N(m)}$ is a subset of the torus centered at the origin

$$\Omega = \{x \in \mathbb{R}^n \ : \ R_i < \|x\|_{\mathbb{R}^n} < R_o\}, \tag{6.185}$$

where the inner radius is given by $R_i = 2\sqrt{2} - 0.3$ and the outer radius is given by $R_o = 2\sqrt{2} + 0.3$. The matrix P in the adaptive law (6.124) is the symmetric positive-definite solution of the Lyapunov equation (6.123) with $Q = I$. For any step $m \in \mathbb{Z}^+$, the greedy basis augmentation is triggered if the tracking error $\|\tilde{x}(t)\|_{\mathbb{X}} \leq c_m \sup_{x \in \Omega_0} \mathcal{P}_m(x)\|f\|_{\mathcal{H}}$ for a period of 20 s to prevent triggering basis augmentation during the transient phase of the control implementation. Figure 6.29 shows an example of the centers used for basis augmentation using the greedy update for triggering events $m_1 < m_2$. Note here that for $m_1 < m_2$, $\Xi_{N(m_1)} \subset \Xi_{N(m_2)}$.

Figures 6.30 and 6.31 show the CLC system's performance with the adaptive rate $\gamma_f = 10^5$. Figure 6.32 shows the system's performance with $\gamma_f = 10^6$. Additional cases are studied in the reference [85], and the reader is referred to the discussion there for the details.

6.7 Concluding Remarks

Given an RKHS-embedded adaptive control problem, this chapter discussed how to employ deterministic data-driven methods to identify the RKHS that best approximates the hypothesis space. Three fundamental architectures were discussed, namely,

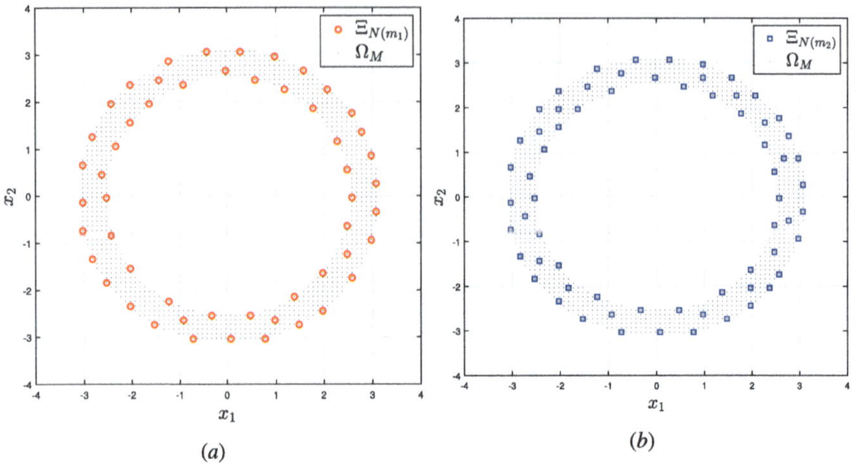

Fig. 6.29 (*a*) Centers selected by the proposed greedy method after the m_1th trigger event. (*b*) Centers selected by the proposed greedy method after the m_2th trigger event with $m_2 > m_1$. Note that the centers are typically placed in the largest empty portions of the approximation region. Additionally, the center sequences are nested, that is, $\Xi_{N(m_1)} \subset \Xi_{N(m_1)} \subset \Omega$

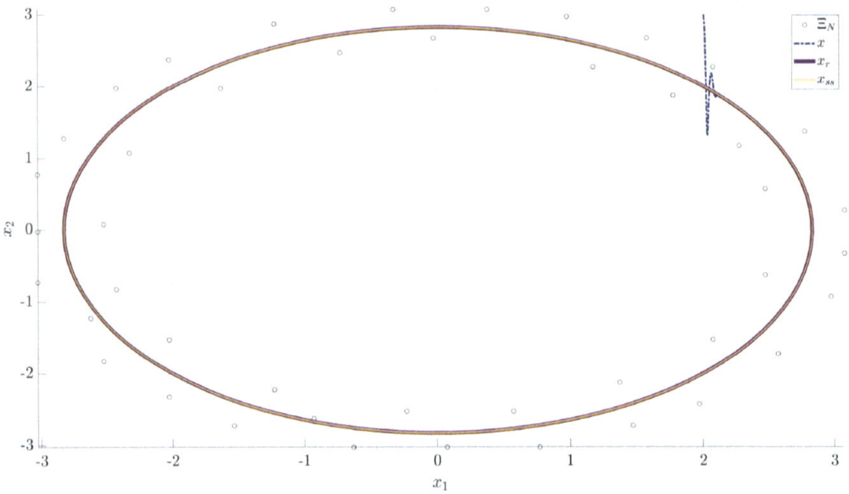

Fig. 6.30 Kernel center locations, closed-loop trajectory, and reference trajectory with adaptive rate $\gamma_f = 10^5$

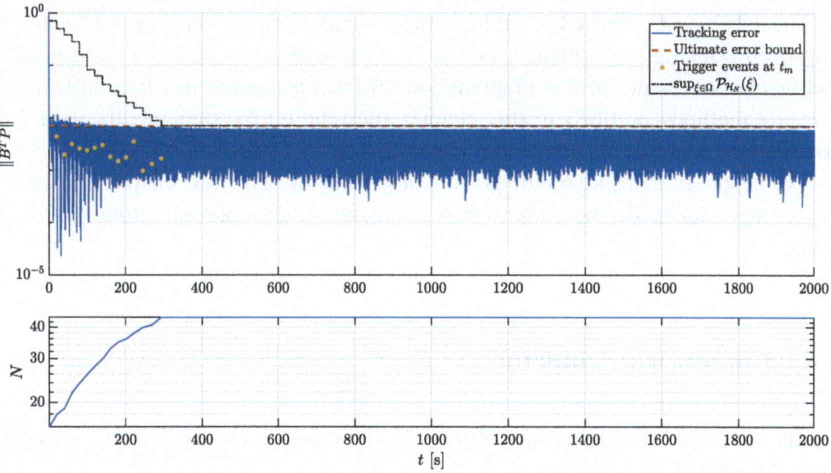

Fig. 6.31 Normalized tracking error with adaptive rate $\gamma = 10^5$

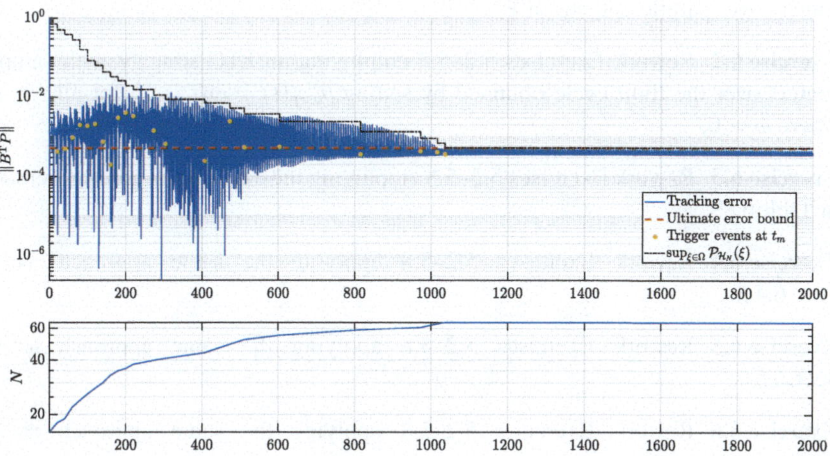

Fig. 6.32 Normalized norm of the tracking error, $\gamma = 10^6$

LTC, SLC, and CLC. The discussion on LTC method comprehended the definition of performance measures for an RKHS-embedded adaptive control system and a discussion on how the distribution of the kernel centers affects our ability to forecast the ultimate bounds on the trajectory tracking error. The SLC method is less mature than the SLC method and was addressed by employing a reinforcement learning method. Future research directions involve designing mechanisms that explicitly prevent Zeno solutions and beating while reducing the extent of the learning phase to the minimum. Finally, a novel CLC architecture was presented. This approach is based on the use of the adaptive bound method and a greedy method to identify the center location through the use of Newton bases.

The LTC, SLC, and CLC architectures blend learning and control stages for improved tracking performance. For each of these architectures, the control stages can be chosen as any of the adaptive control laws presented in Chap. 5. Thus, the learning methods outlined in this chapter show how to systematically model the approximations of the hypothesis space leveraging the properties of RKHSs.

The proposed learning methods were deterministic. The next chapter shows how to leverage stochastic methods to best characterize the approximating hypothesis space.

6.8 Homework Problems

For an improved comprehension of the material exposed in this chapter, we recommend the following problems.

Exercise 6.1 Rework Exercises 5.2–5.8 employing an LTC strategy with sample centers over parallelepipeds in $\mathbb{R}^2$. △

Exercise 6.2 Rework Exercises 5.2–5.8 employing an LTC strategy with sample centers over the limit cycle obtained by setting $\theta_{\text{user}}(t) = \omega(t - t_0)$ for all $t \geq t_0$, where $\omega > 0$. △

Exercise 6.3 Rework Exercises 5.2–5.8 employing the data-driven greedy methods in Table 6.89. △

Exercise 6.4 Rework Exercises 5.2–5.8 applying the same approach as in Sect. 6.4.3. △

Exercise 6.5 Rework Exercises 5.2–5.8 applying the same approach as in Sect. 6.5. △

Exercise 6.6 Rework Exercises 5.2–5.8 applying the same approach as in Sect. 6.6. △

References

1. Farrell, J.A.: Stability and approximator convergence in nonparametric nonlinear adaptive control. IEEE Trans. Neural Netw. **9**(5), 1008–1020 (1998)
2. Chowdhary, G., How, J., Kingravi, H.: Model reference adaptive control using nonparametric adaptive elements. In: AIAA Guidance, Navigation, and Control Conference, p. 5038. MN, Minneapolis (2012)
3. Kurdila, A., Lei, Y.: Adaptive control via embedding in reproducing kernel Hilbert spaces. In: American Control Conference, pp. 3384–3389. IEEE, Florence, Italy (2013)
4. Choi, J.Y., Farrell, J.A.: Nonlinear adaptive control using networks of piecewise linear approximators. IEEE Trans. Neural Netw. **11**(2), 390–401 (2000)

5. Farrell, J.A., Murlis, J., Long, X., Li, W., Cardé, R.T.: Filament-based atmospheric dispersion model to achieve short time-scale structure of odor plumes. Environ. Fluid Mech. **2**(1–2), 143–169 (2002)
6. Stoev, J., Choi, J.Y., Farrell, J.: Adaptive control for output feedback nonlinear systems in the presence of modeling errors. Automatica **38**(10), 1761–1767 (2002)
7. Naicanishi, J., Farrel, J.A., Schaal, S.: Learning composite adaptive control for a class of nonlinear systems. In: International Conference on Robotics and Automation. IEEE, New Orleans, LA, pp. 2647–2652 (2004)
8. Nakanishi, J., Farrell, J.A., Schaal, S.: Composite adaptive control with locally weighted statistical learning. Neural Netw. **18**(1), 71–90 (2005)
9. Zhao, Y., Farrell, J.A.: Self-organizing approximation-based control for higher order systems. IEEE Trans. Neural Netw. **18**(4), 1220–1231 (2007)
10. Farrell, J.A.: Aided NavigationâŁ¯: GPS With High Rate Sensors. McGraw-Hill, New York, NY (2008)
11. Dong, W., Farrell, J.A., Polycarpou, M.M., Djapic, V., Sharma, M.: Command filtered adaptive backstepping. IEEE Trans. Control Syst. Technol. **3**(20), 566–580 (2012)
12. Kingravi, H.A., Chowdhary, G., Vela, P.A., Johnson, E.N.: Reproducing kernel Hilbert space approach for the online update of radial bases in neuro-adaptive control. IEEE Trans. Neural Netw. Learn. Syst. **23**(7), 1130–1141 (2012)
13. Kamalapurkar, R., Rosenfeld, J.A., Dixon, W.E.: State following (StaF) kernel functions for function approximation part II: adaptive dynamic programming. In: American Control Conference, pp. 521–526. IEEE, Chicago, IL (2015)
14. Rosenfeld, J.A., Kamalapurkar, R., Dixon, W.E.: The state following approximation method. IEEE Trans. Neural Netw. Learn. Syst. **30**(6), 1716–1730 (2019)
15. Paruchuri, S.T., Guo, J., Kurdila, A.: Reproducing kernel Hilbert space embedding for adaptive estimation of nonlinearities in piezoelectric systems. Nonlinear Dyn. **101**(2), 1397–1415 (2020)
16. Paruchuri, S.T., Guo, J., Kurdila, A.: Kernel center adaptation in the reproducing kernel hilbert space embedding method. Int. J. Adapt. Control. Signal Process. **36**(7), 1562–1583 (2022)
17. Paruchuri, S.T., Guo, J., Kurdila, A.: Sufficient conditions for parameter convergence over embedded manifolds using kernel techniques. IEEE Trans. Autom. Control **68**(2), 753–765 (2022)
18. Farrell, J.A., Polycarpou, M.M.: Adaptive Approximation Based Control: Unifying Neural, Fuzzy and Traditional Adaptive Approximation Approaches, vol. 48. Wiley, Danvers, MA (2006)
19. Lavretsky, E., Wise, K.: Robust and Adaptive Control: With Aerospace Applications. Springer, London, UK (2013)
20. Narendra, K.S., Annaswamy, A.M.: Stable Adaptive Systems. Dover, New York, NY (1989)
21. Ioannou, P.A., Sun, J.: Robust Adaptive Control. Courier Corporation, Mineola, NY (2012)
22. Krstic, M., Kanellakopoulos, I., Kokotovic, P.: Nonlinear and Adaptive Control Design. Wiley, New York, NY (1995)
23. Sastry, S., Bodson, M.: Adaptive Control: Stability, Convergence and Robustness. Courier Corporation, New York, NY (2011)
24. Slotine, J.-J.E., Li W. et al.: Applied Nonlinear Control, vol. 199, no. 1. Prentice Hall, Englewood Cliffs, NJ (1991)
25. Wendland, H.: Scattered Data Approximation, vol. 17. Cambridge University Press, Cambridge, UK (2004)
26. G. Tao, *Adaptive control design and analysis*. Danvers, MA: Wiley, 2003, vol. 37
27. DeVore, R.A.: Nonlinear approximation. Acta Numerica **7**, 51–150 (1998)
28. Cucker, F., Smale, S.: On the mathematical foundations of learning. Bull. Am. Math. Soc. **39**(1), 1–49 (2002)
29. Binev, P., Cohen, A., Dahmen, W., DeVore, R., Temlyakov, V., Bartlett, P.: Universal algorithms for learning theory part I: piecewise constant functions. J. Mach. Learn. Res. **6**(9) (2005)
30. Temlyakov, V.: Greedy Approximation, vol. 20. Cambridge University Press, Cambridge, UK (2011)

31. Hangelbroek, T., Narcowich, F., Ward, J.: Polyharmonic and related kernels on manifolds: interpolation and approximation. Found. Comput. Math. **12**, 625–670 (2012)
32. Narcowich, F.J., Ward, J.D., Wendland, H.: Sobolev error estimates and a Bernstein inequality for scattered data interpolation via radial basis functions. Constr. Approx. **24**(2), 175–186 (2006)
33. DeVore, R.A., Lorentz, G.G.: Constructive Approximation. Springer-Verlag, Berlin, Germany (1993)
34. Pietsch, A.: Approximation spaces. J. Approx. Theory **32**(2), 115–134 (1981)
35. Traub, J.F., Werschulz, A.G.: Complexity and Information, , vol. 26862. Cambridge University Press, Cambridge, UK (1998)
36. Traub, J.F.: Information-based complexity. In: Encyclopedia of Computer Science, pp. 850–854 (2003)
37. Traub, J.F., Werschulz, A.G.: Information-Based Complexity and Information-Based Optimization, pp. 1603–1608. Springer, Boston, MA, USA (2009)
38. Wenzel, T., Santin, G., Haasdonk, B.: Analysis of target data-dependent greedy kernel algorithms: convergence rates for f-, $f \cdot p$-, and f/p-greedy. Constr. Approx. 1–30 (2022)
39. Wenzel, T., Santin, G., Haasdonk, B.: A novel class of stabilized greedy kernel approximation algorithms: convergence, stability and uniform point distribution. J. Approx. Theory **262**, 105508 (2021)
40. Binev, P., Cohen, A., Dahmen, W., DeVore, R., Petrova, G., Wojtaszczyk, P.: Convergence rates for greedy algorithms in reduced basis methods. SIAM J. Math. Anal. **43**(3), 1457–1472 (2011)
41. DeVore, R., Petrova, G., Wojtaszczyk, P.: Greedy algorithms for reduced bases in Banach spaces. Constr. Approx. **37**(3), 455–466 (2013)
42. Bouland, A., Niu, S., Paruchuri, S.T., Kurdila, A., Burns, J., Schuster, E.: Rates of convergence in a class of native spaces for reinforcement learning and control. IEEE Control Syst. Lett. 1 – 1 (2023)
43. Paruchuri, S.T., Guo, J., Kurdila, A.: Kernel center adaptation in the reproducing kernel hilbert space embedding method. Int. J. Adapt. Control Signal Process. **36**(7), 1562–1583 (2022)
44. Paruchuri, S.T., Guo, J., Kurdila, A.: Kernel center adaptation in the reproducing kernel Hilbert space embedding method. Int. J. Adapt. Control. Signal Process. **36**(7), 1562–1583 (2022)
45. Wang, H., Scurlock, B., Powell, N., L'Afflitto, A., Kurdila, A.J.: MRAC with adaptive uncertainty bounds via operator-valued reproducing kernels. IEEE Control Syst. Lett. **7**, 3771–3776 (2023)
46. Burns, J.A., Guo, J., Kurdila, A.J., Paruchuri, S.T., Wang, H.: Error bounds for native space embedding observers with operator-valued kernels. In: American Control Conference, pp. 4796–4801. IEEE, San Diego, CA (2023)
47. Burns, J., Estes, B., Guo, J., Kurdila, A., Paruchuri, S.T., Powell, N.: Kernel methods for regression in continuous time over subsets and manifolds. Nonlinear Dyn. **111**(14), 13 165–13 186 (2023)
48. Guo, J., Zhang, F., Kurdila, A.J.: Collaborative persistent excitation in RKHS embedded adaptive estimation with consensus. In: American Control Conference, pp. 3388–3393. IEEE, Atlanta, GA (2022)
49. Guo, J., Paruchuri, S.T., Kurdila, A.J.: Partial persistence of excitation in RKHS embedded adaptive estimation. IEEE Trans. Autom. Control **68**(10), 5850–5861 (2023)
50. Guo, J., Kepler, M.E., Tej Paruchuri, S., Wang, H., Kurdila, A.J., Stilwell, D.J.: Adaptive estimation of external fields in reproducing kernel Hilbert spaces. Int. J. Adapt. Control Signal Process. **36**(8), 1931–1957 (2022)
51. Wittwar, D., Santin, G., Haasdonk, B.: Interpolation with uncoupled separable matrix-valued kernels. Dolomites Res. Notes Approx. **11**, 23–39 (2018)
52. Kurdila, A.J., L'Afflitto, A., Burns, J.A., Wang, H.: Nonparametric adaptive control in native spaces: Finite-dimensional implementations (Part II). Ann. Rev. Control **58**, 100969 (2024)
53. Hangelbroek, T., Narcowich, F.J., Ward, J.D.: Kernel approximation on manifolds I: bounding the Lebesgue constant. SIAM J. Math. Anal. **42**(4), 1732–1760 (2010)

54. Fuselier, E., Wright, G.B.: Scattered data interpolation on embedded submanifolds with restricted positive definite kernels: Sobolev error estimates. SIAM J. Numer. Anal. **50**(3), 1753–1776 (2012)
55. Mauroy, A., Mezić, I., Susuki, Y.: The Koopman Operator in Systems and Control: Concepts, Methodologies, and Applications, ser. Lecture Notes in Control and Information Sciences, vol. 484. Springer (2020)
56. Devaney, R.: An Introduction to Chaotic Dynamical Systems. CRC Press, Boca Raton, FL (2018)
57. Walker, J.A.: Dynamical Systems and Evolution Equations: Theory and Applications. Plenum Press, New York (1980)
58. Saperstone, S.H.: Semidynamical Systems in Infinite Dimensional Spaces, vol. 37. Springer, New York, NY (2012)
59. Mischaikow, K., Smith, H., Thieme, H.R.: Asymptotically autonomous semiflows: chain recurrence and lyapunov functions. Trans. Am. Math. Soc. **347**(5), 1669–1685 (1995)
60. Györfy, L., Kohler, M., Krzyzak, A., Walk, H.: A Distribution-Free Theory of Nonparametric Regression. Springer, Berlin, Germany (2002)
61. Barron, A.R., Cohen, A., Dahmen, W., DeVore, R.A.: Approximation and learning by greedy algorithms. Ann. Stat. **36**(1), 64–94 (2008)
62. Pinkus, A.: N-widths in Approximation Theory, vol. 7. Springer, New York, NY (2012)
63. Gao, T., Kovalsky, S.Z., Boyer, D.M., Daubechies, I.: Gaussian process landmarking for three-dimensional geometric morphometrics. SIAM J. Math. Data Sci. **1**(1), 237–267 (2019)
64. Wenzel, T., Santin, G., Haasdonk, B.: Analysis of target data-dependent greedy kernel algorithms: convergence rates for f-, $f \cdot P$- and f/P-Greedy. Constr. Approx. **57**(1), 45–74 (2023)
65. Bouland, A., Niu, S., Paruchuri, S.T., Kurdila, A., Burns, J., Schuster, E.: Rates of convergence in a class of native spaces for reinforcement learning and control. IEEE Control Syst. Lett. **8**, 55–60 (2024)
66. Niu, S., Bouland, A., Wang, H., Fotiadis, F., Kurdila, A., L'Afflitto, A., Paruchuri, S.T., Vamvoudakis, K.G.: Convergence rates of online critic value function approximation in native spaces. IEEE Control Syst. Lett. **8**, 2145–2150 (2024)
67. Kamalapurkar, R., Walters, P., Rosenfeld, J., Dixon, W.: Reinforcement Learning for Optimal Feedback Control. Springer, London, UK (2018)
68. Vamvoudakis, K.G., Wan, Y., Lewis, F.L., Cansever, D.: Handbook of Reinforcement Learning and Control. Springer, New York, NY (2021)
69. Bertsekas, D.: Reinforcement Learning and Optimal Control, vol. 1. Athena Scientific, Nashua, NH (2019)
70. Anderson, R.B., Marshall, J.A., L'Afflitto, A.: Novel model reference adaptive control laws for improved transient dynamics and guaranteed saturation constraints. J. Frankl. Inst. **358**(12), 6281–6308 (2021)
71. Haddad, W.M., Chellaboina, V.: Nonlinear Dynamical Systems and Control: A Lyapunov-Based Approach. Princeton University Press, Princeton, NJ (2008)
72. Bernstein, D.S.: Nonquadratic cost and nonlinear feedback control. Int. J. Robust Nonlinear Control. **3**(3), 211–229 (1993)
73. L'Afflitto, A., Haddad, W.M., Hui, Q.: Optimal control for linear and nonlinear semistabilization. J. Frankl. Inst. **352**(3), 851–881 (2015)
74. Haddad, W., L'Afflitto, A.: Finite-time stabilization and optimal feedback control. IEEE Trans. Autom. Control 1–6 (2015)
75. Haddad, W.M., L'Afflitto, A.: Finite-time partial stability and stabilization, and optimal feedback control. J. Frankl. Inst. **352**(6), 2329–2357 (2015)
76. L'Afflitto, A.: Differential games, partial-state stabilization, and model reference adaptive control. J. Frankl. Inst. **354**(1), 456–478 (2017)
77. Vamvoudakis, K.G., Lewis, F.L.: Online actor-critic algorithm to solve the continuous-time infinite horizon optimal control problem. Automatica **46**(5), 878–888 (2010)
78. Pazouki, M., Schaback, R.: Bases for kernel-based spaces. J. Comput. Appl. Math. **236**(4), 575–588 (2011)

79. Müller, S., Schaback, R.: A newton basis for kernel spaces. J. Approx. Theory **161**(2), 645–655 (2009)
80. Powell, N., Liu, B., Kurdila, A.J.: Koopman methods for estimation of motion over unknown, regularly embedded submanifolds. In: American Control Conference, pp. 2584–2591. IEEE, Atlanta, GA (2022)
81. Schaback, R.: Error estimates and condition numbers for radial basis function interpolation. Adv. Comput. Math. **3**, 251–264 (1994)
82. Guo, J., Zhang, F.: Event-triggered basis augmentation for data-driven adaptive control. In: American Control Conference, pp. 2987–2992. IEEE, San Diego, CA (2023)
83. De Marchi, S., Santin, G.: A new stable basis for radial basis function interpolation. J. Comput. Appl. Math. **253**, 1–13 (2013)
84. Powell, N., Guo, J., Wang, H., L'Afflitto, A., Burns, J., Kurdila, A.: Greedy Native Space Embedding MRAC via Newton Bases (2023). Arxiv.org
85. Powell, N., Wang, H., L'Afflitto, A., Kurdila, A.J.: Computer codes to implement concurrent learning and control architectures (2024). https://lafflitto.com/Codes/Springer_RKHS_Book_Newton_basis_MRAC.zip

Chapter 7
Data-Driven Methods and Adaptive Control: Stochastic Analysis

Chapter 6 gave a detailed account of how contemporary practices in *deterministic* data-driven methods and learning theory can be exploited to define a general nonparametric theory of deterministic learning and adaptive control. The overall goal of Chap. 6 was to develop a general framework that can reduce, or ideally eliminate, the reliance on oracular information of modern parametric adaptive control theory methods. The deterministic analysis in Chap. 6 built such a theory both by introducing and making systematic use of Jackson inequalities of approximation theory, as they naturally arise in native spaces, and by introducing various control architectures for scheduling subspace learning and designing closed-loop control laws for the control phase. This chapter discusses how a stochastic approach can be employed to sample the space of matched functional uncertainties acting on a deterministic, continuous-time set of ordinary differential equations (ODEs). Once again, in this effort, key properties of native spaces will help forecast the ultimate bounds on the trajectory tracking error as a function of how the uncertainty space is sampled.

The majority of recent attempts to exploit RKHSs and explore nonparametric adaptive control methods for ODEs incorporate some type of stochastic analysis of data on the functional uncertainty or the closed-loop system to guarantee controller performance. This stochastic analysis usually employs principles of Gaussian processes (GPs) for the development of the underlying theory. Stochastic analysis has been used to develop different types of nonparametric adaptive control systems based on principles of GPs in references like [1–32]. We would argue that many of these most recent approaches, including those introduced in the past few years, come closer to defining a general theory of nonparametric adaptive control, whereas approaches based on a purely deterministic analysis do not form a cohesive theory yet. Such a predisposition to use GPs is easy to understand. One of the singular achievements during the last decade in the development of statistical and machine learning theory has been the realization that GPs provide systematic, pragmatic, and theoretically rich means of solving many important classical learning problems. Consequently, one may deduce that the prevalence of recent approaches that use GPs to study adaptive control of ODEs in continuous time is predicated on their clear ability to

A. J. Kurdila et al., *Data-Driven, Nonparametric, Adaptive Control Theory*,
Lecture Notes in Control and Information Sciences 495,
https://doi.org/10.1007/978-3-031-78003-5_7

formulate and solve certain learning problems in discrete time that use independent and identically distributed (IID) measurements of the input/output response.

The decision to include a stochastic component in the analysis of adaptive controllers for ODEs has both advantages and disadvantages. On the positive side, the incorporation of probabilistic concepts enables stochastic analysis to formulate and solve other questions of interest beyond those addressed in a solely deterministic analysis. Notions of confidence, reliability, or safety of a control strategy are intuitive and provide valuable insights about the performance of a control strategy. These concepts are incorporated naturally in many recent formulations using GPs to derive adaptive controllers. On the other hand, the performance analysis of a feedback controller that includes stochastic principles is more difficult. A large part of this chapter explores how this analysis becomes more difficult. We also explain how the theoretical guarantees of performance in the stochastic analysis are often more complicated to state, and sometimes less sharp, than those obtained via a deterministic analysis as in Chap. 6. Sometimes, choosing to use error bounds that arise in a stochastic analysis of the nonparametric controller must account for an approximation error term and a stochastic error term. The approximation error term depends on the choice of the hypothesis space, decreases with the number N of parameters that characterize the approximating hypothesis space, and is equivalent to expressions obtained for the error in the deterministic methods studied in Chaps. 5 and 6. The stochastic error term, on the other hand, generally depends on both N and the number of measurements M. It is a classical result that, often, this stochastic term grows with N.

Several examples presented in this chapter emphasize that, at least sometimes, trying to find a role for GPs to build feedback controllers for ODEs in continuous time may rely on assumptions that are difficult to realize in practice or justify theoretically. GPs excel when we study offline regression problems described in terms of IID measurement processes. The use of samples along a trajectory of a deterministic ODE is the foundation of many recent data-driven modeling approaches. For an ODE in continuous time, where it is most natural to collect samples along a trajectory, such samples are not IID. Assuming that the plant model is structured so that input/output samples of the uncertainty are available from IID measurements can be quite counter-intuitive in a setting governed by ODEs. At the very least, we need to fully understand the qualitative, practical, and theoretical differences between using samples collected along the trajectory of a deterministic ODE, in contrast to the situation when we make direct IID measurements of the uncertainty dictated by a random walk on some domain.

This chapter is based on the study of deterministic plant models captured by ODEs in continuous time that depend parametrically on random samples of the matched functional uncertainty. The proposed analysis relies on the tools of conventional deterministic Lyapunov analysis for ODEs in continuous time, combined with certain concentration inequalities for the probabilistic error in the functional uncertainty. This latter strategy encompasses many recent efforts, like those in [2, 33, 34]. An alternative approach can be based on replacing the deterministic ODEs with a random dynamical system and then using stochastic notions of solutions, stability, convergence, and Lyapunov analysis to study performance. This is the approach taken in [35], which is left for future research.

This chapter is organized as follows. Section 7.1 recalls some of the most important symbols and notational conventions used in this chapter. Section 7.2 presents selected notions of probability theory. Section 7.3 presents the distribution-free regression problem through a generic formulation. Successively, Sect. 7.4 explains how the distribution-free regression problem can be solved by employing the empirical risk minimization approach. Results of Sect. 7.4 are further specialized in Sect. 7.5 considering GPs. A useful method to quantify the quality of stochastic methods for approximations, namely the use of concentration inequalities, is presented in Sect. 7.6. These results are applied in Sect. 7.7 to characterize LTC architectures. Section 7.8 presents numerical simulation results employing LTC and SLC architectures. Finally, Sect. 7.9 presents some concluding remarks, and Sect. 7.10 suggests some homework problems.

7.1 Notation, Symbols, and Terminology

Consistently with the previous chapters, a scalar-valued native space is denoted by $\mathcal{H}$, a vector-valued native space is denoted by $\boldsymbol{\mathcal{H}}$, a scalar-valued reproducing kernel is denoted by $\mathfrak{K}$, an operator-valued reproducing kernel is denoted by $\mathcal{K}$, a set of centers is denoted by Ξ_N, and subspaces of approximants are denoted by $\mathcal{H}_N \subseteq \mathcal{H}$ or $\boldsymbol{\mathcal{H}}_{d(N)} \subseteq \boldsymbol{\mathcal{H}}$. We consider plant models in the same form as (6.4),

$$\dot{x}(t) = Ax(t) + B(u(t) + f(x(t))) + \Gamma(t), \quad x(t_0) = x_0, \quad t \geq t_0, \tag{7.1}$$

which we repeat here for convenience. As in the rest of this book, the state vector is denoted by $x(t) \in \mathbb{X} \triangleq \mathbb{R}^n$, the control input is denoted by $u(t) \in \mathbb{U} \triangleq \mathbb{R}^m$, and the unknown unmatched disturbance is denoted by $\Gamma(t) \in L^\infty([t_0, \infty), \mathbb{X})$. Finally, as in the earlier chapters of this book, we assume that (7.1) is forward complete.

The term "prior" is used in two distinct ways, according to the context. In approximation theory, a prior usually takes the form of a smoothness condition on an unknown function that can be used to inform the construction of approximations or to derive more precise bounds on approximation errors. In the context of GPs, a prior is used to distinguish between our initial notion about a GP used to represent a function and the improved posterior estimate that is constructed after making some observations.

As in Chap. 6, we denote a generic Banach space by $\mathbb{B}$ and a generic Hilbert space by $\mathbb{H}$. Despite Chap. 6, where the hypothesis space was a Hilbert space, in this chapter, the hypothesis space may also be a Banach space. Thus, to remark on the fact that hypothesis space may be a Banach space or a Hilbert space, here we denote the hypothesis space by H. As shown in Fig. 6.1, approximating hypothesis spaces are such that

$$H_1 \subset \ldots \subset H_{d(N)} \subset H_{d(N)+1} \subset \ldots \subset H \hookrightarrow \mathbb{B}; \tag{7.2}$$

it is worthwhile comparing (6.3) with (7.2).

Remark 7.1 In this chapter, the approximating hypothesis space is not necessarily a vector-valued native space. For consistency with the notation employed in the previous chapters, the dimension of the approximating space is shown as a subscript. Thus, $H_{d(N)}$ has dimension $d(N)$. Until this generic approximating hypothesis space is explicitly set as an RKHS generated by N centers, the expression $d(N)$ may seem redundant. However, we preferred employing this notation to help making analogies and connecting concepts.

7.2 Elements of Probability Theory

For their relevance to the scope of this chapter, recall the notions of probability space, sample space, and probability measure introduced in Definition 2.65 and the notion of a random variable given in Definition 2.66 in the context of measure theory in Sect. 2.4.5. In this section, we present additional key elements of probability theory.

Given the probability space $(\mathbb{X}, \Sigma, \nu)$, the *probability* $\Pr(A)$ of the event $A \subseteq \mathbb{X}$ is $\nu(A)$. We also write this probability as

$$\Pr(A) \triangleq \nu(A) = \int_A \nu(\mathrm{d}x), \tag{7.3}$$

where the integral is interpreted in the sense of Lebesgue on $\mathbb{R}^n$.

Definition 7.1 *(Distribution of a Random Variable)* The *distribution* λ_f for a random variable $f : \mathbb{X} \to \mathbb{R}$ is the probability measure on $\mathbb{R}$ defined by

$$\lambda_f(O) \triangleq \nu(f^{-1}(O)) \tag{7.4}$$

for each open set $O \subseteq \mathbb{R}$.

Intuitively, the distribution λ_f of a random variable f is the probability measure on the range $\mathbb{R}$ that describes where the values of $f(x)$ accumulate when we select x randomly in $\mathbb{X}$ according to ν. A *random process* is an infinite family of random variables. Examples of random processes include sequences of random variables such as $\{f_k\}_{k\in\mathbb{N}}$ with each $f_k(\cdot) : \mathbb{X} \to \mathbb{R}$, or families of time-varying functions $\{f(t,\cdot)\}_{t\geq 0}$ with each $f(t,\cdot) : \mathbb{X} \to \mathbb{R}$, fixed t.

One of the most popular probability measures is defined over $\mathbb{X} \triangleq \mathbb{R}^n$ and such that $\nu(\mathrm{d}x) = p(x)\mathrm{d}x$, the probability of A is computed as

$$\Pr(A) \triangleq \nu(A) = \int_A p(x)\mathrm{d}x, \tag{7.5}$$

and

$$p(x) \triangleq \frac{e^{-\frac{1}{2}(x-\bar{m})^{\mathrm{T}}\mathbb{K}^{-1}(x-\bar{m})}}{\sqrt{(2\pi)^{-n}\det(\mathbb{K})}}, \quad \text{for all } x \in \mathbb{X} \triangleq \mathbb{R}^n \tag{7.6}$$

denotes the Gaussian *probability density function*. In (7.6), $\bar{m} \in \mathbb{R}^n$ denotes the *mean*, $\mathbb{K} \in \mathbb{R}^{n\times n}$ denotes the symmetric, positive-definite *covariance matrix*, and $\det(\mathbb{K})$ denotes the determinant of $\mathbb{K}$. If the density function (7.6) defines a probability measure on $\mathbb{X} \triangleq \mathbb{R}^n$, then we say that it has a *normal distribution* or *Gaussian distribution* and write $\nu \sim \mathcal{N}(\bar{m}, \mathbb{K})$ for some fixed mean and covariance. If $n = 1$, it is conventional to define the *standard deviation* $\sigma \in \mathbb{R}$ in terms of the covariance so that $\sigma^2 = \mathbb{K} > 0$.

To introduce identically distributed (IID) processes, recall that if $(\mathbb{X}, \Sigma, \nu)$ is a probability space, then two events $A, B \subset \Sigma$ are *independent* if $\Pr(A \bigcap B) = \Pr(A)\Pr(B)$. The random variables $f, g \in \mathbb{X}$ are *independent* whenever $\Pr(\{(f, g) \in A \times B\}) = \Pr(\{f \in A\})\Pr(\{g \in B\})$. We say that f and g are *independent identically distributed* (IID) if their distribution functions λ_f and λ_g, are equal to each other. An alternative definition is the following.

Definition 7.2 (*IID Random Variables on* $\mathbb{R}$) Let $\mathbb{X} \triangleq \mathbb{R}$, let $I \subseteq \mathbb{X}$ be a non-zero measure set, and let y, y_1, y_2 be real-valued random variables over I. We define the cumulative distribution functions as

$$F_y(x) \triangleq \Pr\left(\{y \in I : y \leq x\}\right), \quad \text{for any } y \in I, \tag{7.7a}$$

$$F_{y_1,y_2}(x_1, x_2) \triangleq \Pr\left(\{(y_1 \in I : y_1 \leq x_1) \text{ or } (y_2 \in I : y_2 \leq x_2)\}\right), \quad \text{for any } y_1, y_2 \in I. \tag{7.7b}$$

The random variables y_1 and y_2 are *independent identically distributed* (*IID*) if $F_{y_1}(x) = F_{y_2}(x)$ for all $x \in I$ and $F_{y_1,y_2}(x_1, x_2) = F_{y_1}(x_1)F_{y_2}(x_2)$.

The notion of random IID variables can be extended to the case whereby $\mathbb{X} \triangleq \mathbb{R}^n$ by applying Definition 7.2 component-wise. In this chapter, we denote by ρ a *probability measure* over the measurements space Z. If the random variables $\{z_1, \ldots, z_M\}_{i=1}^M$ are IID, then we define the probability measure ρ^M on Z^M as $\rho^M(\mathrm{d}z) = \rho(\mathrm{d}z_1)\rho(\mathrm{d}z_2) \cdot \ldots \cdot \rho(\mathrm{d}z_M)$ for $z \triangleq \{z_1, \ldots, z_M\} \in Z^M$.

We conclude this introduction to elements of probability theory by introducing concentration inequalities, which play a distinguished role in this chapter. Suppose that an estimate $f_{N,M} \in H_{d(N)}$ of the uncertainty $f \in \mathbb{B}$ depends parametrically on M IID observations $\{\nu_1, \ldots, \nu_M\}$ of a random variable ν. In this case, $f_{N,M}$ can be written as

$$f_{N,M}(x) \triangleq f_N(x; \nu_1, \ldots, \nu_M) \in H_{d(N)} \subset \mathbb{B}. \tag{7.8}$$

If we pick some desired error level $\epsilon_{N,M} > 0$, we can distinguish between a "bad" set of samples that generate large error relative to the target $\epsilon_{N,M}$,

$$\mathscr{B}_{N,M} \triangleq \left\{ z_M \in Z^M : \|f - f_{N,M}\|_{\mathbb{B}} > \epsilon_{N,M} \right\} \tag{7.9}$$

and a "good" set of samples that generate small errors relative to the target $\epsilon_{N,M}$,

$$\mathscr{G}_{N,M} \triangleq \left\{ z_M \in Z^M : \|f - f_{N,M}\|_{\mathbb{B}} \le \epsilon_{N,M} \right\}. \tag{7.10}$$

Let $\Delta_{N,M} \in [0, 1]$ measure the sizes of the bad sets. Concentration inequalities are defined such that

$$\rho^M\left(\mathscr{B}_{N,M}\right) \le \Delta_{M,N}, \tag{7.11a}$$

$$\rho^M(\mathscr{G}_{N,M}) \ge 1 - \Delta_{M,N}, \tag{7.11b}$$

respectively. Intuitively, these inequalities show how the sizes of the good and bad sets vary as the approximation error varies. Often, (7.10) is concisely presented as

$$\rho^M\left(\left\{ z_M \in Z^M : \|f - f_{N,M}\|_{\mathbb{R}} \le \epsilon_{N,M} \right\}\right) \ge 1 - \Delta_{N,M}, \tag{7.12}$$

or equivalently,

$$\Pr\left(\left\{ z_M \in Z^M : \|f - f_{N,M}\|_{\mathbb{R}} \le \epsilon_{N,M} \right\}\right) \ge 1 - \Delta_{N,M}. \tag{7.13}$$

We will find it more useful to present our arguments by making references to the "bad set" (7.9) and the "good set" (7.10) associated with a particular concentration inequality.

7.3 Distribution-Free Regression

The foundations of statistical learning theory were developed between 1960 and 1970 [36]. This discipline was later referred to as *machine learning*, owing to its more recent emphasis on computational methods and their subsequent influence on solving some common problems of artificial intelligence. Canonical problems of learning theory include classification, pattern recognition, and regression, among others. In this section, we focus on the distribution-free regression problem. The roots of its formulation as a learning problem can be found in [36], and a good resource for the myriad of subsequent refinements can be found in [37]. We also follow closely the approximation theoretic analysis of learning problems initiated by Cucker and Smale in [38], which has subsequently been refined by numerous authors [39–44].

The *learning problem*, or *regression problem*, seeks to approximate a function $f : \mathbb{X} \to \mathbb{Y}$, where $\mathbb{X}$ denotes the *input space* and $\mathbb{Y}$ denotes the *output space*. We denote the *input/output space* by $Z \triangleq \mathbb{X} \times \mathbb{Y}$. Approximations are typically constructed from noisy IID observations $\{(x_i, y_i)\}_{i=1}^M \approx \{(x_i, f(x_i))\}_{i=1}^M \subset Z$. We let $Z^M \triangleq \bigcup_{i=1}^M \{(x_i, y_i)\} \subset Z$. The IID observations are generated from random walks defined in terms of an unknown probability measure ρ on Z. In the literature on machine learning, there is considerable diversity in the assumptions on the set $\mathbb{X}$ over which the unknown function f is defined. A common assumption is that the input space $\mathbb{X}$ is a compact subset of $\mathbb{R}^n$. Another important case is when $\mathbb{X}$ is a compact Riemannian manifold.

Assuming that $\mathbb{X}$ is a compact metric space and $\mathbb{Y}$ is a normed vector space, a common case studied in the literature assumes that measurements are given by

$$y_i = f(x_i) + \nu_i \in \mathbb{Y}, \quad i \in \{1, \ldots, M\}, \tag{7.14}$$

where ν_i is a sample of a random variable on $\mathbb{Y}$ and the state x_i is selected randomly in $\Omega \subset \mathbb{X}$ according to a probability measure μ on $\mathbb{X}$. Since the assumptions whereby x_i and ν_i are random variables for any $i \in \{1, \ldots, M\}$ are accepted in the overwhelming majority of works in this area, we refer to them as the "standard assumptions" of statistical data-driven methods. The problem of identifying f from $\{(x_i, y_i)\}_{i=1}^M$ and (7.14) is known as the *distribution-free regression problem* of data-driven methods [37].

The working assumptions of statistical data-driven methods in the standard setting are just a bit more general than that in (7.14). We decompose the unknown measure ρ as

$$\rho(\mathrm{d}x, \mathrm{d}y) = \rho(\mathrm{d}y|x)\mu(\mathrm{d}x), \tag{7.15}$$

where

$$\mu(A) \triangleq \rho(A \times \mathbb{Y}) \tag{7.16}$$

denotes the *marginal measure over* $\mathbb{X}$ for the measure ρ over Z for all ρ-measurable subsets $A \times \mathbb{Y} \subset Z$, and $\rho(\mathrm{d}y|x)$ denotes the conditional probability measure over $\mathbb{Y}$, given $x \in \mathbb{X}$. This decomposition enables us to view the random walk that generates samples $(x_i, y_i) \in Z$ intuitively. First, we choose a sample $x_i \in \mathbb{X}$ randomly according to the probability measure μ over $\mathbb{X}$. Then, we generate the output sample y_i randomly according to the conditional probability measure $\rho(\cdot|x)$ over $\mathbb{Y}$.

With this interpretation of the measurement process, we construct estimates of the uncertainty f that depend on M samples using the setting of an optimization problem. We choose a hypothesis space $H \triangleq H(\mathbb{X}, \mathbb{Y}) \subseteq L^2_\mu(\mathbb{X}, \mathbb{Y})$ that defines the class of functions from which our estimates of the unknown f will be constructed. For this choice of the hypothesis space, we introduce the most popular *ideal risk* or *error functional* given by the quadratic cost

$$J_\rho(g) \triangleq \frac{1}{2}\int_Z \|g(x) - y\|_Y^2 \rho(\mathrm{d}x, \mathrm{d}y) + \frac{1}{2}\lambda\|g\|_S^2, \quad \text{for all } g \in H, \tag{7.17}$$

where S denotes a normed space and $\lambda \geq 0$ denotes a regularization parameter; in Sect. 7.4, we discuss how, if $\lambda \neq 0$, then the normed space S is used to regularize the construction of estimates. The ideal estimate $\hat{f} \in H$ is then defined such that

$$\hat{f} \in \arg\min_{g \in H} J_\rho(g). \tag{7.18}$$

The next result shows that if we choose the hypothesis space as $H = L^2_\mu(\mathbb{X}, \mathbb{Y})$, then it is possible to find the ideal minimizer $\hat{f} \in L^2_\mu(\mathbb{X}, \mathbb{Y})$. This result may be surprising since the probability measure ρ that generates IID samples in Z is unknown.

Theorem 7.1 *Suppose that $f \in L^2_\mu(\mathbb{X}, \mathbb{Y})$, and consider the error functional* (7.17) *with $\lambda = 0$. Then, the minimizer $\hat{f} \in L^2_\mu(\mathbb{X}, \mathbb{Y})$ of J_ρ over $H = L^2_\mu(\mathbb{X}, \mathbb{Y})$ in* (7.18) *is the* regressor function *defined as*

$$\hat{f}(x) = f_\rho(x) \triangleq \int_Y y\rho(\mathrm{d}y|x). \tag{7.19}$$

Proof For all $g \in L^2_\mu(\mathbb{X}, \mathbb{Y})$, it holds that

$$\begin{aligned}
J_\rho(g) &= \frac{1}{2}\int_X \int_Y (\|g(x)\|_Y^2 - 2(g(x), y)_Y + \|y\|_Y^2)\rho(\mathrm{d}y|x)\mu(\mathrm{d}x) \\
&= \frac{1}{2}\int_X \|g(x) - f_\rho(x)\|_Y^2 \mu(\mathrm{d}x) \\
&\quad \underbrace{-\frac{1}{2}\int_X \|f_\rho(x))\|_Y^2 \mu(\mathrm{d}x) + \frac{1}{2}\int_X \int_Y \|y\|_Y^2 \rho(\mathrm{d}y|x)\mu(\mathrm{d}x)}_{J_\rho(f_\rho)} \\
&= \frac{1}{2}\int_X \|g(x) - f_\rho(x)\|_Y^2 \mu(\mathrm{d}x) + J_\rho(f_\rho) = \frac{1}{2}\|g - f_\rho\|_{L^2_\mu(\mathbb{X},\mathbb{Y})} + J_\rho(f_\rho).
\end{aligned}$$

Therefore, if $f_\rho \in L^2_\mu(\mathbb{X}, \mathbb{Y})$, then

$$f_\rho \in \arg\min_{g \in L^2_\mu(\mathbb{X},\mathbb{Y})} J_\rho(g),$$

and the result is proven. ■

Theorem 7.1 characterizes the minimizer of the error functional J_ρ, in case the hypothesis space is chosen as the largest possible, that is, $H = L^2_\mu(\mathbb{X}, \mathbb{Y})$. In particular, this result shows that the problem of constructing an estimate $\hat{f}$ of the function solving the minimization of the quadratic risk (7.17) is equivalent to the approximation of the regressor function f_ρ. Although we know the form of the regressor function $f_\rho(x) \triangleq \int_Y y\rho(\mathrm{d}y|x)$, the measure ρ is unknown. For this reason, the underlying optimization problem is referred to as a *distribution-free learning problem*. In practice, we choose the hypothesis space $H \subseteq L^2_\mu(\mathbb{X}, \mathbb{Y})$ so that the minimizer in (7.18) provides a good approximation of the regressor $f_\rho \in L^2_\mu(\mathbb{X}, \mathbb{Y})$.

The overall aim is to develop a theory and associated algorithms that are robust in the sense that they guarantee measures of performance for a class $C \subset \mathbb{B}$ of regressor functions $f_\rho \in C$ that is as wide as possible. The assumption whereby $f_\rho \in C \subset \mathbb{B}$ is the *approximation theoretic prior.*

7.4 Empirical Risk Minimization

7.4.1 Overview

One of the earlier, and most fundamental, achievements of learning theory was the systematic development of the method of empirical risk minimization (ERM) [36]. The method of ERM was developed to address the fact that the probability measure appearing in the regressor function f_ρ is unknown. Since it is assumed that input/output data samples $\{(x_i, y_i)\}_{i=1}^M$ in Z are obtained by an IID measurement process defined in terms of ρ, we define the *empirical risk* J_M that approximates the ideal quadratic risk J_ρ as

$$J_M(g) \triangleq \frac{1}{M}\sum_{i=1}^{M} \|g(x_i) - y_i\|_Y^2 + \frac{1}{2}\lambda\|g\|_S^2, \tag{7.20}$$

where $\lambda \geq 0$, and the norm in S defines a regularization term when $\lambda \neq 0$. We define the *ideal empirical estimate* $f_M \in H$ using M observations as the solution of the minimization problem

$$f_M \in \arg\min_{g\in H} J_M(g). \tag{7.21}$$

The empirical estimate $f_M \in H$ is defined *ideal* since it is defined in terms of a minimization over the generally infinite-dimensional hypothesis space $H \subset L^2_\mu(\mathbb{X}, \mathbb{Y})$.

To generate practical algorithms, estimates $f_{N,M} \in H_{d(N)}$ of the ideal empirical minimizer $f_M \in H$ are usually defined in terms of some N-dimensional subspace $H_{d(N)}$ of a chosen hypothesis space $H \subseteq L^2_\rho(\mathbb{X}, \mathbb{Y})$. Thus, we seek an estimate $f_{N,M}$ such that

$$f_{N,M} \in \arg\min_{g\in H_{d(N)}} J_M(g). \tag{7.22}$$

In (7.22), $f_{N,M}$ is obtained by searching over the finite-dimensional space $H_{d(N)}$, in contrast to (7.19) and (7.21) that characterize f_ρ and f_M, respectively. We refer to $f_{N,M} \in H_{d(N)} \subseteq H$ as the *realizable empirical estimate* since the optimization problem (7.22) is designed to be solvable via computation. Thus, in approximations constructed from the ERM principle, we want to construct estimates $f_{N,M}$ that attain the minimum for the empirical risk J_M. The additional knowledge that the ideal estimate $f_\rho \in C$ can be used to derive sharper error bounds on performance.

Example 7.1 (Coordinate Realizations) Consider the problem of studying the vector-valued uncertainty $f : \mathbb{X} \to \mathbb{Y}$ with $x \subset \mathbb{R}^n$ and $\mathbb{Y} \triangleq \mathbb{R}^m$. To this goal, let the smoothness space S be an inner product space and choose basis functions with $\phi_l(x) \in \mathbb{R}^m$. Furthermore, let

$$\Phi_N(x) \triangleq \left[\phi_1(x), \ldots, \phi_{d(N)}(x)\right] \in \mathbb{R}^{m\times d(N)}, \tag{7.23a}$$

$$H_{d(N)} \triangleq \text{span}(\Phi_N), \tag{7.23b}$$

$$X_M \triangleq [x_1, x_2, \ldots, x_M]^{\mathrm{T}} \in \mathbb{R}^{M\times n}, \tag{7.23c}$$

$$Y_M \triangleq \left[y_1^{\mathrm{T}}, \ldots, y_M^{\mathrm{T}}\right]^{\mathrm{T}} \in \mathbb{R}^{mM}, \tag{7.23d}$$

$$\Phi_N(X_M) \triangleq \begin{bmatrix} \Phi_N(x_1) \\ \vdots \\ \Phi_N(x_M) \end{bmatrix} \in \mathbb{R}^{mM\times d(N)}, \tag{7.23e}$$

$$\mathbb{S}_N \triangleq [\langle \phi_i, \phi_j\rangle_S] \in \mathbb{R}^{d(N)\times d(N)}. \tag{7.23f}$$

With this notation, the empirical risk for a function

$$g \triangleq \sum_{i=1}^{d(N)} \theta_i \phi_i = \Phi(X_M)\Theta \tag{7.24}$$

of coordinates $\Theta \triangleq \left[\theta_1, \ldots, \theta_{d(N)}\right]^{\mathrm{T}} \in \mathbb{R}^{d(N)}$ can be rewritten as

$$J_M(g) = \Theta^{\mathrm{T}} \left(\frac{1}{M}\Phi_N^{\mathrm{T}}(X_M)\Phi_N(X_M) + \lambda \mathbb{S}\right)\Theta - \frac{2}{M}\Phi_N^{\mathrm{T}}(X_M)Y_M + \frac{1}{M}Y_M^{\mathrm{T}}Y_M. \tag{7.25}$$

This expression for the empirical risk is just a quadratic form in the unknown coefficients, and a local minimum can be calculated by setting the gradient of J_M with respect to Θ equal to zero. In particular, the coordinates Θ are determined by the solution of the matrix equations

$$\left(\Phi_N^{\mathrm{T}}(X_M)\Phi_N(X_M) + \bar{\lambda}\mathbb{S}_N\right)\Theta = \Phi_N^{\mathrm{T}}(X_M)Y_M \tag{7.26}$$

for the new regularization constant $\bar{\lambda} \triangleq M\lambda$. If the basis functions in Φ_N are linearly independent in the smoothness space $\mathbb{S}$, then (7.26) has a unique solution

$$\Theta = \left(\Phi_N^{\mathrm{T}}(X_M)\Phi_N(X_M) + \bar{\lambda}\mathbb{S}_N\right)^{-1} \Phi^{\mathrm{T}}(X_M)Y_M \in \mathbb{R}^{d(N)}, \tag{7.27}$$

which defines the realizable empirical risk estimate $f_{N,M} = \Phi_N\Theta \in H_{d(N)}$. $\triangle$

7.4.2 The Role of the Representer Theorem

Classical tools of RKHS theory allow addressing many important learning problems, in which data is given by discrete IID measurements. This section gives a first hint of the power of RKHS in studying approximations generated by the ERM approach.

An estimate obtained using ERM to approximate the solution of a distribution-free regression problem depends on several key parameters, such as, the choice of the universe Banach space $\mathbb{B}$, the choice of the hypothesis space H, the dimension $d(N)$ of the space of approximants $H_{d(N)}$, the basis functions Φ_N that define $H_{d(N)}$, the number M of IID input/output observations, and the set C that defines the (approximation theory) priors for the regressor $f_\rho \in C \subset \mathbb{B}$. The inventory of learning algorithms derived for different choices of these parameters is vast. Just for scalar-valued uncertainties defined over $\mathbb{X} \triangleq \Omega \subset \mathbb{R}^n$, the performance of learning algorithms for hypothesis spaces such as the r-times continuously differentiable functions, Hölder continuous functions, Sobolev spaces, and Besov spaces have been considered. In many cases, these results for scalar-valued functions over subsets of the Euclidean space have been extended to obtain results for vector-valued uncertainty, or different choices of $\mathbb{X}$ such as Riemannian manifolds. The bases used for constructing finite-dimensional approximations in any of these spaces vary across different approaches. Fourier bases, various finite element bases, piecewise polynomial bases, spline bases, wavelets, and kernel bases are just a few of the commonly used choices. The interested reader should consult [45], the treatise [46], and the more recent work [47] to gain an appreciation of the breadth and generality of specific learning algorithms for nonparametric regression. Learning theory allows assessing the performance of robust and adaptive controllers for plants affected by matched functional uncertainties assumed to lie in some Banach space $\mathbb{B}$, and these bounds on the tracking error are in terms of $d(N)$ and M. The most flexible, and most generally useful, error bounds allow N and M to be chosen independently.

Since $H_{d(N)} \subseteq H \subset L^2_\mu(\mathbb{X}, \mathbb{Y})$, it holds that

$$J_M(f_M) \leq J_M(f_{N,M}), \tag{7.28}$$

so that the realizable empirical estimate $f_{N,M} \in H_{d(N)} \subset H$ cannot perform better than the ideal empirical estimate $f_M \in H$. Indeed, for an arbitrary H and an arbitrary choice of bases $\Phi_N \subset H$, $f_M \neq f_{N,M}$. The next result provides sufficient conditions, which leverage the properties of native spaces, for the realizable empirical estimate to be equal to the ideal empirical estimate.

Theorem 7.2 (Representer Theorem on Scalar-Valued RKHSs, [48, 49, Theorem 8.7]) *Assume the hypothesis space H is a scalar-valued RKHS $\mathcal{H}$ of scalar-valued functions over $\mathbb{X}$ with kernel $\mathfrak{K}$, and assume there exist M samples $\{(x_i, y_i)\}_{i=1}^M$ in the input/output space $Z \triangleq \mathbb{X} \times \mathbb{Y}$. Let $S = \mathcal{H}$ and define*

$$H_M \triangleq \mathcal{H}_M \triangleq \text{span}\left\{\mathfrak{K}_{x_i} : x_i \in \Xi_M, 1 \leq i \leq M\right\}, \tag{7.29}$$

where

$$\Xi_M \triangleq \{x_1, \ldots, x_M\} \subset \mathbb{X}. \tag{7.30}$$

Then,

$$f_M = f_{M,M}, \tag{7.31}$$

where the minimizer f_M *of the ideal empirical optimization problem verifies* (7.21) *and the minimizer* $f_{M,M}$ *of the realizable optimization problem verifies* (7.22).

It is important to note that Theorem 7.2 applies to scalar-valued hypothesis spaces with $N = M$ and, hence, $H_{d(N)} = H_M$. Under the assumptions of Theorem 7.2, the coordinate realization for the minimizer $f_M \triangleq f_{M,M}$ can be deduced from (7.27). Indeed, let $N = M$, $m = 1$, $H = \mathcal{H}$, and $S = \mathcal{H}$. In this case, (7.23) imply that

$$\Phi_M(x) \triangleq \mathfrak{K}_{\Xi_M}(x) = \left[\mathfrak{K}_{x_1}(x), \ldots, \mathfrak{K}_{x_M}(x)\right]^{\mathrm{T}} \in \mathbb{R}^M, \tag{7.32}$$

$$H_{d(N)} \triangleq \mathcal{H}_M = \operatorname{span}(\Phi_M), \tag{7.33}$$

$$Y_M \triangleq [y_1, \ldots, y_M]^{\mathrm{T}} \in \mathbb{R}^M. \tag{7.34}$$

Thus, $f_M = \mathfrak{K}_{\Xi_M}^{\mathrm{T}}\Theta$, where

$$\Theta = \left(\mathbb{K}_N + \bar{\lambda} I_M\right)^{-1} Y_M \in \mathbb{R}^m \tag{7.35}$$

with $\bar{\lambda} \triangleq M\lambda$. It is noteworthy that if $\lambda = 0$ and there is no regularization, then the estimate $f_M = f_{M,M}$ is the orthogonal projection of f onto $H_M = \mathcal{H}_M$, and $f_M = \Pi_M f$.

Theorem 7.2 has been extended in a wide variety of cases; see [50–53]. In particular, proceeding as in [50, 51], Theorem 7.2 can be restated without essential changes for vector-valued learning problems, that is, for $\mathbb{Y}$ being an m-dimensional normed vector space. In this case, we choose

$$d(N) = mM, \tag{7.36a}$$

$$H = \mathcal{H}, \tag{7.36b}$$

$$S = \mathcal{H}, \tag{7.36c}$$

$$\phi_{i,j} \triangleq \mathcal{K}_{x_i} e_j, \quad \text{for } 1 \leq i \leq M, 1 \leq j \leq m, \tag{7.36d}$$

$$\Phi_{d(M)}(x) = \mathcal{K}_{\Xi_M}(x) \triangleq \left[(\mathcal{K}_{x_1} e_1), \ldots, (\mathcal{K}_{x_M} e_m)\right]^{\mathrm{T}} \in \mathbb{R}^{mM}, \tag{7.36e}$$

$$\mathcal{H}_{d(M)} = \operatorname{span}(\Phi_{d(M)}), \tag{7.36f}$$

$$Y_M = [y_1, \ldots, y_M]^{\mathrm{T}} \in \mathbb{R}^{M \times m}. \tag{7.36g}$$

Thus, the ERM process generates coefficients in Θ that have the same form as in (7.35) with $\mathbb{K}_N$ now denoting the Grammian matrix $\mathbb{K}_N \in \mathbb{R}^{mM \times mM}$ of the basis

$\mathcal{K}_{\Xi_M}$ discussed in Sect. 3.8, and we replace the identity matrix I_M with the identity matrix I_{mM}. If $\lambda = 0$, then the minimizer $f_{M,M}$ is the orthogonal projection Π_M of f onto $\mathcal{H}_{d(M)}$.

We have expressed the analysis of the ERM principle in this section under the assumption that the samples $\{(x_i, y_i)\}_{i=1}^M$ are generated randomly according to the probability measure ρ. This is the standard setting of ERM in statistical learning theory. However, Theorem 7.2 holds for any samples $\{(x_i, y_i)\}_{i=1}^M$, whether they are deterministically prescribed or generated randomly according to ρ.

7.4.3 The Classical Bias-Variance Trade-Off

Given the plethora of approaches to derive estimates $f_{N,M}$ via ERM, the challenge for control theory is to discriminate among them and choose learning strategies that are attractive in a real-time control setting. In this section, we begin a study of the structures of errors as they commonly arise in statistical learning theory methods based on ERM. This study will help focusing on the factors contributing to error bounds and influencing the performance bounds derived later in this chapter for families of controllers. Specifically, this section reviews the bias versus variance trade-off.

Let $\mathbb{Y} \triangleq \mathbb{R}$, let $\mathbb{B} \triangleq L^2_\mu(\mathbb{X})$ be the universe space, let $H \subseteq \mathbb{B}$ be the hypothesis space, and let the M samples $\{(x_i, y_i)\}_{i=1}^M$ be IID measurements generated by the probability measure ρ so that (7.15) is verified on $Z \triangleq \mathbb{X} \times \mathbb{Y}$. Furthermore, consider the quadratic error functional (7.17). If the regressor is contained in the uncertainty class C, then, as shown in [45], the error can be decomposed as

$$\|f_\rho - f_{M,N}\|_{L^2_\mu(\mathbb{X})} \leq \|f_\rho - f_{\rho,N}\|_{L^2_\mu(\mathbb{X})} + \|f_{\rho,N} - f_{M,N}\|_{L^2_\mu(\mathbb{X})}, \tag{7.37}$$

where $f_{\rho,N}$ denotes the element in $H_{d(N)} \bigcap C$ that is closest to the regressor function f_ρ in the norm defined over the universe space $\mathbb{B}$. The first term on the right-hand side of (7.37) is known as the *approximation error* or *bias*, and it is a measure of the distance between the regressor f_ρ and its closest approximation in $\mathcal{H}_{d(N)} \bigcap C$. The second term on the right-hand side of (7.37) is commonly referred to as the *probabilistic error*, *stochastic error*, or *variance*. It is worthwhile noting that the approximation error is not a function of M. Typically, the approximation error term converges to zero as $N \to \infty$. The stochastic error increases with N and decreases with M. Figure 7.1 shows a graphical representation of this behavior, which is known as *bias versus variance trade-off*. The theory of inverse methods addresses the bias versus variance trade-off in great detail.

The bias-variance trade-off presented in (7.37) for scalar-valued uncertainties can be extended with relatively minor modifications to vector-valued uncertainties. For example, as pointed out [44, p. 1301] or [37, Th. 13.3], for any linear finite-dimensional vector-valued space $H_{d(N)}$, it holds that

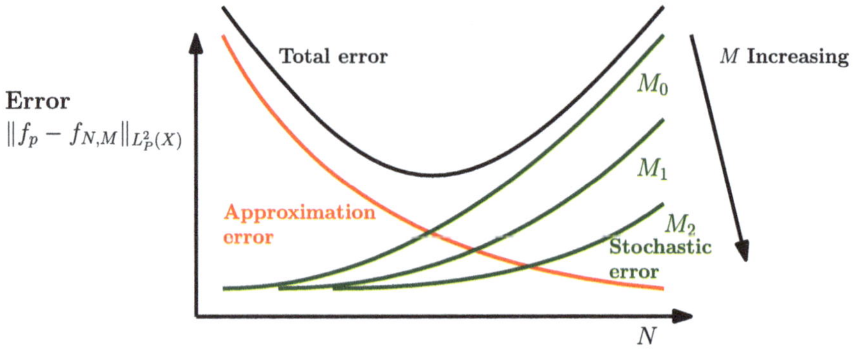

Fig. 7.1 Graphical representation of the classical bias-variance trade-off

$$\mathbb{E}_{\rho^M}\left(\|f_\rho - \mathrm{Tr}(f_{N,M})\|_{L^2_\mu(\mathbb{X})}\right) \leq \underbrace{\inf_{g\in H_{d(N)}} \|g - f_\rho\|_{L^2_\mu(\Omega)}}_{\text{approximation error}} + \underbrace{N\frac{\log(M)}{M}}_{\text{stochastic error}}, \tag{7.38}$$

where the *truncation operator* $\mathrm{Tr} : \mathbb{Y} \to \mathbb{Y}$ is such that $\mathrm{Tr}(g)$ is contained in a compact subset of the output space $\mathbb{Y}$. In (7.38), $\mathbb{E}_{\rho^M}(\cdot)$ denotes the expectation with respect to the probability measure ρ^M on Z^M that generates the IID samples in Z^M.

Bounds of this sort are well-understood [46, 54–56] and are an important classical consideration in assessing the performance of any ERM approximation method. Presently, typical studies of approximation theory and adaptive control as in [57, 58] emphasize parsimony. Even with improvements in onboard processor storage and bandwidth, the dimension $d(N) = mN$ of the space of approximants $H_{d(N)}$ is still rather small, say $O(100)$. In this case, approximants suitable for online computations usually exhibit errors on the left of the bias-variance trade-off curve. We discuss this issue further in the context of deep learning and deep neural networks (DNN) later in this section.

7.4.4 The Curse/Blessing of Dimensionality and Deep Neural Networks

In Sects. 6.4.2.1–6.4.2.3, we discussed bounds on the closed-loop trajectory tracking error that are the first of their kind in the literature on adaptive control systems. For instance, observing (6.45), in Sect. 6.4.2.1, we noted that for smaller values of the user-defined bounds $\varepsilon^* \in (0, 1)$ on the tracking error, larger values of the dimension $n \in \mathbb{N}$ of the state space $\mathbb{X} \triangleq \mathbb{R}^n$ imply that considerably larger values of N, which characterizes the dimensions of the approximating hypothesis space, are required. This effect, known as the *curse of dimensionality*, was determined to be due to the lack of specific information regarding the parallelepiped $\Omega \subset \mathbb{X}$, in which the closed-loop trajectory is assumed to remain. In Sects. 6.4.2.2 and 6.4.2.3, we showed that if we reduced the size of the uncertainty class, by choosing a set $\Omega \subset \mathbb{X}$ that is smaller, then we could alleviate some of the difficulties associated with the curse of dimensionality. In this section, we discuss the curse of dimensionality, and the associated notion of blessing of dimensionality, in further detail as it arises and is addressed in techniques based on learning theory [59–61].

To give a good account of the curse or blessing of dimensionality, it is useful to recount a common way of organizing algorithms in learning theory. The current practice in machine learning theory often discriminates among methods based on the number of variables that characterize the phenomenon at hand in comparison to the number of measurements or samples used to construct estimates [59, 62, 63]. This criterion gives rise to learning problems in the over-parameterized and under-parameterized regimes, which is described next.

For ease of illustration, we initially assume that $m = 1$, so that uncertainties are scalar-valued. We are interested in building a linear prediction $y = x^{\mathrm{T}}\Theta$ from measurements $\{(x_i, y_i)\}_{i=1}^{M} \subset \mathbb{X} \times \mathbb{Y}$ using an ERM approach. In this initial setting, the number of states n can be understood as the number of variables, and M denotes the number of measurements. Let

$$Y_M \triangleq [y_1, \ldots, y_M]^{\mathrm{T}} \in \mathbb{R}^M, \tag{7.39}$$

$$X_M \triangleq [x_1, x_2, \ldots, x_M]^{\mathrm{T}} = \begin{bmatrix} x_{1,1} & \cdots & x_{n,1} \\ \vdots & \ddots & \vdots \\ x_{1,M} & \cdots & x_{n,M} \end{bmatrix} \in \mathbb{R}^{M \times n}. \tag{7.40}$$

If $\lambda = 0$ and there is no regularization, then the empirical risk (7.20) reduces to

$$J_M(g) = \frac{1}{M} (X_M\Theta - Y_M)^{\mathrm{T}} (X_M\Theta - Y_M), \tag{7.41}$$

and (7.27) implies that the vector $\Theta \in \mathbb{R}^n$ is a solution of the matrix equation

$$X_M\Theta = Y_M \in \mathbb{R}^M. \tag{7.42}$$

The study of such a linear matrix equation is a focus of numerical linear algebra and well-known volumes, such [64], which are filled with algorithms for its approximate solution. These methods are generally organized into techniques that are designed for the over-determined problem when $M > n$, the exactly determined problem when $M = n$, and the under-determined problem when $M < n$.

Learning theory provides a venue to generalize the above linear problem and consider nonlinear regression problems via ERM [62, 65]. We can think of the ERM estimate as an approximation of the solution of the linear matrix equation

$$\Phi_M(X_M)\Theta = Y_M. \tag{7.43}$$

In this case, if $\lambda = 0$ and there is no regularization, then the ERM principle generates an approximation based on the minimization of the empirical risk

$$J_M(g) = \frac{1}{M}\left(\Phi_N(X_M)\Theta - Y_M\right)^{\mathrm{T}}\left(\Phi_N(X_M)\Theta - Y_M\right). \tag{7.44}$$

If the matrix $\Phi_N(X_M) \in \mathbb{R}^{mM \times d(N)}$ is tall and narrow, that is, if $mM \gg d(N)$, then we are studying a learning problem in an over-determined regime. On the other hand, if the data matrix is short and wide, which occurs if $mM << d(N)$, then the associated learning problem is an example of an under-determined regime. Recent references like [62, 63, 66, 67] often use the terms under-parameterized and over-parameterized, respectively, instead of the older terms over-determined and under-determined common in numerical linear algebra.

Many older problems in statistical learning theory could be understood as examples in the over-determined regime. Reference [59] refers to these as the "classical world" of learning problems and gives a good summary of some of these cases. They include classical problems of optimization, function approximation, and numerical integration. The data matrices for such problems can arise from sensor data arrays, hyperspectral imaging, or time histories of financial data, for example. For these classical problems, it is therefore common that $mM \gg d(N)$, and historically the curse of dimensionality can be observed in some methods of learning theory. While there are many ways to define the curse of dimensionality, given the aim of this text, we only consider how the term applies to estimates generated via ERM. Perhaps, the simplest definition is the following.

Definition 7.3 *Curse of Dimensionality*) An algorithm that generates the estimate $f_{N,M}$ suffers from the *curse of dimensionality* if we must take $mM \approx O((1/\epsilon)^{d(N)}$ observations to achieve an accuracy of order $O(\epsilon)$.

In this definition, the cost of computing the estimate $f_{N,M}$ is described by the number of observations mM. The computational complexity, which measures the amount of computational work needed to generate estimates of a given accuracy [68], scales very poorly with increases in the dimension $d(N)$ in these cases.

As a specific example of such behavior, the authors in [59] credit Kalman with describing an example of the curse of dimensionality arising in the optimization of a scalar loss function that depends on trajectories having initial conditions in $x_0 \in \Omega \triangleq [0, 1]^n$. The cost of evaluating the scalar cost functional for all initial conditions on a uniform, multidimensional grid of mesh size Δ grows like $O(\Delta^n)$. Using such grids of initial points with the hope of achieving accuracy in approximating the cost function of $O(\Delta)$ becomes infeasible as the dimension n grows. This example of the curse of dimensionality for dynamics problems is commonly encountered in learning problems in the over-determined regime.

It is worthwhile noting the similarities of the argument put forth by Kalman for the computational complexity, and the curse of dimensionality in approximating cost functionals over some parallelepiped domain in $\mathbb{R}^n$. We have given explicit upper bounds on tracking error of the same qualitative nature when the trajectory is assumed to belong to a parallelepiped in $\mathbb{R}^n$.

The emphasis on the over-parameterized regime has resulted in the search for methods that do not exhibit such a catastrophic curse of dimensionality as $d(N)$ increases. The enthusiasm for this pursuit can be attributed also to the success of deep learning methods. Many authors have documented the fact that such methods perform extraordinarily well in empirical studies that involve high-dimensional data. The development of a general explanation or theory that accounts for this progress has lagged somewhat, as noted by [69]. Presently, the development of such an overall theory for deep learning is an active area of research, with many investigators. Notable examples include [66, 69–72], among others.

In this pursuit to achieve algorithms that do well for high-dimensional spaces, several authors have described algorithms that exhibit a so-called "blessing of dimensionality" [59–61, 73–75]. Unsurprisingly, a method exhibits the blessing of dimensionality if it does not exhibit a curse of dimensionality. A common feature of many methods that exhibit a blessing of dimensionality exploit what is known as a concentration of measure guarantees. The basic idea behind the concentration of measure can be described as follows. If A is an event in the probability space Z having a measure ρ, then we introduce the dilation $A_r \triangleq \{z \in Z : \text{dist}(z, A) \leq r\} \supset A$ for the scalar factor $r \geq 0$, where $\text{dist}(\cdot, \cdot)$ denotes the distance from z to the set Z; see (2.73). The concentration of measure occurs whenever $\Pr(A_r)$ grows rapidly with r. There are important problems and methods in learning theory for which $\Pr(A_r)$ grows quickly, even exponentially fast as $r \to \infty$.

An example of such behavior, which is particularly relevant to this book, involves the problem of estimating empirically the mean of a function. Let $F : Z \to \mathbb{R}$ be a Lipschitz continuous function, where $Z \triangleq \mathbb{X} \times \mathbb{Y} \subset \mathbb{R}^{n+1}$, and suppose that the observations $\{z_i\}_{i=1}^M \in Z^M$ are generated as IID samples according to the measure ρ on Z. Theorem 22.10 in [76] gives conditions on ρ so that there is a constant $C > 0$ such that the concentration inequality

$$\Pr\left(\mathscr{B}_M\right) \leq e^{-\frac{CM\epsilon^2}{\|F\|_{C^{0,1}(Z)}}} \tag{7.45}$$

is verified for the "bad set"

$$\mathscr{B}_M \triangleq \left\{ z_M \in Z^M : \frac{1}{M}\sum_{i=1}^{M} F(z_i) - \int F(z)\rho(dz) \geq \epsilon \right\}. \tag{7.46}$$

The probability of the bad set of samples $\mathscr{B}_M$ decreases exponentially fast as the number of samples M used to approximate the mean $\bar{F} \triangleq \int F(z)\rho(dz)$ increases. Concurrently, the size of the complementary set of good samples expands exponentially quickly. Note that (7.45) does not depend on the dimension n of the space that contains $\mathbb{X}$. The authors in [59, 76] shows that this result implies that Lipschitz continuous functions are, probabilistically speaking, very close to constant functions in high dimensions.

It should be noted that many of the standard bounds on the approximation error of ERM are derived using a concentration inequality similar to (7.45). For example, the strategies in [38, 39, 45, 47] employ Bennett's inequality or Hoeffding's inequality, which are quite similar in form to (7.45). In these references, the application of a concentration inequality such as in (7.45) is carried out for

$$F(z) \triangleq (f(x) - y)^2, \tag{7.47}$$

which would yield

$$\Pr\left(\left\{z_M \in Z^M : J_M - J_\rho \geq \epsilon\right\}\right) \leq e^{-\frac{CM\epsilon^2}{\|F\|_{C^{0,1}(Z)}}} \tag{7.48}$$

for $\lambda = 0$. In other words, using a concentration inequality like (7.45) gives us a characterization of how confident we can be that the empirical risk J_M approximates the quadratic error functional J_ρ. Fundamentally, the fact that the concentration inequality (7.45) does not depend on the dimension n of the underlying input space $\mathbb{X} \subset \mathbb{R}^n$ allows developing algorithms that exhibit the blessing of dimensionality.

It is therefore apparent how concentration inequalities provide distinct advantages for problems in high dimensions in learning theory. This chapter will use concentration inequalities developed on native spaces to study nonparametric adaptive control in a stochastic setting. These inequalities are subsequently used in Sect. 7.7 below to derive adaptive control strategies.

7.4.5 The Double Descent Phenomenon

The more recent push to study the over-parameterized regime, wherein $d(N) \gg mM$, resulted in some surprising innovations and insights regarding the efficacy of approximations obtained via ERM over the past five years. One of these latest results is known as the "double descent" phenomenon, which occurs in ERM estimators. If no regularization is used, that is, $\lambda = 0$, then finding the approximants $f_{N,M}$ to interpolate all sample points amounts to finding the coefficients Θ so that the empirical risk (7.44) is identically zero. This is only possible in the over-parameterized regime, where there are enough basis functions and coefficients to interpolate all the samples $\{(x_i, y_i)\}_{i=1}^M$. Because there is no regularization, the classical bias-variance trade-off in ERM might suggest that the error should grow as $d(N)$ becomes sufficiently large. However, while studying methods for the case whereby $d(N) \gg mM$, the authors in [62, 67, 77–81] documented numerous cases in which this is not the case. Instead, the error plots without regularization exhibit a "double descent" phenomenon as shown qualitatively in Fig. 7.2. Historically, these studies have often arisen in studies of the performance of deep learning algorithms or DNNs.

A comparison of errors between the under-parameterized and over-parameterized regimes is a complicated task, and much depends on the design of a particular problem

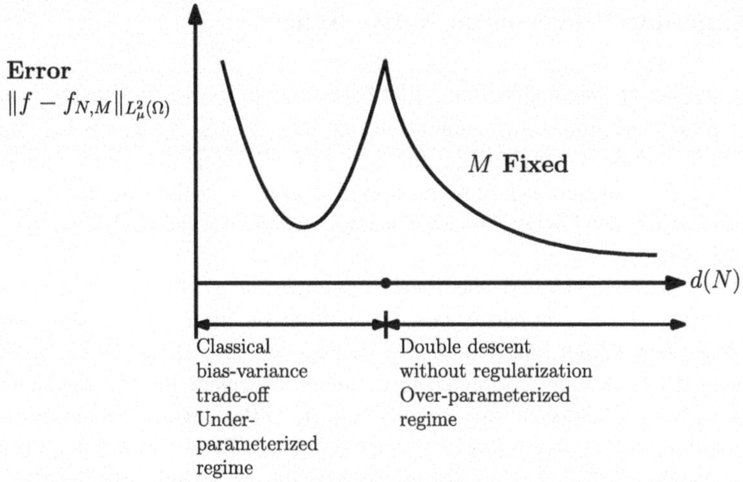

Fig. 7.2 Graphical representation of the double descent phenomenon

at hand when comparing the relative size of error over the two regimes [78]. Despite ample empirical evidence that double descent occurs in ERM, presently, researchers are only just beginning to develop a general theory that explains this observation.

One way to understand the double descent phenomenon is as follows. It can be interpreted as a way to "recover" the approximation error rates guaranteed near the minimum of the classical bias-variance curve as $d(N) \to \infty$. This means that if we have an effective method that balances the approximation error near the minimum of the classical bias-variance plot, we are not guaranteed that letting $d(N) \to \infty$ will yield better results. However, in some cases, the approximation errors appear commensurate to the errors achieved with classical regularization methods as $d(N) \to \infty$. As noted in [69], the provable rates of convergence for DNNs are only better in certain special cases, and, despite the progress in studies like [62, 67, 77–81], it fair to say that much needs yet to be done to have a full theoretical understanding of the double descent behavior.

In conclusion, it is important to understand the implications of the double descent phenomenon for "evaluating spaces and bases" for adaptive control methods. Here, we would only emphasize two observations. Firstly, the empirical reports on the performance of DNNs have often arisen in the context of large dimension $d(N)$ of the space of approximants. Secondly, current limits on computational bandwidth and storage in a real-time feedback control setting raises the question of the suitability of DNN networks of very high dimension $d(N)$.

7.5 Gaussian Processes on Native Spaces

Having discussed key elements of the ERM approach to address the distribution-free regression problem under the assumption that $H \subset L^2_\mu(\mathbb{X}, \mathbb{Y})$ and $\mathbb{B}$ is a Banach space, in this section, we review how the theory of GPs generates approximations of solutions of the distribution-free regression problem, under the assumption that the hypothesis space is an RKHS. Detailed discussions on GPs and RKHS can be found in [48, 82, 83].

A GP is constructed by first selecting both a *covariance kernel* $\mathfrak{K} : \mathbb{X} \times \mathbb{X} \to \mathbb{R}^+$, that is, a real-valued, symmetric, positive-definite function, and a mean function $\bar{m} : \mathbb{X} \to \mathbb{R}$. Any kernel function of positive type $\mathfrak{K} : \Omega \times \Omega \to \mathbb{R}^+$ that defines a scalar-valued RKHS can be selected to define a covariance kernel; see Definition 3.5. The associated Gaussian process, denoted by $GP(\bar{m}, \mathfrak{K})$, is the family of random variables $\{f_x\}_{x \in \mathbb{X}}$. Each random variable f_x for a fixed $x \in \mathbb{X}$ has a Gaussian distribution $\mathcal{N}(\bar{m}(x), \mathfrak{K}(x, x))$ characterized by the covariance $\mathfrak{K}(x, x)$ and mean function $\bar{m}(x) \in \mathbb{R}$. It is conventional to say that the Gaussian process $\{f_x\}_{x \in \mathbb{X}}$ has a *prior distribution* defined by $GP(\bar{m}, \mathfrak{K})$, and this fact is usually expressed by writing

$$f \sim GP(\bar{m}, \mathfrak{K}). \tag{7.49}$$

It is also common to say that we have a *prior belief* about the Gaussian process $f \sim GP(\bar{m}, \mathfrak{K})$. In these cases, $\sigma^2(x) = \mathfrak{K}(x, x)$ for all $x \in \mathbb{X}$. However, our prior belief $f \sim GP(\bar{m}, \mathfrak{K})$ is often just an initial guess about how the values of $f(\cdot)$ vary when sampled. Such an initial guess may be improved upon by leveraging additional information. Such improved estimates are generally referred to as *posterior estimates*. The probabilistic distribution of these posterior estimates is referred to as *posterior beliefs* about the Gaussian process $GP(\bar{m}, \mathfrak{K})$ that characterizes f.

Given a finite collection of centers $\{x_1, \ldots, x_M\} \subset \mathbb{X} \triangleq \mathbb{R}^n$ and the Gaussian process $\{f_x\}_{x \in \mathbb{X}}$, we can always construct the vector-valued random variable $\left[f_{x_1}(\cdot), \ldots, f_{x_M}(\cdot)\right]^{\mathrm{T}} \in \mathbb{R}^M$. It is a remarkable fact, which is the topic of many introductory accounts of GPs [83], that this vector-valued random variable has a normal distribution denoted by $\mathcal{N}(\bar{m}_M, \mathbb{K}_M)$ with a mean vector

$$\bar{m}_M = [\bar{m}_1, \ldots, \bar{m}_M]^{\mathrm{T}} \triangleq [\bar{m}(x_1), \ldots, \bar{m}(x_M)]^{\mathrm{T}}, \tag{7.50}$$

and covariance matrix

$$\mathbb{K}_M \triangleq \mathbb{K}(x_1, \ldots, x_M) \triangleq \begin{bmatrix} \mathfrak{K}(x_1, x_1) & \mathfrak{K}(x_1, x_2) & \cdots & \mathfrak{K}(x_1, x_k) \\ \mathfrak{K}(x_2, x_1) & \mathfrak{K}(x_2, x_2) & \cdots & \mathfrak{K}(x_2, x_k) \\ \vdots & \cdots & \ddots & \vdots \\ \mathfrak{K}(x_M, x_1) & \mathfrak{K}(x_M, x_2) & \cdots & \mathfrak{K}(x_M, x_M) \end{bmatrix}. \tag{7.51}$$

A good example of how this construction can be extended from $\mathbb{X} \triangleq \mathbb{R}^n$ to other structures, such as Riemannian manifolds, can be found in [84].

Remark 7.2 (Comment on Notation) In the previous chapters, we denote the centers defining an RKHS as $\xi_i \in \mathbb{X}$ for any $i \in \mathbb{N}$, and the collection of centers by Ξ. In this chapter, centers are points at which the uncertain function f is evaluated giving some output y_i for some $i \in \mathbb{N}$. Thus, to preserve the canonical notation (x_i, y_i) to capture the input-output pair, in this chapter, centers are denoted by x_i for any $i \in \mathbb{N}$. However, the collection of centers is still denoted by Ξ.

Suppose now that we collect M IID data

$$z_M \triangleq \{(x_i, y_i)\}_{i=1}^M = \{(x_i, f(x_i))\}_{i=1}^M \subset (\mathbb{X} \times \mathbb{Y})^M \triangleq Z^M$$

of the inputs and outputs of a GP $f \sim GP(\bar{m}, \mathfrak{K})$, for some choice of the mean function $\bar{m} : \mathbb{X} \to \mathbb{R}$ and covariance kernel $\mathfrak{K} : \mathbb{X} \times \mathbb{X} \to \mathbb{R}$. Let Y_M be given by (7.39) and, employing the same notation as in Chap. 3, let

$$\mathbb{K}_M \triangleq \begin{bmatrix} \mathfrak{K}(x_1, x_1) & \mathfrak{K}(x_1, x_2) & \cdots & \mathfrak{K}(x_1, x_M) \\ \mathfrak{K}(x_2, x_1) & \mathfrak{K}(x_2, x_2) & \cdots & \mathfrak{K}(x_2, x_M) \\ \vdots & \cdots & \ddots & \vdots \\ \mathfrak{K}(x_M, x_1) & \mathfrak{K}(x_M, x_2) & \cdots & \mathfrak{K}(x_M, x_M) \end{bmatrix} \in \mathbb{R}^{M \times M}, \tag{7.52}$$

$$\mathcal{H}_M \triangleq \operatorname{span}\left\{\mathfrak{K}_{x_i} : x_i \in \Xi_M, 1 \le i \le M\right\}, \tag{7.53}$$

$$\mathfrak{K}_{\Xi_M}(x) \triangleq \left[\mathfrak{K}_{x_1}(x), \ldots, \mathfrak{K}_{x_M}(x)\right]^{\mathrm{T}} \in \mathbb{R}^M; \tag{7.54}$$

see (3.6), (3.8), and (3.55). Remarkably, the Gaussian process $f \sim GP(\bar{m}, \mathfrak{K})$ is such that the output prediction

$$y(x) \triangleq f_M(x|z_M), \quad \text{for any } x \in \mathbb{X} \triangleq \mathbb{R} \tag{7.55}$$

has Gaussian distribution $\mathcal{N}(\bar{m}_M, \mathfrak{K}_M)$, where

$$\bar{m}_M(x) \triangleq \mathfrak{K}_{\Xi}^{\mathrm{T}}(x)\mathbb{K}_M^{-1} Y_M, \tag{7.56a}$$

$$\mathfrak{K}_M(x, y) \triangleq \mathfrak{K}(x, y) - \mathfrak{K}_{\Xi_M}^{\mathrm{T}}(x)\mathbb{K}_M^{-1}\mathfrak{K}_{\Xi_M}(y), \tag{7.56b}$$

$$\sigma_M(x) \triangleq \sqrt{\mathfrak{K}_M(x, x)}. \tag{7.56c}$$

Thus, the unknown GP has a *prior Gaussian distribution* $\mathcal{N}(\bar{m}, \mathfrak{K})$, and the output prediction $y(x|z_M)$ has a *posterior Gaussian distribution* $\mathcal{N}(\bar{m}_M, \mathfrak{K}_M)$.

In light of (3.68), (3.74), and (3.75), we can correlate (7.56a) and (7.56b) to the interpolation and projection problems discussed earlier in Sect. 3.5. Indeed, the mean function $\bar{m}_M$ can be then interpreted in terms of the orthogonal projection of f onto $\mathcal{H}_M$, and the covariance kernel $\mathfrak{K}_M$ can be related to the power function $\mathcal{P}_M$ of the subspace $\mathcal{H}_M$ in the RKHS $\mathcal{H}$. Specifically,

$$\bar{m}_M(x) = (\Pi_M f)(x), \quad \text{for all } x \in \mathbb{X}, \tag{7.57}$$

$$\mathfrak{K}_M(x, x) = \mathcal{P}_M^2(x). \tag{7.58}$$

Suppose that we make M noisy IID observations

$$y_i = f(x_i) + \epsilon, \quad i \in \{1, \dots, M\}, \tag{7.59}$$

where $f \sim GP(\bar{m}, \mathfrak{K})$ and the real-valued noise is distributed as a zero-mean Gaussian process with $\epsilon \sim \mathcal{N}(0, \sigma^2)$. In this case, the output prediction $y(x) \triangleq f_M(x|z_M)$ has the Gaussian distribution $GP(\bar{m}_M, \mathfrak{K}_M)$ with mean function $\bar{m}_M : \mathbb{X} \to \mathbb{R}$, covariance kernel $\mathfrak{K}_M : \mathbb{X} \times \mathbb{X} \to \mathbb{R}$, and standard deviation $\sigma_M : \mathbb{X} \to \mathbb{R}$ given by

$$\bar{m}_M(x) \triangleq \mathfrak{K}^{\mathrm{T}}_{\Xi_M}(x)(\mathbb{K}_M + \sigma^2 I_M)^{-1} Y_M, \quad \text{for all } x \in \mathbb{X}, \tag{7.60}$$

$$\mathfrak{K}_M(x, y) \triangleq \mathfrak{K}(x, y) - \mathfrak{K}^{\mathrm{T}}_{\Xi_M}(x)(\mathbb{K}_M + \sigma^2 I)^{-1} \mathfrak{K}_{\Xi_M}(y), \tag{7.61}$$

$$\sigma_M(x) = \sqrt{\mathfrak{K}_M(x, x)}. \tag{7.62}$$

Several algorithms and publicly available computer codes enable a recursive approach to generate predictions of a GP using M samples. In the following, we present an example of a recursive approach; the interested reader is referred to [83] for a detailed account of the derivation of these results. Suppose that we make an additional measurement (x_{M+1}, y_{M+1}). In this case,

$$\mathbb{K}_{M+1} = \begin{bmatrix} \mathbb{K}_M + \sigma^2 I_M & \mathfrak{K}_{\Xi_M}(x_{M+1}) \\ \mathfrak{K}^{\mathrm{T}}_{\Xi_M}(x_{M+1}) & \mathfrak{K}(x_{M+1}, x_{M+1}) \end{bmatrix} \tag{7.63}$$

and

$$\mathbb{K}^{-1}_{M+1} = \begin{bmatrix} \left(\mathbb{K}_M + \sigma^2 I_M\right)^{-1} & \left(I_M + \lambda_{M+1} \mathfrak{K}^{\mathrm{T}}_{\Xi_M}(x_{M+1}) \mathfrak{K}^{\mathrm{T}}_{\Xi_M}(x_{M+1}) \mathbb{K}^{-1}_M\right) - \lambda_{M+1} \mathbb{K}^{-1}_M \mathfrak{K}_{\Xi_M} \\ -\lambda_{M+1} \mathfrak{K}^{\mathrm{T}}_{\Xi_M}(x_{M+1}) \mathbb{K}^{-1}_M & \lambda_{M+1} \end{bmatrix}, \tag{7.64}$$

where

$$\lambda_{M+1} \triangleq \left(\mathfrak{K}(x_{M+1}, x_{M+1}) - \mathfrak{K}^{\mathrm{T}}_{\Xi_M}(x_{M+1}) \left(\mathbb{K}_M + \sigma^2 I_M\right)^{-1} \mathfrak{K}_{\Xi_M}(x_{M+1}) \right)^{-1}. \tag{7.65}$$

Equation (7.64) is the inversion formula of a matrix in terms of the inverse of a submatrix. These recursive formulas provides for an efficient update of the formula for the prediction $y(x|Y_{M+1})$ once the inversion, or preferably the factorization, of $\mathbb{K}_M$ is complete.

7.6 Concentration Inequalities in Native Spaces

In this section, we discuss how to characterize the quality of approximations using concentration inequalities of the form

$$\Pr\left(\left\{\|f - f_{N,M}\|_{L^\infty(\Omega,\mathbb{U})} \geq \epsilon_{M,N}\right\}\right) \leq \Delta_{M,N} \tag{7.66}$$

for the universe $\mathbb{B} \triangleq L^\infty(\Omega, \mathbb{U})$ and some $\Delta_{M,N} \in [0, 1]$, and

$$\Pr\left(\left\{\|f - f_{N,M}\|_{\mathcal{H}} \geq \epsilon_{M,N}\right\}\right) \leq \Delta_{M,N} \tag{7.67}$$

for $\mathbb{B} \triangleq \mathcal{H}(\Omega, \mathbb{U})$. These concentration inequalities will play a pivotal role in assessing ultimate bounds on the trajectory tracking error for closed-loop plant models governed by adaptive laws similar to those discussed in Chap. 5 and with learning schemes similar to those discussed in Chap. 6.

7.6.1 Concentration Inequalities for Scalar-Valued Uncertainties

7.6.1.1 Results on the Space of Bounded Functions

In this section, we derive a concentration inequality for the scalar-valued uncertainty $f : \Omega \to \mathbb{R}$ assuming that $\mathbb{B} \triangleq L^\infty(\Omega)$ and $\Omega \subset \mathbb{X} \triangleq \mathbb{R}^n$ is compact. This concentration inequality is a special form of the concentration inequality in (7.66), and it has been used for the stochastic analysis of recent adaptive control methods in [12, 33, 85–87]. For the statement of this result, let $f \sim GP(\bar{m}, \mathfrak{K})$ be uncertain, and assume that the Gaussian process is defined in terms of a scalar-valued continuous reproducing (Mercer) kernel $\mathfrak{K} : \Omega \times \Omega \to \mathbb{R}$ that is bounded on the diagonal and defines the RKHS space $\mathcal{H}$ over Ω. In this case, it holds that

$$\mathcal{H} \hookrightarrow C_b(\Omega) \hookrightarrow L^2_\mu(\Omega); \tag{7.68}$$

see (3.38). There exists a self-adjoint, positive, and compact operator $L_{\mathfrak{K}} : L^2_\mu(\Omega) \to L^2_\mu(\Omega)$ such that

$$(L_{\mathfrak{K}} g)(x) \triangleq \int_\Omega \mathfrak{K}(x, \xi) g(\xi) \mu(d\xi), \quad \text{for all } x \in \Omega, g \in L^2_\mu(\Omega); \tag{7.69}$$

see (3.41) and Theorem 3.10. Finally, it follows from (3.50) that

$$\mathcal{H} \subset \overline{\mathscr{R}(\mathcal{I})} \subset L^2_\mu(\Omega), \tag{7.70}$$

where $I : \mathcal{H} \to L^2_\mu(\Omega)$ denotes the canonical injection of $\mathcal{H}$ into $L^2_\mu(\Omega)$ and $\overline{\mathscr{R}(I)}$ denotes the $L^2_\mu(\Omega)$-closure of the span of the family of eigenvectors $\{e_i\}_{i\in\mathbb{N}}$ of $L_{\mathfrak{K}}$. In [33], the space $\overline{\mathscr{R}(I)} \supset \mathcal{H}$ is referred to as the *sample space of the Gaussian process* $GP(\mu, \mathfrak{K})$, and its properties have been studied in some settings recently such as in [88, 89].

Let $\Omega \subset \mathbb{X} \triangleq \mathbb{R}^n$ be a compact set. Define $N_c(R, \Omega)$ to be the minimal number of open balls of radius $R > 0$ needed to cover Ω. Equivalently, $N_c(R, \Omega)$ denotes the minimal number of points $\{x_j \in \Omega : j = 1, 2, \dots, M\}$ such that every other point $y \in \Omega$ is within a distance R from one of them. This number $N_c(R, \Omega)$ always exists since Ω is compact. Indeed, the collection of balls $\mathscr{O}_R = \{U_x = \mathcal{B}_R(x)\}$ is an open cover of Ω since Ω is compact and Definition 2.23 implies there exists a finite subcover of $\mathscr{O}_R$ given by $\mathscr{O}_R^M = \left\{U_{x_j} = \mathcal{B}_R(x_j),\ j = 1, 2, \dots, M\right\}$. Thus, $N_c(R, \Omega)$ exists.

Theorem 7.3 ([33, Theorem 3.1]) *Let $\Omega \subset \mathbb{R}^n$ be compact, let $N_c(R, \Omega)$ denote the covering number of the set Ω by balls in $\mathbb{R}^n$ of radius $R > 0$, and let the continuous scalar-valued reproducing kernel $\mathfrak{K} : \Omega \times \Omega \to \mathbb{R}$ be Lipschitz continuous on $\Omega \times \Omega$ with the Lipschitz constant $\ell_{\mathfrak{K}}$. Assume that the unknown function $f \in \overline{\mathscr{R}(I)}$ is Lipschitz continuous with constant ℓ_f and a sample of the Gaussian process $GP(0, \mathfrak{K})$, and M IID observations $z_M \triangleq \{(x_i, y_i)\}_{i=1}^M \subset Z^M$ are generated from*

$$y = f(x) + \epsilon, \quad \textit{for all } x \in \Omega, \tag{7.71}$$

where $\epsilon \sim \mathcal{N}(0, \sigma^2)$. Then, the posterior mean function $\bar{m}_M$ and standard deviation σ_M of a Gaussian process conditioned on the training data $\{(x_i, y_i)\}_{i=1}^M$ are continuous with Lipschitz constant $\ell_{\bar{m}_M}$ and modulus of continuity ω_{σ_M} over Ω such that

$$\ell_{\bar{m}_M} \le \ell_{\mathfrak{K}} \sqrt{M} \left\| (\mathbb{K}_M + \sigma^2 I_M)^{-1} Y_N \right\|, \tag{7.72}$$

$$\omega_{\sigma_M}(R) \le \sqrt{2R\ell_{\mathfrak{K}} \left(1 + M \|\mathbb{K}_M + \sigma^2 I_M\| \max_{\xi,\eta\in\Omega} \mathfrak{K}(\xi, \eta)\right)}. \tag{7.73}$$

Finally, the probabilistic bound

$$\Pr\left(|f(x) - \bar{m}_M(x)| \le \sqrt{\beta(R)}\sigma_M(x) + \gamma(R),\ \textit{for all } x \in \Omega\right) \ge 1 - \delta \tag{7.74}$$

is verified for

$$\beta(R) \triangleq 2\log\left(\frac{N_c(R, \Omega)}{\delta}\right), \tag{7.75}$$

$$\gamma(R) \triangleq \left(\ell_{\bar{m}_M} + \ell_f\right) R + \sqrt{\beta(R)}\omega_\sigma(R). \tag{7.76}$$

The probabilistic bound (7.74) departs slightly from the standard form of the concentration inequality in (7.11a) since the event in the parentheses in (7.74) is defined pointwise. However, (7.74) can lead to a concentration inequality of the form in (7.11a) in a straightforward way, as shown by the corollary below.

Corollary 7.1 *Assume that conditions of Theorem 7.3 are verified. Then, it holds that*

$$\Pr\left(\left\{\|f - \bar{m}_M\|_{L^\infty(\Omega)} \leq \epsilon_{M,M}, \textit{ for all } x \in \Omega\right\}\right) \geq 1 - \delta, \tag{7.77}$$

where

$$\epsilon_{M,M} \triangleq \sqrt{\beta(R)} \sup_{\xi \in \Omega} \sigma_M(\xi) + \gamma(R). \tag{7.78}$$

Proof Note that the inequality in (7.74) can be written as

$$\rho^M \underbrace{\left(z_M \in Z^M : |f(x) - \bar{m}_M(x)| \leq \epsilon_{M,M}, \text{ for all } x \in \Omega\right)}_{\mathscr{G}_{PW} \subset Z^M} \geq 1 - \delta, \tag{7.79}$$

where the measure ρ^M on Z^M is the product measure of $\rho(\mathrm{d}z_M) \triangleq \rho(\mathrm{d}z_1)\rho(\mathrm{d}z_2) \cdot \ldots \cdot \rho(\mathrm{d}z_M)$, ρ denotes the probability measure on Z that generates the IID samples $z_M \triangleq \{(x_i, y_i)\}_{i=1}^M$, and

$$\bar{m}_M(x) \triangleq \bar{m}_M(x|z_M), \tag{7.80}$$

$$\sigma_M(x) \triangleq \sigma_M(x|z_M). \tag{7.81}$$

The term $\mathscr{G}_{PW}$ in (7.79) denotes the event in Z^M that characterizes the set whose samples, with high probability, generate small errors; the subscript "PW" refers to the fact that the inequality that defines $\mathscr{G}_{PW}$ holds pointwise for all $x \in \Omega$. Now, introduce the event

$$\mathscr{G}_\infty \triangleq \left\{ z_M \in Z^M : \|f - \bar{m}_M\|_{L^\infty(\Omega)} \leq \sqrt{\beta(R)} \sup_{\xi \in \Omega} \sigma_M(\xi) + \gamma(R) \right\} \subset Z^M. \tag{7.82}$$

To prove that $\mathscr{G}_{PW} \subseteq \mathscr{G}_\infty$, suppose that $z_M \in \mathscr{G}_{PW}$. Then,

$$|f(x) - \bar{m}_M(x)| \leq \sqrt{\beta(R)}\sigma_M(x) + \gamma(R), \quad \text{for all } x \in \Omega. \tag{7.83}$$

By first taking the supremum on the right-hand side of (7.83), we know that

$$|f(x) - \bar{m}_M(x)| \leq \sqrt{\beta(R)} \sup_{\xi \in \Omega} \sigma_M(\xi) + \gamma(R), \quad \text{for all } x \in \Omega, \tag{7.84}$$

and subsequently taking the supremum on the left-hand side of (7.84), yields

$$\|f - \bar{m}_M\|_{L^\infty(\Omega)} \leq \sqrt{\beta(R)} \sup_{\xi \in \Omega} \sigma_M(\xi) + \gamma(R). \tag{7.85}$$

Thus, $\mathscr{G}_{PW} \subset \mathscr{G}_\infty$, and the result is proven since $\Pr(\mathscr{G}_\infty) \geq \Pr(\mathscr{G}_{PW})$. ■

The concentration inequality (7.77) is the only example in this text that addresses the problem of generating approximations that "escape the native space," that is, the case wherein the uncertainty lies outside a native space, which serves as hypothesis space, but lies within the given universe. This is a topic of much interest in recent studies of the theory of approximations in RKHSs; see, for example, [90–95] for some particular studies that illustrate the analysis for certain types of Sobolev–Matérn kernels over subsets of Euclidean spaces or compact Riemannian manifolds. The authors in [88, 96] also studied this topic in certain power spaces.

Theorem 7.3 is stated for a scalar-valued uncertainty f, and it was extended in [34] to uncertainties contained in the Cartesian product space $\boldsymbol{\mathcal{H}} \triangleq \mathcal{H}^m$ by substantially reproducing component-wise the arguments in [33]. The analysis of the extension in [34] is nontrivial, also in its statement. Such complex descriptions of the error should be compared to the concise statements in the deterministic case in Chaps. 5 and 6. Often, as we have shown in the deterministic analysis, there are large collections of native spaces wherein we obtain deterministic error bounds that are simple expressions in terms of the fill distance when the error is of the order $O(h^s_{\Xi_N,\Omega})$.

In general, ERM approaches generate approximations of regressors that are indexed by the dimensionality N of the subspace $H_{d(N)}$ of the approximations and the cardinality M of the information used to construct estimates. In Theorem 7.3 and Corollary 7.1, however, M denotes both the number of observations used to generate the approximation $\bar{m}_M$ and the dimension of the space of approximants $\mathcal{H}_M$. Thus, the algorithms and learning approximations generated via Gaussian process methods captured by Theorem 7.3 and Corollary 7.1 are such that $N = M$. This means that $\bar{m}_M$ is a special kind of the general estimate $f_{N,M}$ with $N = M$, that is,

$$\bar{m}_M \triangleq f_{M,M}. \tag{7.86}$$

However, in general, it is more advantageous to choose N and M independently.

It is common practice to modify the basic estimate $\bar{m}_M$ using outlier rejection methods, basis rejection methods, or other methods to modify the pool of randomly selected bases. The authors in [97–100] provide examples of algorithms that modify the basic random process for the selection of bases. The concentration inequality (7.77) relies on retaining all the bases generated by the first M samples, strictly speaking. More nuanced versions would be required for estimates that feature such basis adaptation. Concentration inequalities for such modifications of the basic estimates obtained using IID samples for GPs are needed. These results, however, have not yet been systematically developed to the authors' knowledge.

Theorem 7.3 and Corollary 7.1 do not illustrate how the quality of the approximation inequalities (7.74) or (7.79) decreases for less smooth functions in the sample space near $L^2_\mu(\Omega)$. To comprehend this limitation, assume that the native space $\mathcal{H}$ is the Sobolev space $W^{s,2}(\Omega)$ contained in the Lipschitz functions $C^{0,1}(\Omega)$. In this case, it holds that

$$\ldots \hookrightarrow W^{s+1,2}(\Omega) \hookrightarrow \mathcal{H} \equiv W^{s,2}(\Omega) \hookrightarrow W^{s-1,2}(\Omega) \subset \ldots \subset L^2_\mu(\Omega). \tag{7.87}$$

As shown in [90, 92–94], applications of Theorem 2.8 generate approximations from $\mathcal{H}_N$ such that

$$\|f - f_N\|_{L^2_\mu(\Omega)} \leq C h^s_{\Xi_N,\Omega} \|f\|_{W^{s,2}(\Omega)}, \quad \text{for all } f \in W^{s,2}(\Omega), \tag{7.88}$$

and the larger the smoothness $s > 0$, the faster approximations of functions converge. Thus, the knowledge that the unknown function f lies in the hypothesis spaces $W^{s,2}(\Omega)$ defines the approximation theory priors that allow forecasting convergence rates as functions of the smoothness of the function. The simple relationship (7.68) does not provide such an insight. Later in this chapter, we will discuss a simple way to achieve bounds with approximation theoretic priors in the scale of spaces

$$\mathcal{H}^2(T_{\mathfrak{K}}) \hookrightarrow \mathcal{H}^s(T_{\mathfrak{K}}) \subset \cdots \subset \mathcal{H}^0(T_{\mathfrak{K}}) \equiv \mathcal{H} \hookrightarrow L^2_\mu(\Omega), \tag{7.89}$$

where $T_{\mathfrak{K}}$ is defined in (3.40) and $\mathcal{H}^s(T_{\mathfrak{K}})$ denotes the power space for the smoothness index s introduced in Sect. 3.4.5.2.

7.6.1.2 Results on a Native Space

In this section, we derive a concentration inequality for the scalar-valued uncertainty $f : \Omega \to \mathbb{R}$ assuming that $\mathbb{B} \triangleq \mathcal{H}(\Omega)$ and $\Omega \subset \mathbb{X} \triangleq \mathbb{R}^n$ is compact. This concentration inequality is a special form of the concentration inequality in (7.67). We introduce a simple way to generate concentration inequalities for scalar-valued native spaces that allow for $N \neq M$, and discuss how these inequalities give error guarantees explicit in approximation theoretic priors defined in terms of a smoothness index s. Leveraging the results in [101, 102], these concentration inequalities are not given by a Gaussian process but by a stochastic gradient descent method.

The scalar-valued native space $\mathcal{H}$ is defined in terms of a strictly positive Mercer kernel $\mathfrak{K}(\cdot,\cdot) : \Omega \times \Omega \to \mathbb{R}$ that is uniformly bounded on the diagonal, that is, $\mathfrak{K}(x, x) \leq \bar{\mathfrak{K}} < \infty$ for all $x \in \Omega$. Consider the linear, compact, self-adjoint, and nonnegative integral operator $T_{\mathfrak{K}} : \mathcal{H} \to \mathcal{H}$ given by

$$(T_{\mathfrak{K}} g)(x) \int_\Omega \mathfrak{K}(x, \xi) f(\xi) \mu(d\xi), \quad \text{for all } x \in \Omega,\ g \in \mathcal{H}; \tag{7.90}$$

see (3.40), and recall Remark 3.2 to appreciate the difference between (7.90) and (7.69). As discussed in Remark 3.4, it is possible to construct an $\mathcal{H}$-orthogonal basis $\{\psi_i\}_{i=1}^\infty \subset \mathcal{H}$ of $(\mathcal{N}(T_{\mathfrak{K}}))^\perp \subseteq \mathcal{H}$ from the eigenvectors $\{\psi_i\}_{i\in\mathbb{N}}$ of the operator $T_{\mathfrak{K}}$.

By construction, the operators $L_{\mathfrak{K}}$ and $T_{\mathfrak{K}}$ have the same eigenvalues. Furthermore, the eigenvectors $\{e_i\}_{i\in\mathbb{N}}$ of $L_{\mathfrak{K}}$ and $\{\psi_i\}_{i\in\mathbb{N}}$ of $T_{\mathfrak{K}}$ are such that

$$\psi_i \triangleq \sqrt{\lambda_i} e_i. \tag{7.91}$$

Thus, we define the power spaces $\mathcal{H}^s(T_{\mathfrak{K}})$, which give the sequence of continuous embeddings

$$\mathcal{H}^{s_2}(T_{\mathfrak{K}}) \hookrightarrow \mathcal{H}^{s_1}(T_{\mathfrak{K}}) \subset (\mathcal{N}(T_{\mathfrak{K}}))^{\perp} \equiv \mathcal{H}^0 \subset \mathcal{H} \tag{7.92}$$

for $s_2 > s_1 > 0$; see (3.54).

References [102, 103] describe various algorithms for constructing approximations of an unknown function $f \in \mathcal{H}^s(T_{\mathfrak{K}})$ that are based on sequential improvements of estimates along the trajectory of a random walk on $Z \triangleq \mathbb{X} \times \mathbb{Y}$ determined by the probability measure $\rho(dz) \triangleq \rho(dy|x)\mu(dx)$. The initial estimate is chosen as $f^{(0)} \triangleq 0 \in \mathcal{H}$. At step $i+1$ of the random walk in Z determined by the measure ρ, the estimate $f^{(i+1)} \in \mathcal{H}$ is generated from the input/output observation $(x_i, y_i) \in Z$ and the previous estimate $f^{(i)} \in \mathcal{H}$ from the recursion

$$f^{(i+1)} = a_i f^{(i)} + b_i \mathfrak{K}_{x_i} \left(y_i - f^{(i)}\right) \in \mathcal{H}, \tag{7.93}$$

where the learning coefficents $a_i, b_i \in \mathbb{R}^+$ are defined in a different ways in [101, 102]. In this book, we let

$$a_i \triangleq \frac{i+i_0+1}{i+i_0}, \quad \text{for } i = 0, 1, \ldots, \tag{7.94a}$$

$$a_i b_i \triangleq \frac{a}{(i+i_0)^{(s+1)/(s+2)}} \tag{7.94b}$$

for some choices of the constants a and m_0 described in [101, 102] and omitted for brevity. The constant a_i, b_i in (7.94) are derived from [102], but are expressed in terms of the notation of [101, Sect. 3]

Theorem 7.4 ([101, 102, Theorem 3.1]) *Consider the stochastic recursion* (7.93). *If* $f \in \mathcal{H}^s(T_{\mathfrak{K}})$ *for some* $0 < s \leq 2$, *then there is a constant* $0 < C \triangleq C(s, \|f\|_{\mathcal{H}^s}, \delta)$ *such that*

$$\Pr\left\{\|f - f^{(i)}\|_{\mathcal{H}} \leq \frac{C}{(i+i_0)^{s/(s+2)}}\right\} \geq 1 - \delta. \tag{7.95}$$

It follows from the stochastic recursion in (7.93) that

$$f^{(i+1)} \in \text{span}\left\{f^{(i)}, \mathfrak{K}_{x_i}\right\}, \quad \text{for each } i \in \mathbb{N}, \tag{7.96}$$

which implies that the estimate $f_M \triangleq f^{(M+1)}$ resides in the finite-dimensional space

$$\mathcal{H}_M \triangleq \text{span}\left\{\mathfrak{K}_{x_i} : 1 \le i \le M\right\}. \tag{7.97}$$

Similarly to the results in Sect. 7.6.1, Theorem 7.4 holds for $N = M$. The right-hand side of (7.95) holds for any uncertainty f in the power space $\mathcal{H}^s \triangleq \mathcal{H}^s(T_{\mathfrak{K}}) \subset \mathcal{H}$ with smoothness index $s \in (0, 2]$. This means that the rate of convergence is described in terms of approximation theoretic priors given by the scale of power spaces,

$$\mathcal{H}^2(T_{\mathfrak{K}}) \hookrightarrow \mathcal{H}^s(T_{\mathfrak{K}}) \hookrightarrow \mathcal{H}^0(T_{\mathfrak{K}}) \triangleq (\mathcal{N}(T_{\mathfrak{K}}))^{\perp} \hookrightarrow \mathcal{H} \tag{7.98}$$

for $s \in (0, 2]$. This result gives a refined estimate of how fast approximations of the uncertainty converge as the smoothness of the function f varies. In particular, approximations converge faster for smoother uncertainty as s increases. As already discussed, the concentration inequalities (7.74) and (7.79) do not allow considerations on convergence based on the smoothness of the uncertainty. On the other hand, (7.95) holds only for spaces contained in the RKHS $\mathcal{H}$. The results in Sect. 7.6.1 give a universal bound for all functions in the sample space of the GP determined by the kernel $\mathfrak{K}$.

7.6.1.3 A Simple Modifications for $N \neq M$

The dimension $N = M$ of the space $\mathcal{H}_M$ in the concentration inequalities grows with the number M of samples of the recursion. With the high bandwidth of measurement processes for contemporary instrumentation, the dimension of the space $\mathcal{H}_M$ can grow very rapidly. In many situations, the number of available measurements may be much larger than the dimension we would like to use for the space $\mathcal{H}_N$ underlying the adaptive control scheme. A possibility is to decimate the available pool of samples or reduce the sampling rate of the measurement process.

In the remainder of this section, we review a simple method that allows a judicious choice of the space $\mathcal{H}_N$ used in the control phase for which $N \ll M$. The idea is that while M may become large, we project the estimate provided by the mean function $f_M \triangleq \bar{m}_M$ onto a different space of approximants, $\mathcal{H}_N$, determined by some deterministically prescribed set of centers $\Xi_N \subset \Omega$. Typically, we choose the centers in Ξ_N to have small fill distances in Ω. The space $\mathcal{H}_N$ is chosen so that $d(N) \triangleq \dim(\mathcal{H}_N)$ is small enough for online implementations.

For the statement of the next result, let $X_M \subset \Omega$ denote the collection of states visited by a random walk that chooses $\{(x_i, y_i)\}_{i=1}^M$ generated by the measure ρ on Z. As in Chaps. 3, 5, and 6, let Π_N denote the $\mathcal{H}$-orthogonal projection onto $\mathcal{H}_N$, and let $\mathcal{P}_N$ denote the power function of $\mathcal{H}_N$ in $\mathcal{H}$. Finally, assume that $\mathbb{B} \triangleq L^\infty(\Omega)$.

Corollary 7.2 *Assume that the hypotheses of Theorem 7.4 are verified with $f \in \mathcal{H}^s(T_{\mathfrak{K}})$ and some $s \in (0, 2]$, and define the approximations*

$$f_M \triangleq f^{M+1} \in \mathcal{H}_M, \tag{7.99}$$

$$f_{N,M} \triangleq \Pi_N f_M. \tag{7.100}$$

Then,

$$\Pr\left(\left\{\|f - f_{N,M}\|_{L^\infty(\Omega)} \leq \epsilon_{N,M}\right\}\right) \geq 1 - \delta, \tag{7.101}$$

where $C > 0$ is defined as in Theorem 7.4 and

$$\epsilon_{N,M} \triangleq \sup_{\xi \in \Omega} \mathcal{P}_N(\xi)\|f\|_{\mathcal{H}} + \frac{C}{(M + i_0)^{s/(s+2)}}. \tag{7.102}$$

Furthermore,

$$\Pr\left(\left\{\|f - f_{N,M}\|_{\mathcal{H}} \leq \epsilon_{N,M}\right\}\right) \geq 1 - \delta. \tag{7.103}$$

Proof It holds that

$$\begin{aligned} \|f - f_{N,M}\|_{\mathcal{H}} &\leq \|(I - \Pi_N)f\|_{\mathcal{H}} + \|\Pi_N(f - f_M)\|_{\mathcal{H}} && (7.104)\\ &\leq \underbrace{\|(I - \Pi_N)f\|_{\mathcal{H}}}_{\text{approximation error}} + \underbrace{\|f - f_M\|_{\mathcal{H}}}_{\text{stochastic error}}, && (7.105) \end{aligned}$$

where $f_{N,M}$ and f_M depend on the samples $z_m \in Z^M$ of a random variable, but $\Pi_N f$ does not. Furthermore,

$$|E_x(f - f_{N,M})| \leq |E_x(I - \Pi_N)f| + |E_x\Pi_M(f - f_M)|, \quad \text{for all } x \in \Omega. \tag{7.106}$$

Since the kernel $\mathfrak{K}(x, x) \leq \bar{\mathfrak{K}}^2$ is bounded on the diagonal, that is, $\|E_x\| \leq \bar{\mathfrak{K}}$, and $\|\Pi_N\| \leq 1$, we deduce that

$$|E_x(f - f_{N,M})| \leq \mathcal{P}_N(x)\|f\|_{\mathcal{H}} + \|f - f_M\|_{\mathcal{H}}, \tag{7.107}$$

and it follows from (7.105) that

$$\|f - f_{N,M}\|_{\mathcal{H}} \leq \sup_{\xi \in \Omega} \mathcal{P}_N(\xi)\|f\|_{\mathcal{H}} + \|f - f_M\|_{\mathcal{H}}. \tag{7.108}$$

This inequality holds for any sample $z_M \in Z_M$. Whenever the sample z_M is contained in the good set (7.10) of the concentration inequality, we have

$$\|f - f_{N,M}\|_{\mathcal{H}} \leq \underbrace{\sup_{\xi \in \Omega} \mathcal{P}_N(\xi)\|f\|_{\mathcal{H}}}_{\text{Approximation error}} + \underbrace{\frac{C}{(i + i_0)^{s/(s+2)}}}_{\text{Stochastic error}}. \tag{7.109}$$

Since this inequality holds for every possible sample in the good set, this inequality holds with probability $1 - \delta$, and (7.103) is verified. A similar reasoning yields (7.101). ■

In practice, Corollary 7.2 can be employed to deterministically choose the dimension of the control phase for N different than the number of samples M collected randomly during the learning phase. The techniques discussed in Chap. 3 to bound $\mathcal{P}_N$ in terms of the fill distance using Table 3.3 can be applied to bound the approximation error in (7.109). The approximation bound (7.109) and its counterpart on $L^2_\mu(\mathbb{X})$ are in the same form as (7.37). The approximation error term is the only term that appears in the purely deterministic studies in Chaps. 5 and 6.

7.6.2 Concentration Inequalities for Vector-Valued Uncertainty

The previous section summarized concentration inequalities that can be used in the stochastic analysis of control systems when the uncertainty is scalar-valued. In this section, we consider the case wherein the uncertainty is vector-valued.

The error bound in (7.109) is decomposed into the approximation error and the stochastic error. In the general theory of ERM, when the hypothesis space is a generic Banach space, advanced methods seek to balance the approximation and stochastic error terms. These terms are balanced by deriving relationships between the dimension of the space of approximants $d(N)$ and the number of samples mM. That is, $d(N)$ is defined in terms of mM to create estimates that lie at the minimum depicted in Fig. 7.1. For details, see, for example, [44, 104]. Alternatively, the regularization parameter λ in ERM is chosen as a function of the number of samples M to obtain a guarantee of errors that lie at the minimum of the curve shown in Fig. 7.1. Thus, regularization can have the same qualitative effect as choosing the dimensions of subspaces to equilibrate the error in the bias-variance trade-off. For details, see [96, 105], which derive concentration inequalities for scalar-valued uncertainties and vector-valued uncertainties, respectively. The next result provides a special case of the results provided in [105].

Theorem 7.5 ([105, Corollary 2]) *Let $\Omega \subset \mathbb{R}^n$ be compact and let the IID samples $\{(x_i, y_i)\}_{i=1}^M$ be generated in terms of the probability measure ρ on $Z \triangleq \mathbb{X} \times \mathbb{Y}$. Consider the scalar-valued RKHS $\mathcal{H}$ defined in terms of the Mercer kernel $\mathfrak{K} : \Omega \times \Omega \to \mathbb{R}^+$ that is uniformly bounded on the diagonal, and let the operator kernel $\mathcal{K} \triangleq \mathfrak{K} I_m$ define the vector-valued RKHS $\boldsymbol{\mathcal{H}}$. Assume that $\boldsymbol{\mathcal{H}}$ is equivalent to the vector Sobolev space $\mathbf{W}^{s,2}(\Omega, \mathbb{R}^m)$, that is, $\boldsymbol{\mathcal{H}} \approx \mathbf{W}^{s,2}(\Omega, \mathbb{R}^m)$, for a smoothness index $s > n/2$, and that the regressor function satisfies the regularity condition $f_\rho \in \mathbf{W}^{r,2}(\Omega, \mathbb{R}^m)$ for some smoothness index $0 < r < 2s$. Then, for $t \in [0, r]$, for the sequence of regularization parameters $\lambda_M = M^{\frac{s}{r+n/2}}$, for all $\tau > \log(5)$, and for all sufficiently large $M \geq 1$, there is a constant $C > 0$ independent of M and τ such* *that the solution f_M of ERM given by* (7.35) *for the parameters in* (7.36) *guarantees the concentration inequality*

$$\Pr\left(\left\{\|f_\rho - f_M\|_{W^{t,2}(\Omega)} \leq \epsilon_{M,M}\right\}\right) \geq 1 - \delta \tag{7.110}$$

with $\delta \triangleq 5e^{-\tau}$, where

$$\epsilon_{M,M} \triangleq C\tau^2 M^{-\frac{r-t}{r+n/2}}. \tag{7.111}$$

The inequality in (7.110) has the advantage that it applies whenever we use the scalar-valued Sobolev–Matérn kernels, and it gives a fine description of how the approximation theoretic prior on the smoothness r of the regressor influences the rate of convergence of approximations in the scale of Sobolev spaces

$$W^{t,2}(\Omega) \hookrightarrow W^{r,2}(\Omega) \hookrightarrow W^{2s,2}(\Omega) \hookrightarrow L^2_\mu(\Omega, \mathbb{R}^m). \tag{7.112}$$

It should still be noted, however, that the results in [96, 105] still constrain the dimensions of the subspace of approximants to $d(N) = mM$. This can be a limitation in applications to adaptive control, where the number of offline measurements during the learning phase can considerably exceed the dimension of the space of approximants used during the adaptive control phase. If the number of observations is large in the learning phase and we do not want to be constrained to using M basis functions, then the same procedure as in Sect. 7.6.1.3 can be applied to obtain the alternative concentration inequality

$$\Pr\left(\left\{\|f_\rho - f_{N,M}\|_{W^{t,2}(\Omega)} \leq C\left(\sup_{\xi\in\Omega} \mathcal{P}_N(\xi)\|f\|_{\mathcal{H}} + \tau^2 M^{-\frac{r-t}{r+n/2}}\right)\right\}\right) \geq 1 - \delta. \tag{7.113}$$

This inequality is based on the properties of the power function $\mathcal{P}_N$ of $\mathcal{H}$; see Sect. 3.8.

7.7 LTC Architectures with Stochastic Subspace Learning

The results produced in the previous sections can be used to analyze nonparametric adaptive control methods in light of stochastic data-driven methods. As an example, in this section, we develop a stochastic analysis based on GPs for LTC architectures discussed in Sect. 6.4.2.4.

The learning phase solves an offline problem of statistical and machine learning theory to generate the realizable, empirical approximation $f_{N,M} \in H_{d(N)}$ of the matched parametric uncertainty f in (7.1) that depends on M IID observations $z_M \triangleq \{(x_i, y_i)\}_{i=1}^M \in Z^M$ of the uncertainty. These observation are generated by the random walk determined by the unknown probability measure ρ on Z. The control phase, which can be performed by applying any adaptive control system discussed in Chap. 5, now depends on the particular sample $z_M \in Z^M$ used to construct $f_{N,M}$ via statistical and machine learning theory. Thus, we have

$$\Xi_M \triangleq \Xi_M(z_M), \tag{7.114a}$$

$$H_{d(N)} \triangleq H_{d(N)}(z_M), \tag{7.114b}$$

$$\Phi_N(x) \triangleq \Phi_N(x; z_M), \quad \text{for all } x \in \Omega, \tag{7.114c}$$

$$f_{N,M}(x) \triangleq f_{N,M}(x; z_M), \quad \text{for all } x \in \Omega. \tag{7.114d}$$

For notational brevity, we suppress the symbol z_M when it denotes parametric dependence whenever possible, and recall the functional dependency on the sample values only when it leads to a clearer understanding of a result.

We limit our discussion to an analysis of the results obtained by applying the dead-zone modification of MRAC; see Sect. 5.4.1. This choice is due to the relative simplicity of the proof of this approach. The same ideas can be extended to any of the adaptive control systems presented in Chap. 5. For the statement of the next theorem, which is the main result of this section we assume a concentration inequality of the form

$$\Pr\left(\left\{\|f - f_{N,M}\|_{\mathbb{B}} \le \epsilon_{N,M}\right\}\right) \ge 1 - \Delta_{N,M} \tag{7.115}$$

with $f_{N,M} \in \mathcal{H}_{d(N)}$, $\Delta_{N,M} \in [0, 1]$, and the universe $\mathbb{B}$ is a generic Banach space.

Theorem 7.6 (ERM+Dead-Zone) *Consider the plant model* (7.1) *and assume that the matched nonparametric uncertainty f is an element of an RKHS defined in terms of the continuous operator-valued kernel $\mathcal{K} : \Omega \times \Omega \to \mathcal{L}(\mathbb{U})$ such that $\|\mathcal{K}(x, x)\|_{\mathcal{L}(\mathbb{U})} \le \bar{\mathcal{K}}$ for all $x \in \Omega$ and some constant $\bar{\mathcal{K}} > 0$. Assume that the concentration inequality* (7.115) *holds for either $\mathbb{B} \triangleq L^\infty(\Omega, \mathbb{U})$ or $\mathbb{B} \triangleq \mathcal{H}$. Assume that the conditions of Theorem 5.5 are verified employing the randomly selected basis Φ_N associated with $f_{N,M}$ for a dead-zone size*

$$d \ge d_{N,M} \triangleq \frac{2\|PB\|}{\lambda_{\min}(Q)} \epsilon_{N,M}. \tag{7.116}$$

If the trajectory $x_{N,M}(t) \triangleq x_{N,M}(t; \Xi_M, H_{d(N)})$ remains in the set Ω for all $t \ge t_0$, then, with probability $1 - \Delta_{N,M}$, the tracking error $e_{N,M}(t) \triangleq x_{N,M}(t) - x_r(t)$ converges to a closed ball of radius $O(d)$ centered in the origin as $t \to \infty$. Furthermore, the ultimate bound on the trajectory tracking error decreases in N, M like $O(\epsilon_{N,M})$.

Proof The approach to this proof is based on modifying the strategy presented in Sect. 6.4.2.4 for deterministic offline approximation of the uncertainty. Recall that, in the statistical learning theory scenario, it holds that

$$f_{N,M}(x) \triangleq f_{N,M}(x; z_M) \tag{7.117}$$

for the samples $z_M \in Z^M$. The performance of our adaptive control strategy will depend on whether $z_M \in \mathscr{G}_M \subset Z_M$ or $z_M \in \mathscr{B}_M \subset Z_M$, where

$$\mathscr{B}_M \triangleq \left\{ z_M \in Z^M : \|f - f_{N,M}\|_{L^\infty(\Omega,\mathbb{U})} > \epsilon_{N,M} \right\}, \tag{7.118}$$

$$\mathscr{G}_M \triangleq \left\{ z_M \in Z^M : \|f - f_{N,M}\|_{L^\infty(\Omega,\mathbb{U})} \leq \epsilon_{N,M} \right\}. \tag{7.119}$$

If $z_M \subset \mathscr{G}_M$, then $f_{N,M}(x) \triangleq f_{N,M}(x; z_M)$ and

$$\|f - f_{N,M}(\cdot; z_M)\|_{L^\infty(\Omega,\mathbb{U})} \leq \epsilon_{N,M}, \quad \text{for all } x \in \Omega. \tag{7.120}$$

For this fixed $z_M \in \mathscr{G}_M$, we proceed as in Sect. 6.4.2.4 and define the online estimation error

$$\tilde{f}_{N,M}(t, \cdot) \triangleq f_{N,M} - \hat{f}_N(t, \cdot), \tag{7.121}$$

where $\hat{f}_N(t, \cdot)$ verifies (5.81c). Following the same steps as in the proof of Theorem 5.5, it follows from (7.101) that

$$\begin{aligned}
\dot{V}(t) &\leq -\langle Qe_N(t), e_N(t)\rangle_{\mathbb{X}} + 2\|PB\|\|e_N(t)\|_{\mathbb{X}}\|E_{x_N(t)}(f - f_{N,M})\|_{\mathbb{U}} \\
&\leq -\langle Qe_N(t), e_N(t)\rangle_{\mathbb{X}} + 2\|PB\|\|e_N(t)\|_{\mathbb{X}}\|f - f_{N,M}\|_{L^\infty(\Omega,\mathbb{U})} \\
&\leq -\langle Qe_N(t), e_N(t)\rangle_{\mathbb{X}} + 2\|PB\|\|e_N(t)\|_{\mathbb{X}}\epsilon_{N,M} \\
&\leq -\lambda_{\min}(Q)\|e_N(t)\|_{\mathbb{X}}\left(\|e_N(t)\|_{\mathbb{X}} - \frac{2\|PB\|}{\lambda_{\min}(Q)}\epsilon_{N,M}\right).
\end{aligned} \tag{7.122}$$

Alternatively, if we apply the concentration inequality (7.103), we obtain the pointwise error

$$\|f - f_{N,M}\|_{L^\infty(\Omega,\mathbb{U})} \leq \|f - f_{N,M}\|_{\mathcal{H}} \tag{7.123}$$

and resume (7.103). In conclusion, since we have assumed that the fixed set of samples is contained in $\mathscr{G}_M$, we can proceed exactly as in the proof of the deterministic dead-zone modification of MRAC to conclude that there is a time $T > t_0$ such that $e_N(t; z_M) \triangleq \|x_N(t; z_M) - x_r(t)\|_{\mathbb{X}} \leq d$ for all $t \geq T$. ■

The convergence analysis of Theorem 7.6 holds with probability $(1 - \Delta_{M,N})$. In practice, we choose $\Delta_{N,M} << 1$, so that the conclusions of this theorem hold with high probability $(1 - \Delta_{N,M})$. Note that the best performance is obtained if we can select the dead-zone $d \approx d_{N,M}$, so that the asymptotic tracking error scales like $O(\epsilon_{N,M})$. This general form of the stochastic analysis of the dead-zone method depends on the general form of the concentration inequality in (7.115).

In Sect. 7.6.2, three different choices are summarized. These choices give rise to the following corollaries, whose proofs are relatively intuitive and omitted for brevity.

Corollary 7.3 *Suppose that the hypotheses of Theorem 7.6 hold with $m = 1$ so the uncertainty is scalar-valued, and suppose that the hypotheses of Corollary 7.1 hold. Let $\epsilon_{M,M}$ be defined as in Theorem 7.2, let $d_{N,M}$ in (7.116) be defined for $N = M$ in terms of $\epsilon_{M,M}$, and let the dead-zone width be given by $d = (1 + \eta d_{M,M})$ for a small constant $\eta > 0$. Then, the ultimate bound on the trajectory tracking error is such that*

$$\|x_{M,M}(t) - x_r(t)\|_{\mathbb{X}} \lesssim \sqrt{\beta(R)} \sup_{\xi \in \Omega} \sigma_M(\xi) + \gamma(R) \tag{7.124}$$

with probability at least $(1 - \delta)$ for all $t \geq T$ and some $T > t_0$.

Corollary 7.4 *Suppose that the hypotheses of Theorem 7.6 hold with $m = 1$, and suppose that the hypotheses of Corollary 7.2 hold. Let $\epsilon_{N,M}$ be defined as in Theorem 7.2, let $d_{N,M}$ in (7.116) be defined in terms of $\epsilon_{N,M}$, and let the dead-zone width be given by $d = (1 + \eta)d_{N,M}$ with $\eta > 0$ small. Then, the ultimate bound on the trajectory tracking error is such that*

$$\|x_r(t) - x_{N,M}(t)\|_{\mathbb{X}} \lesssim \sup_{\xi \in \Omega} \mathcal{P}_N(\xi)\|f\|_{\mathcal{H}} + \frac{C}{(M + i_0)^{s/(s+2)}} \tag{7.125}$$

with probability at least $(1 - \delta)$ for all $t \geq T$ and some $T > t_0$.

Corollary 7.5 *Suppose that the hypotheses of Theorems 7.5 and 7.6 hold. Let $\epsilon_{M,M}$ be defined as in Theorem 7.5, let $d_{N,M}$ in (7.116) be defined for $N = M$ in terms of $\epsilon_{M,M}$, and let the dead-zone width be given by $d = (1 + \eta)d_{M,M}$ where $d_{M,M}$ is defined as in (7.116) for $\epsilon_{N,M} = \epsilon_{M,M}$. Then, the ultimate bound on the trajectory tracking error is such that*

$$\|x_r(t) - x_{M,M}(t)\|_{\mathbb{X}} \lesssim C\tau^2 M^{-\frac{r-t}{r+n/2}} \tag{7.126}$$

with probability at least $(1 - \delta)$ for $\delta = 5e^{-\tau}$ and for all $t \geq T$ and some $T > t_0$.

Corollary 7.3 yields setting the dimension N of $\mathcal{H}_N = \mathcal{H}_M$ to be equal to the number of samples M, that is, $N = M$. Corollary 7.5 applies to the vector-valued case with RKHS $\mathcal{H}_{d(N)}$ and $d(N) = mM$; see Remark 6.2 on the notation used in this book for the subscripts of RKHSs. Thus, the number of measurements and the dimension of the space of approximants are not independent. Corollary 7.4 yields for the approximant $f_{N,M} \in \mathcal{H}_N$ using M measurements, and one can set $M \neq N$. Typically, we expect $M > N$. Corollary 7.3 holds for uncertainties in the sample space of $GP(\mathfrak{K}, \bar{m})$, with $f \in \overline{\mathscr{R}(\mathcal{I})}$ and (7.70) being verified. This uncertainty escapes the native space $\mathcal{H}$, which means that the uncertainty class is larger than the one studied in Chaps. 5 and 6. This approach does not describe how the performance of the closed-loop controller varies for uncertainties that satisfy approximation theoretic priors, i.e., varying smoothness in the sample space. Corollaries 7.4 and 7.5 describe

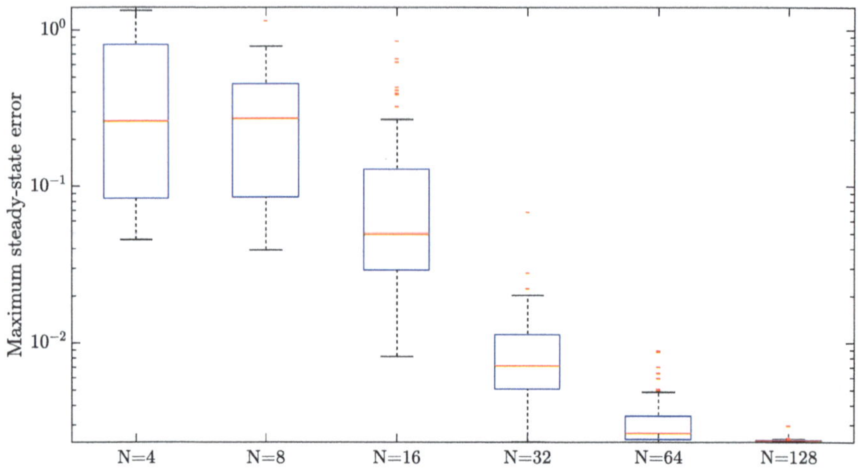

Fig. 7.3 Boxplot of steady-state error with Wendland C^2 kernel and σ-modification

how approximation theoretic priors in terms of a smoothness index s affect the performance. In this sense, these corollaries give a finer resolution of performance for uncertainty in a smoothness scale, compared to Corollary 7.3.

The study of SLC architectures in a stochastic setting is omitted for brevity. However, by proceeding as in this section, similar conclusions can be attained. This task is left to the interested reader as a research problem.

7.8 Numerical Examples

In this section, we review the performance of LTC and SLC control architectures when the learning phase generates estimates of the uncertainty based on GPs.

7.8.1 GPs for LTC Architectures

Let us revisit Example 5.4.1.5 in light of the results presented in this chapter. Assume that we are given samples $\{(x_i, y_i)\}_{i=1}^M \subset \Omega \times \mathbb{U}$, where $\Omega \subset \mathbb{X}$ and $\mathbb{U} \triangleq \mathbb{R}^m$. These are IID samples of a random walk on $Z \triangleq \Omega \times \mathbb{U}$ defined in terms of the probability measure ρ on Z such that (7.15) is verified. The learning phase builds the approximation f_M of the regressor function f_ρ by choosing f_M to be the posterior mean function $\bar{m}_M$ for the GP representing f given in (7.60). Finally, the subspace learning algorithm L in the LTC architecture chooses the set of centers Ξ_M used to represent the posterior mean function $\bar{m}_M$,

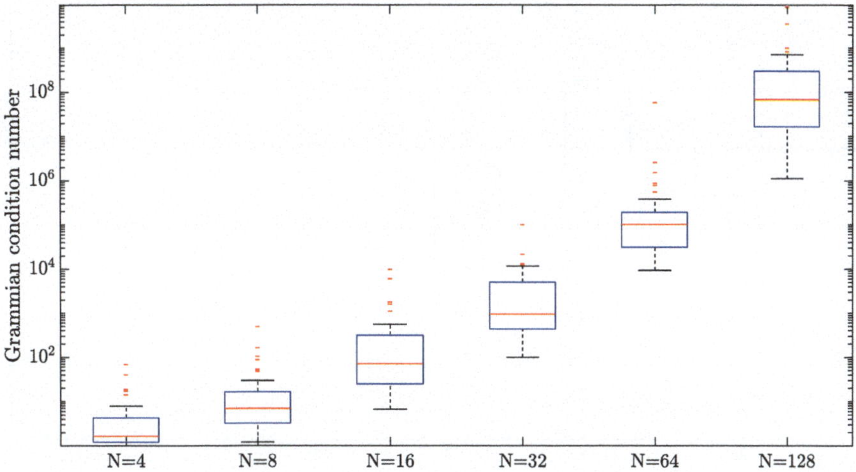

Fig. 7.4 Boxplot of the Grammian condition number with Wendland C^2 kernel and σ-modification

We choose the initial conditions $\hat{f}_M(t,\cdot)$ for the online estimate during the control phase as

$$\hat{f}_M(t,\cdot) = \bar{m}_M. \tag{7.127}$$

The adaptive laws during the control phase are the same adaptive laws used in the deterministic setting of Chap. 5. However, since the centers Ξ_M are selected in a random walk, the state variables during the control phase now depend parametrically on the sample of the random walk. Hence, the probabilistic analysis of this chapter can be applied to assess the ultimate bounds on the trajectory tracking error (Fig. 7.5).

The marginal probability measure μ on $\Omega \subset \mathbb{X}$, which generates the centers in $\Xi_M \subset \Omega$, is the uniform probability measure over Ω. To define statistics over trials, 50 simulations are conducted with M randomly generated basis centers. In other words, the centers' locations vary across each run with M fixed as the basis centers are randomly generated. To highlight the effect of kernel function choice and robust modification, the Wendland C^2 kernel and Gaussian kernels are chosen. Both kernels share the same hyperparameter value $\alpha = 1$ from Table 3.1.

Two additional robust methods are also considered in this stochastic study, namely the σ-modification and the bounding control method. For the σ-modification, we set $\sigma = 0.01$. For the bounding control method, we set the buffer zone value $\varepsilon = 0.1$ and bounding radius $R_{\text{bounding}} = 2$. The simulation time is set to 500 s to assess the steady-state behavior with a good confidence. The rest of the test parameters are kept the same as in Example 5.4.1.5.

Figures 7.3, 7.4, 7.5, 7.6, 7.7, and 7.8 show the primary results of this simulation. As the number of random basis centers populate and extend throughout the domain, the maximum steady-state error decreases and its discrete variance over

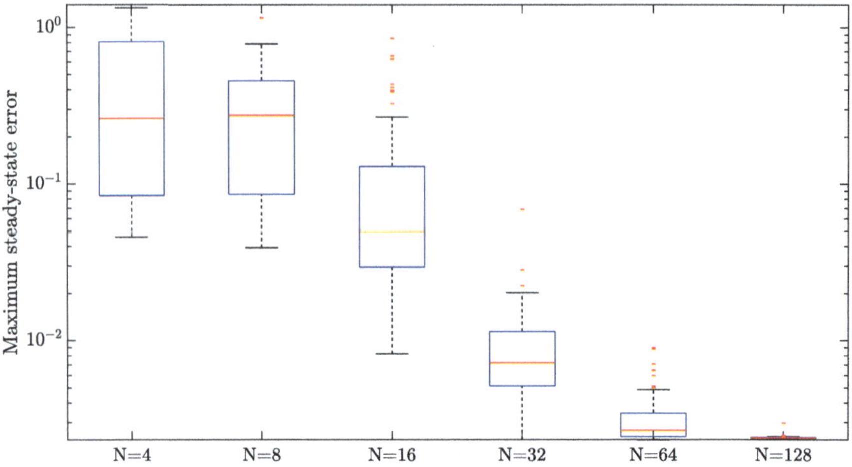

Fig. 7.5 Boxplot of steady-state error with Gaussian kernel and σ-modification

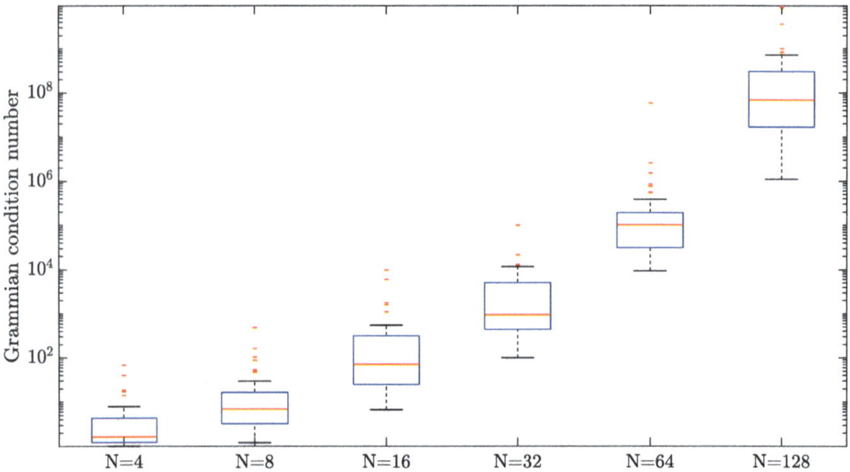

Fig. 7.6 Boxplot of Grammian condition number with Gaussian kernel and σ-modification

trials decreases, as expected from the theory. The average condition number of the Grammian matrix increases and follows the same general trends as observed in Chap. 5 for the deterministic studies. However, the discrete variance of the condition number of the Grammian does not decrease as M increases, in contrast to the steady-state error. The logarithm of the discrete standard deviation about the discrete mean is fairly consistent as M varies. Intuitively, for large values of M, many outliers have a condition number much greater than the average; in Chap. 5 we discussed how poor conditioning of the Grammian can inject numerical noise in the closed-loop governing equations. The deleterious effects of these outliers on the performance

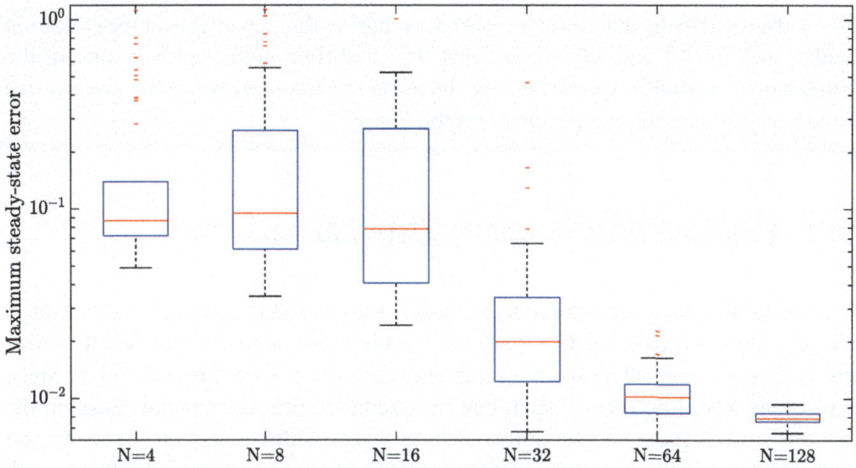

Fig. 7.7 Boxplot of steady-state error with Wendland C^2 kernel and the bounding control

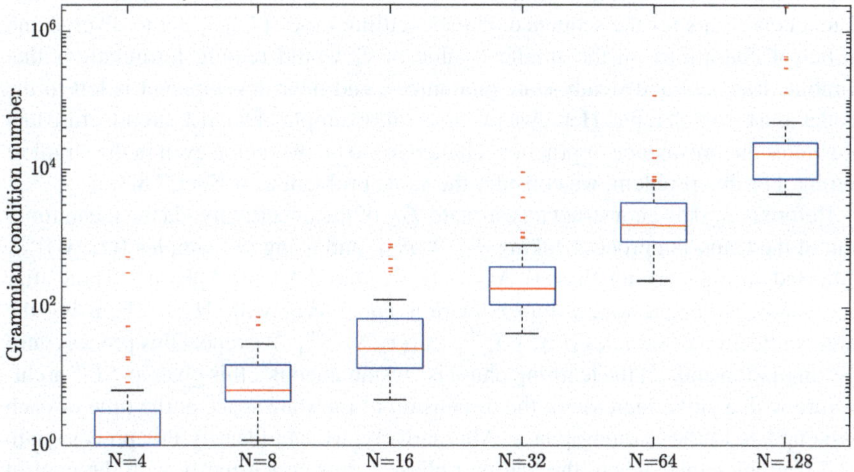

Fig. 7.8 Boxplot of Grammian condition number with Wendland C^2 kernel and the bounding control

should be studied more closely in further works on adaptive control methods defined in terms of Gaussian process adaptive methods.

Perhaps unsurprisingly, a careful study of the discrete average of the closed-loop controller performance for each of the three methods in this stochastic setting resembles the corresponding results for each method in Chap. 5. For example, the maximum steady-state error of the test cases using the Gaussian kernel is a bit lower and exhibits a smaller variance compared to cases with the Wendland C^2 kernel, especially when the number of basis functions is low. A comparison of the condition number growth in the stochastic setting also reflects some of the qualitative results observed in the deterministic case. Comparing the growth rate of the condition number in Figs. 7.4 and 7.6, we see that the condition number growth rate of the Wendland C^2 kernel is approximately linear on the log plot, while the growth rate of the Gaussian kernel is superlinear on the log plot.

7.8.2 Gaussian Processes for SLC Architectures

Because an SLC architecture can be obtained from a modification of the LTC architecture, it may be enticing to repeat the learning and control processes to define new and better controllers by applying Theorem 7.6 or Corollaries 7.3–7.5. Since there exists a settling time T such that the predicted ultimate boundedness on the tracking error is met, we can simply collate several sufficiently long learning and control phases together to achieve better tracking. If we have a finite number of such switches, the performance after the last switching time is guaranteed by our theory.

However, as usual, such a strategy might be far from a practical approach. Sufficient conditions for the sequence of such settling times $\{T_i\}_{i\in\mathbb{N}}$ not to diverge and sufficient conditions on the smallest value of T_i would require formulations that combine transient and steady-state guarantees, and their development is left to the reader as a research topic. Here, we just give one example of an SLC architecture that illustrates the prevalence of practical challenges to be overcome even in the simplest setting. For this problem, we consider the same problem as in Sect. 7.8.1.

Before $t_0 = 0$, we construct an estimate f_{M_0} of the uncertainty via the mean function of the Gaussian process, taking $f_{M_0} \triangleq \bar{m}_{M_0}$ and using the samples $\{(x_i, y_i)\}_{i=1}^{M_0}$ collected during learning phase 0. At $t = t_0$, we start the control phase 0. Then, after T seconds, we create a new approximation $f_{M_1} = \bar{m}_{M_1}$ with $M_1 > M_0$, using the larger collection of samples $\{(x_i, y_i)\}_{i=1}^{M_1} \supset \{(x_i, y_i)\}_{i=1}^{M_0}$. We repeat this process until the simulation ends. If the learning phase is "instantaneous", this gives an SLC architecture with a spike increase in the dimension of the state space at the time of each basis update at the learning phase. Alternatively, we can identify this process with a CLC architecture, where the learning phase i runs concurrently with the control phase $i - 1$.

In our specific example, we consider five learning and control phases. The learning phases create approximations of the uncertainty using the posterior mean of the Gaussian process for $M_i = 10, 20, \ldots, 100$ samples. The time T between basis updates is 500 s, and the total time of the simulation is 5000 s. To compare the effect of the kernel function choice, we consider the Wendland C^2 kernel and the Gaussian kernel with the same hyperparameter of $\alpha = 1$ in Table 3.1. The rest of the simulation parameters such as learning rate and tolerances remain the same as in the last example.

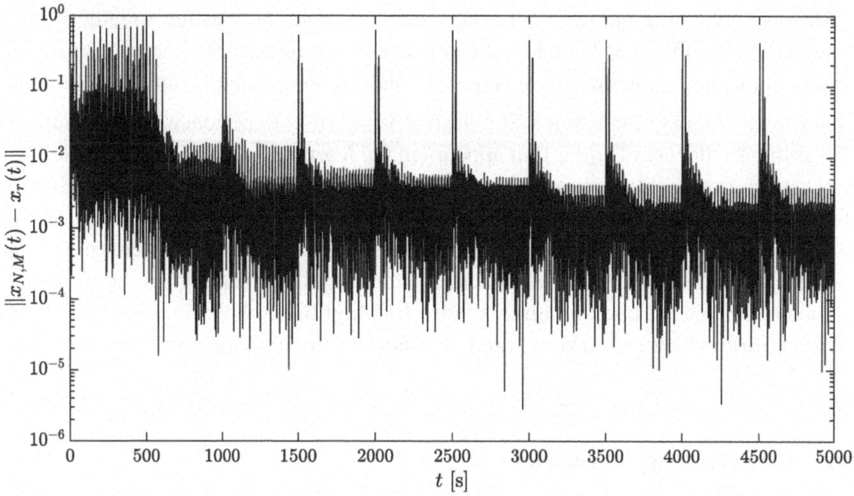

Fig. 7.9 Norm of the tracking error as a function of time employing an SCL architecture

Figures 7.9 and 7.10 show that qualitatively speaking, adding more basis improves the performance of the SLC architecture. As more basis centers are added, the tracking error eventually reaches a plateau and stops improving. This behavior is qualitatively what one would expect as the approximation errors become commensurate in magnitude with other sources of external noise and numerical noise.

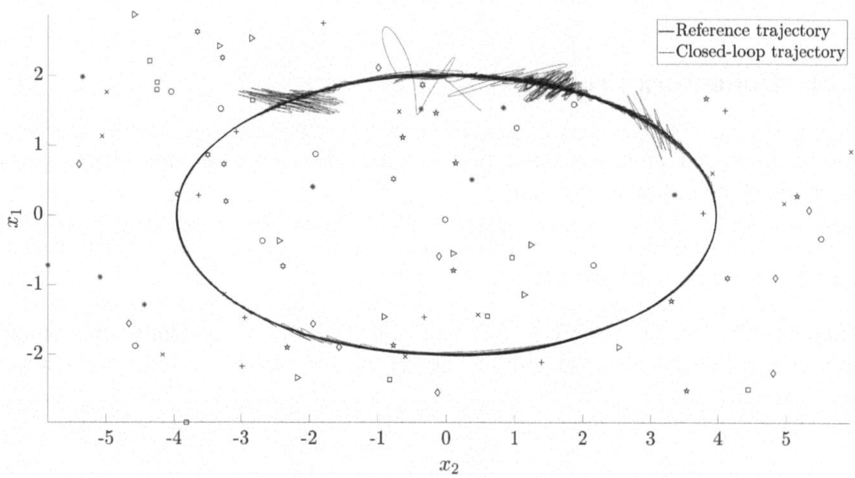

Fig. 7.10 Phase plot of the reference signal, trajectory, and basis centers of different epochs. Different epochs of basis center addition are marked with different shapes

However, the simulation results show that just naively "pasting together" LTC architectures to obtain SLC architectures can yield some new challenges. While the initial conditions at the switching times are chosen so that state continuity is maintained over the switching times, the derivatives $\dot{\hat{f}}(t, \cdot)$ experience discontinuities. This is due to the term $\Pi_{M_i}\mathfrak{K}_x$ that appears in the learning laws for the function estimates just before control phase M_i, and the term $\Pi_{M_{i+1}}\mathfrak{K}_x$ that appears immediately after switching. We attribute this undesirable behavior to the sudden change in the adaptive law dimension. The dimension of the adaptive law for $\hat{f}_N$ is increased along with the number of basis centers. This results in a significant change in the $\hat{f}_N$ gradient, which spikes the tracking error. Figure 7.10 highlights the random nature of the basis addition which explains the random nature of the tracking error improvement.

7.9 Concluding Remarks

This chapter presented the use of stochastic data-driven methods to approximate matched functional uncertainties affecting deterministic plant models in the form of ODEs. Exploiting state-of-the-art concepts in GPs on native spaces, we deduced ultimate bounds on the trajectory tracking error dynamics of adaptive control systems, employing the dead-zone modification of MRAC as a running example. Numerical examples involving LTC and SLC architectures showcased the applicability of the proposed results and highlighted some future research challenges. This chapter closes the technical part of this book. The next chapter will provide an overview of the results presented in this work and will outline future research directions.

7.10 Homework Problems

For an improved comprehension of the material exposed in this chapter, we recommend the following problems.

Exercise 7.1 Rework Exercise 6.1 with sample centers randomly distributed around the vertices parallelepipeds in $\mathbb{R}^2$. △

Exercise 7.2 Rework Exercise 6.2 with sample centers randomly distributed around the limit cycle obtained by setting $\theta_{\text{user}}(t) = \omega(t - t_0)$ for all $t \geq t_0$, where $\omega > 0$. △

References

1. Liu, M., Chowdhary, G., Da Silva, B.C., Liu, S.-Y., How, J.P.: Gaussian processes for learning and control: a tutorial with examples. IEEE Control Syst. Mag. **38**(5), 53–86 (2018)
2. Boffi, N.M., Tu, S., Slotine, J.-J.E.: Nonparametric adaptive control and prediction: theory and randomized algorithms. J. Mach. Learn. Res. **23**(1), 12 841–12 886 (2022)
3. Chen, K., Yi, J., Song, D.: Gaussian-process-based control of underactuated balance robots with guaranteed performance. IEEE Trans. Robot. **39**(1), 572–589 (2022)
4. Umlauft, J., Beckers, T., Kimmel, M., Hirche, S.: Feedback linearization using Gaussian processes. In: Conference on Decision and Control, pp. 5249–5255. IEEE, Melbourne, Australia (2017)
5. Umlauft, J., Pöhler, L., Hirche, S.: An uncertainty-based control lyapunov approach for control-affine systems modeled by Gaussian process. IEEE Control Syst. Lett. **2**(3), 483–488 (2018)
6. Umlauft, J., Hirche, S.: Feedback linearization based on Gaussian processes with event-triggered online learning. IEEE Trans. Autom. Control **65**(10), 4154–4169 (2019)
7. Jagtap, P., Pappas, G.J., Zamani, M.: Control barrier functions for unknown nonlinear systems using Gaussian processes. In: Conference on Decision and Control, pp. 3699–3704. IEEE, Jeju Island, Republic of Korea (2020)
8. Beckers, T., Umlauft, J., Kulic, D., Hirche, S.: Stable Gaussian process based tracking control of Lagrangian systems. In: Conference on Decision and Control, vol. 5. IEEE, Melbourne, Australia (2017)
9. Beckers, T., Kulić, D., Hirche, S.: Stable Gaussian process based tracking control of Euler-Lagrange systems. Automatica **103**, 390–397 (2019)
10. Fan, D.D., Nguyen, J., Thakker, R., Alatur, N., Agha-mohammadi, A.-A., Theodorou, E.A.: Bayesian learning-based adaptive control for safety critical systems. In: International Conference on Robotics and Automation, pp. 4093–4099. IEEE, Paris, France (2020)
11. Dhiman, V., Khojasteh, M.J., Franceschetti, M., Atanasov, N.: Control barriers in bayesian learning of system dynamics. IEEE Trans. Autom. Control **68**(1), 214–229 (2023)
12. Gahlawat, A., Zhao, P., Patterson, A., Hovakimyan, N., Theodorou, E.: L1-GP: L1 adaptive control with bayesian learning. In: Learning for Dynamics and Control. PMLR, pp. 826–837 (2020)
13. Helwa, M.K., Heins, A., Schoellig, A.P.: Provably robust learning-based approach for high-accuracy tracking control of Lagrangian systems. IEEE Robot. Autom. Lett. **4**(2), 1587–1594 (2019)
14. König, C., Turchetta, M., Lygeros, J., Rupenyan, A., Krause, A.: Safe and efficient model-free adaptive control via bayesian optimization. In: International Conference on Robotics and Automation, pp. 9782–9788. IEEE (2021)
15. Capone, A., Hirche, S.: Backstepping for partially unknown nonlinear systems using Gaussian processes. IEEE Control Syst. Lett. **3**(2), 416–421 (2019)
16. Westenbroek, T., Fridovich-Keil, D., Mazumdar, E., Arora, S., Prabhu, V., Sastry, S.S., Tomlin, C.J.: Feedback linearization for uncertain systems via reinforcement learning. In: International Conference on Robotics and Automation, pp. 1364–1371. IEEE, Paris, France (2020)
17. Sinha, R., Harrison, J., Richards, S.M., Pavone, M.: Adaptive robust model predictive control with matched and unmatched uncertainty. In: American Control Conference, pp. 906–913. IEEE, Atlanta, GA (2022)
18. Jiao, J., Capone, A., Hirche, S.: Backstepping tracking control using Gaussian processes with event-triggered online learning. IEEE Control Syst. Lett. **6**, 3176–3181 (2022)
19. Lederer, A., Capone, A., Umlauft, J., Hirche, S.: How training data impacts performance in learning-based control. IEEE Control Syst. Lett. **5**(3), 905–910 (2020)
20. Arabi, E., Yucelen, T., Gruenwald, B.C., Fravolini, M., Balakrishnan, S., Nguyen, N.T.: A neuroadaptive architecture for model reference control of uncertain dynamical systems with performance guarantees. Syst. Control Lett. **125**, 37–44 (2019)

21. Sforni, L., Notarnicola, I., Notarstefano, G.: Learning-driven nonlinear optimal control via Gaussian process regression. In: Conference on Decision and Control, pp. 4412–4417. IEEE, Austin, TX (2021)
22. Nayyer, S.S., Hozefa, J., Gunjal, R., Bhil, S., Wagh, S., Singh, N.: Passivity and immersion (P&I) approach with Gaussian process for stabilization and control of nonlinear systems. IEEE Access **10**, 132 621–132 634 (2022)
23. Chen, K., Yi, J., Song, D.: Gaussian processes model-based control of underactuated balance robots. In: International Conference on Robotics and Automation, pp. 4458–4464. IEEE, Montreal, CA (2019)
24. Joshi, G., Chowdhary, G.: Adaptive control using Gaussian-process with model reference generative network. In: Conference on Decision and Control, pp. 237–243. IEEE, Miami Beach, FL (2018)
25. Grande, R.C., Chowdhary, G., How, J.P.: Experimental validation of Bayesian nonparametric adaptive control using Gaussian processes. J. Aerosp. Inf. Syst. **11**(9), 565–578 (2014)
26. Ignatyev, D.I., Shin, H.-S., Tsourdos, A.: Sparse online Gaussian process adaptation for incremental backstepping flight control. Aerosp. Sci. Technol. **136**, 108157 (2023)
27. Chowdhary, G., Kingravi, H.A., How, J.P., Vela, P.A.: Bayesian nonparametric adaptive control using Gaussian processes. IEEE Trans. Neural Netw. Learn. Syst. **26**(3), 537–550 (2015)
28. Chang, A.H., Hubicki, C.M., Aguilar, J.J., Goldman, D.I., Ames, A.D., Vela, P.A.: Learning to jump in granular media: Unifying optimal control synthesis with Gaussian process-based regression. In: International Conference on Robotics and Automation, pp. 2154–2160. IEEE, Bay Sands, Singapore (2017)
29. Grande, R.C., Walsh, T.J., Chowdhary, G., Ferguson, S., How, J.P.: Online regression for data with changepoints using Gaussian processes and reusable models. IEEE Trans. Neural Netw. Learn. Syst. **28**(9), 2115–2128 (2016)
30. Wajid, R., Awan, A.U., Zamani, M.: Formal synthesis of safety controllers for unknown stochastic control systems using Gaussian process learning. In: Learning for Dynamics and Control Conference, pp. 624–636. PMLR, Stanford, CA (2022)
31. Carè, A., Carli, R., Dalla Libera, A., Romeres, D., Pillonetto, G.: Kernel methods and Gaussian processes for system identification and control: A road map on regularized kernel-based learning for control. IEEE Control Systems Magazine, vol. 43, no. 5, pp. 69–110 (2023)
32. Y. He, Y. Zhao, Adaptive robust control of uncertain euler–lagrange systems using Gaussian processes. In: IEEE Transactions on Neural Networks and Learning Systems (2022)
33. Lederer, A., Umlauft, J., Hirche, S.: Uniform error bounds for Gaussian process regression with application to safe control. Advances in Neural Information Processing Systems, vol. 32 (2019)
34. Gahlawat, A., Lakshmanan, A., Song, L., Patterson, A., Wu, Z., Hovakimyan, N., Theodorou, E.: RL1-GP: safe simultaneous learning and control (2020). arXiv:2009.03864
35. Chowdhary, G., How, J., Kingravi, H.: Model reference adaptive control using nonparametric adaptive elements. In: AIAA Guidance, Navigation, and Control Conference, MN, Minneapolis, p. 5038 (2012)
36. Vapnik, V.: Statistical Learning Theory. Wiley, New York, NY (1998)
37. L. Györfi, M. Kohler, A. Krzyzak, H. Walk et al., A Distribution-free Theory of Nonparametric Regression, vol. 1. Springer, New York, NY (2002)
38. Cucker, F., Smale, S.: On the mathematical foundations of learning. Bull. Am. Math. Soc. **39**(1), 1–49 (2002)
39. Cucker, F., Zhou, D.X.: Learning Theory: An Approximation Theory Viewpoint. Cambridge Press, Cambridge, UK (2007)
40. Binev, P., Cohen, A., Dahmen, W., DeVore, R., Temlyakov, V., Bartlett, P.: Universal algorithms for learning theory part I: piecewise constant functions. J. Mach. Learn. Res. **6**(9) (2005)
41. Binev, P., Cohen, A., Dahmen, W., DeVore, R.: Universal algorithms for learning theory. Part II: piecewise polynomial functions. Const. Approx. **26**, 127–152 (2007)

42. Kerkyacharian, G., Picard, D.: Entropy, universal coding, approximation, and bases properties. Const. Approx. **20**, 1–37 (2003)
43. Binev, P., Bonito, A., DeVore, R., Petrova, G.: Optimal learning. Calcolo **61**(1), 15 (2024)
44. DeVore, R., Kerkyacharian, G., Picard, D., Temlyakov, V.: Approximation methods for supervised learning. Found. Comput. Math. **6**, 3–58 (2006)
45. DeVore, R.A.: Nonlinear approximation. Acta Numer. **7**, 51–150 (1998)
46. Györfy, L., Kohler, M., Krzyzak, A., Walk, H.: A Distribution-free Theory of Nonparametric Regression. Springer, Berlin, Germany (2002)
47. Temlyakov, V.: Greedy Approximation. Cambridge University Press, Cambridge, UK, vol. 20 (2011)
48. Paulsen, V.I., Raghupathi, M.: An Introduction to the Theory of Reproducing Kernel Hilbert Spaces, vol. 152. Cambridge University Press, Cambridge, UK (2016)
49. Schölkopf, B., Smola, A.J.: Learning with Kernels: Support Vector Machines, Regularization, Optimization, and Beyond. MIT press, Cambridge, MA (2002)
50. Micchelli, C.A., Pontil, M.: On learning vector-valued functions. Neural Comput. **17**(1), 177–204 (2005)
51. Minh, H.Q.: Operator-valued bochner theorem, fourier feature maps for operator-valued kernels, and vector-valued learning (2016). arXiv:1608.05639
52. Bohn, B., Griebel, M., Rieger, C.: A representer theorem for deep kernel learning. J. Mach. Learn. Res. **20**(64), 1–32 (2019)
53. Diwale, S., Jones, C.: A generalized representer theorem for hilbert space-valued functions (2018). *arXiv preprint* arXiv:1809.07347
54. A. Kirsch, An Introduction to the Mathematical Theory of Inverse Problems, vol. 120. Springer, Berlin, Germany (2011)
55. Kress, R., Maz'ya, V., Kozlov, V.: Linear Integral Equations, vol. 82. Springer (1989)
56. Engl, H.W., Hanke, M., Neubauer, A.: Regularization of Inverse Problems, vol. 375. Springer, New York, NY (1996)
57. Lavretsky, E., Wise, K.: Robust and Adaptive Control: With Aerospace Applications. Springer, London, UK (2013)
58. Farrell, J.A., Polycarpou, M.M.: Adaptive Approximation Based Control: Unifying Neural, Fuzzy and Traditional Adaptive Approximation Approaches, vol. 48. Wiley, Danvers, MA (2006)
59. Donoho, D.L.: High-dimensional data analysis: the curses and blessings of dimensionality. AMS Math. Challenges Lect. **1**(2000), 32 (2000)
60. Gorban, A.N., Tyukin, I.Y.: Blessing of dimensionality: mathematical foundations of the statistical physics of data. Philosop. Trans. R. Soc. A: Math. Phys. Eng. Sci. **376**(2118), 20170237 (2018)
61. Gorban, A.N., Makarov, V.A., Tyukin, I.Y.: High-dimensional brain in a high-dimensional world: blessing of dimensionality. Entropy **22**(1), 82 (2020)
62. Belkin, M., Hsu, D., Ma, S., Mandal, S.: Reconciling modern machine-learning practice and the classical bias-variance trade-off. Nat. Acad. Sci. **116**(32), 15 849–15 854 (2019)
63. Liu, C., Zhu, L., Belkin, M.: Loss landscapes and optimization in over-parameterized non-linear systems and neural networks. Appl. Comput. Harm. Anal. **59**, 85–116 (2022)
64. Golub, G.H., Van Loan, C.F.: Matrix Computations. JHU Press, Baltimore, MD (2013)
65. Liu, F., Shi, L., Huang, X., Yang, J., Suykens, J.A.: Analysis of regularized least-squares in reproducing kernel Krein spaces. Mach. Learn. **110**, 1145–1173 (2021)
66. Grohs, P., Voigtlaender, F.: Proof of the theory-to-practice gap in deep learning via sampling complexity bounds for neural network approximation spaces. Found. Comput. Math. **24**(4), 1085–1143 (2024)
67. Belkin, M., Ma, S., Mandal, S.: To understand deep learning we need to understand kernel learning. In: International Conference on Machine Learning, pp. 541–549. PMLR, Stockholm, Sweden (2018)
68. Traub, J.F., Werschulz, A.G.: Complexity and Information, vol. 26862. Cambridge University Press, Cambridge, UK (1998)

69. Daubechies, I., DeVore, R., Foucart, S., Hanin, B., Petrova, G.: Nonlinear approximation and (deep) ReLU networks. Const. Approx. **55**(1), 127–172 (2022)
70. Yarotsky, D.: Error bounds for approximations with deep ReLU networks. Neural Netw. **94**, 103–114 (2017)
71. Adcock, B., Dexter, N.: The gap between theory and practice in function approximation with deep neural networks. SIAM J. Math. Data Sci. **3**(2), 624–655 (2021)
72. Gribonval, R., Kutyniok, G., Nielsen, M., Voigtlaender, F.: Approximation spaces of deep neural networks. Const. Approx. **55**(1), 259–367 (2022)
73. Tan, J., LeJeune, D., Mason, B., Javadi, H., Baraniuk, R.G.: A blessing of dimensionality in membership inference through regularization. In: International Conference on Artificial Intelligence and Statistics, pp. 10 968–10 993. PMLR, Valencia, Spain (2023)
74. Tan, H.Y., Mukherjee, S., Tang, J., Schönlieb, C.-B.: Blessing of dimensionality for approximating sobolev classes on manifolds (2024). arXiv:2408.06996
75. Hu, Z., Shukla, K., Karniadakis, G.E., Kawaguchi, K.: Tackling the curse of dimensionality with physics-informed neural networks. Neural Netw. **176**, 106369 (2024)
76. Villani, C.: Optimal Transport: Old and New, vol. 338. Springer (2009)
77. Allen-Zhu, Z., Li, Y., Song, Z.: A convergence theory for deep learning via over-parameterization. In: International Conference on Machine Learning, pp. 242–252. PMLR, Long Beach, CA (2019)
78. Dar, Y., Muthukumar, V., Baraniuk, R.G.: A farewell to the bias-variance tradeoff? An overview of the theory of overparameterized machine learning (2021). arXiv:2109.02355
79. Liu, C., Zhu, L., Belkin, M.: Toward a theory of optimization for over-parameterized systems of non-linear equations: the lessons of deep learning, vol. 7 (2020). arXiv:2003.00307
80. Liu, Z., Tegmark, M.: Machine learning conservation laws from trajectories. Phys. Rev. Lett. **126**, 180604 (2021)
81. Simon, J.B., Karkada, D., Ghosh, N., Belkin, M.: More is better: when infinite overparameterization is optimal and overfitting is obligatory. In: International Conference on Learning Representations, Kigali, Rwanda, (2023)
82. Berlinet, A., Thomas-Agnan, C.: Reproducing Kernel Hilbert Spaces in Probability and Statistics. Springer, New York, NY (2011)
83. Williams, C.K., Rasmussen, C.E.: Gaussian Processes for Machine Learning, vol. 2, no. 3. MIT Press, Cambridge, MA (2006)
84. Gao, W., Wang, W., Zhu, H., Huang, G., Wu, D., Du, Z.: Robust radiation sources localization based on the peak suppressed particle filter for mixed multi-modal environments. Sensors **18**(11), 3784 (2018)
85. Gahlawat, A., Lakshmanan, A., Song, L., Patterson, A., Wu, Z., Hovakimyan, N., Theodorou, E.A.: Contraction $\mathcal{L}_1$-adaptive control using Gaussian processes. In: Learning for Dynamics and Control, pp. 1027–1040. PMLR (2021)
86. Lederer, A., Conejo, A.J.O., Maier, K.A., Xiao, W., Umlauft, J., Hirche, S.: Gaussian process-based real-time learning for safety critical applications. In: International Conference on Machine Learning, pp. 6055–6064. PMLR, Virtual (2021)
87. Lederer, A., Umlauft, J., Hirche, S.: Uniform error and posterior variance bounds for Gaussian process regression with application to safe control (2021). arXiv:2101.05328
88. Steinwart, I.: Convergence types and rates in generic Karhunen-Loeve expansions with applications to sample path properties. Potential Anal. **51**(3), 361–395 (2019)
89. Kanagawa, M., Hennig, P., Sejdinovic, D., Sriperumbudur, B.K.: Gaussian processes and kernel methods: a review on connections and equivalences (2018). arXiv:1807.02582
90. Wendland, H.: Scattered Data Approximation, vol. 17. Cambridge University Press, Cambridge, UK (2004)
91. Narcowich, F.J., Ward, J.D., Wendland, H.: Sobolev error estimates and a Bernstein inequality for scattered data interpolation via radial basis functions. Const. Approx. **24**(2), 175–186 (2006)
92. Hangelbroek, T., Narcowich, F.J., Ward, J.D.: Kernel approximation on manifolds I: bounding the Lebesgue constant. SIAM J. Math. Anal. **42**(4), 1732–1760 (2010)

93. Hangelbroek, T., Narcowich, F., Ward, J.: Polyharmonic and related kernels on manifolds: interpolation and approximation. Found. Comput. Math. **12**, 625–670 (2012)
94. Hangelbroek, T., Rieger, C.: Extending error bounds for radial basis function interpolation to measuring the error in higher order sobolev norms. Math. Comput. (2024)
95. Narcowich, F., Ward, J., Wendland, H.: Sobolev bounds on functions with scattered zeros, with applications to radial basis function surface fitting. Math. Comput. **74**(250), 743–763 (2005)
96. Fischer, S., Steinwart, I.: Sobolev norm learning rates for regularized least-squares algorithms. J. Mach. Learn. Res. **21**(205), 1–38 (2020)
97. Csató, L., Opper, M.: Sparse on-line Gaussian processes. Neural Comput. **14**(3), 641–668 (2002)
98. Titsias, M.: Variational learning of inducing variables in sparse Gaussian processes. In: Artificial Intelligence and Statistics, pp. 567–574. PMLR, Clearwater Beach, FL (2009)
99. Bui, T.D., Nguyen, C., Turner, R.E.: Streaming sparse Gaussian process approximations. Advances in Neural Information Processing Systems, vol. 30 (2017)
100. Uhrenholt, A.K., Charvet, V., Jensen, B.S.: Probabilistic selection of inducing points in sparse Gaussian processes. In: Uncertainty in Artificial Intelligence, pp. 1035–1044. PMLR (2021)
101. Griebel, M., Oswald, P.: Convergence analysis of online algorithms for vector-valued kernel regression (2023). arXiv:2309.07779
102. Tarrès, P., Yao, Y.: Online learning as stochastic approximation of regularization paths: optimality and almost-sure convergence. IEEE Trans. Inf. Theory **60**(9), 5716–5735 (2014)
103. Griebel, M., Rieger, C., Zwicknagl, B.: Multiscale approximation and reproducing kernel Hilbert space methods. SIAM J. Numer. Anal. **53**(2), 852–873 (2015)
104. DeVore, R., Kerkyacharian, G., Picard, D., Temlyakov, V.: Mathematical methods for supervised learning. IMI Preprints **22**, 1–51 (2004)
105. Leiter, C., Lertvittayakumjorn, P., Fomicheva, M., Zhao, W., Gao, Y., Eger, S.: Towards explainable evaluation metrics for machine translation. J. Mach. Learn. Res. **25**(75), 1–49 (2024)

Chapter 8
Conclusion

This book presented a self-contained overall approach to the problem of controlling deterministic, nonlinear ordinary differential equations affected by functional uncertainties. The proposed framework is novel because it does not require characterizing the functional uncertainties affecting the plant model using a regressor vector or some equivalent tool. The uncertainty is assumed to lie in a reproducing kernel Hilbert space (RKHS). Classical robust and adaptive control techniques, such as control Lyapunov functions, backstepping, and model reference adaptive control (MRAC) have been extended to this new framework, which enables capturing considerably broader classes of uncertainties than in the classical formulation. The problem of identifying and approximating the hypothesis space, that is, the RKHS of functional uncertainties, has been addressed to minimize the burden on the user. Thus, deterministic and stochastic data-driven methods have been discussed to define the location of the centers that characterize the approximating hypothesis space and characterize the controller's performance. The exposition of these results, that are truly novel in the context of robust and adaptive control theory, required a systematic exposition of fundamental elements of real and functional analysis, theory of RKHSs, and dynamical systems theory.

The proposed material can be used by researchers and more experienced graduate students for self-study and as a textbook for an advanced control system design course. Such a course is currently offered at the authors' institution and is intended for students, who already took a graduate class in nonlinear systems theory and a course in adaptive control systems. The computer codes available at [1–3] are meant to help the reader apply some key results and help their comprehension of several theoretical notions. Although a sequential approach to the study of the proposed material is recommended, this approach may be tedious. In this case, the reader is encouraged to study this book from Chap. 5 and consult Chaps. 2–4 according to needs. To ease this approach, throughout this text, abundant references are made to earlier results, especially those in the first four chapters. Furthermore, a rich index

A. J. Kurdila et al., *Data-Driven, Nonparametric, Adaptive Control Theory*,
Lecture Notes in Control and Information Sciences 495,
https://doi.org/10.1007/978-3-031-78003-5_8

aims to help the reader to readily identify where concepts are presented for the first time.

Assigning homework and project problems for courses as advanced as a course based on this text may appear difficult. Homework problems are recommended at the end of each chapter and may inspire additional assignments. Instructors are encouraged to challenge students to customize the computer codes provided in [1, 2], which are needed to solve some of the recommended homework assignments, to problems of increasing complexity. The authors applied with success and low computational times the proposed theoretical results on problems as complex as the design of control systems for quadcopters; this material will be soon disclosed in future publications. Problems as complex as this may serve as group projects. Additional homework problems can be designed to challenge the student's comprehension of RKHSs by analyzing in a discursive manner the mathematical concepts exposed in Chapters 2 and 3 and numerically approximating functions using RKHSs. Worked-out examples and problems from classical textbooks such as [4–7] can be leveraged to strengthen the student's understanding of the real and functional analysis. Classical nonlinear dynamical systems analysis and synthesis problems, such as those suggested in [8, 9], can be employed to motivate students in their comprehension of the material in Chap. 4. The numerous examples presented in [10] are strongly recommended to readers with minimal or no experience in classical adaptive control systems design.

8.1 Summary of Noteworthy Features

In this text, many new results have distinct advantages over the standard approaches to classical robust and adaptive control theories. In the following, we highlight some of these advantages, which have been remarked on throughout this book as well.

8.1.1 Nonparametric Adaptive Control via Operator Kernels

Chapter 5 shows how the most popular robust and parametric adaptive control approaches can be lifted to a nonparametric setting while leveraging the RKHS embedding strategy. All of these modifications in this text are explicit in the orthogonal projection operator that maps functions in the hypothesis space onto a suitably chosen finite-dimensional space used to build approximations of the uncertainty. This book is the first general treatment of nonlinear robust and adaptive control theory based on vector-valued uncertainties in an RKHS defined by general operator kernels. The proposed results provide performance bounds on the ultimate tracking error, which are explicit in the dimension of the RKHS employed to approximate the ideal hypothesis space of functional uncertainties. To the authors' knowledge, there is no equivalent result in the classical parametric approach. For these essential properties and the key role played by the location of the centers and the choice of the kernel functions defining the RKHS of functional uncertainties, we view the process of basis selection as an integral part of the design of robust and adaptive control methods.

8.1.2 Functional Uncertainty Classes and AAO Controllers

Chapter 5 introduced nonparametric functional uncertainty classes that are typically larger than the uncertainty classes defined in terms of a finite number of real parameters and proved uniform ultimate boundedness of the proposed robust and adaptive controllers despite uncertainties in these classes. Chapter 6 introduced the notion of asymptotically approximation theory nearly optimal (AAO) and ϵ-AAO adaptive control schemes. These are new concepts for evaluating the performance of various proposed nonparametric adaptive control schemes relative to the best offline approximation error. The developments in Chap. 6 culminated in the description of AAO and ϵ-AAO performance for the adaptive control laws introduced in Chap. 5. Tables 6.1 and 6.2 provided the user with quick guidelines on the performance levels to expect for each control technique. These tables should be compared, for example, to Tables 9.1–9.3 of [11] or Table 11.1 of [10], which summarize classical methods of parametric adaptive control theory. The tables in these classical references summarize the structure of standard robust modification methods but do not describe the associated ultimate tracking performance measures as functions of the number of parameters used to approximate the space of uncertainties.

8.1.3 Deterministic and Stochastic General Concentration Inequalities

Chapters 6 and 7 described how principles of deterministic and stochastic machine learning theory in general and data-driven methods in particular can be employed to forecast the performance of nonparametric adaptive controllers and guide the design of algorithms for choosing the location of the kernel centers. In Chap. 6, the learn-then-control (LTC), the switched learn-then-control (SLC), and the concurrent learn-then-control (CLC) strategies were introduced. Within the LTC framework, bounds on the performance of nonparametric adaptive control systems were given according to multiple choices of centers' distribution. In the context of SLC, we provided an example of how the actor-critic approach to the optimal control problem can be employed to define the learning stage. Finally, a CLC method based on a greedy search algorithm was discussed and the corresponding bounds on the trajectory tracking error have been derived.

Chapter 7 leveraged a body of literature that is broader than that exploited in Chap. 6 for the characterization of functional uncertainties. The analysis proposed in Chap. 7 extended approaches first encountered in specific cases in references like [12–14] for Gaussian process (GP) methods to design a *general* strategy for nonparametric adaptive control theory. In particular, Chap. 7 showed how common GP techniques, which provide concentration inequalities for the posterior mean of the GP, can be used to derive probabilistic ultimate tracking performance bounds. Chapter 7 also described a general process by which standard concentration inequalities can be modified to estimate the functional uncertainties that depend both on the deterministically defined space of approximants and the posterior mean function defined in terms of the number of measured samples of the functional uncertainty. The analysis in this approach generates concentration inequalities in both the number of parameters characterizing the approximating hypothesis space and the number of measurements of the functional uncertainty. Existing techniques require that these two sets of parameters are identical, with consequential limitations. The approach in Chap. 7 allowed quantifying explicitly the trade-off between approximation error and probabilistic error, which is a classical topic of learning and approximation theory. This analysis of the contributions of approximation error and stochastic error is not found in the existing literature on adaptive control approaches based on GPs.

8.2 Selected Future Research Directions

This work aspires to introduce the control community to a new strategy for addressing control problems in the presence of functional uncertainties. Indeed, the authors envision how the ideas presented in this book can be extended to any existing class of control systems, eventually leading to a new branch within the domain of nonlinear control systems theory that focuses on nonparametric uncertainties. This text, however, only represents an initial effort to build on the vast array of powerful existing methods in parametric robust and adaptive control theory. In the following, we summarize some substantial future research topics that, in our opinion, should be studied to extend the proposed framework and provide novel concrete performance guarantees in the presence of functional uncertainties.

8.2.1 The Search for Tighter Bounds

Classical approaches to the parametric adaptive control problem, such as [10, 15] and the numerous works inspired by them, generally make a universal assumption about the size of the approximation error. In this book, assumptions are made to exploit the tools provided by the RKHS setting to analyze an adaptive controller's performance. In particular, the proposed results make systematic use of Jackson inequalities that arise naturally in an RKHS setting and are obtained either using the power function of the many zeros theorem. Still, the resulting bounds on the controller's performance are typically very conservative. This fact is more apparent in some of the robust modifications of MRAC, like the σ-modification in the RKHS setting, but it can also be observed in almost all simulations. One reason for this fact is that the functional uncertainty is usually "excessively smooth" in the native space. For these reasons, the general strategy in this text would benefit greatly if tighter bounds could be derived, or if more precise explanations of how excess smoothness in the functional uncertainty $f \in \mathcal{H}$ affects controller's performance.

8.2.2 General, Systematic Methods for SLC and CLC Architectures

Although the study of LTC systems has been particularly detailed, the study of SLC and CLC architectures in this book has been rather limited. The study of learning strategies in the context of RKHS-embedded robust adaptive controllers is still at its infancy and was summarized for the most part in Chaps. 6 and 7 of this book. It is apparent how there is a need for the development of further *general* methods that further relax or eliminate the reliance on oracles that choose the hypothesis space and its approximations without insight into their rationale. Additional challenges lie in the concatenation of learning and control stages in LTC and SLC strategies. We suppose that Newton bases, employed in the context of the CLC framework, will provide fruitful results. Developing a rich theory that holds in general for a broad family of robust adaptive control modifications, would be an important accomplishment and a step toward a useful general theory for SLC and CLC architectures.

8.2.3 Guarantees of Transient Performance

Although the mathematical techniques, theory, and algorithms described in this manuscript constitute a first step toward a general theory of learning and nonparametric adaptive control, all the proposed results considered in this book employ the ultimate bounds on the tracking error as the sole measure of the performance of any controller. Numerical examples have shown that the performance of the closed-loop plant trajectory in the transient stage must also be given priority. In general, devising such guarantees of transient and steady-state performance is one of the highest goals of adaptive control theory. However, this objective may become exceptionally important in nonparametric adaptive control strategies that change the dimensions of subspaces of approximants. A potential approach lies in the use of barrier Lyapunov functions and the extension of results such as those in [16] to the RKHS setting.

8.2.4 Safety and Control Barrier Functions

Many of the results on nonparametric adaptive control presented in this book assume that the closed-loop plant trajectory lies in some compact subset of the state space at all times. Being able to identify the smallest compact subset such that this assumption is verified is usually a difficult, if not impossible, task. However, the proposed ultimate bounds on the tracking error are usually explicit functions of this compact subset of the state space, and the smaller is this set, the tighter are the guaranteed ultimate bounds on the tracking error. In practice, one can only hope to choose such a set that is not too large.

The state-limiting arguments in Chap. 12 of [10] or in Sect. 7.2.4 of [15] may be employed in the future within the nonparametric setting to overcome this limitation of the proposed arguments. Alternatively, barrier Lyapunov functions [16–19] may be extended to the RKHS embedding framework presented in this book. This topic is currently under study by the authors and will soon be discussed in future publications.

8.2.5 Scaling of Complexity Estimates for Controller Performance

Chapters 5 and 6 presented explicit bounds on the performance of nonparametric controllers for a wide variety of choices of RKHSs. The subset of the state space in which the centers of the kernels are collocated has been proven to play a key role in the computational complexity of the proposed control systems. Although centers can be naively distributed over regular grids that cover parts of the state space, in which the closed-loop trajectory is expected to dwell, this distribution of centers scales poorly with the dimension of the state space. For problems involving a significant number of states, more careful choices of the subset of the state space in which centers can be located should be made. Thus, in Chap. 6, we outlined several choices that are valid alternatives to the center distribution over grids. For instance, we analyzed how to distribute centers along limit cycles known a priori and limit sets unknown a priori. Furthermore, we presented multiple greedy methods to distribute centers according to needs as the dynamical process unfolds. These approaches allow the proposed control systems to scale considerably better with the dimension of the state space, but are limited to smaller, yet infinite dimensional, nonparametric uncertainty classes. An important future development of the theory in this book would be a systematic study of the trade-off between the size of the functional uncertainty class and the rate of growth of the computational complexity for classes of adaptive controllers.

8.2.6 Deep Neural Networks and RKHSs

If we polled researchers, mathematicians, and scientists in general about exciting new topics of research in statistical and machine learning theory, the list would most probably include techniques for approximation, including the RKHS setting, estimation methods based on GPs, and methods of deep learning and deep neural networks (DNN). This book presented several results about approximation and estimation techniques based on GPs that are relevant to nonparametric control. However, the only discussion on DNNs in this text is limited to our analysis of the double descent phenomenon. A reason for the lack of methods based on DNN in this book is due to the early stage of development in efforts to derive rigorous approximation error bounds for DNNs. Reference [20] provides a good overview of the state-of-the-art on DNNs.

As noted throughout this book, the sharpest bounds on the closed-loop system trajectory tracking error are provided in the case of scalar-valued RKHSs, and, hence, in the case of scalar-valued control inputs, or in the case of vector-valued RKHSs deduced as the Cartesian product of scalar-valued RKHSs. However, there is an immense, and substantially rich, class of vector-valued RKHSs, which is not reducible to the Cartesian product of scalar-valued RKHSs. For these hypothesis spaces, this book provides considerably fewer results. A reason for this disparity is that RKHS theory provides considerably fewer results for generic vector-valued RKHSs. We argue that the situation is even worse in the theory of DNN to characterize uncertainties. While some progress has been made in characterizing the expressivity of DNNs, there is a general lack of concrete realizable algorithms with associated sharp error estimates. Some studies proved good approximations for various multilayer DNNs, but well-defined methods to construct such estimates are still lacking.

Another reason for the lack of results in this text for controllers defined in terms of DNN approximants is that, historically, the empirical results that show the potential of DNNs have arisen in the search for algorithms that perform well for problems of very high dimensions. The problems considered in this book, however, are usually concerned with relatively small numbers of parameters to characterize approximations of the hypothesis space. The literature on classical parametric adaptive control is limited to problems in which the number of parameters characterizing the space of functional uncertainties, i.e., the dimension of the regressor vector or some equivalent tool, is considerably smaller than dimensions of typical DNNs used in offline learning methods.

Quite recently, in [21], a representer theorem has been derived. This result is expressed in terms of multilayer DNN approximants constructed from an RKHS. An important next step for the general approaches in this book, which can be interpreted as certain types of single-layer neural networks, would be to extend the results in [21]. This extension would be nontrivial because approximants would no longer be linear in the parameters. Furthermore, although [21] provides a solid first step toward the use of multilayer approximants, simple approximation methods with associated rates of convergence of multilayer architectures in an RKHS are not yet known.

8.2.7 Escaping the Native Space

Throughout this book, we assumed that the functional uncertainty affecting a nonlinear plant model lies in a Hibert space that serves as the hypothesis space. Special emphasis was given to problems in which the hypothesis space is an RKHS. However, very little attention was given to the complex problem in which the functional uncertainty lies in a Banach space, in which the hypothesis space is embedded, but does not belong to the hypothesis space. The only results pertaining to this case, known as escaping the native space, is Corollary 7.1, which is based on the concentration inequality in Theorem 7.3. However, as explained in the discussion following Theorem 7.3, this bound does not give a precise description of how approximation theoretic priors affect the rate of convergence for uncertainty outside the native space. Extending the approach in this book to allow for larger uncertainty classes would be an important improvement to the current setting. A general theory would require concentration inequalities that include the effects of information-theoretic priors and hold outside the hypothesis space.

8.3 Concluding Remarks

The future research directions recommended in Sect. 8.2 are just some of the major research initiatives that might be undertaken, taking this text as an initial condition. The authors hope that the interested reader will feel free to reach out and contact us for collaborations on these topics, or on other topics reading this text.

References

1. Wang, H., L'Afflitto, A., Kurdila, A.J.: Computer codes to implement control systems based on RKHS-embedded robust control Lyapunov functions (2024). https://lafflitto.com/Codes/Springer_RKHS_Book_RCLF_Codes.zip
2. Wang, H., L'Afflitto, A., Kurdila, A.J.: Computer codes to implement RKHS-embedded model reference adaptive control systems (2024). https://lafflitto.com/Codes/Springer_RKHS_Book_MRAC_Codes.zip
3. Powell, N., Wang, H., L'Afflitto, A., Kurdila, A.J.: Computer codes to implement concurrent learning and control architectures (2024). https://lafflitto.com/Codes/Springer_RKHS_Book_Newton_basis_MRAC.zip
4. Kreyszig, E.: Introductory Functional Analysis with Applications, vol. 17. Wiley (1991)
5. Naylor, A.W., Sell, G.R.: Linear Operator Theory in Engineering and Science. Springer Science and Business Media, New York, NY (1982)

6. Conway, J.: A Course in Functional Analysis, Series Graduate Texts in Mathematics. Springer, New York, NY (1994)
7. Heil, C.: Introduction to Real Analysis. Springer, New York, NY (2019)
8. Haddad, W.M., Chellaboina, V.: Nonlinear Dynamical Systems and Control: A Lyapunov-Based Approach. Princeton University Press, Princeton, NJ (2008)
9. Khalil, H.K., Grizzle, J.W.: Nonlinear Systems, vol. 3. Prentice Hall, Upper Saddle River, NJ, (2002)
10. Lavretsky, E., Wise, K.: Robust and Adaptive Control: With Aerospace Applications. Springer, London, UK (2013)
11. Ioannou, P.A., Sun, J.: Robust Adaptive Control. Courier Corporation, Mineola, NY (2012)
12. Boffi, N.M., Tu, S., Slotine, J.-J.E.: Nonparametric adaptive control and prediction: theory and randomized algorithms. J. Mach. Learn. Res. **23**(1), 12 841–12 886 (2022)
13. Lederer, A., Umlauft, J., Hirche, S.: Uniform error bounds for Gaussian process regression with application to safe control. Adv. Neural Inf. Process. Syst. **32** (2019)
14. Gahlawat, A., Lakshmanan, A., Song, L., Patterson, A., Wu, Z., Hovakimyan, N., Theodorou, E.: RL1-GP: safe simultaneous learning and control (2020). arXiv:2009.03864
15. Farrell, J.A., Polycarpou, M.M.: Adaptive Approximation Based Control: Unifying Neural, Fuzzy and Traditional Adaptive Approximation Approaches, vol. 48. Wiley, Danvers, MA (2006)
16. Anderson, R.B., Marshall, J.A., L'Afflitto, A.: Novel model reference adaptive control laws for improved transient dynamics and guaranteed saturation constraints. J. Frank. Inst. **358**(12), 6281–6308 (2021)
17. L'Afflitto, A.: Barrier lyapunov functions and constrained model reference adaptive control. IEEE Control Syst. Lett. **2**(3), 441–446 (2018)
18. L'Afflitto, A., Mohammadi, K.: Robust observer-based control of nonlinear dynamical systems with state constraints. J. Frank. Inst. **354**(16), 7385–7409 (2017)
19. L'Afflitto, A., Blackford, T.A.: Constrained dynamical systems, robust model reference adaptive control, and unreliable reference signals. Int. J. Control **93**(5), 1039–1052 (2020)
20. Daubechies, I., DeVore, R., Foucart, S., Hanin, B., Petrova, G.: Nonlinear approximation and (deep) ReLU networks. Const. Approx. **55**(1), 127–172 (2022)
21. Bohn, B., Griebel, M., Rieger, C.: A representer theorem for deep kernel learning. J. Mach. Learn. Res. **20**(64), 1–32 (2019)

Index

A. J. Kurdila et al., *Data-Driven, Nonparametric, Adaptive Control Theory*, Lecture Notes in Control and Information Sciences 495,
https://doi.org/10.1007/978-3-031-78003-5

The manufacturer's authorised representative in the EU is Springer Nature Customer Service Centre GmbH, Europaplatz 3, 69115 Heidelberg, Germany. If you have any concerns regarding our products, please contact ProductSafety@springernature.com

Printed and bound by CPI Group (UK) Ltd, Croydon, CR0 4YY
30/09/2026
02225568-0011